THE PENGUIN DICTIONARY OF

BOTANY

Consultant Editor STEPHEN BLACKMORE

General Editor ELIZABETH TOOTILL

PENGUIN BOOKS

Penguin Books Ltd, Harmondsworth, Middlesex, England
Penguin Books, 40 West 23rd Street, New York, New York 10010, U.S.A.
Penguin Books Australia Ltd, Ringwood, Victoria, Australia
Penguin Books Canada Ltd, 2801 John Street, Markham, Ontario, Canada L3R 1B4
Penguin Books (N.Z.) Ltd, 182–190 Wairau Road, Auckland 10, New Zealand

First published 1984
Published simultaneously by Allen Lane

Copyright © Market House Books Ltd, 1984
All rights reserved

Made and printed in Great Britain by
Richard Clay (The Chaucer Press) Ltd, Bungay, Suffolk

Set in Times by
Market House Books Ltd, Aylesbury

REFERENCE
N 010831
0416058
R 581.03

THE LIBRARY
THE COLLEGE
SWINDON

Except in the United States of America,
this book is sold subject to the condition
that it shall not, by way of trade or otherwise,
be lent, re-sold, hired out, or otherwise circulated
without the publisher's prior consent in any form of
binding or cover other than that in which it is
published and without a similar condition
including this condition being imposed
on the subsequent purchaser

This book has been prepared by
Market House Books Ltd, Aylesbury

Consultant Editor: Stephen Blackmore, B.Sc., Ph.D.

General Editor: Elizabeth Tootill, B.Sc., M.Sc.

Contributors: Richard A. Blackman, M.Sc., D.I.C.
Jonathan Y. Clark, B.Sc.
Mary Clarke, B.Sc.
E. K. Daintith, B.Sc.
M. R. Ingle, B.Sc., Ph.D., M.I.Biol., F.L.S.
Sheila C. A. Jones, B.Sc.
Sabina G. Knees, B.Sc., M.Sc.
Lynne Mayers, B.Sc., D.T.A.
Ian Andrew Solway, B.Sc.
Elizabeth Tootill, B.Sc., M.Sc.

PREFACE

The Penguin Dictionary of Botany encompasses in over 3,000 entries all the major fields of pure and applied plant science, including taxonomy and classification, anatomy and morphology, physiology, biochemistry, cell biology, plant pathology, genetics, evolution, and ecology. It also covers selected items from such related fields as agricultural botany, horticulture and microbiology, and includes entries on certain laboratory equipment and techniques. The dictionary should prove invaluable to 'A' level and undergraduate students of botany and biology and to naturalists, geographers, and others studying or working in related fields. It is also hoped it will be of use to any with a general interest in the plant world.

The trivial, rather than systematic, names of organic compounds are used throughout. Table 2 in the Appendix lists the recommended chemical names of some of the commoner organic compounds. No attempt has been made to include named species or genera of plants since the numbers involved make this beyond the scope of the dictionary. However, higher ranks of plant groups (divisions, classes and the more important orders) are included, as are some of the larger families of flowering plants. The vernacular names of these are given where appropriate. Plant taxonomy is in a considerable state of flux and many new and often radically different classification systems have recently been published. Comprehensive coverage of these would heavily bias the book towards taxonomy to the detriment of other subjects. Hence only the better known taxa are included.

Bacteria and viruses are included, although neither would normally be classified in the plant kingdom. Justification for this lies in their considerable importance in plant ecology and pathology.

ELIZABETH TOOTILL, 1983

NOTES ON THE USE OF THIS DICTIONARY

The entries in this dictionary are ordered strictly alphabetically. However, two points should be noted. Where numbers occur within a term, these are treated as though they precede the letter A. Consequently, **C_3 plant** is to be found at the beginning of letter C and not at cp. Both numbers and Greek letters that precede a term, on the other hand, are ignored for alphabetization purposes. Hence **2,4-D** appears at the beginning of D and **β-oxidation** is to be found at oxidation.

The dictionary incorporates a comprehensive cross-referencing system to guide the user to other entries that contain further relevant information. Cross references are indicated in the text by an asterisk before the appropriate word or are listed at the end of the entry. Synonyms appear as simple cross references to the better known name for the term, where they follow the headword in brackets. Terms, such as DNA and ATP, that are more often encountered in their abbreviated form, are defined at the appropriate abbreviation.

A

ABA *See* abscisic acid.

abaxial Describing the side of leaves, petals, etc. facing away from the stem or main axis, i.e. the lower surface. *Compare* adaxial. *See also* dorsal.

abiogenesis The development of living organisms from nonliving matter as envisaged in the modern theory of the *origin of life.

abiotic factors The nonliving components of the environment that directly affect plant and animal life, such as water, carbon dioxide, oxygen, and light. Abiotic factors include *climatic, *edaphic, and *physiographic factors. *Compare* biotic factors.

abortive transduction *See* transduction.

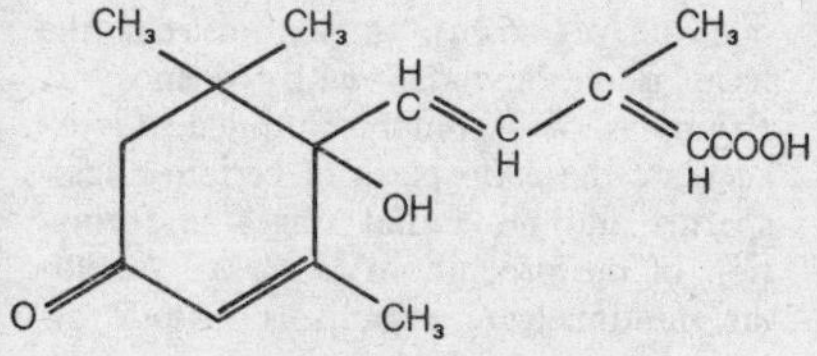

Abscisic acid.

abscisic acid (ABA, dormin) A *growth substance that exerts numerous different, mainly inhibitory, effects on the growth and development of many species. It is a sesquiterpenoid, formula $C_{15}H_{20}O_4$, (*see* diagram) and, like gibberellin, may be synthesized from mevalonic acid. It is active, possibly in association with gibberellic acid, in the promotion of leaf and fruit *abscission and the control of *dormancy. It prevents cell elongation and shoot growth and also inhibits seed germination and some tropic responses. At physiological concentrations, ABA is not toxic to plants, which can, if necessary, remove its effects by converting it to an *abscisyl glucoside* by linking it to a glucose residue. A large proportion of ABA is synthesized in the chloroplasts. The rate of synthesis increases dramatically when the plant is under stress, especially from water shortage. ABA overrides the normal diurnal pattern of stomatal opening and closure and causes the stomata to close during the day. This response decreases water loss by transpiration in times of drought.

abscission The controlled shedding of a part, such as a leaf, fruit, or flower, by a plant. The process is usually associated with a decline in auxin level within the organ to be detached. An *abscission zone* of tissue often forms at the point of separation, which is normally at the base of a petiole or pedicel. A thin plate of cells, the *abscission* (or *separation*) *layer*, forms within the abscission zone. The pectic acid in the cell walls of the abscission layer is converted to pectin, resulting in a softening and weakening of the region. The organ is then easily dislodged from the plant by wind, heavy rain, etc. Abscisic acid, which promotes leaf senescence, may also play a part in abscission. Ethylene has been shown to accelerate the abscission of senescent leaves. Precocious abscission of fruits (*fruit drop) is a common phenomenon.

absorption 1. The uptake into a plant of water, solutes, or other substances by either active or passive means. Entry invariably involves movement across cellular membranes. *Active absorption*, for example the uptake of a solute against an osmotic gradient, involves expenditure of energy. An example of *passive absorption* is the intake of water by plant roots, which is controlled by the rate of transpiration.

2. The retention of radiant energy by the pigments of a plant. About 80% of the incident visible light and about 10% of the infrared radiation falling on a leaf is absorbed. Generally, less than 2% of this is used in photosynthesis.

absorption spectrum The pattern of bands or lines obtained by passing white light through a selectively absorbing substance into a spectroscope. It is specific for any one compound and gives a characteristic profile when plotted against wavelength. Absorption spectra have been important in the study of photosynthesis: a comparison of the absorption spectra of the various photosynthetic pigments with the *action spectrum of photosynthesis shows which pigments are contributing absorbed light energy to the photosynthetic process.

acaulescent Describing plants that have no stem or an extremely short stem, such as tufted or rosette plants.

accessory cell *See* subsidiary cell.

accessory chromosome *See* B-chromosome.

accessory pigments Pigments other than chlorophyll *a* found in photosynthetic cells. They include the carotenes and the xanthophylls, together known as the carotenoids, chlorophylls *b*, *c*, and *d*, and the phycobilins. The latter are found only in blue-green and red algae. The composition of accessory pigments in algae is used as a taxonomic character.

The accessory pigments function as secondary absorbers of light in regions of the visible spectrum not covered by chlorophyll *a*. The light energy that they absorb must be transferred to chlorophyll *a* before it can be used in the photosynthetic process. As energy transfer from one molecule to another can only occur from a shorter wavelength absorbing form to a longer wavelength absorbing form, all accessory pigments have absorption maxima at shorter wavelengths than chlorophyll *a*. Energy is passed from chlorophyll *a* to the reaction centre pigments, P680 and P700. Some accessory pigments may have a protective function, preventing photooxidation of the cell's chlorophyll at high light intensities.

accessory transfusion tissue In certain gymnosperms, e.g. *Cycas*, *transfusion tissue that extends through the mesophyll from the midrib to the margin.

acellular Describing tissues, organs, or organisms consisting of a mass of protoplasm not divided by cell walls into discrete units. The term may be used of relatively large uninucleate organisms, such as the alga *Acetabularia*, to distinguish them from comparatively less specialized and smaller unicellular organisms. More often acellular structures are multinucleate and result from free nuclear division with no accompanying cell wall formation. When describing multinucleate organisms, tissues, or cells the term is synonymous with *coenocytic. Examples of acellular multinucleate tissues are the endosperm of certain angiosperms and the initial stages in formation of the proembryo in *Cycas*. Acellular multinucleate organisms include algae of the order Siphonales.

acervulus A small disc-shaped mass of conidiophores that erupts through the epidermis of plants infected by fungi of the order Melanconiales.

acetaldehyde (ethanal) An intermediate in the conversion of pyruvic acid to ethanol, the final stage of glycolysis under anaerobic conditions in plants. The conversion of acetaldehyde to carbon dioxide and ethanol is energy requiring, involving oxidation of a reduced molecule of NAD. Acetaldehyde is also involved in the synthesis and breakdown of the amino acid threonine.

acetic acid (ethanoic acid) A weak organic acid, formula CH_3COOH. Acetic acid can be used as an alternative carbon source by certain algae (e.g. *Chlamydomonas mundana* and species of *Chlorella*), while green algae of the genus *Chlamydobotrys* are totally dependent on acetic acid as a carbon source. In combination with coenzyme A (*see* acetyl CoA) acetic acid plays a central role in aerobic energy metabolism.

aceto-orcein A stain used in the preparation of root-tip or anther squashes for chromosome examination. The material to be stained is placed in the aceto-orcein (which is acidified with hydrochloric acid) and heated at 60°C for 15 minutes. The material is then removed and mounted in acetic acid.

acetyl CoA (acetyl coenzyme A) A compound consisting of acetyl combined through a sulphur bridge with *coenzyme A. The formation of acetyl CoA is an energy-requiring reaction, involving the conversion of ATP to AMP and pyrophosphate. Acetyl CoA plays a central role in intermediary metabolism. It is a product of the degradation of fatty acids, carbohydrates, and some amino acids. It is an essential precursor in the TCA and glyoxylate cycles and is the starting point for the synthesis of fatty acids, terpenes, and some amino acids.

achene Any simple one-seeded indehiscent dry fruit that develops from a monocarpellary ovary. The *caryopsis, *cypsela, and *samara are all types of achene. *Compare* nut.

acicular Needle shaped, for example the acicular crystals that, closely packed together, form a raphide, and the leaves of pine (*Pinus*) trees.

acid-fast stain A widely applied stain in bacteriology that is used to identify organisms that can retain a dye on washing with acid alcohol. One technique is to stain a bacterial smear with hot carbol fuchsin and, after rinsing with water, expose the smear to concentrated hydrochloric acid dissolved in water or ethanol. Following a second rinsing, the smear is counterstained with methylene blue. Acid-fast organisms retain the red colour of the carbol fuchsin while other organisms appear blue. Examples of acid-fast bacteria are species of the genus *Mycobacterium*.

acid rain Rain with a high level of acidity caused by pollution by oxides of nitrogen and sulphur produced by coal and oil combustion. The oxides combine with water in the atmosphere forming acids such as nitric and sulphuric acids. Acid rain has caused the deaths of countless living organisms both by producing a high level of acidity and by leaching metal ions from the soil into rivers and lakes. The disastrous effects have been extensively documented in Scandinavia and North America. Recently, the Warren Spring preliminary report has revealed a similar level of pollution in Scotland and parts of northern England.

acid soil *See* pedalfer.

acid stain *See* staining.

ACP *See* acyl carrier protein.

acquired characteristic A characteristic of a living structure or biochemical system that has been brought about by environmental factors acting during the life of the organism. For example, plants receiving fertilizer applications may grow taller and more vigorously than before and pot plants moved to a shady position may become etiolated. It is a fundamental tenet of Neo-Darwinism that such variations are not inherited by the succeeding generation. *See* Lamarckism.

Acrasiomycetes (cellular slime moulds) A class of the *Myxomycota containing organisms that feed on soil bacteria. It includes some 25 species divided among

10 genera. Cellular slime moulds are made up of an aggregation of *myxamoebae forming a *pseudoplasmodium. These produce uncovered masses of walled spores or pseudospores by a process different from that of other fungi. The cellular slime moulds are often classified as an order (Acrasiales) of the Myxomycetes.

acrocarpous Describing mosses in which the main axis is terminated by the development of the reproductive organs so subsequent growth is sympodial. In such mosses the main axis is almost always erect. *Compare* pleurocarpous.

acrocentric Describing a chromosome in which the centromere is not centrally placed, giving two arms of different lengths. *Compare* metacentric, telocentric.

acronematic flagellum *See* whiplash flagellum.

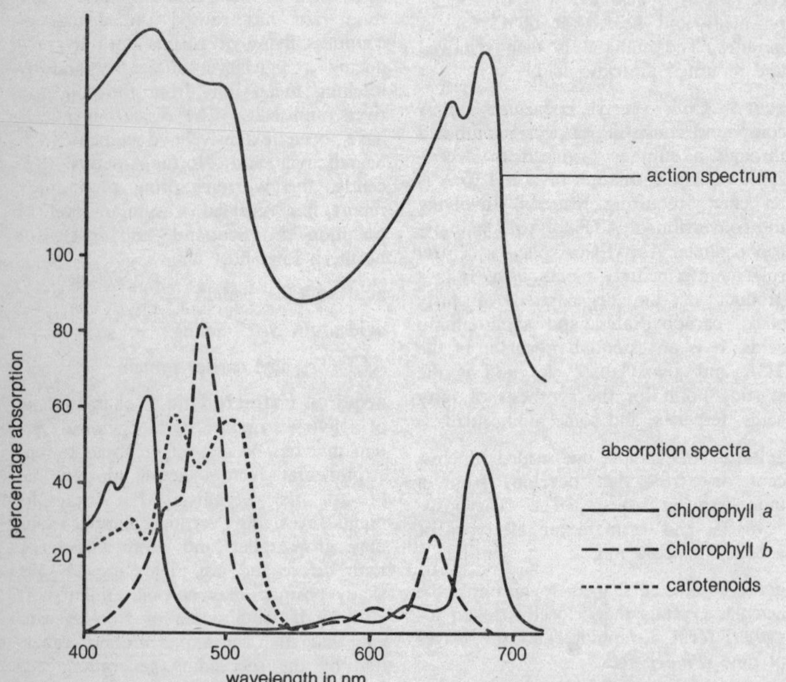

A comparison of the action spectrum of photosynthesis with the absorption spectra of chlorophylls *a* and *b* and the carotenoids.

acropetal Describing movement, differentiation, etc. occurring from base to apex. For example, the development of flowers is usually acropetal with the calyx being formed first and the gynoecium last. Movement of water through the plant is acropetal. *Compare* basipetal.

acrostichoid condition The situation found in certain ferns, e.g. *Platycerium*, in which the sporangia are formed all over the undersurface of a fertile frond rather than in specialized organs or sori.

actinodromous Describing a form of leaf venation in which three or more primary veins originate at the base of the lamina and run out towards the margin. The leaves of sycamore (*Acer pseudoplatanus*) are an example. In traditional terminology such venation is termed palmate or digitate. *See illustration at* venation.

actinomorphy *See* radial symmetry.

Actinomycetales An order of bacteria containing forms that develop a branched mycelium reminiscent of fungal growth forms, though on a far smaller scale (actinomycete hyphae rarely exceed 1.5 μm in diameter). Most are saprophytic though some are pathogenic.

actinostele A *protostele in which the central core of xylem is star shaped or somewhat lobed as viewed in transverse section. This type of stele is exhibited in various species of *Psilotum* and *Lycopodium* and in the roots of higher plants. Depending on the number of xylem lobes, usually between two and eight, the stele is described as diarch, triarch, etc., and polyarch if the number of lobes exceeds eight. *See illustration at* stele.

action spectrum A plot of the rate of a reaction, e.g. the phototropic response or photosynthesis, at different wavelengths of light. The action spectrum for photosynthesis (*see* diagram) is obtained by measuring the photosynthetic yield for a given amount of light incident upon the plant over a range of wavelengths. For green plants the action spectrum shows that chlorophyll is the pigment responsible for photosynthesis, since peak photosynthetic activity occurs

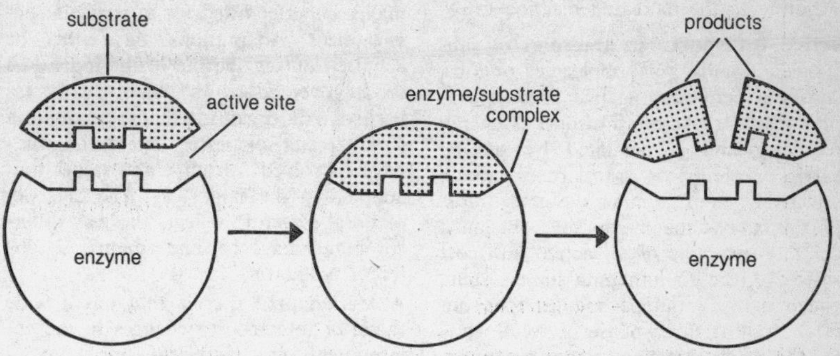

Schematic representation of the induced fit hypothesis.

at the absorption peaks of chlorophylls *a* and *b*.

For some algae the action spectrum shows peak photosynthetic activity at the absorption peak of one of the accessory pigments. In this situation although chlorophyll is still used for transfer of light energy into the photosynthetic process, it is not the primary light-absorbing pigment.

active site The site at which the substrate of an enzyme is bound during catalysis. Once bound, the substrate reacts to form a product or products, which are then released from the active site. The conformation and charge distribution of the active site is carefully tailored to accept only a specific substrate or class of substrates. Within the active site there are areas concerned with binding the substrate and other areas that create conditions conducive to the catalysis of substrate to products.

The theory that best explains how the active site and the substrate interact is the *induced fit hypothesis* (*see* diagram). This considers the active site to be flexible and to adjust its conformation in response to the presence of a substrate molecule. The resulting distortion of both substrate and enzyme puts a strain on the substrate and is one factor contributing to the increased reaction rate.

active transport The transport of substances, usually polar molecules or ions, across a membrane against a concentration gradient. Active transport is energy requiring and is mediated by specific carrier proteins or translocases, which selectively bind a substrate and transport it across the membrane. The most common example of an active transport system is the sodium/potassium system, which actively pumps sodium ions out of the cell at the expense of ATP, while at the same time carrying potassium ions into the cell. The resulting concentration gradient can be very high; in the freshwater alga *Nitella clavata* the ratio of potassium inside the cell to that outside is 1065:1.

acuminate Gradually narrowing to a point, as do certain leaves.

acute Having a sharp but not extended point, as certain types of leaf.

acyl carrier protein (ACP) A low-molecular-weight conjugated protein. It forms a complex with the enzymes of fatty acid synthesis and binds the acyl group of the growing fatty acid molecule, thus bringing enzymes and substrate together. The prosthetic group of ACP, 4′-phosphopantetheine, is identical to the acyl-carrying portion of *coenzyme A. *See also* fatty acid metabolism, malonyl ACP.

acylglycerol (glyceride) An ester of the alcohol glycerol in which a hydroxyl group is replaced by a fatty acid. Acylglycerols are one of the four main types of complex *lipids. Often all three of the glycerol hydroxyl groups are esterified giving a *triacylglycerol. Some diacylglycerols contain a sugar attached to the unesterified hydroxyl group. These glycosyldiaclyglycerols are commonly termed *glycolipids.

adaptation A modification to an organism, or a feature of an organism, that makes it better fitted for a particular environment. Adaptations may either be acquired during the life of the individual or, if governed by the genotype, be inherited. All organisms show adaptation to a greater or lesser extent, i.e. they have evolved from ancestors not adapted in the same way. The diversity of floral structure reflects the way different plants have become adapted to different pollinators.

A less adapted species that has a wide range of tolerance to changes in the environment may be better able to colonize new areas or survive changes in its own habitat. Highly adapted species are less likely to survive sudden environmental changes but more likely to thrive in their particular niche. Weeds are exceptions to this generalization, having evolved adaptations that specifically enable them to colonize disturbed ground.

adaptive enzyme *See* induced enzyme.

adaptive radiation The evolution of a number of different groups of organisms from a common ancestral group, each adapted to life in different environments. Examples are the various species of the rubiaceous genus *Hedyotis*, which occupy different niches in the Hawaiian Islands. *See also* speciation.

adaxial Describing the side of lateral organs facing towards the stem or main axis, i.e. the upper surface. *Compare* abaxial. *See also* ventral.

adelphous Describing an androecium in which the filaments are fused. *Compare* syngenesious. *See also* monadelphous, diadelphous, polyadelphous.

adenine A nitrogenous base, more correctly described as 6-aminopurine, derived from amino acids and sugars and found in all living organisms. Adenine is a constituent of *DNA, *RNA, *ATP, *NAD, and *FAD. Many cytokinins are derivatives of adenine and adenine itself shows cytokinin activity.

adenosine A combination of adenine with D-ribose, a pentose sugar. The two molecules are linked by a β-glycosidic bond and together form a *nucleoside. Adenine may alternatively combine with 2-deoxy-D-ribose, in which case the nucleoside is called *deoxyadenosine*. The latter is the form found in DNA, while adenosine is found in RNA, ATP, ADP, and AMP. Certain derivatives of adenosine (e.g. isopentenyl adenosine or *IPA) are cytokinins. *See also* nucleotide.

adenosine diphosphate *See* ADP.

adenosine monophosphate *See* AMP.

adenosine triphosphate *See* ATP.

adnate Describing unlike organs that are joined together, such as stamens fused with the petals. *Compare* connate.

ADP (adenosine diphosphate) A diphosphorylated nucleoside with the same structure as adenosine monophosphate but with a second phosphate linked through a *high-energy phosphate bond to the first. In photophosphorylation ADP combines with another phosphate molecule to form ATP. This conversion is a major mechanism for the conservation of the light energy absorbed in photosynthesis. In some plants ADP acts as a sugar carrier, forming such molecules as ADP-glucose and ADP-sucrose. These ADP-sugars can then be oxidized to give energy or used in the synthesis of starch and ascorbic acid. *See* nucleoside diphosphate sugars.

adventitious Describing organs that arise in unexpected positions, such as roots growing from a leaf.

adventive embryony (adventitious embryogenesis) The formation of an embryo in some position other than within an embryo sac. It is a form of *apomixis. Adventive embryos may develop from somatic cells of the nucellus or chalaza (*apospory) and may occur together with a normal zygotic embryo. One seed can contain several embryos, as is often the case in *Citrus* seeds. Alternatively an embryo may develop from the unreduced egg cell (*diplospory).

aeciospore (aecidiospore) A dikaryotic spore formed by certain rust fungi in a small cup-shaped sorus, the *aecium* or *aecidium*. The aeciospores of *Puccinia graminis* are formed on the lower surfaces of barberry leaves in clusters of aecia, this stage of the life cycle commonly being called the cluster-cup stage. The aeciospores cannot reinfect barberry but will germinate on certain grass or cereal leaves where the subsequent infection shortly gives rise to *uredospores. *See* heteroecious.

aerenchyma Tissue with numerous large intercellular spaces. Aerenchyma is common in the cortex of the roots and stems of many aquatic plants, where it facilitates oxygenation of the roots and may increase buoyancy.

aerial root A *root that arises above soil level. The term is usually applied to the tangled masses of roots developed by epiphytes, which hang down in the moist air. The root epidermis develops a sheath of dead empty cells, the *velamen, that helps absorb water from the atmosphere. *See also* pneumatophore, climbing root, prop root, buttress root.

aerobe An organism capable of living only in the presence of air or free oxygen, the oxygen being needed for *aerobic respiration. All plants, except certain bacteria and fungi, are aerobes. In contrast to most animals they do not die instantly when deprived of oxygen but can continue to respire anaerobically for a time. *Compare* anaerobe.

aerobic respiration *Respiration involving the absorption of free oxygen and the complete oxidation of the organic starting materials. The oxygen is used as the terminal electron acceptor and the ultimate products are water and carbon dioxide. Approximately 700 kcals are released when a molecule of glucose is completely oxidized in aerobic respiration. This compares to only about 72 kcals obtained when a glucose molecule is broken down by *anaerobic respiration to alcohol and carbon dioxide.

aerophore *See* pneumatophore.

aerotaxis A form of *chemotaxis in which the cell or organism moves in response to oxygen.

aerotropism A form of *chemotropism in which oxygen is the orientating factor.

aestivation (prefloration) The arrangement of perianth parts within an immature flower bud. The classification of aestivation types is based mainly on the extent of overlap of perianth segments. *Compare* vernation.

after-ripening The processes that must be undergone by certain seeds after harvest before germination can take place. Although the embryo appears fully mature the seed remains dormant until the after-ripening period is completed. It has been suggested that after-ripening involves the breakdown of some growth inhibitor or the formation of a growth-promoting substance. However since after-ripening processes often occur in dry seeds, in which metabolic activity is very low, it would appear that the changes are not necessarily metabolic in nature. An after-ripening requirement is exhibited by many cereals and grasses and prevents the seed from germinating in the ear under moist conditions.

agamospermy *See* apomixis.

agar A mucilaginous carbohydrate that is used in the form of a gel as a microbiological supporting medium. It is obtained from certain red algae, notably *Gelidium* species. The gel is produced by dissolving agar crystals in boiling water and then allowing the mixture to cool. It may be sterilized by placing in an autoclave. A nutrient agar gel has various nutrients added to it. Agar is widely used because it is resistant to attack by almost all microorganisms, and, as it does not melt until 100°C, incubation of the selected culture can take place at high temperatures. When agar is used in a Petri dish it is known as an *agar plate.*

Agaricales An order of the *Hymenomycetes containing the *mushrooms and toadstools. The 3250 species and 220 genera are divided into two families: the Agaricaceae, in which the hymenium is borne on the surface of gills; and the Boletaceae, in which the hymenium lines fleshy pores. Both families contain many edible species, including the cultivated mushroom *Agaricus bisporus*, which is the most widely eaten fungus in the UK.

aggregate fruit A fruitlike structure that has developed from the carpels of a single flower and is composed of a number of separate fruits. It may be an aggregation of achenes (as in *Anemone*), berries (as in *Actaea*), drupelets (as in *Rubus*), follicles (as in *Delphinium*), or samaras (as in *Liriodendron*). The term *etaerio* is often used to mean any aggregate fruit but its use is sometimes restricted to a collection of drupelets.

aglycone *See* glycoside.

Agonomycetales (Mycelia Sterilia) An order of the *Hyphomycetes containing those mycelial imperfect fungi that do not produce spores, although some develop sclerotia. It contains about 200 species and 28 genera including *Rhizoctonia*, certain species of which cause *damping-off disease, and *Sclerotium*.

air bladder **1.** A swollen air-filled region of the thallus seen in many algae of the order Fucales, notably *Fucus vesiculosus* (bladder wrack). The air bladders serve to increase buoyancy and may also play a part in respiration.
2. Any of the air-filled protuberances that develop by separation of the exine layers in the pollen grains of certain gymnosperms, e.g. *Pinus*. They give the pollen a characteristic winged appearance and assist the wind dispersal of the pollen.
See also bladder.

air chamber Any of the air-filled cavities beneath the upper epidermis of the gametophyte thallus in many liverworts of the order Marchantiales. Such air chambers increase the surface area of the internal photosynthetic cells and so facilitate gaseous exchange. At the same time they maintain a humid atmosphere around these cells, thus reducing transpiration. Connection of the air chambers with the external atmosphere is facilitated by pores in the upper surface of the thallus.

air layering A form of vegetative propagation often used to multiply greenhouse and indoor plants. A branch is stimulated to produce roots while still attached to the parent plant by making a shallow cut or removing a narrow ring of bark just below a bud. The cut is dusted with rooting powder containing synthetic auxins and kept open by inserting sphagnum moss The whole area is then kept moist by wrapping the stem in moss and surrounding this with a plastic sleeve. Root formation is slow and may take up to two years.

air plant *See* epiphyte.

akinete A thick-walled resting spore that is formed during unfavourable conditions by certain blue-green algae. Akinetes are highly resistant to temperature extremes and desiccation. On germination a *hormogonium may be formed.

alanine A simple nonpolar amino acid in which the R group is CH_3 (*see illustration at* amino acid). Alanine is formed in a transaminase reaction in which glutamine donates an amino group to pyruvic acid to form alanine and α-ketoglutaric acid. This reaction is reversible; alanine can be deaminated to pyruvate and subsequently oxidized in the TCA cycle. Along with aspartic acid, serine, glutamic acid, and glycine, alanine is one of the early products of photosynthesis.

albedo The ratio of the intensity of light reflected from a surface to the intensity of light received. All leaves reflect or transmit infrared radiation but different species reflect different amounts of visible radiation. Thus xerophytic plants tend to reflect more light than mesophytic plants.

albumin Any of a class of simple low-molecular-weight proteins found in many plants. An example is leucosin, an albumin found in wheat. Albumins have no prosthetic group and only one amino-acid chain. They are often found in association with globulins.

albuminous cell 1. A specialized parenchyma cell, physiologically and anatomically associated with a sieve cell in gymnosperm phloem. Albuminous cells are analogous to the companion cells of angiosperm phloem but unlike companion cells they do not usually arise from the same mother cell as the sieve element. **2.** An albumen-containing cell, found in certain seeds.

albuminous seed *See* endosperm.

alburnum *See* sapwood.

alcian blue A stain that enhances the preservation of cell surface materials, especially mucopolysaccharides. It is usually used during fixation with glutaraldehyde for transmission electron microscopy.

alcoholic fermentation A form of *anaerobic respiration in which glucose is broken down to form ethanol and carbon dioxide. It is carried out by yeasts and various other fungi and by certain bacteria. Fermentation takes place outside the organism and is catalysed by enzymes of the zymase complex. These are either secreted by living cells or released on cell death. Fermentation usually stops due to cell poisoning when the alcohol level reaches about 15%. The process is central to the brewing, wine-making, and baking industries. Since free oxygen is not available as a hydrogen acceptor acetaldehyde is used instead. Pyruvic acid, formed by glycolysis, is broken down to acetaldehyde and carbon dioxide. The acetaldehyde is then reduced by $NADH_2$ to form ethanol and NAD. The process yields about 72 kcals from each glucose molecule. This is only about 10% of the energy that would be released by complete oxidation of glucose, as in *aerobic respiration.

glyceraldehyde: CHO–HCOH–CH_2OH
erythrose: CHO–HCOH–HCOH–CH_2OH
ribose: CHO–HCOH–HCOH–HCOH–CH_2OH
arabinose: CHO–HOCH–HCOH–HCOH–CH_2OH
xylose: CHO–HCOH–HOCH–HCOH–CH_2OH
glucose: CHO–HCOH–HOCH–HCOH–HCOH–CH_2OH
mannose: CHO–HOCH–HOCH–HCOH–HCOH–CH_2OH
galactose: CHO–HCOH–HOCH–HOCH–HCOH–CH_2OH

Some common aldoses.

aldose Any monosaccharide with the carbonyl (CO) group on the terminal carbon, so forming an aldehyde (CHO) group. The simplest aldose is the three-carbon sugar *glyceraldehyde. Other aldoses include glucose, galactose, and mannose, which are aldohexoses, and ribose, arabinose, and xylose, which are aldopentoses (*see* diagram). *Compare* ketose.

aleurone layer (proteinaceous endosperm layer) The proteinaceous outermost layer of the *endosperm in seeds of the Gramineae (grasses) and Polygonaceae. The cells of the aleurone layer contain *aleurone grains*, which store protein for later use by the embryo.

algae An extremely diverse group consisting predominantly of aquatic plants showing relatively little differentiation of tissues and organs as compared to bryophytes and tracheophytes. The group includes both prokaryotic and eukaryotic organisms ranging from unicells through colonial and filamentous forms to parenchymatous seaweeds over 50 m long. Some 23 500 species of eukaryotic algae and some 1450 species of prokaryotic algae have been described. Adaptation to an aquatic environment has led to the development of many distinctive biochemical traits. Algal sex organs are usually unicellular.

Formerly the algae were placed with the bacteria and fungi in the division *Thallophyta. Increasing appreciation of the very basic differences between such organisms and between the algal groups themselves has led most systematists to consider each group as a separate class or division. The groups are classified according to their pigments, food reserves, cell-wall materials, number and types of flagella, and ultrastructural details. *See* Cyanophyta, Chrysophyta, Xanthophyta, Haptophyta, Bacillariophyta, Chlorophyta, Charophyta, Euglenophyta, Dinophyta, Cryptophyta, Phaeophyta, Rhodophyta.

algal bloom *See* bloom.

alginic acid A carbohydrate polymer consisting of D-mannuronic acid and L-glucuronic acid units. It is found principally in the cell walls of brown algae where it functions as an ion exchange agent. Alginic acid derivatives have commercial value: they are used in the making of fireproof and disposable fabrics, as emulsifying agents, in some types of surgical suture, and as stabilizers in food.

alien *See* exotic.

Alismatidae (Alismidae) A subclass of the monocotyledons containing aquatic or semiaquatic herbaceous plants. General features of the group are: unfused carpels, in contrast to other monocotyledons in which the gynoecium is usually syncarpous; trinucleate pollen, as opposed to the binucleate pollen common to most other monocotyledons; absence of a starchy endosperm; and absence of vessels in the stem and often also in the roots. There are usually two subsidiary cells associated with each stoma. Four orders are commonly recognized: the Alismatales (or Alismales); the Hydrocharitales; the Najadales; and the Triuridales. The Triuridales is sometimes placed in the subclass Liliidae. The families of the Alismatidae are usually small, some containing only one genus (e.g. Najadaceae: *Najus*, naiads; Ruppiaceae: *Ruppia*, ditch grasses; Scheuchzeriaceae: *Scheuchzeria*) while the Butomaceae includes only one species, *Butomus umbellatus* (flowering rush). The most important family is the Potamogetonaceae (pondweeds).

alkaline soil *See* pedocal.

alkaloids A class of nitrogen-containing usually basic plant products, which are often poisonous. Many alkaloids, e.g. morphine, codeine, nicotine, and cocaine, have been utilized in medicine and other fields. Their natural functions are not well understood. Some, through their bitterness and toxicity, may serve to protect the plants from herbivorous animals. Others are thought to be involved in nitrogen metabolism (e.g. nicotine is involved in the absorption of nitrate through plant roots). Many are thought to be simply end products of nitrogen metabolism, which are stored in leaves, fruits, and flowers before being discarded. Certain families, e.g. the Solanaceae, Leguminosae, and Papaveraceae, contain many alkaloidal species while monocotyledons rarely produce alkaloids. Most groups of lower plants contain alkaloidal species though alkaloids have not been found in the algae.

Three classes of alkaloids have been recognized. The *true alkaloids* have a nitrogen-containing heterocyclic nucleus, examples being the *isoquinoline alkaloids. The *protoalkaloids* lack a heterocyclic ring and are usually simple amines. Examples are mescaline and ephedrine. Some of the protoalkaloids may be precursors of true alkaloids. Both the true alkaloids and the protoalkaloids are derived from amino acids. The *pseudoalkaloids* are not derived from amino acids but from such compounds as terpenes, purines, and sterols. Pseudoalkaloids include theobromine and caffeine, which are both methylated purines. *See also* indole alkaloids, pyridine alkaloids, piperidine alkaloids, tropane alkaloids.

allele (allelomorph) A form in which a gene may occur. Different alleles of a gene give rise to different expressions of a character. Hence alleles for 'green' and 'yellow' are alternative expressions of the gene governing the characteristic for seed colour. Alleles of a gene always occupy the same site (locus) on homologous chromosomes. A diploid organism whose cells contain two identical alleles is said to be *homozygous. One with two different alleles at a locus is said to be *heterozygous. The allele expressed in a heterozygous organism depends on the dominance relationship between the two alleles. If one is expressed to the exclusion of the other it is described as *dominant and the latter *recessive. If the heterozygote is intermediate in appearance between the two homozygotes, the alleles are said to exhibit *incomplete dominance.

allelopathy The release of a chemical by a plant that inhibits the growth of nearby plants and thus reduces competition. For example, pines (*Pinus*) produce substances that kill any seedlings of the same species (autotoxicity or autoallelopathy) growing too close to the parent plant. Allelopathy may also be indirect, if one plant inhibits the growth of a second plant or microorganism that itself is essential to the growth of a third plant. Allelopathic substances have been shown to be responsible for various changes during plant successions. For example, in a field succession the pioneer weed stage is replaced by annual grasses because the weeds produce substances inhibitory to other weeds.

In arid or semiarid habitats the allelopathic substances are often volatile terpenes released from the leaves. For example, in the Californian chaparral aromatic shrubs, such as *Artemisia* and *Salvia*, release oils (e.g. camphor) that inhibit the germination of herbaceous plants. In wetter areas various phenolic compounds have been shown to be allelopathic. *See also* phytoalexin.

allogamy (cross fertilization, exogamy) Fusion of female and male gametes derived from genetically dissimilar individuals of the same species. It promotes the recombination of variable genetic material so that the total population shows greater variation than an *autogamic population and has greater adaptive potential. Various methods have developed in the plant kingdom to encourage or necessitate allogamy (*see* dichogamy, dioecious, monoecious, incompatibility, chasmogamy, heterostyly). When cross fertilization is obligate, useful genetic traits may be diluted and isolated individuals cannot reproduce.

allometric growth The growth of part of an organism as compared to the whole. It is rare for all parts of a plant to be growing at the same rate. Parts that grow faster than the whole are said to exhibit *positive allometry*, for example, a newly formed inflorescence would usually be growing faster than the vegetative structures. Other parts or organs may show *negative allometry*, as when a meristem releases growth substances that reduce the growth rates of other neighbouring meristems.

allopatric Describing a population or species that is unable to breed with a related group of organisms because of geographical separation by distance or by natural barriers, such as water, mountain ranges, deserts, etc. If two groups that have been separated in such a way develop different adaptations such that they would not be able to interbreed even if the barriers broke down then *allopatric speciation* is said to have occurred. *Compare* sympatric. *See also* vicariance.

allopolyploidy A form of *polyploidy that results from the combination of sets of chromosomes from two or more different species. A (diploid) interspecific hybrid is normally sterile because there is only one of each kind of chromosome per cell. Thus at meiosis no bivalents are formed and any resulting gametes are usually inviable as they contain either too many or too few chromosomes. However if some unreduced diploid gametes form and fuse then the resulting tetraploid will usually be fertile as its nuclei will contain pairs of homologous chromosomes. An allotetraploid may cross with a diploid to form a sterile triploid hybrid, which, if it produces unreduced triploid gametes, may give rise to a fertile allohexaploid. If an allohexaploid crosses with a diploid an allo-octaploid could arise in a similar fashion, and so on to higher levels of allopolyploidy. An allopolyploid is often intermediate in appearance between both parental species and cannot reproduce with either. Hence it may merit the status of a new species. Allopolyploidy has been described as instant evolution. Many crop plants are believed to have originated in this way. Wheat (*Triticum aestivum*), for example, is an allohexaploid. It has a chromosome number of 42 and is probably derived from the three species *T. monococcum*, *Aegilops speltoides*, and *A. squarrosa*, each with 14 chromosomes.

If at meiosis an allopolyploid forms only bivalents, i.e. it acts like a diploid, then it is termed an *amphidiploid. Raphanobrassica* (*see* allotetraploid) is an example. However if certain of the chromosomes from the two parent species are sufficiently similar (*see* homoeology) then multivalents may be seen at meiosis. When this occurs the allopolyploids are termed *segmental allopolyploids*. An example is the allotetraploid *Primula kewensis* derived from a cross between *P. floribunda* and *P. verticillata*. Sometimes segmental allopolyploids show reduced fertility as compared to amphidiploids and it is clear that in some allopolyploids there are mechanisms that prevent the formation of multivalents. Wheat, for example, normally behaves as an amphidiploid but if chromosome V of the B genome (the genome derived from *A. speltoides*) is missing then multivalents are formed.

allosteric enzyme A regulatory enzyme responsive to alterations in the metabolic state of a cell or tissue. Its catalytic activity is modified by the noncovalent binding of a specific metabolite at a site (*see* allosteric site) other than the active site. The most common type of allosteric enzyme is one found at the beginning of a multienzyme sequence that is inhibited specifically by the end product of the reaction sequence (*feedback inhibition). However, the activity of an allosteric enzyme is not always decreased by the action of a modulator; some allosteric modulators increase the enzyme's activity. Allosteric

enzymes may also have more than one modulator.
Reactions involving allosteric enzymes are always irreversible in the cell, as it would be impossible to regulate a reversible reaction. Allosteric enzymes are often more complex structurally than other enzymes; all known allosteric enzymes have at least two protein subunits.

allosteric site A regulatory site on an allosteric enzyme where a specific effector or modulator can reversibly bind. When bound, the modulator either activates (a positive modulator) or inhibits (a negative modulator) the enzyme by changing its shape. Some allosteric enzymes have a site for only one modulator (monovalent enzymes) while others have several allosteric sites (polyvalent enzymes).

allosyndesis Pairing of chromosomes derived from different species. It is seen in some diploid hybrids and in segmental allopolyploids (*see* allopolyploidy). *See also* homoeology.

allotetraploid An allopolyploid that has originated from an interspecific hybrid through the formation and fusion of unreduced diploid gametes. An example is *Raphanobrassica*, which is obtained by crossing *Raphanus sativus* (radish) with *Brassica oleracea* (cabbage). The F_1 is sterile but diploid gametes are sometimes formed, which on selfing give a fertile tetraploid F_2. *See* allopolyploidy.

alluvial soil A type of azonal *soil formed on the flood plains of river valleys and at river mouths (alluvial fans or deltas). New material is successively deposited on the surface when the land is subjected to flooding. In tropical regions such areas may be used as paddy fields, and in arid regions the extent of the alluvial soils may be artificially increased by irrigation methods, as along the banks of the Nile. The polders of the Netherlands are reclaimed marine alluviums.

alpha-naphthol test (Molisch's test) A standard procedure for detecting the presence of carbohydrates in solution. A small amount of alcoholic alpha-naphthol is added to the test solution in a test tube. Concentrated sulphuric acid is then poured slowly down the side of the tube. A violet ring forming at the junction of the liquids indicates a positive reaction.

alpine A major regional community (*biome) of vegetation in high mountainous regions and on high-level plateaus. The plants generally grow in thin stony soil and are subjected to high light intensity and high wind speeds. The climate differs from that of the *tundra as there is daylight in winter and darkness in summer, more precipitation and wind, and a higher degree of solar radiation, but no permafrost. The number of plant species, although limited, is greater than that found in the tundra. Certain drought-tolerant species (e.g. grasses, sedges, mosses, and lichens) are common to both communities. In temperate latitudes there are many brightly coloured flowers in summer on the alpine pastures.
The vegetation changes as the altitude increases due to the associated drop in temperature. The vegetation also differs between north- and south-facing slopes. The lower limit of the alpine zone (*see* tree line) varies in different mountain regions and also varies with the wetness of the locality and the mass of the mountain range.

alternate Describing a form of leaf arrangement in which there is one leaf at each node (*see illustration at* phyllotaxis). This pattern is found in most plants. *Compare* opposite, whorled.

alternate host A plant, other than the main host, on which a pathogen or pest can live. Alternate hosts, which are often weeds, can provide a means for the pathogen to survive when its main host is not available. For example, downy mildew of beet (*Peronospora farinosa*) can overwinter in wild beet. Hawthorn (*Crataegus*) is an alternate host and important inoculum source for the fireblight bacterium (*Erwinia amylovora*), which infects pears and apples. Many *rusts also overwinter on alternate hosts. Certain insect pests overwinter as eggs on alternate hosts, e.g. blackfly (*Aphis fabae*) migrates to the spindle tree (*Euonymus europaeus*) in the autumn. Virus diseases may be found in weed hosts near crops. The term *alternative host* may be used when the pathogen has a number of different hosts. Control of the alternate hosts can be an important way of reducing inoculum sources of some diseases.

alternation of generations The occurrence of two or more reproductive forms during the life cycle of an organism. In plants it usually includes an asexual and a sexual phase resulting in alternating diploid and haploid individuals in the life cycle. In bryophytes the dominant generation is the *gametophyte, whereas in the pteridophytes and seed plants it is the *sporophyte. If the two generations are markedly different the life cycle is termed *heteromorphic whereas if they are similar it is termed *isomorphic. *See also* haplontic, diplontic, haplobiontic, diplobiontic.

alternative host *See* alternate host.

amensalism Any association between two organisms in which one is harmed by the activities of the other. It includes parasitism and also the various ways in which plants protect themselves from pests and predators. Associations between microorganisms in which one is adversely affected are termed *antibiosis*. *See also* commensalism, mutualism.

Amentiferae *See* Hamamelidae.

amino acid A molecule containing both carboxylic acid and amino groups and having the general formula $RCHNH_2CO$ OH. The nature of the R group varies widely, from a hydrogen atom in glycine to aromatic and heterocyclic ring structures in such amino acids as tyrosine and tryptophan. Amino acids can be classified as nonpolar, polar uncharged, acidic, or basic, according to the nature of the R group (*see* tables).

Amino acids are the basic structural units of proteins with some 20 amino acids commonly occurring in proteins. In addition to these there are a few unusual amino acids that occur only in a few proteins and over 200 nonprotein amino acids that have been isolated from various plant sources. In many cases the function of these nonprotein amino acids is unclear but some are intermediates in the synthesis of commoı amino acids while others may have protective or storage functions.

Because they contain both acidic and basic groups, amino acids will react with both acids and bases. Consequently the charge on an amino acid varies with its pH. Each amino acid has a specific pH, known as its isoelectric point, at which the net charge on the molecule is zero. All naturally occurring amino acids, except glycine, are optically active due to the asymmetry of the α carbon atom. Most amino acids in nature are in the L form, although some D-amino acids are found in bacterial cell walls.

amino acid sequencing The determination of the amino acid sequence of a protein. The sequencing of homologous proteins from different species has been used as a method of determining phylogenetic relationships. It is assumed that the number of differences in the sequence may be related to the length of time since the different species diverged from a common ancestor. Phylogenetic trees have been constructed from data on the sequences of cytochrome *c* and

amino acid	symbol	R group
alanine	ala	$-CH_3$
valine	val	$-CH{<}\genfrac{}{}{0pt}{}{CH_3}{CH_3}$
leucine	leu	$-CH_2-CH{<}\genfrac{}{}{0pt}{}{CH_3}{CH_3}$
isoleucine	ile	$-\underset{\underset{CH_3}{\vert}}{CH}-CH_2-CH_3$
proline	pro	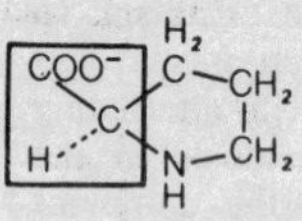
methionine	met	$-CH_2-CH_2-S-CH_3$
phenylalanine	phe	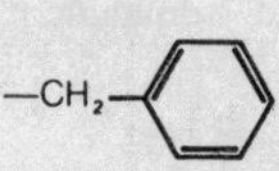
tryptophan	trp	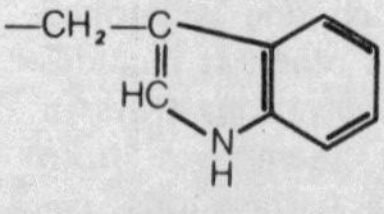

amino acids with nonpolar R groups

Structures of the 20 amino acids commonly found

structure is given, the remainder of the

amino acid	symbol	R group
glycine	gly	$-H$
serine	ser	$-CH_2-OH$
threonine	thr	$-\overset{OH}{\underset{H}{\vert C \vert}}-CH_3$
cysteine	cys	$-CH_2-SH$
asparagine	asn	$-CH_2-C\overset{NH_2}{\underset{O}{\lessgtr}}$
glutamine	gln	$-CH_2-CH_2-C\overset{NH_2}{\underset{O}{\lessgtr}}$
tyrosine	tyr	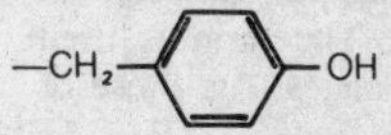 $-CH_2-C_6H_4-OH$

amino acids with uncharged polar R groups

amino acid	symbol	R group	
aspartic acid	asp	$-CH_2-C\overset{O^-}{\underset{O}{\lessgtr}}$	acidic amino acids
glutamic acid	glu	$-CH_2-CH_2-C\overset{O^-}{\underset{O}{\lessgtr}}$	
lysine	lys	$-CH_2-CH_2-CH_2-CH_2-\overset{+}{N}H_3$	basic amino acids
arginine	arg	$-CH_2-CH_2-CH_2-NH-\underset{\overset{\Vert}{\overset{+}{N}H_2}}{C}-NH_2$	
histidine	his	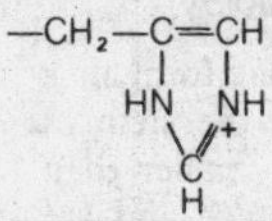 $-CH_2-C{=}CH$ $HN\quad NH$ $\overset{+}{C}H$	

amino acids with charged polar R groups

in proteins. In all except proline, where the complete molecule has the structure

$$HOOC-\overset{H}{\underset{NH_2}{\vert C \vert}}-$$

plastocyanin. These do not always correspond with information derived from other sources. *See also* DNA hybridization.

amino sugar A monosaccharide with an amino (NH_2) group in place of one of the hydroxyl (OH) groups. The most commonly occurring amino sugars are glucosamine and galactosamine. Amino sugars are components of glycoproteins.

amitosis Nuclear division by constriction into two parts without the appearance of chromosomes. The nuclear membrane does not break down and a spindle is not formed. Amitosis occurs in some primitive unicellular algae and has also been observed in the formation of endosperm tissue.

amoeboid Resembling an amoeba in form and movement. The term is used to describe certain gametes, e.g. those of the Zygnemaphyceae.

AMP (adenosine monophosphate) A phosphorylated nucleoside consisting of the purine adenine and the sugar ribose phosphorylated in the 5′ position. AMP is involved in the regulation of glycolysis and gluconeogenesis, promoting the formation of fructose bisphosphate from fructose 6-phosphate (i.e. promoting glycolysis) while inhibiting the back reaction to fructose bisphosphate.

amphicribal Describing a *concentric vascular bundle that has the phloem surrounding the xylem. *Compare* amphivasal.

amphidiploid *See* allopolyploidy.

amphigastria The leaves that form in a row on the undersurface of the stem of a leafy liverwort. They are smaller than the leaves on the upper surface and are often only seen just below the apex, having been shed lower down.

amphimixis True sexual reproduction involving the fusion of two gametes. *Compare* apomixis.

amphiphloic Having phloem arranged on both sides of the xylem as, for example, in a solenostele.

amphiphloic siphonostele *See* solenostele.

amphithecium The outer layer of the young sporophyte in bryophytes, giving rise to the capsule wall, which, in many Bryales, differentiates a *peristome. In *Sphagnum* and *Anthoceros*, the sporogenous tissue develops from the amphithecium, whereas in other bryophytes it develops from the *endothecium.

amphivasal Describing a *concentric vascular bundle in which the xylem surrounds the phloem. *Compare* amphicribal.

amplexicaul Describing a sessile leaf in which the base of the lamina clasps the stem at the node as, for example, the upper leaves of henbit (*Lamium amplexicaule*).

amylase The *hydrolase enzyme that catalyses the hydrolysis of the $\alpha(1-4)$ glucosidic bonds in starch. Amylase occurs in two forms, designated α and β.

β-amylase only attacks the nonreducing ends of a starch molecule, successively hydrolysing alternate $\alpha(1-4)$ linkages and releasing maltose molecules. β-amylase is found in germinating seeds and is important for the production of malt in the brewing industry. Amylose is completely degraded to maltose by β-amylase but amylopectin is only partially broken down because the β-amylase cannot attack $\alpha(1-4)$ linkages beyond the first branch on each chain. A separate enzyme, $\alpha(1-6)$-glucosidase, exists to hydrolyse the $\alpha(1-6)$ branching linkages of amylopectin.

α-amylase differs from β-amylase in that it can attack $\alpha(1-4)$ bonds within the starch molecule. It thus degrades amylopectin more completely than β-amylase.

amylopectin A polysaccharide that, with *amylose, makes up starch. It consists of glucose units linked by $\alpha(1-4)$ glucosidic bonds with branches formed by $\alpha(1-6)$ bonds. The molecular weight of amylopectin may be as high as 10^7. It can form either colloidal or micellar solutions with water. *See also* amylase.

amylose A polysaccharide that, with *amylopectin, makes up starch. Amylose is an unbranched chain of glucose molecules linked by $\alpha(1-4)$ glucosidic bonds. The molecular weight of amylose can vary from a few thousand to over half a million. In water amylose has a micellar structure and the amylose chains form helical coils. *See also* amylase.

amylum *See* starch.

anabolism The metabolic synthesis of complex molecules from simpler ones. It requires an input of chemical energy, which is provided by ATP. The Calvin cycle is an example of an anabolic pathway. *Compare* catabolism.

anaerobe An organism that can live in the absence of free oxygen. Obligate or strict anaerobes cannot live in the presence of free oxygen. Facultative or indifferent anaerobes can grow in the presence of oxygen but do not use it. Such organisms include the denitrifying bacteria and lactic acid bacteria. The term facultative anaerobe may also be used of such organisms as yeasts that can grow under anaerobic conditions but given free oxygen will use it to oxidize the products of anaerobic respiration. They thus grow better with free oxygen. However yeasts and similar organisms could equally well be described as facultative aerobes. *Compare* aerobe.

anaerobic respiration Any of various catabolic pathways by which chemical energy is obtained from organic compounds in the absence of free oxygen. The glycolytic pathway (*see* glycolysis) and *alcoholic fermentation are the two main examples. Other types include *mixed lactic fermentation and various other bacterial fermentations in which the end products include propionic acid, butyric acid, and acetone.

anagenesis The degree or rate of evolutionary divergence.

analogous Describing organs, often similar in appearance, that carry out similar functions but have different origins. For example, phyllodes are analogous to leaf blades but are derived from petioles. *Compare* homologous. *See also* convergent evolution.

anaperturate Describing a pollen grain without any type of aperture.

anaphase The stage following metaphase in nuclear division, during which separation of either chromatids or homologous chromosomes commences. In *mitosis and in the second division of *meiosis daughter chromatids are seen to move apart towards opposite poles of the spindle. Although the nature of the forces operating to initiate this separation is as yet unknown, two processes within the spindle, both probably involving microtubule activity, are believed to contribute to chromatid movement. The first is the elongation of the spindle fibres in the equatorial region between the chromatids. The spindle eventually doubles in length. The second is the shortening of the fibres attached to the *kinetochores in the centromeric regions. This is thought to be due to depolymerization of tubulin in the constituent microtubules at the ends nearest to the poles.

In anaphase of the first division of meiosis, the homologous chromosomes of each bivalent, each with a complete centromere, become separated and move towards opposite poles of the spindle.

anatomy Plant structure, or the study thereof, with an emphasis on tissues and their component cells in the interior of the plant body. *Compare* morphology.

anatropous Describing the form of ovule orientation in the ovary in which the funiculus has lengthened and the ovule turned through 180° so that the micropyle is folded over and lies near the base of the funiculus (*see illustration at* ovule). This arrangement is the most common. *Compare* campylotropous, orthotropous.

Andreaeales *See* Musci.

androdioecious Describing plant species in which male and hermaphrodite flowers are borne on separate individuals, as in *Caltha* (marsh marigolds). *Compare* andromonoecious, gynodioecious.

androecium The male component of a flower, made up of several *stamens and sometimes also staminodes. The androecium usually surrounds the *gynoecium, although the exact arrangement or position may not be symmetrical in the more advanced forms. The androecium is described as *apostemonous* when the stamens are separate, *monadelphous* when the filaments are all fused, and *syngenesious* when the anthers are fused. If the stamens form two groups, the members of each group being joined by their filaments, the androecium is *diadelphous* and if three or more groups are formed in this way, *polydelphous*. *Petalostemonous* describes the condition where the filaments are joined to the petals. The androecium is represented in the floral formula by the letter A.

androgynophore (androphore) An extension of the receptacle between the petals and the stamens on which the androecium and gynoecium are borne. It is seen in many members of the Capparaceae.

andromonoecious Describing plant species in which male and hermaphrodite flowers are borne separately on the same individual, as in *Aesculus hippocastanum* (horse chestnut). *Compare* androdioecious, gynomonoecious.

androphore *See* androgynophore.

androspore A specialized zoospore produced in the Oedogoniales. It does not participate directly in the fertilization process but swims towards and attaches itself to a female filament. Here it germinates to form a microfilament, which in turn liberates spermatozoids from the upper disc-shaped cells (antheridia). These then fertilize the female gamete.

anemophily (wind pollination) Pollination by pollen carried on the wind. Wind-pollinated flowers often have reduced sepals and petals and often appear before the leaves. This helps ensure that the stigmas are effectively positioned for pollen interception and the stamens are free to release their pollen. The stamens often have very long filaments while the styles may be long and feathery. As female and male parts compete for room they are often found in different flowers on the same plant, as in hazels (*Corylus*), or on separate plants, as in black bryony (*Tamus communis*).

The pollen of wind-pollinated plants needs to be light and smooth surfaced and is only released from the anthers on warm dry days. In catkin-bearing species it can be stored in saucer-like bracts until disturbed and transported on a windy day. Wind-transmitted pollen has a typical diameter of 20–30 μm and may be carried thousands of miles but will normally travel less than 1 km. Because air movement is random pollen needs to be produced in vast quantities (a hazel catkin may produce 4 000 000 pollen grains). This is wasteful of the plant's resources and is thought by some to be the most primitive form of pollination. Others consider that the surviving species using this method show evidence of secondary modification from entomophilic forms. *Compare* entomophily, hydrophily.

anergized culture *See* habituation.

aneuploidy A condition in which not all the chromosomes are present in equal numbers and hence the total number is not an exact multiple of the haploid set. It occurs when chromosomes fail to separate at meiosis (*see* nondisjunction), so a gamete may either lack one chromosome altogether or have an additional copy. On fertilization the resulting zygote may thus have only one homologue of a given chromosome, and is described as a *monosomic, or it may have three homologues, and is called a *trisomic. If both gametes lack the same chromosome then the zygote is said to be *nullisomic*. Nullisomics are often inviable. Alternatively if both gametes contain the same additional chromosome, the zygote is termed *tetrasomic*. If there are missing or additional copies of two chromosomes the zygote is described as double monosomic or double trisomic respectively.

Aneuploidy results in unusual segregation ratios. It has been most closely studied in the thorn apple (*Datura stramonium*) where 12 types of trisomics have been recognized (one for each of the 12 different chromosomes), each producing a different mutant phenotype.

aneuspory The production, through a modification of the meiotic process, of an unusual number of spores (usually two) instead of the four normally formed from each spore mother cell. It is seen in the formation of megaspores in dandelions (*Taraxacum*) where, after the first meiotic division the chromosomes stay in the one cell forming a *restitution nucleus*. The second meiotic division gives rise to two cells each with an unreduced number of chromosomes. One of these develops parthenogenetically into an embryo. Crossing over and hence reassortment of the genes can occur during the first meiotic division. This accounts for some of the variation found in apomictic complexes that have arisen by aneuspory.

Angiospermae (flowering plants) A class of vascular plants, all characteristically bearing seeds within enclosing carpellary structures. The sporophyte is the dominant generation and is either herbaceous or woody, the woody habit being considered more primitive. The reproductive axis and its associated, often brightly coloured, sepals and petals, is called a flower. The gametophyte is reduced to the female embryo sac and the male pollen grain. The pollen does not germinate directly on the ovule, as in the *Gymnospermae, but on a specialized extension of the carpel, the stigma. The male gametes, unlike certain gymnosperm gametes, are never flagellate. Double fertilization to form a zygote and a triploid endosperm nucleus is characteristic.

Secondary vascular tissue is usually but not always present. The xylem contains vessels, except in certain primitive woody forms, and the phloem has distinct companion cells associated with the sieve tube elements.

Angiosperms are the most advanced, most abundant, and most widely distributed vascular plants. The group contains some 250 000 species and is subdivided, on the basis of the number of cotyledons in the embryo, into the *Monocotyledonae and the *Dicotyledonae. Beyond these groups further subdivision into superorders and orders is based mainly on the structure of the flower and especially on the form, number, and arrangement of the stamens and carpels. Different classifications recognize various numbers of orders and the names and contents of these often differ widely between various authorities.

From fossil pollen evidence it would seem the angiosperms appeared at the beginning of the Cretaceous. They had replaced the gymnosperms as the dominant vegetation by the second half of the Cretaceous period. This may have been due in part to the relatively rapid life cycle of angiosperms in which seed set occurs days or weeks after flowering whereas in gymnosperms the period between pollination and seed set is at least a year. The radial symmetry of angiosperm seeds suggests the group may have evolved from the Pteridospermales. However numerous groups, including the Coniferales, Gnetales, Bennettitales, and Cycadales, have been postulated as angiosperm ancestors. These have all been discounted for various reasons and the origin of the angiosperms remains obscure. *See also* Magnoliophyta.

Angstrom Symbol: Å. A former unit of length equal to one thousandth of a micrometre or one tenth of a nanometre, i.e. 10^{-10} metre.

aniline stain Any of a group of dyes derived from aniline that are used to stain biological material. Aniline sulphate and aniline hydrochloride both stain lignin yellow. Aniline blue is often used as a counterstain with safranin.

anisogamy The production or fusion of motile gametes that differ in size. This condition is found in various algae and fungi. *Compare* isogamy. *See also* oogamy.

annual A plant that germinates from a seed, grows, flowers, produces seeds, and then dies within a single year. Examples are marigold (*Calendula officinalis*) and poppies (*Papaver*). *Compare* ephemeral, biennial, perennial.

annual ring The increment of secondary xylem added to the wood of a plant in a single year. In transverse section this often appears as one or more rings due to the seasonal variation in tracheary element diameter. *See also* growth ring, ring-porous wood.

annular thickening A type of secondary wall patterning in tracheary elements in which the secondary cell wall is laid down in rings (*see illustration at* tracheary elements). Annular thickening is common in tracheary elements that have not yet finished elongating, for example those in protoxylem. *Compare* pitted thickening, reticulate thickening, scalariform thickening, spiral thickening.

annulus A band or circle of tissue, especially: **1.** The ring of differentially thickened cells that encircles the sporangium of certain ferns. It aids spore dispersal by inducing tension in the sporangial wall as the water in its cells evaporates. This causes the rupture of the *stomium and the top of the sporangium gradually curls back. As the tension increases a certain stage is reached when the water in the cells vaporizes, so releasing the tension and causing the top of the sporangium to spring back. Such movements serve to release the spores.
2. (velum) The remnants of the ruptured partial veil that encircle the stalk of a mushroom or toadstool.

anomocytic *See* subsidiary cell.

anther The apical portion of a *stamen, which produces the microspores or pollen grains. An anther normally comprises four *pollen sacs (but only two in the Malvaceae) arranged in two groups or lobes joined by the connective tissue to the filament. If the anther is attached to the filament on the dorsal surface allowing it to pivot it is called a *versatile* anther. If it is attached dorsally but there is no movement it is termed a *dorsifixed* anther and if attached at the base it is a *basifixed* anther.
The wall of the anther lobes consists of an outer epidermis below which is the *endothecium. This surrounds an inner

nutritive layer or *tapetum. Within the cavity (*loculus*) of the lobes pollen mother cells undergo meiosis to form tetrads of pollen grains. The individual pollen sacs are joined by zones of parenchyma tissue. *Compare* ovule.

anther culture The culture of excised anthers on sterile nutrient medium. If anthers are taken from a plant at a certain stage of development and cultured under appropriate conditions then embryoids and subsequently haploid plantlets may be induced to form from the pollen grains. In some species it is possible to culture isolated pollen grains. This precludes the possibility that any resulting plantlets might be derived from somatic tissue rather than a pollen grain. In culturing isolated pollen grains, anther tissue may have to be used as nurse tissue. If the ploidy level of plantlets derived from pollen grains is doubled using colchicine then completely homozygous diploid plants can be obtained. *Compare* ovule culture.

antheridial cell The cell from which the antheridium develops. In seed plants, it is the generative cell in the pollen grain, which divides to provide the two sperm cells in the pollen tube. *Compare* prothallial cell.

antheridiophore An upright structure consisting of a stalk and cap that bears the antheridia in certain liverworts of the Marchantiales. The antheridia are borne in pits on the upper surface of the cap. *Compare* archegoniophore.

antheridium The male sex organ of the lower plants. In the algae and fungi it is unicellular, whereas in the bryophytes and pteridophytes it may be multicellular and surrounded by a sterile jacket. It usually produces numerous small motile gametes. *Compare* oogonium, archegonium.

antherocyte (spermatocyte) A cell that differentiates into an *antherozoid without further cell division. The antherozoid is usually released from the antherocyte after it has been discharged from the antheridium.

antherozoid (spermatozoid) A motile male gamete produced by lower plants and some gymnosperms, which moves by means of flagella. In lower plants antherozoids are released from an antheridium but in the gymnosperms they are formed in the pollen tube prior to fertilization. Most antherozoids consist of an elongated nucleus contained within a ribbon-like cell. This form enables them to penetrate the narrow neck of an archegonium. The number of flagella borne by an antherozoid may be used as a diagnostic character.

anthesis The period from flower opening to fruit set.

Anthocerotae (hornworts, horned liverworts) A class of the *Bryophyta containing a single order, Anthocerotales, and some five genera. Members of this order are distinguished from those of liverwort orders, with which they are sometimes classified, by their long-lived axial sporophytes. These contain photosynthetic tissue and stomata and occasionally persist after the death of the thallose gametophyte. Most species are also unusual in having a single pyrenoid-containing chloroplast in each cell, a feature reminiscent of the algae.

anthochlor pigments A group of yellow flavonoid flower pigments containing the *chalcones and *aurones. They turn red on exposure to ammonia. This reaction distinguishes them from the yellow carotenoid pigments, which do not react in this way.

anthocyanescence The development of red pigments as a symptom of disease as, for example, seen in peach leaf curl caused by the fungus *Taphrina deformans*.

anthocyanin Any of a group of glycoside pigments formed by the addition of sugars and other residues to an *anthocyanidin* precursor (usually pelargonidin, delphinidin, or cyanidin). A very large number of these pigments have been characterized, all of them either red, blue, or violet. They are sap soluble, and occur in the cell sap of flowers, fruits, stems, and leaves.

anthoxanthin Any of a class of yellow or cream glycoside plant pigments normally consisting of a glucose molecule attached to a flavone or xanthone molecule.

anthracnose A fungal plant disease in which the characteristic symptoms are limited lesions, *necrosis, and *hypoplasia. Anthracnose diseases are generally caused by one of the *Melanconiales (e.g. *Colletotrichum lindemuthianum*, bean anthracnose; *C. coffeanum*, coffee anthracnose or coffee-berry disease; and *Gloeosporium limetticola*, lime anthracnose). The fungi causing anthracnose diseases produce numerous spores that are spread by rain. High humidity is required for infection and the diseases are most destructive when there is some water-soaking of tissues. Anthracnose diseases also produce symptoms, including scab, leaf-spots, and blight, that are not exclusive to the group.

antibiosis *See* amensalism.

antibiotic Any substance produced by a microorganism that inhibits the growth of another microorganism. Antibiotics are widely used as drugs to combat bacterial diseases. Examples are penicillin, obtained from the mould fungus *Penicillium notatum* and active against staphylococcal infections and many other gram-positive bacteria, and streptomycin, obtained from the actinomycete bacterium *Streptomyces griseus* and used to treat tuberculosis. Biosynthesis of antibiotics may be from amino acids (e.g. penicillin), sugars (e.g. streptomycin), or from acetate or propionate (e.g. tetracyclines). Commercial production is usually by large scale culture of the appropriate organism though some simple antibiotics, e.g. chloramphenicol, are cheaper to produce by artificial synthesis.

Antibiotics have proved useful research tools. Those that inhibit protein synthesis have been used to investigate ribosome structure and function. Most antibiotics inhibit protein synthesis on the 70S ribosomes of prokaryotes but not the 80S ribosomes of eukaryotes. The susceptibility of mitochondrial and chloroplast ribosomes to antibiotics is taken as further evidence that these organelles are derived from endosymbiotic prokaryotic organisms (*see* serial endosymbiotic theory). Some antibiotics, e.g. cycloheximide, inhibit protein synthesis by 80S ribosomes but not 70S ribosomes. *See also* phytoalexin, allelopathy.

antibody *See* serology.

anticlinal At right angles to the surface. The *anticlinal wall* of a cell is thus arranged perpendicular to the surface of the plant body. An *anticlinal division* results in the formation of anticlinal walls between daughter cells. Such a division enables a tissue to increase its circumference, thus keeping pace with any increase in girth of the organ. In cylindrical organs, such as stems and roots, the term *radial* may be used in place of anticlinal, especially when describing cell walls. *Compare* periclinal.

anticodon A sequence of three nucleotides on transfer RNA that is complementary to a sequence (the *codon) on messenger RNA, to which it temporarily binds during protein synthesis. A given molecule of transfer RNA will possess a specific anticodon that only complexes with one particular amino acid. It is this absolute correspondence between an amino acid, an anticodon, and a codon, that enables the type and sequence of amino acids in a protein to be determined precisely.

antigen *See* serology.

antipodal cells The haploid cells, usually three in number, found in the embryo sac at the opposite end to the micropyle. They are derived by mitotic divisions of the *megaspore and have no distinct cell wall. They take no part in the fertilization process and their function is unknown. At fertilization they may disintegrate or multiply and enlarge.

antithetic *See* heteromorphic.

antitranspirant A compound that reduces transpiration, either by closing the stomata or by depositing a film over the stomata. Antitranspirants are sprayed on some crops and ornamentals, but their use also limits photosynthesis and affects various other metabolic processes.

aperturate Describing a pollen grain having one or more apertures (areas where the exine is either thinner or absent). If the apertures are colpi the pollen is termed *colpate* and if they are pores, *porate*.

aphid Any insect of the family Aphididae (greenfly, plant lice) of the insect order Hemiptera. Aphids are small plant bugs that feed by sucking plant juices. Many species are pests in their own right, such as the blackfly (*Aphis fabae*) on broad beans. Other species are chiefly important as vectors of plant virus diseases, such as cucumber mosaic virus, which is transmitted by several species of aphid (including *Aphis gossypii*). *Macrosiphum avenae* and *Rhopalosiphum* species both transmit barley yellow dwarf virus of cereals and grasses.

Aphyllophorales (Polyporales) An order of the *Hymenomycetes containing fungi that produce a large basidiocarp, which rarely contains gills. It contains over 1000 species in some 375 genera. Four main groups are recognized depending on the structure of the basidiocarp: the polypores or bracket fungi, which grow out in a bracket-like manner from both living and dead wood (e.g. dryad's saddle, *Polyporus squamosus*); the club fungi, in which the fruiting body consists of a number of finger or clublike projections (e.g. *Clavulinopsis helvola*); fungi in which there are many tooth or spinelike projections from the basidium (e.g. hedgehog fungus, *Hydnum erinaceus*); and fungi in which the basidium is more or less flattened (e.g. earth fan, *Telephora terrestris*). These groups have been classified as families but it is now realized that these are artificial.

Apiaceae *See* Umbelliferae.

apical dominance The inhibition of the development of some or all of the lateral buds by the terminal (apical) bud of a shoot. Removal of the terminal bud releases some of the lateral buds from inhibition. This implies that a substance produced at the apex, most probably auxin, is responsible for the inhibition, though its method of action is unclear. Apical dominance is not so marked when nutrients are plentiful, suggesting that available nutrients are first delivered to the terminal bud and only those in excess of requirements reach the lateral buds. Cytokinins have also been shown to promote the growth of lateral buds.

apical meristem The *meristem at the tip of a stem or root that gives rise to primary tissues and is responsible for increase in length rather than girth of the axis. *Compare* lateral meristem. *See also* ground meristem, procambium, protoderm, histogen theory, tunica-corpus theory.

apical placentation (pendulous placentation, suspended placentation) A form of *placentation, found in ovaries containing only one ovule, in which the placenta develops at the top of the ovary.

apiculate Having a small broad point at the apex.

aplanetic Describing organisms in which there is no motile stage.

aplanospore A nonmotile spore as produced, for example, by fungi in the Zygomycotina and by certain algae in the Chlorophyta and Chrysophyta. *Compare* zoospore.

apocarpous Describing a *gynoecium in which the *carpels are free, as in buttercups (*Ranunculus*). This type of arrangement is thought to be more primitive than a *syncarpous gynoecium.

apoenzyme The catalytically inactive protein portion of an enzyme that remains when the prosthetic group or *cofactor has been removed. Examples of enzymes needing both apoenzyme and cofactor for catalytic activity include: alcohol dehydrogenase, which requires zinc ions; kinases, which require magnesium or manganese ions; and the cytochromes, which require ferrous, Fe (II), or ferric, Fe(III), ions. *See also* holoenzyme.

apogamy The development of the sporophyte directly from the gametophyte without the formation of gametes. The resulting sporophyte therefore has the same chromosome number as the gametophyte. Although there is no nuclear alternation between the generations the morphological differences persist. The phenomenon is seen in certain ferns, fungi, and algae. If male gametes are produced they are redundant although they have been shown to be capable of functioning in certain species. For example, the male fern (*Dryopteris borreri*) produces antherozoids that can fertilize the female gametes of related ferns.
Apogamy often occurs when the gametophyte has been produced by *apospory. It may also be induced by ageing or chemical agents. The term may also refer to the development of an unreduced diploid cell of the embryo sac into an embryo without fertilization occurring (i.e. parthenogenesis).

apomixis Any form of *asexual reproduction, including vegetative propagation. The term is very often used in a narrower sense to mean the production of seeds without fertilization occurring. In this restricted sense the term is synonymous with *agamospermy*. Seed production by apomixis may occur either by the formation of a diploid embryo (*see* adventive embryony) or embryo sac by a somatic cell, or by suppression or modification of the meiotic process to give unreduced megaspores (*see* diplospory, aneuspory). Development into a mature diploid embryo can then proceed without fertilization (*see* parthenogenesis, pseudogamy).
Apomixis is usually associated with polyploidy. An organism that reproduces by apomixis is termed an *apomict*. Facultative apomicts, e.g. the cinquefoils (*Potentilla*), can reproduce both sexually and apomictically. In such species the incidence of apomixis may be affected by environmental factors, such as photoperiod. Obligate apomicts can only reproduce apomictically. Often these are triploids or pentaploids that cannot produce viable pollen. Plants that form apomictic complexes are notoriously difficult to classify. Some apomictic races are so constant that they have been given taxonomic status as species (for example, almost 400 species of *Rubus* have been recognized in Britain).

apomorphy In discussions of phylogeny or cladistics, a derived or advanced character state. Apomorphies shared by different taxa are termed *synapomorphies* (*compare* autapomorphy). Only synapomorphies whose origin can be traced back to a recent common ancestor (i.e. homologous character states) are of use in constructing phylogenies or cladograms. Analogous 'synapomorphies' may arise by convergent evolution and if identified as such are ignored. *Compare* plesiomorphy.

apophysis The slightly swollen region between the seta and the capsule in a moss sporophyte. Its cells are rich in chloroplasts, there are numerous intercellular spaces, and the epidermis contains stomata, similar in structure to those of vascular plants. The apophysis is thus an active photosynthetic region and helps to nourish the developing sporogonium.

apoplast The continuum of nonprotoplasmic matter, such as cell walls and intercellular material, throughout a plant. The movement of water in nonvascular tissues is principally through the apoplast as its resistance to flow is approximately 50 times less than that of the *symplast.

apospory The development of the gametophyte from the sporophyte without meiosis and spore production, so that the gametophyte has the same number of chromosomes as the sporophyte. If fertile gametes are formed by such a gametophyte then a sporophyte with twice the original number of chromosomes is produced. A polyploid series may be built up in this way. Apospory can be induced artificially in certain ferns, e.g. the lady fern (*Athyrum filix-femina*), by pinning segments of frond to damp sand. Prothalli then arise from buds on the frond.
Apospory may also be used to describe the condition in angiosperms in which a diploid embryo forms from a cell of the nucellus or chalaza and megaspore formation is bypassed (*see* adventive embryony). *Compare* apogamy.

apothecium The disc- or cup-shaped ascocarp characteristic of discomycete fungi (excepting the Tuberales). The tips of the asci are freely exposed. Most lichens contain discomycete fungi and such lichen fungi also form apothecia.

apposition The laying down of layers of *cellulose on the inner surface of a plant cell wall to form the secondary cell wall. The process normally occurs once extension of the cell wall is completed and it serves to strengthen the overall cell structure. *Compare* intussusception.

appressorium A hyphal structure, formed by many parasitic fungi, that serves to effect penetration of the host epidermis. It is flattened and closely pressed to the outer surface of the epidermis. From the undersurface a narrow *infection hypha* or *penetration tube* is pushed through into the cell or cell spaces below. This then expands into a haustorium or develops into hyphae.

aquatic *See* hydrophyte.

araban (arabinan) A polysaccharide in which the major monosaccharide subunit is the pentose sugar arabinose. Arabans are found with pectic substances in mature primary cell walls and in hemicelluloses and gums.

arabinose A pentose sugar of the aldose group commonly found in plants. It often occurs in the polymerized form *araban. *See illustration at* aldose.

arboretum An area devoted to the cultivation of a wide selection of all the woody plants (trees, shrubs, vines, etc.) that may be grown in a particular climatic region. Arboreta are maintained both as centres of research and as educational and recreational areas. *See also* botanic garden.

arbuscular-vesicular mycorrhiza *See* vesicular-arbuscular mycorrhiza.

archegoniophore An upright structure consisting of a stalk and cap that bears the archegonia in certain liverworts. In *Marchantia* sterile raylike structures radiate from the cap giving it the appearance of a chimney sweep's brush. In *Reboulia* the archegoniophore is umbrella shaped. The archegonia are borne in groups on the lower surface of the cap, the groups being separated by involucral scales termed *perichaetia.* The stalk of the archegoniophore does not elongate until fertilization has occurred. Thus the antherozoids can swim to the archegonia before the water film between the thallus and archegoniophore is disrupted. *Compare* antheridiophore.

archegonium The female sex organ of the bryophytes, pteridophytes, and most gymnosperms. It is normally produced within the maternal tissue and is related to the development of the terrestrial habit. It is made up of a narrow neck and a swollen base (or venter) that contains the female gamete. *Compare* antheridium.

archesporium The tissue that gives rise to the spore mother cells.

arctic *See* tundra.

Arecaceae *See* Palmae.

Arecidae A subclass of the monocotyledons containing both herbaceous and arborescent (the palms) plants. Members of the Arecidae differ from other monocotyledons in having broad net-veined petiolate leaves. Their numerous small flowers are usually unisexual and grouped into an inflorescence subtended by a spathe. Four orders are commonly recognized: the Arecales, which contains one family, the Arecaceae or *Palmae (palms); the Cyclanthales, which also contains one family, the Cyclanthaceae; the Arales, which contains the two families Lemnaceae (which includes the duckweeds), and Araceae (which includes the arums or aroids); and the Pandanales, which comprises the one family Pandanaceae (which includes the screw pines). In some classifications the order Typhales is placed in the Arecidae, while in others it is allocated to the Commelinidae. The Typhales includes two families, the monotypic Typhaceae, containing the cattails and reedmace bulrush (*Typhus*), and the monotypic Sparganiaceae, containing the bur-reeds (*Sparganum*).

areole 1. A sunken cushion representing a condensed lateral shoot from which spines, branches, and flowers arise in cacti. Areoles may occur either singly on tubercles (e.g. in *Mammillaria zeilmanniana*) or in rows along raised ridges.
2. Any area outlined on a surface, e.g. a segment of leaf lamina surrounded by veins. A surface divided into areoles, e.g. lichen thalli split by cracks into roughly hexagonal areas, is described as *areolate.*

arginine A basic amino acid with the formula $H_2NC(:NH)NH(CH_2)_3CH(NH_2)COOH$ (*see illustration at* amino acid). It is important in histone proteins, which are rich in this amino acid.

aril A fleshy or hairy outgrowth of a seed or fertilized ovule, commonly derived from the funiculus or hilum. It may be regarded as a modified outer integument that only becomes conspicuous following fertilization. The brightly coloured mace surrounding the seed of nutmeg (*Myristica fragrans*) is an example. In the white water lily (*Nymphaea alba*), in which seed dispersal is by water, the seed is covered by a spongy aril that helps keep the seed afloat. The red fleshy cup surrounding the seed of yew (*Taxus baccata*) is also an aril. *See also* caruncle.

aristate Describing a structure, such as a glume or lemma, bearing an awn. *Compare* bearded.

arithmetic mean *See* mean.

arthrospore *See* oidium.

ascocarp The fruiting body of all ascomycete fungi except the Hemiascomycetes. It consists of an aggregation of hyphae surrounding the asci. The various types of ascocarp are used in dividing the Ascomycotina into classes. *See* cleistothecium, apothecium, perithecium, pseudothecium.

ascogenous hypha Any of a number of small multinucleate branches that develop from an ascogonium. Ascogenous hyphae give rise to asci by *crozier formation.

ascogonium The large coiled multinucleate female gametangium of certain fungi in the Ascomycotina. It is fertilized either by antheridial contents or by spermatia and gives rise to ascogenous hyphae.

Ascomycotina (sac fungi) A subdivision of the *Eumycota, most of whose members have a mycelial septate thallus though some (the yeasts) are unicellular. It contains about 16 500 species in almost 2000 genera. Sexual reproduction is by ascospores, with eight characteristically developing within an ascus. It is divided into the classes: *Hemiascomycetes; *Plectomycetes; *Pyrenomycetes; *Discomycetes; Laboulbeniomycetes, which are ectoparasites of arthropods; and *Loculoascomycetes. The sac fungi are often placed in the class Ascomycetes, and divided into the two subclasses Hemiascomycetidae (or Protoascomycetidae) and Euascomycetidae. About half the ascomycetes have developed a special mode of nutrition in which algal cells are incorporated in the thallus (*see* lichen).

ascorbic acid (vitamin C) A lactone of a *sugar acid that is found in high concentrations in certain fruits and green vegetables. It acts as a cofactor in the hydroxylation of proline to hydroxyproline. However its function in this reaction can be replaced by other compounds and it is not clear why it is essential to the growth of some vertebrates. *See* vitamin.

ascospore One of usually eight haploid spores characteristically formed inside an ascus. The two nuclei of the dikaryotic ascus initially fuse and undergo meiosis to produce four haploid daughter nuclei. These then undergo one mitotic division resulting in eight haploid nuclei. The cytoplasm of the ascus is then isolated around each nucleus to form the ascospores.

ascus A specialized cell in fungi of the Ascomycotina in which ascospores develop. If it is cylindrical, as is usually the case, then the ascospores are violently discharged. However, in some ascomycetes (e.g. the Endomycetales and Eurotiales) the ascus is a globular sac and in these fungi the ascospores are released passively. The tip of the ascus may possess a cap, the operculum, or it may simply have a terminal pore. This characteristic is important in separating the operculate Pezizales from the inoperculate Helotiales. In most ascomycetes the ascus wall is a single layer (unitunicate) but in the Loculoascomycetes it consists of two layers (bitunicate).

asexual reproduction (*apomixis) The formation of new individuals from the parent without the fusion of gametes. This may be achieved by *budding, *fission, or *fragmentation in the lower plant forms, or by *spore formation or *vegetative reproduction in the higher plants. Individuals so formed have a genetic constitution identical to that of the parent. *Compare* sexual reproduction. *See also* clone.

asparagine An uncharged polar amino acid with the formula $NH_2COCH_2CH(NH_2)COOH$ (*see illustration at* amino acid). Asparagine is formed by the ATP-assisted addition of ammonia to aspartic acid. The reverse of this reaction (but without ATP formation) is the route of asparagine breakdown.

Because ammonia is extremely toxic to living cells, it is used to synthesize *glutamine and asparagine, in which form it can be stored for later use in amino acid synthesis.

aspartic acid An acidic amino acid with the formula $HOOCCH_2CH(NH_2)COOH$ (*see illustration at* amino acid). It is formed by a transamination, in which the amino group of glutamic acid is transferred to oxaloacetic acid so forming aspartic acid. Breakdown is by the reverse reaction, followed by further oxidation in the TCA cycle.

Aspartate is an important precursor of nitrogenous compounds. The amino acids leucine, isoleucine, threonine, methionine, and lysine all derive from aspartic acid, and it is necessary for synthesis of purines, pyrimidines, and porphyrins. Aspartate also acts as an amino-group donor in several reactions.

Aspergillales *See* Eurotiales.

association A large climax community named after the dominant types of plant species, e.g. deciduous-forest association, heath association. Most associations have more than one dominant plant. *See also* consociation.

assortive mating Breeding that is nonrandom. If similar phenotypes breed together it is *positive assortive mating* and leads to *inbreeding. If dissimilar phenotypes breed together it is *negative assortive mating* and leads to *outbreeding. The pollinating mechanism often has a major influence on the breeding patterns found. Thus in insect-pollinated plants, the pollinator may preferentially take pollen to flowers of a certain colour or with certain markings. In wind-pollinated plants crossing is more likely to be random but there is a tendency for plants of the same height to cross. *See also* self incompatibility.

aster A starlike arrangement formed by fibrils radiating from a *centriole. Asters become conspicuous at the poles of the *spindle as cell division commences in some primitive plant cells.

Asteraceae *See* Compositae.

Asteridae (Sympetalae) A subclass of the dicotyledons containing mostly herbaceous plants having flowers with fused petals. They have few stamens and usually only two carpels and the seeds are surrounded by only one integument (*unitegmic*). The following seven orders are generally recognized: Gentianales (Contortae), including the Loganiaceae or Buddlejaceae (e.g. buddleias), Gentianaceae, Apocynaceae (e.g. oleanders), and Asclepiadaceae (e.g. milkweeds); Polemoniales, including the Convolvulaceae (e.g. bindweeds), Polemoniaceae (e.g. phlox), and Boraginaceae (e.g. forget-me-nots); Lamiales, including the Verbenaceae and *Labiatae; Scrophulariales, including the *Scrophulariaceae (foxglove family), Gesneriaceae (e.g. gloxinias), and Acanthaceae; Campanulales (Campanales), including the Campanulaceae (bellflower family) and Lobeliaceae; Dipsacales, including the Caprifoliaceae (e.g. honeysuckles); and Asterales, comprising the *Compositae. Members of the Polemoniales, Lamiales, and Scrophulariales are sometimes grouped in the Tubiflorae.

Three further orders are sometimes recognized: Plantaginales (e.g. plantains) often included in the Scrophulariales; Rubiales, often included in the Gentianales; and Calycerales, often included in the Dipsacales.

The affinities of the important family *Solanaceae are uncertain, some placing it in the Polemoniales and others in the Scrophulariales.

astrosclereid (asterosclereid, star sclereid) A relatively short sclerenchyma cell (*sclereid), differing from a *brachysclereid by its often conspicuously branched shape. Astrosclereids are usually present singly or in small groups and are often found in the mesophyll of leaves, where they act as a strengthening agent.

atactostele A stele, typical of monocotyledon stems, in which the vascular bundles are arranged more or less irregularly in the ground tissue (*see illustration at* stele). *Compare* eustele.

atmometer *See* potometer.

ATP (adenosine triphosphate) The triphosphorylated form of adenosine, similar in structure to AMP but with three phosphates, linked by *high-energy phosphate bonds. It is the major energy-transferring molecule in all biological systems. The energy produced by photophosphorylation and metabolic oxidation reactions is used to form ATP from ADP and inorganic phosphate. ATP then provides energy for biosynthesis, active transport of ions and metabolites, and other energy-requiring processes. It is also essential for the synthesis of the nucleic acids DNA and RNA.

Besides its ubiquitous role as an energy-transferring molecule, ATP is involved in the regulation of sugar oxidation, being an inhibitor of the conversions of pyruvic acid to acetyl CoA and of fructose 6-phosphate to fructose bisphosphate.

atropous *See* orthotropous.

auricle A small earlike projection from the base of a leaf or petal. Auricles are seen at the base of the leaf blade in grasses (Gramineae). *See also* ligule.

aurones One of the two groups of *anthochlor flower pigments. They are probably formed by oxidation of *chalcones. Aurones are found in many members of the Compositae, especially the genus *Coreopsis*.

autapomorphy Any derived character state (*see* apomorphy) only possessed by members of one particular taxon. Autapomorphies distinguish a taxon from other related taxa. The number of autapomorphies possessed by one taxon as compared to another taxon assumed to be derived from the same common ancestor, provides a measure of anagenesis.

autecology The study of a single species and its relationship with the environment. It involves the investigation of the life cycle of the organism, recording the effects of the nonliving and living factors in the environment at each stage. Quantitative records are kept, particularly those relating to variations in numbers within and between populations. *Compare* synecology.

author In taxonomy, the person who published the first valid name of a taxon. The name of the author, or a recognized abbreviation of it, should follow the name of the taxon for it to be complete. Many plant taxa are followed by L. (for Linnaeus). If the rank of a taxon at the genus level or below is changed but the name or epithet retained then the name of the original author is also retained in brackets before the name of the person who made the change. For example, *Medicago polymorpha* var. *orbicularis* L. was raised to the species level by Allioni and is thus named *Medicago orbicularis* (L.) All. A similar procedure is followed if a taxon at the subgeneric level is transferred to another taxon but retains its epithet.

If a name proposed by one person but not published by him is published by another and ascribed to the first author then the taxon is followed by the name of the first author followed by ex (described by) and then the name of the publishing author. For example, the Corsican speedwell is *Veronica repens* Clarion ex DC. (DC. for De Candolle). The citation of authors in this way helps other workers find the original descriptions and type specimens.

autoclave An apparatus within which very high temperatures are produced by steam under pressure. It is used to sterilize laboratory equipment and is essentially a large pressure cooker.

autodiploid A gamete containing a diploid rather than haploid number of chromosomes. Such gametes may arise either through faulty meiosis or from tetraploid tissues. They may fuse with another such gamete to give a tetraploid zygote or may develop parthenogenetically. The term is also applied to diploid plants that have been obtained by doubling the chromosomes of a haploid plant. Such plants are completely homozygous.

autoecious Describing a rust fungus in which the various spore forms are all developed on the same host. An example is *Puccinia menthae*, mint rust. *Compare* heteroecious.

autogamy (self fertilization, endogamy) Fusion of female and male gametes derived from genetically similar sources, usually the same individual. This mechanism restricts genetic variability of the population but can stabilize selected traits and ensures that isolated individuals have the opportunity to reproduce. Autogamy may be found in species where cross pollination cannot be assured, as in pioneer populations (many weeds are autogamic) or where insect vectors may be rare (as in the tundra ecosystems). Autogamy is thought to be derived from *allogamy by the overcoming of self-incompatibility systems. Total autogamy is rare in the plant kingdom except in species showing *cleistogamy. In most cases autogamy is facultative and may be triggered by adverse climatic conditions, such as high humidity or intense cold, or by failure to achieve cross fertilization. *See also* homogamy.

autolysis The process of self-digestion undergone by organelles or cells when their useful life is completed. It is effected by enzymes, primarily hydrolytic in character, that are produced by the cytoplasm. The process is believed to be under the control of growth substances. The products of the digestion are subsequently reabsorbed by the surrounding cells. *See also* autophagy, deliquescence.

autonomic movement A movement of a plant or plant part in response to a stimulus generated within the plant. Autonomic movements include the beating of cilia and flagella, protoplasmic streaming, circumnutation, and the movement of chromosomes during nuclear division. *Compare* paratonic movement.

autophagy A process utilized by certain cells to digest worn out or superfluous cell organelles. An *autophagic vacuole* forms around a portion of cytoplasm containing one or more unwanted organelles. Hydrolytic enzymes are secreted into the vacuole by the surrounding cytoplasm and digestion occurs followed by the subsequent reabsorption of the breakdown products. *See also* autolysis.

autopolyploidy A form of *polyploidy that results from the multiplication of chromosome sets from a single species. Autopolyploids may arise either through a failure in the mitotic process (*see* syndiploidy), or in the meiotic process, resulting in the formation of diploid rather than haploid gametes. Autopolyploids often resemble their diploid parents, except that they may grow more slowly and flower later. Their cells are usually larger and the epidermis thicker. Autopolyploids with an uneven number of chromosome sets (autotriploids, autopentaploids) are often sterile because their gametes contain unbalanced numbers of chromosomes. Autotetraploids also show reduced fertility, the degree depending on whether any trivalents or univalents are formed at meiosis and also on the orientations

taken up by the multivalents at metaphase. The patterns of inheritance, although they follow Mendelian laws, are very complex in autopolyploids because of the increased number of genotypes that may exist for any given locus with two alternative alleles. Thus for the two alleles A and a, a diploid can have three possible genotypes, AA, Aa, and aa, while an autotetraploid can have five, AAAA, AAAa, AAaa, Aaaa, and aaaa. In addition, while the frequencies of the gametes A and a can usually be predicted in diploids, the frequencies of the gametes AA, Aa, and aa of an autotetraploid are hard to predict because segregation can occur at either the first or second meiotic division.
Compare allopolyploidy.

autoradiography A technique for detecting the distribution of radioactive isotopes previously incorporated into cells. A photographic emulsion is placed over a thin piece of squashed or sectioned tissue in the dark. The developed film (*autoradiograph*) shows the location of the radioactive substance in the form of concentrated dark patches. *See also* isotopic tracer.

autosome Either of any pair of chromosomes that do not play a principal role in sex determination. The term is used of plant chromosomes when discussing dioecious species that have cytologically identifiable *sex chromosomes.

autotetraploid *See* autopolyploidy.

autotroph An organism that needs only simple inorganic compounds to grow. Thus carbon dioxide or carbonates serve

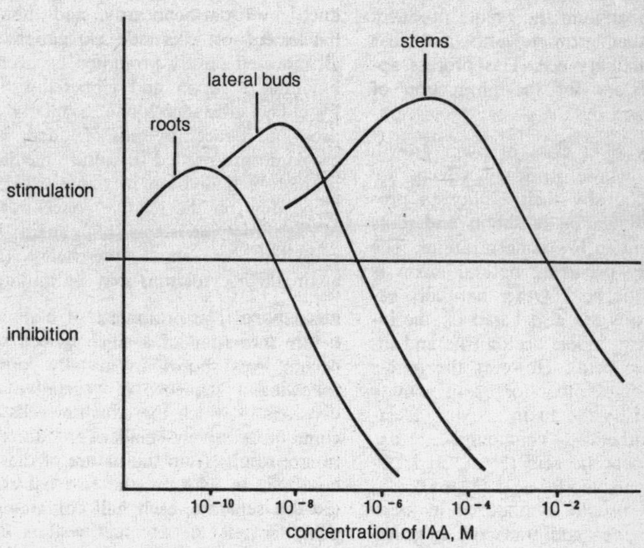

The effects of auxin concentration on the growth of different organs.

as the carbon source and can be built up into complex organic molecules using light (*see* phototroph) or chemical (*see* chemotroph) energy. Simple inorganic nitrogen compounds are utilized in the synthesis of proteins. Most plants and some bacteria are autotrophic. *Compare* heterotroph.

auxanometer An apparatus that is designed to measure the growth in length of plant organs. The organ is attached by a taut piece of cotton to the end of the shorter arm of a pivoted lever. Any growth results in the upward movement of the arm, which is magnified and recorded by the other longer arm of the lever as its tip moves across a calibrated scale, or as it traces the movement on a slowly rotating drum (*kymograph).

auxiliary cell A specialized cell in the gametophyte thallus of some red algae, e.g. *Polysiphonia*, situated adjacent to the carpogonium. Following fertilization of the carpogonium the zygote produces filaments that grow towards and fuse with the auxiliary cell. This process appears necessary for the production of carpospores.

auxin Any of a class of plant growth substances whose principal effects are brought about by their ability to promote the elongation of shoots and roots when present in low concentrations. The most widely occurring natural auxin is *indole acetic acid. Other naturally occurring auxins are also based on the indole ring (e.g. indole acetonitrile and indole pyruvic acid). However the indole group is not essential for auxin activity as is shown by the auxin activity of certain synthetic compounds, e.g. *naphthaleneacetic acid (NAA) and 2,4-dichlorophenoxyacetic acid (*2,4-D).

Auxins are usually synthesized in meristematic regions and transport is generally from apex to base, i.e. basipetal. The concentration of auxin is very important in determining the nature of the growth response and the optimum auxin concentration differs for different organs (*see* diagram). The inhibition of growth at higher auxin concentrations may be due to the auxin-promoted synthesis of *ethylene, which inhibits cell elongation. Such inhibitory effects have been exploited in the production of herbicides based on 2,4-D and *2,4,5-T.

Auxins also promote root initiation, and certain synthetic auxins (e.g. NAA and indole butyric acid or IBA) are widely used as rooting compounds. Abscission has been correlated with low levels of auxin in the organ concerned and auxins have thus been used to prevent premature fruit drop. Other phenomena in which auxins have been implicated include *apical dominance, phototropism (*see Avena* curvature test), and epinasty. Many effects are brought about by the combined action of auxin with other growth substances. For example, the stimulation of cambial activity, the induction of parthenocarpy, and the enhancement of internode elongation are all more effectively promoted by a combination of auxin and gibberellin than by either substance alone. Similarly appropriate concentrations of auxin and cytokinin are needed in culture media to promote cell division in tissue explants. Depending on the relative concentration of each, root meristems (high auxin: low cytokinin) or shoot meristems (low auxin: high cytokinin) may be initiated.

auxospore The protoplast of a diatom before formation of a siliceous cell wall occurs. Auxospores are usually formed following a sequence of vegetative cell divisions in which the daughter cells become progressively smaller. The decrease in size results from the nature of diatom cell division whereby the two halves of the cell separate, each half containing a daughter cell. A new half wall is then formed, the edges of which fit within those of the old half wall in a manner resembling a Petri dish. Thus the diatoms decrease in size until a certain limit is reached, at which point auxospores are released. The naked auxospores are free to expand before forming a cell wall. Auxospores may also form after sexual fusion or by apogamy or parthenogenesis.

auxotroph A type of *biochemical mutant that can grow normally if provided with a supplement in its diet to replace the substance it has lost the ability to make. For example, certain mutants of the mould *Neurospora crassa* need the amino acid arginine in the culture medium to grow normally (the unmutated mould can synthesize all the amino acids it requires from ammonia). By supplying the auxotrophic mould with various precursors of arginine it can be established which of the enzymes in the pathway to arginine is missing.

available space theory The theory that the origin of new leaf primordia at a stem apex is governed physically by the amount of space between existing primordia. It is postulated that a new primordium only arises at a point where a certain minimum amount of space has become available. This is usually between the penultimately formed primordium and the one formed previous to that. The pattern that results from such restrictions is the same as would occur if inhibitors were released by older primordia (*see* repulsion theory). *See also* phyllotaxis.

***Avena* curvature test** A quantitative method used to correlate the degree of curvature obtained in a plant part, after the application of an external stimulus, with the amount of endogenous growth substance (*auxin) required to bring it about.

The tips of *Avena* (oat) coleoptiles are used as a standard. These are removed and placed on an agar block and auxin allowed to diffuse out for a predetermined time. The auxin-impregnated agar is then used to measure the curvature produced when pieces are placed acentrally on decapitated coleoptiles. The response is plotted against the auxin concentration in the block. Over a certain range, increases in auxin concentration result in a regular increase in the degree of bending in the coleoptile, but the linear response is lost at higher levels. Only the straight line section of the graph is useful for assays. By noting the angle of curvature produced by a given stimulus under otherwise controlled standard conditions, one can find the quantity of auxin involved in producing the response. *See also* tropism.

awn A stiff bristle-like projection, usually at the tip of an organ. The glumes and lemmas of grasses (Gramineae) commonly possess awns, as do some fruits, and less commonly leaves. Cereals and grasses having ears or spikes covered in awns are termed awned or bearded. An individual organ bearing an awn is termed aristate. An awn may act to bury a fruit in the soil by uncoiling in damp conditions and is so doing pushing the fruit into the ground.

axenic culture (pure culture) A culture consisting of only one type of organism or cell. Axenic cultures are derived from pure parent cultures and grown under sterile conditions. Such uncontaminated cultures are necessary in investigating the properties of a particular microorganism or cell type and in maintaining cultures of microorganisms over a long period of time.

axial Describing cells derived from the *fusiform initials, which are elongated parallel to the long axis of an organ. The *axial system* refers to all these cells collectively.

axil The upper angle formed by the junction of a leaf or similar organ with the stem. Organs in the axil, such as flowers, inflorescences, meristems, and buds are termed *axillary* or lateral.

axile placentation A form of *placentation in which the placentae arise along the central axis of the ovary. It is seen in compound ovaries that are divided by septa into a number of locules, as in *Hyacinthus*.

axoneme The fibrillar component of *flagella and cilia.

azonal soil *See* soil.

azygospore *See* parthenospore.

B

baccate Shaped like a berry as, for example, the arils of yew (*Taxus baccata*).

Bacillariophyta (diatoms) A division of unicellular or colonial nonflagellate *algae having delicately sculptured silica cell walls divided into two overlapping halves (*see* frustule). Asexual reproduction is by fission and *auxospore formation. There are two classes, the radially symmetrical or *centric* diatoms and the bilaterally symmetrical or *pennate* diatoms. The cell walls are extremely resistant to decay and form deep deposits of diatomaceous earth or kieselguhr in lake and ocean beds. Diatoms are sometimes considered as a class, the Bacillariophyceae, of the division *Chrysophyta.

bacillus Any bacterium that is rod shaped and more or less straight. *Compare* coccus, spirillum, vibrio.

backcross A cross between an individual and one of its parents. If the organism is short lived this may not be a practical possibility, in which case the term refers to a cross between an individual and an organism genetically identical to one of its parents. A backcross performed to ascertain the genotype of an individual is termed a *test cross. *See also* backcrossing.

backcrossing A technique used in plant breeding to introduce a desirable gene into a cultivated variety. Unlike the *pedigree method of plant breeding, the aim of a backcrossing programme is not to create entirely new varieties but to modify existing varieties.

The cultivated variety (the recurrent parent) is crossed with the donor parent, which may be quite useless agriculturally except for its possession of one particularly valued gene (e.g. for disease resistance). The progeny from this cross, which contain 50% of the donor genetic material, are screened for the character and those possessing disease resistance are crossed back to the recurrent parent. The progeny of this cross – the first backcross generation (B_1) – now contain 25% of the genetic material of the donor. The plants are again screened, the resistant plants again backcrossed to the recurrent parent, and this process is repeated until about the seventh or eighth backcross generation, by which time less than 0.25% of the donor genetic material remains. At this stage the B_7 or B_8 generation plants are selfed (crossed with each other) to produce plants homozygous for disease resistance; these may be identified by a *test cross.

The process described above assumes the allele for disease resistance is dominant. If it were recessive then it is necessary to alternate backcrossing with selfing of the backcross generations. Backcrossing is more efficient with self-pollinating species but the method is not necessarily limited to self pollinators.

back-scattered electron imaging In scanning electron microscopy, the use of a special detector to produce an image from the back-scattered electrons given off from below the surface of the specimen rather than the secondary electrons used in conventional imaging. The image obtained consequently includes subsurface information.

bacteria A group containing all the prokaryotic organisms (*see* prokaryote) with the exception of the blue-green algae. These however are included with the bacteria as Cyanobacteria in many classifications. Bacteria are found in virtually all habitats and, in terms of distribution and number, are the most successful of life forms. Many live saprophytically in the soil and are important as decomposers in the carbon cycle. Others, the nitrifying and denitrifying bacteria, are important in the nitrogen cycle. Some are pathogenic, causing rots, cankers, blights, and galls in plants and numerous diseases in animals, e.g. anthrax, brucellosis, and syphilis. Bacteria are important in many industrial processes. Fermenting bacteria, e.g. *Acetobacter*, *Acetomonas*, and *Lactobacillus* are widely used in the food industry. Cheese making and the production of silage are dependent on bacterial activity. Bacteria were once placed with the fungi in the class *Schizomycetes. They are now often classified as a separate division or kingdom, Bacteria or Prokaryota. The numerous different kinds of bacteria are classified on the basis of shape, size, growth form, staining and serological reactions, metabolic activity, and motility. *See* Eubacteriales, Pseudomonadales, Actinomycetales, Chlamydobacteriales, Beggiatoales, Myxobacteriales, Spirochaetales, Mycoplasmatales, Rickettsiales.

bacterial transformation The incorporation of genetic material into a bacterium from DNA in the surrounding medium. The phenomenon was discovered by F. Griffith (1928) who converted (transformed) rough-coated *Diplococcus pneumoniae* into a smooth-coated strain by mixing it with extracts from the latter under appropriate conditions. Later O. T. Avery, C. M. MacLeod, and M. McCarty (1944) showed that the substance responsible for the transformation was DNA. The latter experiment is often quoted as one of the first pieces of evidence that DNA, not protein, is the genetic material.

bacteriochlorophyll Any of the forms of chlorophyll found in photosynthetic bacteria. Bacteriochlorophyll *a*, found in all bacteria, differs from the chlorophyll of plants in being a tetrahydroporphyrin rather than a dihydroporphyrin (i.e. it has more hydrogen atoms in pyrrole rings 2 and 4 of the chlorophyll molecule). It also differs in its absorption spectrum having three peaks, at 365, 605, and 770 nm. Some bacteria contain additional forms of bacteriochlorophyll (Bchl) designated Bchl *b*, Bchl *c*, and Bchl *d*.

bacteriophage (phage) Any virus that infects bacteria. Bacteriophages are often named according to the bacteria they infect, e.g. actinophages, which infect bacteria of the order Actinomycetales, and coliphages, which infect various strains of *Escherichia coli*. All RNA phages bring about the death of the host cell by lysis. Some DNA phages (the temperate phages) become integrated in the host DNA and are replicated with it. Some temperate phages, e.g. the lambda virus, have been shown to transfer genes from one bacterium to another in a process termed *transduction.

bacteroid A *Rhizobium* bacterium in a legume root nodule after it has undergone an increase in size (to about forty times its original size) and a change in shape to a characteristic X- or Y-shaped cell. Bacteroids lose their capacity for autonomous growth and are dependent on the host. It is in this metamorphosed state that the bacteria carry out *nitrogen fixation.

Balbiani ring *See* chromosome puff.

ballistospore A spore that is violently projected, e.g. the basidiospores of the Hymenomycetes.

balsam *See* resin.

Bangioideae (Bangiophycidae) *See* Rhodophyta.

BAP (6-benzylaminopurine, benzyladenine) A synthetic *cytokinin.

Barfoed's reagent A mixture of copper (II) acetate and acetic acid, used to show the presence of strongly reducing sugars in solution. After boiling, monosaccharides cause the formation of a red precipitate of copper(I) oxide. Disaccharides are not such powerful reducing agents and will not show a positive reaction.

bark All the tissues, collectively, lying outside the vascular cambium in the stems and roots of plants showing secondary growth, i.e. the primary and secondary phloem, the cortex, and the periderm. The term is also used in a more restricted sense to mean the tissue arising to the outside of the phellogen, i.e. the *phellem, when this is exposed by the sloughing off of the epidermis. The bark of different trees can be very distinctive and its characteristics are used to aid identification. In some species the same phellogen is active each year and a thick layer consisting solely of phellem is formed (e.g. oak, beech), but in most species a new phellogen arises annually in the cortex below: the bark thus consists of both phellem and dead cortex and is termed *rhytidome*. As the thickness of the bark increases the outer layers may either become fissured (e.g. elm) or be shed as scales (e.g. plane) or rings (e.g. birch).

bar of Sanio *See* crassula.

basal body *See* kinetosome.

basal placentation A form of *placentation, found in ovaries containing only one ovule, in which the placenta develops at the base of the ovary. It is seen in bistorts (*Polygonum*).

base analogue A chemical that resembles a naturally occurring *purine or *pyrimidine base to the extent that it may be incorporated into a DNA molecule during replication. However such analogues are less specific in their pairing properties. For example, an analogue of adenine may pair with cytosine as well as thymine. Over two replications this would result in a point mutation, with a cytosine:guanine pair replacing an adenine:thymine pair on one of a pair of homologous chromosomes. Base analogues may also be incorporated into RNA during RNA synthesis, e.g. thiouracil and fluorouracil can replace uracil. Such changes may have various physiological effects. *See* 5-bromouracil.

base number *See* genome.

base pairing The bonding relationship between the bases in the nucleotides of DNA and RNA. The relationship is highly specific, such that adenine (A) in one strand of DNA only pairs with thymine (T) in the complementary strand (or with uracil (U) in RNA). Similarly guanine (G) only pairs with cytosine (C). The A:T and G:C base pairs of complementary polynucleotide chains are held together by hydrogen bonds. Although these are only weak bonds, large numbers are formed between the two polynucleotide strands of DNA so that the double helix as a whole is extremely stable. The specific pairing of A:T (or A:U in RNA) and G:C means that the genetic material is accurately replicated from one generation to another. It also means that molecules of messenger RNA are accurately transcribed from the genetic material and subsequently accurately translated to protein.

base ratio A comparison of the molar quantities of one base to another base in DNA, or of one base pair with another base pair. The observation that the amount of adenine (A) always equals the amount of thymine (T) and similarly the amount of guanine (G) equals that of cytosine (C) provided one of the first indications that adenine always pairs with thymine and guanine with cytosine in the complementary polynucleotide chains. A comparison of base pair ratios, A+T:G+C, or percentage G+C of the total, shows a constant figure within a species but differences between species.

base triplet hypothesis *See* genetic code.

basic number *See* genome.

basic stain *See* staining.

basidiocarp The reproductive organ of all basidiomycete fungi (except the Ustilaginales and Uredinales), which bears the basidia (*see* basidium). It may be *mushroom shaped, bracket shaped, club shaped, or a hollow sphere or cylinder. Much of the basidiocarp is composed of sterile pseudoparenchymatous tissue. The basidia are borne in a fertile layer, the *hymenium.

Basidiomycotina A subdivision containing the most advanced fungi of the *Eumycota, which are characterized by the production of *basidia. The mycelium, which may be uninucleate or dikaryotic, is typically divided by inflated septa with a central pore (*see* dolipore septum). In dikaryotic mycelia, characteristic *clamp connections are seen. The Basidiomycotina contains over 12 000 species in some 900 genera and is divided into the three classes *Teliomycetes, *Hymenomycetes, and *Gasteromycetes. In some classifications these fungi are grouped in the class Basidiomycetes, which is then divided into the subclasses Heterobasidiomycetidae, in which the basidium is deeply divided or septate (e.g. rusts and smuts), and Homobasidiomycetidae, in which the basidium is a single large cell.

basidiospore A haploid spore, four of which are characteristically borne on the outer surface of a *basidium.

basidium The spore-bearing structure of fungi in the Basidiomycotina. Each basidium usually bears four basidiospores on its outer surface, often on projections called sterigmata. The basidium is usually a single cell but in some basidiomycetes (e.g. Ustilaginales and Uredinales) it is segmented. The basidium is initially binucleate. These nuclei fuse, undergo meiosis, and form four haploid daughter nuclei, which subsequently become the basidiospores. In most basidiomycetes the basidia form a fertile layer, the hymenium.

basifixed Describing an anther that is joined to the filament at its base. *Compare* dorsifixed, versatile.

basipetal Describing movement, differentiation, etc. occurring from apex to base. For example, the direction of proto- and metaxylem formation is from the apex or nodes downwards, while the transport of auxin in the stem is generally towards the base. *Compare* acropetal.

bast *See* phloem.

Baumgrenze *See* tree line.

B-chromosome (accessory chromosome, supernumerary chromosome) Any of a number of small chromosomes present in some organisms in addition to the fixed number of stable chromosomes characteristic of the species (the A-chromosomes). They are of widespread occurrence and have been reported in over 80 genera and 20 families of plants. The numbers found may differ between members of a species and even within a single individual, with gametic nuclei often having larger numbers. B-chromosomes are composed largely of heterochromatin and in many cases appear to have little effect. However when present in large numbers they may be deleterious.

bearded 1. Having many awns as, for example, the ears of the bearded fescue (*Festuca ambigua*).
2. Having any other kind of stiff hairs as, for example, the collection of hairs on the lower petals of the flag iris (*Iris pseudacorus*).

Beaumont period Forty-eight hours during which the minimum temperature is 10°C and the relative humidity is 75% or more. These periods are the basis of the system used in England to forecast an outbreak of late blight of potatoes (*Phytophthora infestans*).
Sporulation of the potato blight fungus and subsequent invasion of potato leaves is favoured by moist weather. By examining meteorological data A. Beaumont worked out that blight could be expected ten days after a 'Beaumont period'. Careful monitoring of local meteorological data enables the advisory services to record Beaumont periods and give farmers accurate advice on when to spray their potato crops.

beet sugar *See* sucrose.

Beggiatoales An order of filamentous or unicellular bacteria that move by gliding over the substrate, hence their common name gliding bacteria. Many of its members (the filamentous sulphur bacteria) grow by oxidizing sulphides. The genus *Beggiatoa* has certain features in common with the blue-green alga *Oscillatoria*. It has been suggested that some of the Beggiatoaceae are colourless blue-green algae.

Benedict's reagent A mixture of copper(II) sulphate in solution together with a filtered mixture of sodium citrate and sodium carbonate. The reagent is used to detect the presence of reducing sugars in solution. After boiling, a high concentration of reducing sugars in the test solution causes the formation of a red precipitate, while a lower concentration results in the formation of a yellow precipitate. It is a more sensitive test than Fehling's solution.

Bennettitales (Cycadeoidales) An extinct order of gymnosperms known from fossils extending throughout the Mesozoic era. They resembled cycads in habit and in leaf structure but can be distinguished from fossil cycads by certain features of the leaf epidermis, notably the possession of a *syndetocheilic stomatal complex as compared to the *haplocheilic stomata of the cycads. Like the cycads, they may have originated from the Pteridospermales.

Benson–Calvin–Bassham cycle The cycle in photosynthetic carbon dioxide fixation that regenerates the primary carbon dioxide acceptor. It is named after the three discoverers but is commonly abbreviated to *Calvin cycle.

benthos The organisms that live on a lake or sea bed. The plants (*phytobenthos*) that grow below low-tide mark include members of the flowering plant family Zosteraceae, such as eelgrasses (*Zostera*), and certain red algae, such as dulse (*Rhodymenia palmata*). A variety of plants grow rooted in the *littoral zone of lakes and ponds, including water lilies (*Nuphar*, *Nymphaea*), pondweeds (*Potamogeton*), and water milfoils (*Myriophyllum*).

benzyladenine *See* BAP.

6-benzylaminopurine *See* BAP.

berry A many-seeded fleshy indehiscent fruit. The epicarp usually forms a tough outer skin, especially in the *pepo and *hesperidium and the mesocarp becomes massive and fleshy. The epicarp and mesocarp may be highly coloured to attract the animals that act as agents of dispersal. Examples are the tomato and grape. *Compare* drupe. *See illustration at* fruit.

betacyanin *See* betalain.

betalain A class of nitrogen-containing pigments containing the red *betacyanins* (e.g. betanidin, the beetroot pigment) and the yellow *betaxanthins* (e.g. indicaxanthin from the fruits of *Opuntia ficus-indica*). They replace and perform the functions of the floral pigment anthocyanin in certain plants. The possession of betalains has been used as a taxonomic character. Its occurrence is restricted to a group of families (nine or ten depending on the classification system used) which, with the exception of the Cactaceae, have all traditionally been placed in the order Centrospermae (or Caryophyllales). The Cactaceae are sometimes included in the Centrospermae on the basis of their resemblance to members of the Aizoaceae (mesembryanthemum family) but others have placed them in the monotypic order Cactales (or Opuntiales). The occurrence of betalains in cacti is strong evidence for their inclusion in the Centrospermae. Interestingly one family, the Caryophyllaceae, normally included in the Centrospermae, lacks betalains.

betaxanthin *See* betalain.

Bial's reagent A mixture of orcinol in concentrated hydrochloric acid with 10% iron(III) chloride solution. It is used to detect the presence of pentose sugars in a test solution. A green colour after boiling indicates a positive reaction.

bicollateral bundle A *vascular bundle in which the phloem occurs both internal and external to the xylem, as in the stems of some dicotyledons, e.g. members of the Solanaceae. The presence of bicollateral bundles distinguishes solanaceous plants from members of the Scrophulariaceae. *Compare* collateral bundle, concentric bundle.

biennial A plant that takes two years to complete its life cycle. It grows vegetatively in the first year and the photosynthates are stored in perennating organs. The stored food is used to produce foliage leaves, flowers, and seeds the following year. The plant then dies. Some important crops are biennials, e.g. carrot (*Daucus carota*) and parsnip (*Pastinaca sativa*). Certain garden flowers that are in fact perennials are more successfully grown as biennials, e.g. wallflower (*Cheiranthus cheiri*) and *Antirrhinum*. *Compare* annual, ephemeral, perennial.

biflavonyls *See* flavonoids.

bigeneric Describing a hybrid derived from an intergeneric cross, e.g. *Triticale* derived from a cross between *Triticum* and *Secale*.

bilateral symmetry The arrangement of parts in an organ or organism such that it can only be split into similar halves along one given plane. Thus most leaves can only be divided into similar halves by cutting along the midrib. Bilateral symmetry in flowers is usually termed *zygomorphy*. The flowers of relatively advanced angiosperm families, e.g. Scrophulariaceae, are often zygomorphic. In a floral formula, zygomorphy is represented by the symbol ·|· or ↑.

biliprotein *See* phycobilin.

binary fission *See* fission.

binomial nomenclature A system of naming species using a generic name and a specific epithet, established by the Swedish botanist Carl von Linné (also known by the Latinized form of his name, Carolus Linnaeus) in 1753. Previous to this the scientific names of species consisted of a short Latin description. This was too cumbersome as a name yet too short for a proper description. By separating the procedures of naming and describing, Linnaeus established the foundations of present-day nomenclature.

The generic name is a noun often based on a classical name as, for example, *Endymion* and *Narcissus*. The name can also be honorific, commemorating the plant's discoverer, such as *Saintpaulia* (African violets), named after Baron Walter von Saint Paul-Illaire. Alternatively, the generic noun may refer to the locality in which the plant was first found, *Ligusticum* (lovage) from Liguria (northwest Italy), for example. Specific names are adjectival and often describe a particular feature of a plant, such as leaf shape or flower colour (e.g. *longifolius*, *albus*). Like generic names they may also be commemorative or geographical (e.g. *wilsonii*, *lusitanica*). *See* table 1 in the appendix for the meanings of some common specific epithets. Binomials are always Latinized and are usually printed in italic script, the generic name beginning with a capital letter while the specific epithet is lower case throughout (e.g. *Primula vulgaris*). When naming a species the rules of the *International Code of Botanical Nomenclature must be followed.

bioassay (biological assay) A quantitative assessment of the effect of a substance on a living organism by comparison with the effects of a similar substance of known concentration. For example, certain fungi are used to find the concentration of a particular vitamin, such as thiamin, in a substance as the growth of the fungus is directly proportional to the concentration of the vitamin. Bioassays have been used extensively to measure the effects of different auxins and gibberellins on plant growth (*see Avena* curvature test).

biochemical genetics The branch of genetics concerned with inheritance at the molecular level. It includes the study of the structure of DNA, its replication, the genetic code and transcription and translation in protein synthesis, and the regulation of gene expression.

biochemical mutant An organism possessing a mutation that causes a particular enzyme either not to be synthesized or to be defective in some respect. The reaction that it catalyses therefore does not take place, and, if this reaction is an intermediate step in a metabolic pathway, the subsequent steps in the pathway are also prevented. Such mutants can be induced by various methods, such as irradiation with X-rays, and have proved particularly useful in elucidating the stages of a number of pathways. *See also* auxotroph.

biochemical taxonomy *See* chemotaxonomy.

biochemistry The study of *metabolism in prokaryotic and eukaryotic cells. As a discrete subject biochemistry has only existed for about 50 years, during which time a huge body of knowledge has accumulated. Originally biochemistry was concerned with cataloguing and investigating the biological occurrence and enzymatic reactions of a large number of organic compounds. However unifying patterns and principles have gradually emerged from the scattered body of facts and hypotheses.

bioenergetics The study of energy transfer in living organisms. Almost all the energy used on Earth comes directly or indirectly from the sun. The way in which it is converted in cells depends on the principles and laws of thermodynamics. The first law concerns the relationship between work, heat, and internal energy; it is equivalent to the statement that energy can be converted from one form into another, but cannot be created or destroyed. Thus, when light energy falls on a plant some is reflected, some absorbed as heat, and some converted into chemical energy in glucose by *photosynthesis, but the total amount of energy is constant. In fact energy transfers in living organisms tend to be inefficient. About 2% of the energy in incident radiation is converted into chemical energy. Heterotrophic organisms obtain energy from the nutrients they ingest; about 10–20% of the energy is passed on at each link in the *food chain, which tends to have three or four links only.

The second law of thermodynamics concerns the way in which energy is transferred or converted. It states that any conversion of energy from one form into another involves some dissipation of energy as unavailable heat energy. The availability of energy in a system is determined by its entropy, which is a measure of randomness or disorder. In any process the total entropy increases (an alternative statement of the second law), and there is a constant degradation of energy in the universe into energy that is unavailable for work. The growth of organisms is characterized by decreases of entropy in the sense that disordered systems (e.g. CO_2 and O_2 gas) are converted into ordered chemical structures. However, the *total* entropy of the organism and its surroundings always increases in any change. All chemical reactions are driven by a decrease of free energy.

biogenesis The theory that new life arises only from preexisting life and never from nonliving matter. *Compare* spontaneous generation. *See also* origin of life.

biological assay *See* bioassay.

biological clock An internal timing mechanism possessed by some organisms and responsible for the periodic or cyclical triggering of certain physiological responses. Biological clocks are often controlled by growth substances. Examples include the occurrence of flowering when a particular day length is achieved and the breaking of dormancy when conditions become suitable for seedling development. *See also* circadian rhythm.

biological control The control of pests and diseases by making use of their natural enemies or by artificially upsetting their life cycle. For example, the moth *Cactoblastis cactorum* has been used in Australia to help control the prickly pear cactus (*Opuntia vulgaris*). The caterpillars bore into the pads of the cactus, severely affecting its growth. Experiments have been conducted on controlling water weeds, such as water hyacinth (*Eichhornia*), by spraying them with disease spores. Populations of damping-off fungi in the soil can be reduced by the addition of organic supplements to encourage the growth of antagonistic saprophytic microorganisms. *See also* chemosterilant, trap crop.

bioluminescence The emission of light from living organisms. It results either from internal chemical reactions or from the reemission of absorbed energy as radiation. In plants, luminescence is probably the result of inefficiencies during oxidation-reduction reactions. Some of the energy released excites a molecule to a high-energy state and when the molecule returns to the ground state visible light is emitted. Luminescence is exhibited by certain fungi, including some of the genus *Mycena*, and also certain bacteria and planktonic algae.

biomass (standing crop) The total weight or volume of either all the living organisms or of one species present at any one time in a community. *See also* pyramid of biomass.

biome (biotic region) Any of a group of major regional terrestrial communities with its own type of climate, vegetation, and animal life. Biomes are not sharply separated but merge gradually into one another. Examples include tundra, temperate deciduous forest, and desert. *See also* ecotone.

biosphere The zone, including the earth's surface (and surface water), the adjacent atmosphere, and the underlying crust, where life can exist. The earth is usually considered as having three spherical zones: the solid lithosphere, the liquid hydrosphere, and the gaseous atmosphere. The biosphere encompasses parts of all three of these zones. Its limits are not well defined; most living organisms exist in a region extending about 100 m into the atmosphere and about 150 m below the surface of the oceans. The term can also include parts of the earth and atmosphere that depend on the present or past existence of living organisms, e.g. coal deposits or atmospheric oxygen. The biosphere can be regarded as a single *ecosystem and is often called the *ecosphere*.

biosynthetic pathway A series of enzymatic reactions in which more complex molecules are built from simpler ones. Biosynthetic pathways are energy requiring, this energy usually being supplied either as phosphate bond energy by ATP or as reducing power by NADPH. Three stages of biosynthesis can be recognized. The first stage is concerned with the manufacture of simple organic molecules from inorganic molecules like carbon dioxide and water. These few simple molecules are then converted in the second stage to building blocks for the many biological macromolecules. Finally macromolecules are synthesized from their constituent subunits. The major biosynthetic pathways in plants are shown in the diagram.

biosystematics The study of variation and relationship in populations rather than individuals. The genetic and evolu-

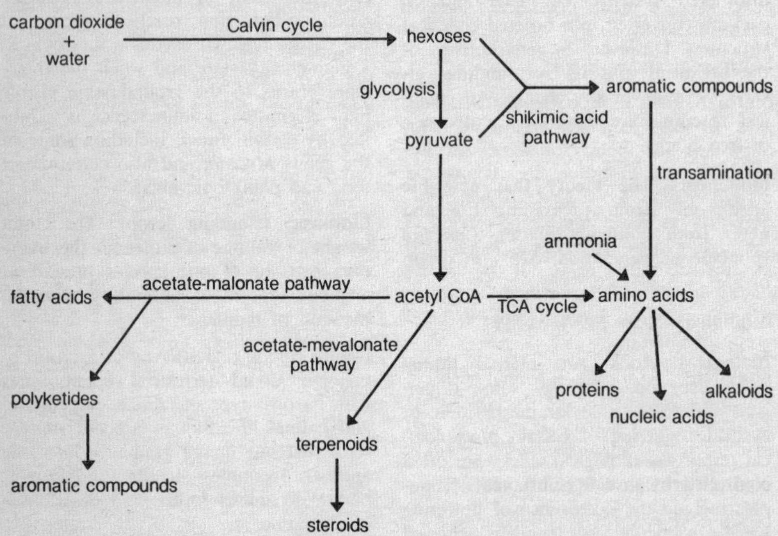

The main biosynthetic pathways in plants.

tionary nature of groups is assessed by the examination of cytology, comparative morphology, and ecology. Biosystematics is often described as the taxonomic application of genecology or, occasionally, as experimental taxonomy. *See also* systematics, taxonomy.

biota The flora and fauna either of a particular region or of a given geological period.

biotechnology The use of organisms (usually microorganisms), or the enzymes they produce, in industrial processes. Biotechnology has potential applications in food production, waste recycling, alternative energy schemes, medicine, and a number of other spheres. *See also* genetic engineering, single-cell protein, enzyme technology.

biotic climax *See* climax.

biotic factors The living components of the environment that by their activities affect the life of an organism. *Compare* abiotic factors.

biotic region *See* biome.

biotin A heat-stable coenzyme that acts as a carrier of carboxyl groups in many important metabolic reactions. An example is the oxidative decarboxylation of pyruvate to acetyl CoA, in which biotin acts as an intermediate carrier of the carboxyl group before it is released as free carbon dioxide. It is usually tightly bound to the enzyme for which it is the cofactor. Biotin is one of the B group of vitamins.

bipinnate Describing a pinnate leaf in which the leaflets themselves are further subdivided in a pinnate fashion, as seen in the sensitive plant (*Mimosa pudica*).

bird pollination *See* ornithophily.

bird's-nest fungi *See* Nidulariales.

biuret test A test for proteins in which sodium hydroxide is added to the test solution, followed by the careful addition of drops of 1% copper(II) sulphate solution. A violet colour indicates a positive result and is caused by the presence of peptide groups (NH–CO) in the proteins or peptides, which form a purple copper(II)–peptide complex. Free amino acids will not give a positive result.

bivalent A pair of homologous chromosomes united by chiasmata. Bivalents are most clearly seen during late prophase of meiosis 1.

black earth *See* chernozem.

bladder Any inflated organ, such as the inflated calyx of the bladder campion (*Silene vulgaris*) or the inflated fruit of the bladder nut (*Staphylea pinnata*). *See also* air bladder, utricle.

blade *See* lamina.

blanket bog *See* bog.

Blastomycetes A class of the *Deuteromycotina containing the imperfect yeasts. It includes about 200 species in some 20 genera and is divided into the Cryptococcales, which reproduce by budding and do not form ballistospores; and the Sporobolomycetales, which reproduce by budding and ballistospores. Most of the Cryptococcales are imperfect states of the *Endomycetales. The order includes the parasitic genus *Candida*, which causes thrush in man. The Sporobolomycetales (the mirror or shadow yeasts) are imperfect states of basidiomycetes, such as the *Tremellales. Many (e.g. *Sporobolomyces* and *Tilletiopsis*) are found as epiphytes on leaves.

blastospore A spore produced by budding. Blastospores are seen, for example, in the Taphrinales, in which they are formed by the budding of ascospores.

blending inheritance The intermingling of the characteristics of the parents so that the offspring are intermediate in form between their parents. It is usually

seen when a characteristic is controlled by many genes, as for example, height or yield. If a characteristic is governed by a single gene then blending inheritance is only seen when the alleles show *incomplete dominance.
It was thought that all characteristics become blended in the offspring. Until the rediscovery of Mendel's work this was the main criticism of Darwin's theory of evolution by natural selection, for the variation upon which it depended would always be lost by such a process. Mendel's demonstration of the particulate nature of inheritance removed this obstacle.

blepharoplast *See* kinetosome.

blight A plant disease in which leaf damage is sudden and serious. Many pathogens can cause blight-type symptoms when conditions are suitable but only a few cause such devastating symptoms on a regular basis: these include the term 'blight' in their common names. Examples include *late* and *early blight* of potatoes (caused by the fungi *Phytophthora infestans* and *Alternaria solani* respectively), and *fireblight of pears (caused by the bacterium *Erwinia amylovora*).

bloom A visible increase in the numbers of a species, usually an algal species, in the plankton. A bloom of diatoms is often seen in the spring, which decreases later in the year, probably as available silica is used up. Sustained algal blooms may lead to the eutrophication of a lake.

blue-green algae *See* Cyanophyta.

body cell A fertile cell formed along with the infertile stalk cell after mitotic division of the *generative cell in the gymnosperm pollen tube. It divides into two *sperm cells.

bog A region of badly drained permanently wet land that is subject to high rainfall and has a persistently moist atmosphere. These conditions result in a *climax community with no trees. The most common plants are bog mosses (*Sphagnum*), cotton grasses (*Eriophorum*), ling (*Calluna vulgaris*), cross-leaved heath (*Erica tetralix*), bog myrtle (*Myrica gale*), rushes (*Juncus*), and sedges (*Carex*). Bogs are commonly found in upland and western areas of temperate regions. There are three main types.
Blanket bog consists of acid peat formed from the remains of the bog plants. It is variable in thickness and the high acidity (pH 3.0–4.5) is caused by the continuous flow of water through the peat. This washes out any bases and prevents minerals from the underlying rocks reaching the plant roots. Thus the organic acids are not neutralized and the peat is poor in nutrients (oligotrophic). Blanket bog can form over limestone.
Raised bogs develop from *fens. Rain water leaches nutrients from the upper layers of the fen peat making it acid. This favours the growth of bog plants and prevents the growth of trees. As the bog plants become established more peat forms, especially in the central regions, resulting in the centre being raised above the periphery.
Valley bogs form in regions, such as the glens of western Scotland, receiving run-off and spring water from surrounding mountains.

bolting The premature production of flowers and seeds. It is a problem in biennial crops, such as sugar beet, which show considerably reduced yields if they 'run to seed' in the first growing season.

bomb calorimeter An apparatus that is used to measure the quantity of heat produced when a known amount of carbohydrate, fat, protein, or some mixture of these is burned. It consists of a strong cylindrical steel chamber, resistant to high pressures, surrounded by a jacket containing a known volume of water. The weighed foodstuff is placed in the chamber, oxygen is pumped in under pressure, and the food is ignited electrically. The rise in temperature of the surrounding water gives a measure of the amount of heat generated by the oxidized food.

borax carmine A dye used to stain nuclei in the preparation of biological material for microscopy.

Bordeaux mixture A mixture of copper sulphate and calcium hydroxide used as a fungicide. It was first used by P. M. A. Millardet in France in 1883–85 to control vine downy mildew (*Plasmopara viticola*). A common mixture is 40 g copper sulphate and 40 g calcium hydroxide in 5 l water.

bordered pit A *pit possessing an extension of the secondary cell wall, i.e. a border, arching over part of the pit cavity. In many gymnosperms, bordered pits also possess a thickening of the primary wall material, termed a *torus, in the central part of the pit membrane, the remaining unthickened part being termed the *margo*. Bordered pits mainly occur in vessel elements, tracheids, and fibres in the xylem, but may also occur in some extraxylary sclerenchyma cells. Pit pairs may be *half-bordered*, in which case one member of a pit pair is bordered and the other is not. This is sometimes the case in pits linking parenchyma cells and tracheary cells. *Compare* simple pit.

border parenchyma *See* bundle sheath.

boreal forest *See* forest.

boron Symbol: B. A metalloid element, atomic number 5, atomic weight 10.81, found in very low concentrations in plant tissues. Its role is uncertain though boron deficiency quickly results in changes in cell membrane function and in cell extension. Various plant diseases, e.g. heart rot of beets and alfalfa yellows, have been attributed to boron deficiency. It has been suggested that a boron–sugar complex is involved in the movement of substrates through the phloem, though this idea is now largely rejected. Pollen germination and pollen tube growth are both greatly enhanced by addition of borate to the growth medium. Boron has been implicated in the uptake of calcium and appears necessary for the proper development of apical meristems.

bostryx A coiled helicoid cyme (*see* monochasium) in which the overall structure resembles a spring. Examples are the inflorescences of *Hypericum* species. *Compare* rhipidium.

botanical **1.** Any insecticide derived from plants, such as rotenone from roots of *Derris* and pyrethrum from *Chrysanthemum* flowers.
2. In horticulture, cultivated plants of an unimproved wild species rather than a cultivated variety.

botanic garden An area in which a wide range of plants are grown for scientific, educational, and aesthetic purposes. Such collections aim either to include representative species of genera from all over the world or to specialize in one or more regions or types of vegetation. Large greenhouses are often maintained to provide appropriate growing conditions for exotic plants. As well as maintaining a large and varied collection of living plants, botanic gardens often also include *herbaria, research laboratories, and, increasingly, *gene banks. In the past botanic gardens have been responsible for the introduction of many ornamental and crop plants into regions where they were previously unknown. The larger botanic gardens also finance plant-collecting expeditions to remote areas where the vegetation remains largely undescribed and uncollected.

boundary layer The layer of liquid or gas next to a solid surface, which flows more slowly than that further away from the surface. A boundary layer of air exists at a leaf surface, its width decreasing with increasing wind speed. It is one of the factors that controls the rate of diffusion of water vapour from a leaf. Transpiration is thus greater from a leaf margin, where the boundary layer is narrower and offers less resistance, than from the centre of the lamina. *See also* essential oil.

brachysclereid (stone cell) A relatively short, more or less isodiametric, sclerenchyma cell (*sclereid). Brachysclereids are usually present singly or in small groups, although they occur in large numbers in some tissues, such as the fleshy part of the pear fruit.

bracket fungi *See* Aphyllophorales.

bract A leaflike organ subtending an inflorescence. Bracts are sometimes brightly coloured and petal-like, as in poinsettia (*Euphorbia pulcherrima*). The glumes, lemmas, and paleae of grass spikes are also examples of bracts. *Compare* bracteole. *See also* involucre.

bracteole A leaflike organ subtending a flower in an inflorescence that is itself subtended by a *bract.

bract scale The structure that subtends the ovuliferous scale in the female strobilus of gymnosperms.

bradytelic In *chronistics, describing a species that has remained more or less unchanged for millions of years. Such species may appear primitive; however when they first appeared they may have represented the end of an increasing trend towards specialization in a particular group. Certain of the present-day gymnosperms, such as *Welwitschia*, *Ginkgo*, and *Podocarpus* are considered bradytelic. The evidence for this comes partly from fossil studies, and also from investigations of the karyotype. Such species have strikingly asymmetrical karyotypes, a feature generally considered to be advanced.

brand spore *See* chlamydospore.

Brassicaceae *See* Cruciferae.

Braun-Blanquet scale A method of describing an area of vegetation devised by J. Braun-Blanquet in 1927. It is used to survey large areas very rapidly. Two scales are used. One consists of a plus sign and a series of numbers from 1 to 5 denoting both the numbers of species and the proportion of the area covered by that species, ranging from + (sparse and covering a small area) to 5 (covering more than 75% of the area). The second scale indicates how the species are grouped and ranges from Soc. 1 (growing singly) to Soc. 5 (growing in pure populations). The information is obtained by laying down adjacent quadrats of increasing size. One of a number of variations of Braun-Blanquet's method is the *Domin scale*, which is more accurate as there are more subdivisions of the original scale. The Braun-Blanquet scale also included a five-point scale to express the *degree of presence* of a plant. For example, 5 = constantly present in 80–100% of the areas; 1 = rare in 1–20% of the areas.

breathing root *See* pneumatophore.

bright-field illumination The normal method of illumination in light microscopy, in which the specimen appears dark against a bright background. Several methods of setting up the condenser lenses and light source are used, of which the two most common are critical illumination and Kohler illumination. *Compare* dark-ground illumination.

brochidodromous *See* camptodromous.

5-bromouracil (5-BU) An analogue of the naturally occurring pyrimidine thymine (5-methyl uracil), in which bro-

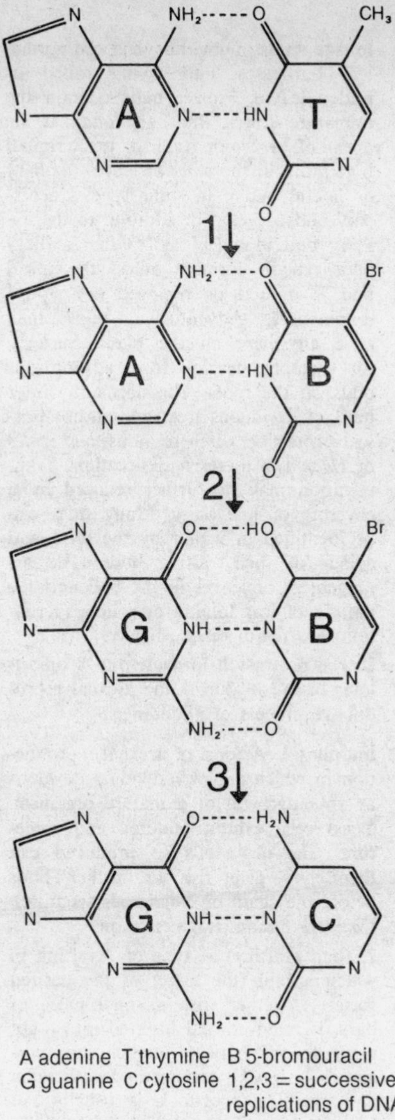

A adenine T thymine B 5-bromouracil
G guanine C cytosine 1,2,3 = successive replications of DNA

The figure shows how 5-bromouracil may be incorporated in place of thymine and by mispairing result in an A:T to G:C mutation.

mine replaces the methyl group on carbon five of the pyrimidine ring. During chromosome replication the analogue may be incorporated into DNA in place of thymine, but it is less specific in its pairing properties. Over a series of replications a T:A nucleotide pair could thus eventually be replaced by a C:G pair resulting in a point mutation (*see* diagram). The ability of 5-bromouracil to produce point mutations without significant chromosomal damage has made it a useful tool in genetic research. Other base analogues with similar effects include 2-aminopurine and 2,6-diaminopurine, which are analogues of adenine, and 5-bromodeoxycytidine, which is an analogue of cytosine.

brown algae *See* Phaeophyta.

brown earth A type of zonal acid *soil (*pedalfer) characteristic of broad-leaved deciduous forest, such as that of Western Europe. The rainfall causes some leaching of lime from the A horizon, which is grey-brown in colour. The temperature is high enough for the rapid decomposition of organic matter resulting in the formation of mild *humus (mull). The B horizon is dark brown because of the accumulation of various iron and other compounds. Many of these deciduous-forest lands have been cleared for cultivation. The natural fertility resulting from decomposed leaf litter is thus no longer available and both liming and manuring are necessary. *Compare* podsol.

Bryales *See* Musci.

bryokinin A growth substance found in certain mosses, e.g. *Funaria*, that stimulates the formation of buds and subsequently leafy plants from the filamentous protonema.

Bryophyta A division of nonvascular plants, mainly terrestrial in habitat containing about 25 000 species. It comprises the three classes *Hepaticae (liverworts), *Musci (mosses), and *Anthocerotae (hornworts). Bryophytes are generally small low-growing plants, in most cases susceptible to desiccation and hence limited to damp or humid environments. Their life cycle shows a heteromorphic alternation of generations with the haploid gametophyte, which may be homo- or heterothallic, the dominant generation. The ephemeral sporophyte is partly or completely parasitic on the gametophyte and consists solely of a stalk bearing the spore capsule.

Certain similarities between mosses and algae, especially between algal filaments and the moss *protonema, suggest that bryophytes evolved from algae, most probably green algae, since these also share similar photosynthetic pigments, food reserves, and cell-wall constituents. Bryophytes however exhibit considerably greater morphological differentiation than the algae and differ also in producing aerial spores and having enclosed sex organs. Nevertheless they still lack roots and water is needed for the dispersal of antherozoids and for fertilization.

Despite their primitive characteristics, bryophytes are the dominant vegetation in certain areas, notably the bogs of temperate latitudes.

Bryopsida *See* Musci.

Bryopsidophyceae A class of the *Chlorophyta consisting predominantly of macroscopic algae with coenocytic thalli. Certain of these algae contain distinctive carotenoid and xanthophyll pigments and reserve carbohydrates and include xylose, mannose, and callose among their cell-wall components. The class includes the orders *Cladophorales, Acrosiphonales, Sphaeropleales, Siphonocladales, Dasycladales, and *Caulerpales.

bud 1. An undeveloped condensed region of a shoot consisting of a short stem terminated by a meristem and, in foliage buds, numerous leaf primordia, leaf buttresses, and young rolled or folded leaves. Flower buds contain the immature flower. Buds are found at the apex of a shoot (apical or terminal buds) and in the axils of leaves (axillary or lateral buds). In some species accessory buds develop in addition to the axillary bud in a leaf axil. Often axillary buds remain dormant unless the apical bud is injured or removed (*see* apical dominance). Adventitious buds may arise anywhere on the plant. Suckers, for example, develop from adventitious buds on the roots. The dormant winter buds of deciduous trees and shrubs possess protective, often resinous, *bud scales* or *cataphylls* to resist desiccation. Transpiration may be further reduced by a covering of fine hairs. Many trees can be identified in winter by the form and colour of their resting buds. The arrangement of leaves in the bud and the pattern of leaf folding or rolling (vernation) are useful diagnostically.

2. The protrusion formed from a unicellular organism during the asexual reproductive process of *budding.

budding 1. A form of asexual reproduction in which a new individual develops as an outgrowth of a mature organism. Yeast cells exhibit budding and sometimes the new cells so produced can themselves give rise to further buds before the chain of individuals separates. *Compare* fission, fragmentation.

2. (bud grafting) A type of *grafting in which a bud (the *scion*) of the desired variety, together with a small piece of bark, is removed and inserted into a slit made in the bark of the chosen rootstock (the *stock*), and secured with raffia or tape. The rootstock is usually cut back and any buds removed. The technique is widely used to propagate rosebushes. Bush roses are produced by inserting buds as near to the ground as possible while standard roses are produced by inserting buds higher up on the stem of the rootstock.

bud grafting *See* budding.

buffer A chemical solution that counteracts small changes in pH when acids or alkalis are added to it. Buffers play an important role in cells and tissues, which usually function best at or near neutrality (pH 7) since changes in pH adversely affect metabolic processes. Examples of buffers in cells are phosphates, borates, and bicarbonates (hydrogen carbonates).
Buffers are used in microscopy to reduce the production of artefacts by maintaining the specimen at its original pH during fixation. Sorenson's phosphate buffer and sodium cacodylate are two of the commonest buffers for electron microscope preparations of plant material.

bulb A fleshy underground perennating organ formed by many members of the Liliaceae and Amaryllidaceae. It is a highly modified shoot, the bulk of which is made up of colourless swollen scale leaves or leaf bases. The central apical bud contains the immature foliage leaves, the future flower, and rudimentary adventitious roots at its base. It is surrounded by numerous layers of fleshy scales, which may be complete modified leaves or the leaf bases of previous years' foliage leaves. The first sign of growth is the rapid elongation of the adventitious roots. When these have established themselves in the soil the apical bud sprouts and the foliage leaves and inflorescence emerge, growing at the expense of the food reserves in the scale leaves. In some bulbs, e.g. tulip, the inflorescence develops from the apex of the bud and further apical growth is consequently prevented. Photosynthates from the foliage leaves are passed down to one or more lateral buds in the axils of the scale leaves. This is an example of sympodial growth. In other bulbs, e.g. daffodil, the inflorescence develops in the axil of one of the foliage leaves and the apex of the bud remains within the bulb. It persists from year to year, each year giving rise to foliage leaves and a lateral inflorescence. Propagation may be effected by the expansion of a lateral bud in the axil of the outermost scale leaf. Bulbs such as daffodil show monopodial rather than sympodial growth.
The food reserves of the bulb may be starch (as in tulip) or sugars (as in onion). The outermost leaves do not contain food but are thin, brown, and scaly and protect the bulb. *Compare* corm.

bulbil A small bulb that develops from an aerial bud. Bulbils are easily detached and function as a means of vegetative propagation. They may form from lateral buds, as in the lesser celandine (*Ranunculus ficaria*), or develop in place of flowers, as in many species of *Allium*. Certain forms of apomixis give rise to bulbils, as seen in the lesser bulbous saxifrage (*Saxifraga cernua*). The term is also applied to various outgrowths formed by lower plants that become detached and develop into new plants. For example, the fern *Asplenium bulbiferum* produces bulbils on the upper surface of its fronds and *Lycopodium selago* has bulbils in the axils of the uppermost leaves.

bulliform cell Any of the large cells that occur in longitudinal rows in the leaf epidermis of certain grasses. They are believed to be involved in the unrolling of young leaves and in the rolling and unrolling movements of the leaf in response to water status.

bundle cap One or more layers of sclerenchyma or thickened parenchyma cells seen at the xylem and/or phloem poles of a vascular bundle.

bundle sheath A region surrounding the small vascular bundles in the leaves of vascular plants. It usually consists of parenchyma cells but may sometimes consist largely of sclerenchyma. In angiosperms, the bundle sheath is a type of endodermis, which may be differentiated as a starch sheath. The cells of the bundle sheath have their long axes parallel to the direction of the vascular bundle. In dicotyledons, the parenchymatous bundle sheath is sometimes termed *border parenchyma*. *Bundle sheath extensions*, in which the cells of the bundle sheath extend to one or both epidermal layers, are present in many monocotyledons. They are believed to be concerned with conduction or support. *See also* Kranz structure.

bunt A covered *smut disease of wheat, sometimes called *stinking smut* because of the smell of rotten fish given off by the spore masses. The disease is caused by *Tilletia caries*. Dwarf bunt of wheat is caused by *T. contraversa*. Infected grains are transformed into balls of bunt spores. At threshing the spores are released and contaminate healthy seed as well as the combine harvester and grain-handling machinery. Contaminated seed in turn gives rise to more infected plants.

bur (burr) A type of *pseudocarp in which the fruit is surrounded by a persistent barbed involucre as, for example, in burdocks (*Arctium*) and cocklebur (*Xanthium*). Often the term refers to any barbed fruits, e.g. those of cleavers (*Galium aparine*) and enchanter's nightshade (*Circaea lutetiana*) both of which have hooks borne directly on the pericarp. Burs cling to passing animals and are thus widely dispersed.

Burgundy mixture A copper fungicide, similar to but stronger than *Bordeaux mixture and used especially against rusts. It contains 60 g sodium carbonate and 50 g copper sulphate in 5 l water.

buttress root A form of *prop root that is asymmetrically thickened to give a planklike outgrowth on the upper side, providing extra support for the tree. Buttress roots are common in many tropical trees, e.g. certain figs (*Ficus*).

C

C_3 plant Any plant that produces, as the first step in photosynthesis, *phosphoglyceric acid, which contains three carbon atoms. Most plants of temperate regions are C_3 plants. They exhibit *photorespiration and are relatively inefficient photosynthetically compared to *C_4 plants. They generally have lower carbon dioxide fixation rates and higher *compensation points than C_4 plants.

C_4 plant Any plant that produces, as the first step in photosynthesis, *oxaloacetic acid, which contains four carbon atoms (*see* Hatch–Slack pathway). Over 100 species of C_4 plants have been identified, most of which are tropical. Examples include maize, sugar cane, sorghum, Bermuda grass, and many desert plants. C_4 plants require 30 molecules of ATP and 24 molecules of water to synthesize a molecule of glucose (*C_3 plants need only 18 molecules of ATP and 12 molecules of water). However C_4 plants produce more glucose for a given leaf area than C_3 plants and consequently grow more quickly. They can also continue to photosynthesize at high light intensities and low carbon dioxide concentrations (*see* compensation point) and, most significantly, do not exhibit *photorespiration. *Compare* crassulacean acid metabolism. *See also* Kranz structure.

Cactaceae A large family of dicotyledonous xerophytic plants, the cacti. It includes over 2000 species in about 87 genera; some authorities however recognize many more, over 300 genera being listed in some classifications. The cacti are limited in distribution to the drier regions of South and Central America and central and southern North

America. Some have become naturalized elsewhere, e.g. prickly pears (*Opuntia*) in Australia and the Mediterranean, and *Rhipsalis*, which is probably naturalized rather than native, in southern Africa.

Cacti are succulents and, being leafless (the leaves are reduced to spines), are a particularly distinctive group of plants. It is this that has led to their popularity as unusual pot plants. Photosynthesis is carried out by the green expanded stems, which show various advanced characteristics in the arrangement of vascular tissues and many other features. The flowers however, which are borne singly directly on the stem, are comparatively primitive having large numbers of stamens, petals, sepals, and bracts arranged spirally. The fruit is a berry.

Very few cacti are of use to man, except as ornamentals, though prickly pears are grown for their fruits in California, Mexico, and some Mediterranean countries.

caducous Describing parts that fall off easily or at an early stage, such as protective nonphotosynthetic stipules.

Caenozoic *See* Cenozoic.

caespitose Having a tufted form of growth, as have many grasses, e.g. *Deschampsia caespitosa* (tufted hair-grass).

Cainozoic *See* Cenozoic.

Calamitales *See* Sphenopsida.

calcareous soil *See* rendzina, terra rossa.

calcicole Describing a plant that flourishes in lime-rich soil. The presence of such a plant is so characteristic of such conditions that it can be used as an indicator. An example of a calcicole plant is the common spotted orchid (*Dactylorhiza fuchsii*). *Compare* calcifuge.

calcifuge Describing a plant that grows in lime-deficient soils, e.g. loams or clays. Tormentil (*Potentilla erecta*) is an example of a calcifuge plant. *Compare* calcicole.

calcisol *See* rendzina.

calcium Symbol: Ca. A soft metal element, atomic number 20, atomic weight 40.08. It is essential to plant growth due to its role, in the form of calcium pectate, in the development of the middle lamella. Calcium ions increase the rigidity of the cell wall by cross-linking the carboxyls of adjacent chains of pectic acid. Calcium is also found as crystals of calcium oxalate in many plant cells. Calcium deficiency is first apparent in the younger parts of the plant because there is little or no translocation of calcium from mature tissues to the growing points.

Calcium is easily leached from soils resulting in an acid soil. This may be countered by adding calcium carbonate, calcium hydroxide (slaked lime), or calcium sulphate (gypsum) to the soil. These compounds also help improve soil structure (*see* liming).

callose A structural polysaccharide found in the sieve plates of the phloem of higher plants, consisting of glucose residues linked through $\beta(1-3)$ glycosidic bonds. It is deposited steadily throughout the growing season eventually causing blockage of the pores of mature sieve tubes. The functions of blocked tubes are taken over by new phloem formed from the cambium. In some species the blockage is reversible, the callose being hydrolysed in the spring. Callose may be laid down very rapidly in response to injury of the phloem.

Callose was the first structural plant polysaccharide to be synthesized in vitro, using an enzyme isolated from bean (*Phaseolus aureus*) seedlings.

Calvin cycle

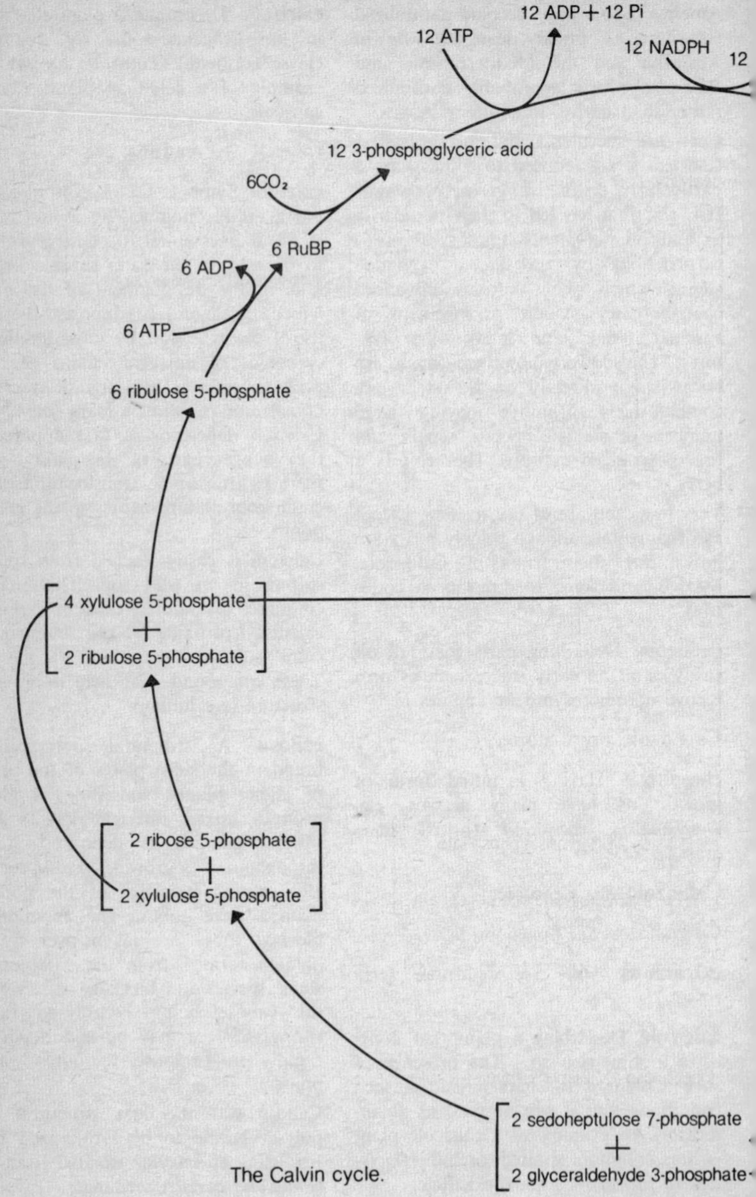

The Calvin cycle.

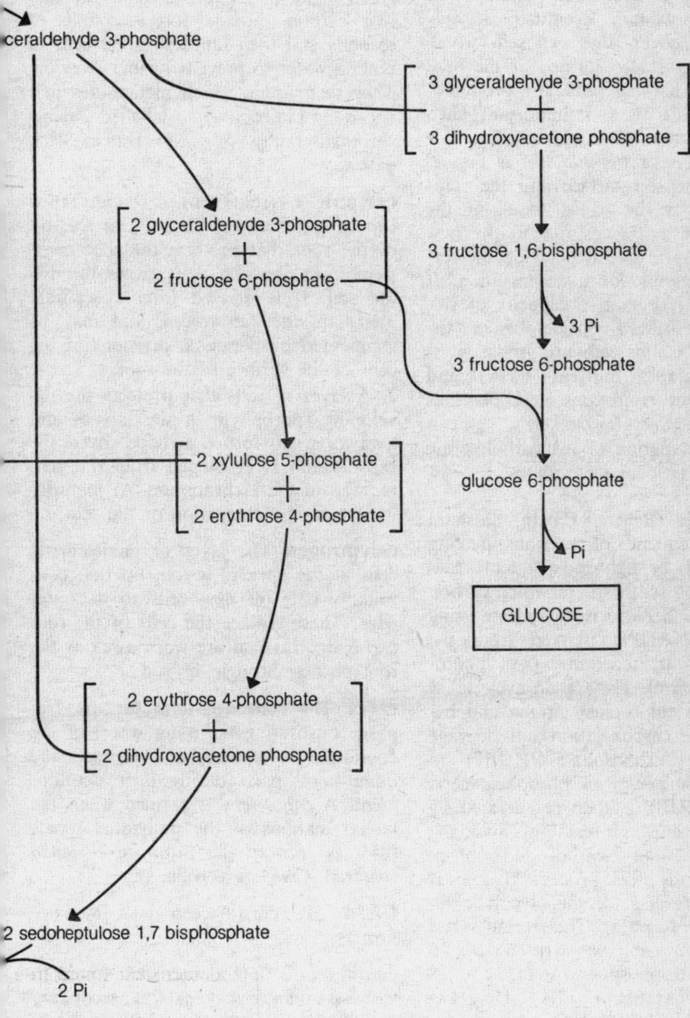

ceraldehyde 3-phosphate
3 glyceraldehyde 3-phosphate
+
3 dihydroxyacetone phosphate
2 glyceraldehyde 3-phosphate
+
2 fructose 6-phosphate
3 fructose 1,6-bisphosphate
3 Pi
3 fructose 6-phosphate
2 xylulose 5-phosphate
+
2 erythrose 4-phosphate
glucose 6-phosphate
Pi
GLUCOSE
2 erythrose 4-phosphate
+
2 dihydroxyacetone phosphate
2 sedoheptulose 1,7 bisphosphate
2 Pi

callus **1.** A mass of parenchymatous cells that forms at a wounded surface, for example where a branch has been cut off a tree. The callus tissue is produced by the cambium and initially forms a ring of thickening (*wound wood*) around the wound. Eventually it may completely cover the exposed wood. Callus tissue is also formed at the base of cuttings before roots are produced. Similarly callus tissue is important when propagating plants by leaf cuttings. For example, when a *Begonia* leaf is pegged down on the soil surface and the veins are cut, it is the callus tissue at the wound sites that gives rise to the new plantlets. Grafts also depend on the formation of callus for a successful graft union. The regenerative capacity of callus tissue is utilized considerably in *tissue culture. Callus cells are grown *in vitro* on suitable nutrient media and under correct conditions new plantlets will develop. *See also* canker.
2. An accumulation of the carbohydrate *callose on phloem sieve plates.

Calvin cycle (Benson–Calvin–Bassham cycle) The sequence of reactions, making up the dark or light-independent reactions of photosynthesis, in which carbon dioxide is reduced to carbohydrate using ATP and NADPH derived from the light-dependent reactions (*see* photophosphorylation). The Calvin cycle takes place in the chloroplast stroma and begins with the carboxylation and cleavage of ribulose bisphosphate (RUBP) to form two molecules of phosphoglyceric acid. The RUBP is then regenerated by a complex series of reactions involving 3-, 4-, 5-, 6-, and 7-carbon sugar phosphates. During this process glucose is formed by reversal of the glycolytic reactions (*see* diagram). The overall series of reactions can be written as: 6 ribulose 1,5-bisphosphate + $6CO_2$ + 18 ATP + 12NADPH + $12H^+$ + $12H_2O$ $\rightarrow$ 6 ribulose 1,5-bisphosphate + glucose + $18P_i$ + 18 ADP + $12NADP^+$. The glucose is subsequently converted to starch, cellulose, and other polysaccharides.
Calvin and his associates worked out this cycle of reactions by illuminating green algae in the presence of radioactive carbon dioxide for a couple of seconds and then immersing the cells in boiling water to prevent further reaction. They then found which metabolites first became radioactively labelled using chromatography. *See also* Hatch–Slack pathway.

calyptra **1.** (root cap) A hood-shaped cap of parenchymatous cells at the tip of the root that protects the root meristem from abrasion as it grows through the soil. It is derived from a separate meristem, the *calyptrogen*, and may be augmented by periclinal divisions of the cells at the surface of the root.
2. A layer of cells that protects the developing sporophyte in the mosses and liverworts. It forms a hood over the capsule and is developed from the venter wall of the archegonium. At maturity it ruptures by elongation of the seta.

calyptrogen The layer of meristematic cells at the apex of a root tip that continually cuts off new cells to its outer edge. These replace the cells of the root cap (calyptra) that are worn away as the root pushes through the soil.

calyx The collective term for the *sepals, constituting the outer whorl of the *perianth. It encloses and protects the other floral parts during their development. A *calyx tube* is formed when the lateral margins of the individual sepals fuse, as seen in the primrose (*Primula vulgaris*). *Compare* corolla.

CAM *See* crassulacean acid metabolism.

cambium A *lateral meristem found in vascular plants that exhibit secondary growth. It gives rise to secondary tissues mostly by *periclinal divisions of *initials. There are two cambia, the *vascular cambium and the *phellogen (cork cambium).

Cambrian The earliest period of geological time in the Palaeozoic era from about 570 to 500 million years ago. Cambrian rocks contain abundant fossils of the marine organisms of that period, i.e. algae and such invertebrate animals as trilobites and brachiopods. Colonization of the land had not yet occurred. It is believed that the climate was uniformly warm. *See* geological time scale.

CAM plant Any of a group of succulent plants that employ *crassulacean acid metabolism (CAM) for fixing atmospheric carbon dioxide.

camptodromous Describing a form of leaf venation in which there is a single primary vein and the secondary veins arising from it curve upwards towards the leaf margin. The secondary veins may join together in a series of conspicuous marginal loops (*brochidodromous* venation, e.g. in wild cherry, *Prunus avium*) or may remain separate (*eucamptodromous* venation, e.g. in downy rose, *Rosa tomentosa*). *See illustration at* venation.

campylodromous Describing a form of leaf venation in which several primary veins originate at the base of the lamina and run in downward-curving arches towards the apex, as seen in black bryony (*Tamus communis*). *See illustration at* venation.

campylotropous Describing a form of ovule orientation in which the ovule develops horizontally and the funiculus appears to be attached half way between the chalaza and the micropyle (*see illustration at* ovule). This arrangement is not very common but may be found in the Malvaceae and Caryophyllaceae. *Compare* anatropous, orthotropous.

Canada balsam A yellowish resin obtained from the fir *Abies balsamifera* that is used as a *mounting medium for microscope slides. It has similar optical properties to glass.

cane sugar *See* sucrose.

canker A plant disease in which there is a well defined area of necrosis of the cortical tissue, which becomes surrounded by layers of callus tissue. Cankers occur mainly on woody stems but may also appear on herbaceous plants. For example, in apple canker, caused by *Nectria galligena*, the fungus penetrates the apple stem through a wound or leaf scar. As the fungus grows into the stem the tree produces callus tissue to seal off the infected zone. Further penetration by the fungus results in the production of more callus tissue, or wound wood, forming the characteristic canker.

cap cell A cell, seen in filaments of algae in the order Oedogoniales, that has one to several 'caps' of cell wall material at its anterior end. These result from a distinctive form of cell division that involves the formation of rings of new cell wall material prior to nuclear division. *See illustration at* Oedogoniales.

capillarity (capillary action) The rise or fall of liquids within narrow tubes as a result of the *surface tension of the liquid. Capillarity is also responsible for the formation of drops, bubbles, etc., and for the retention of water droplets on plants following dew formation, mist, rain, and guttation. The supply of water to all parts of a plant is due, in large part, to capillary action. Water will rise against gravity providing that the forces between the surfaces of the water and the solid it is in contact with are greater than the cohesive forces within the body of water.

capitulum An inflorescence consisting of a head of small closely packed stalkless flowers or florets arising at the same level on a flattened axis. The whole is surrounded or subtended by an *involucre of bracts and simulates, in appearance and function, a single large flower (*see illustration at* racemose inflorescence). The capitulum is typical of members of the family Compositae. Capitula are often made up of two distinct types of floret: *disc florets*, in which the corolla tube terminates in five short teeth; and *ray florets*, in which the tube is extended into a conspicuous strap. When both types of floret are present the disc florets form the centre of the capitulum and the ray florets are arranged around the edge, giving a daisy-like flower. Some composites have ray florets only, e.g. dandelions (*Taraxacum*) and chicory (*Cichorium intybus*), while others have only disc florets, e.g. thistles (*Carduus*, Cirsium) and cornflower (*Centaurea cyanus*).

capsid The protein coat of a virus, which surrounds and protects its nucleic acid. The nature of the capsid determines the virus' host range and the efficiency of its infectivity. Viruses can be identified by the serological reaction of the capsid proteins.

capsomere Any of the identical polypeptide subunits making up the *capsid protein of a virus.

capsule **1.** Any dry dehiscent fruit derived from two or more many-seeded fused carpels (*see illustration at* fruit). Capsular fruits are classified by the nature of dehiscence and the number of carpels in each fruit. For example, if dehiscence is along the dorsal suture the capsule is *loculicidal*, as in willow herbs (*Epilobium*), and if along the septa, *septicidal*, as in St John's worts (*Hypericum*).

2. The structure containing the spores of the *sporophyte generation in mosses, liverworts, and ferns, the wall of which may be uni- or multicellular. The capsule is borne on the end of a stalk and may rupture by a variety of mechanisms.

3. A transparent layer of gelatinous or mucilaginous material that envelops some bacterial cells. The polysaccharides of the capsule are often characteristic of a species. Polypeptides may also be present. In some species the capsules hold together a number of bacteria, forming a structure termed a *zoogloea*.

carbohydrate (saccharide) A polyhydroxy aldehyde or ketone or a derivative of such a compound. Many carbohydrates have the structural formula $(CH_2O)_n$, which suggests that they were originally hydrates of carbon. The simplest carbohydrates are the *monosaccharides or simple sugars. They are single polyhydroxy aldehyde (*see* aldose) or ketone (*see* ketose) units, most commonly containing five or six carbons. *Oligosaccharides are more complex carbohydrates containing between two and ten monosaccharide units. *Polysaccharides contain many monosaccharide units in long linear or branched chains. They function as energy storage molecules and as structural elements in the plant body.

carbon Symbol: C. An amorphous or crystalline element, atomic number 6, atomic weight 12.01, essential to life as it forms the backbone of most organic molecules. It is particularly suited for this role since it readily forms covalent bonds with other carbon atoms and with such elements as hydrogen, oxygen, nitrogen, and sulphur. Thus many different types of functional groups can be incorporated into organic molecules. Being a light element, these bonds are particularly stable as the strength of a covalent bond increases with decreasing atomic weight. In addition, many different three-dimensional structures can be formed by carbon–carbon bonding due to the tetrahedral arrangement of electron pairs around singly bonded carbon.

All organic carbon is ultimately derived from atmospheric *carbon dioxide fixed during *photosynthesis (*see* carbon cycle).

carbon cycle The circulation of carbon between living organisms and the environment. Atmospheric carbon dioxide is fixed as carbohydrates during photosynthesis and animals obtain carbohydrates by eating green plants. Carbon dioxide is returned to the atmosphere by respiration and by the burning of fossil fuels such as coal, oil, and peat.

carbon dioxide A heavy gas, formula CO_2, that makes up 0.033% by volume of the atmosphere. It is the basic carbon source of all life, being converted to carbohydrates by *photosynthesis. The concentration of CO_2 remains relatively constant despite increased burning of fossil fuels and deforestation practices. This is due to the buffering capacity of the oceans, which contain large amounts of dissolved CO_2. The extreme solubility and rapid diffusion of CO_2 in water is a major factor in the maintenance of aquatic life.

The rate of photosynthesis can be increased by raising CO_2 levels. This has been put to practical use in the cultivation of some glasshouse crops. However yield increases are not as great as might be expected since increased CO_2 concentrations also increase stomatal closure.

Carboniferous The second period of the Upper Palaeozoic era from about 345 to 280 million years ago. During this period the climate was generally warm and humid and there were great forests and swamps, dominated by arborescent lycopsids and sphenopsids, from which the coal measures were formed. In the US the term Carboniferous is not widely used because rocks of this age in North America can be separated into two distinct types. American geologists refer to rocks of the Lower Carboniferous (345–310 million years ago) as *Mississippian* and of the Upper Carboniferous (310–280 million years ago) as *Pennsylvanian.* In North America only the Pennsylvanian rocks contain coal. Carboniferous fossil remains include mosses and liverworts, and herbaceous lycopsid fossils similar to present-day *Lycopodium* and *Selaginella* are abundant. Some of the arborescent lycopsids, as represented by *Lepidodendron* (*see* Lepidodendrales), appear to have had a rudimentary ovule. The sphenopsids are represented by the now extinct arborescent Calamitales (giant horsetails) and the extinct herbaceous Sphenophyllales. *Equisetites hemingwayi* in the Upper Carboniferous is one of the few herbaceous forms resembling the present-day *Equisetum.* The ferns were well represented as were the gymnosperms with fossils of the *Pteridospermales, *Cordaitales, and *Coniferales all being found. *See* geological time scale.

carboxydismutase *See* ribulose bisphosphate carboxylase.

carboxylase Any enzyme that catalyses the transfer or incorporation of carbon dioxide into a substrate molecule. Carboxylation reactions usually require energy in the form of ATP or a related compound; a coenzyme, usually biotin, is also a part of most carboxylases. An example is pyruvate carboxylase, which catalyses the formation of oxaloacetate from pyruvic acid. As with many carboxylases, this is an allosteric enzyme, its activity being increased by the binding of acetyl CoA. *See also* ribulose bisphosphate carboxylase, phosphoenolpyruvate carboxylase.

carboxylic acid Any organic acid having one or more –COOH groups. The intermediates of the TCA cycle are important di- and tricarboxylic acids, while acetic acid is a monocarboxylic acid. Carboxylic acids in metabolic pathways are often phosphate or thioesters of the

acid, e.g. acetyl CoA is a thioester of acetic acid.
Some higher plant cells contain high concentrations of certain carboxylic acids, for example citrus fruits are rich in citric acid, and oxalic acid is present in high concentrations in rhubarb and tobacco leaves. In the Crassulaceae and some other succulents and semisucculents there is a large diurnal fluctuation in carboxylic acid levels. This is due to incorporation of carbon dioxide into malic acid in the dark (*see* crassulacean acid metabolism).

carcerulus A type of *capsule (fruit) that breaks up on maturity into one-seeded segments or nutlets, as in the Labiatae.

cardiac glycoside Any of a group of steroid *glycosides characterized by their stimulative effects on the heart. Cardiac glycosides resemble *saponins in being diterpenes and by foaming in solution. They differ however in having a lactone ring attached at carbon 17 of the *steroid nucleus. An important commercial source is the foxglove (*Digitalis purpurea*), which yields digitonin.

carina *See* keel.

carinal canal (protoxylem lacuna) A longitudinal channel in the stem internode of *Equisetum* (horsetail) and some of its fossil relatives, positioned radially opposite a raised stem ridge (*carina*). Each carinal canal is arranged on the inner side of a vascular bundle. The carinal canals apparently result from breakdown of *protoxylem and are believed to function in water conduction. *Compare* vallecular canal.

carnivorous plant *See* insectivorous plant.

carotene (carotin) Any of a class of orange, red, or yellow pigments all of which are hydrocarbons containing eight *isoprene subunits and hence have the formula $C_{40}H_{56}$. The most commonly occurring carotenes are α- and β-carotene (*see also* vitamin A). Another carotene, lycopene, gives tomato fruits their red colour. Carotenes are accessory pigments in most photosynthetic cells, contributing to the process of gathering light energy. However they are relatively inefficient as light absorbers and are thought to have a second function of protecting the other photosynthetic pigments from photooxidation by ultraviolet light. *Xanthophylls are oxygenated carotenes. *See* carotenoids.

carotenoids Yellow, orange, or red fat-soluble pigments found in all photosynthesizing cells, where they act as accessory photosynthetic pigments. Their absorption spectrum suggests they may also be involved in the phototropic response. They are also found in other organs, e.g. roots, notably carrot roots, and petals. There are two groups of carotenoids; the *carotenes, which are hydrocarbons, and the *xanthophylls, which are oxygenated derivatives of the carotenes.
During leaf senescence the carotenoids do not break down as quickly as the chlorophylls and their colours are temporarily revealed.

carotenol *See* xanthophyll.

carotin *See* carotene.

carpel The structure that bears and encloses the ovules in flowering plants. It normally comprises the *ovary, *style, and *stigma. The typical enclosed carpel of the angiosperms has probably evolved from an open carpel bearing megasporangia on the margins, and is thought by many to be homologous with the megasporophyll of pteridophytes and the ovuliferous scale of gymnosperms.
In more primitive angiosperms there are often large numbers of unfused carpels in the gynoecium. In advanced forms the number of carpels is reduced and they tend to be fused and inserted asymmetrically.

carpogonium The female sex organ in the red algae, which has a swollen basal portion containing the egg and a number of hypogynous cells, and an elongated terminal projection or trichogyne, which attracts the male gamete. It develops by the expansion of a single cell.

carposporangium A *sporangium that develops directly from the zygote in situ in some of the red algae, and which is dependent on the parental tissue for nutrition.

carpospore A spore formed following division of the zygote in red algae. It may be haploid if formed following meiosis as, for example, in *Porphyra*. On germination it then gives rise to a haploid gametophyte or occasionally a short-lived juvenile form termed the *Chantransia* stage. The carpospores of the more advanced red algae, e.g. *Polysiphonia*, are diploid and give rise to a diploid plant similar in form to the haploid generation.

carr *See* fen.

caruncle A horny outgrowth near the hilum of a seed that is formed from the integuments. The warty outgrowth covering the hilum and micropyle of the seed of castor oil (*Ricinus communis*) is a caruncle. Caruncilate seeds are also found in other genera of the Euphorbiaceae, e.g. *Euphorbia* and *Jatropha*. *Compare* aril.

Caryophyllidae (Centrospermae) A subclass of the dicotyledons containing mostly herbaceous plants usually with bisexual flowers. The seeds often contain perisperm and *betalain alkaloids are usually present. Four or five orders are usually recognized: the Caryophyllales (or Chenopodiales); the Polygonales, containing the one family Polygonaceae (including buckwheat and sorrel); the Plumbaginales, including the one family Plumbaginaceae; the Batales, comprising the one species *Batis maritima* (saltwort), often placed in the Caryophyllales; and the Theligonales, the one genus of which, *Theligonum*, has been variously placed in the Caryophyllales or Haloragales (subclass Rosidae). The Caryophyllales is divided into some 10–16 families, the more important of which are the *Cactaceae (cactus family), the Aizoaceae (including the mesembryanthemums), the Caryophyllaceae (carnation family), and the Chenopodiaceae (beet family).

caryopsis (grain) A fruit that resembles an *achene except that the seed wall fuses with the carpel wall during embryo development. The caryopsis is typical of cereals and grasses.

Casparian strip (Casparian band) A band of suberized and/or lignified wall material in the radial (anticlinal) and transverse walls of cells of the *endodermis. It ensures that water and solutes pass through the living protoplast of the endodermal cells, rather than through the cell walls, thus facilitating selective filtering of the sap solutes and control of the rate of flow.

catabolism The metabolic breakdown of complex molecules to simpler ones. For example, respiration involves various catabolic reactions that degrade sugars to carbon dioxide and water and release energy, which is stored as ATP. *Compare* anabolism.

catalyst A compound that enhances the rate of a chemical reaction without itself being used up in the process. It achieves this by combining transiently with the reactants to form a transitional complex having a lower activation energy than that of the uncatalysed reaction. Biological catalysts are known as *enzymes.

cataphyll *See* bud.

catkin A pendulous inflorescence modified for wind pollination. It is a loose *spike made up of numerous sessile usually unisexual flowers. The calyx and corolla are normally reduced or absent to allow maximum air circulation around the flower and the catkin itself develops in an exposed position on the plant. Male catkins shed vast amounts of light dry pollen. Female catkins usually have long hairy styles and stigmas to enhance pollen interception. Catkins are formed by many tree species, e.g. the male flowers of the Betulaceae (birches, hazels, alders, hornbeams) are always aggregated into catkins. *See* anemophily.

caudex 1. The trunk of a palm or tree fern.

2. The persistent swollen stem base of certain herbaceous perennials.

Caulerpales (Siphonales) An order of the *Bryopsidophyceae containing algae that lack any dividing walls. They also differ in containing α-carotene in addition to β-carotene. In some classifications these acellular algae are placed in the three orders: Derbesiales, Codiales (including *Bryopsis* and *Codium*), and Caulerpales (including *Caulerpa* and *Halimeda*).

caulid The 'stem' of a moss or liverwort.

cauliflory The production of flowers on tissue that has been secondarily thickened, such as the branches or trunks, rather than at apical meristems. Flowers arising in this way normally develop from suppressed side shoots, but more infrequently they may be formed from the phellem. This trait is found most frequently in species that make up the lower canopy of tropical forests, e.g. cocoa (*Theobroma cacao*). It is believed in some cases to be associated with pollination or fruit dispersal by bats.

cauline Pertaining to the stem, especially applied to leaves arising on the upper part of the stem, as compared to the *radical leaves.

cavitation The formation of vapour-filled cavities in a flowing liquid due to a decrease in pressure. It is the main cause of disruption of water columns in the xylem when a plant is suffering water stress.

Caytoniales An extinct order of gymnosperms known from fossils of the Jurassic period. Their habit is uncertain but it is known they had compound leaves consisting of three to six leaflets with reticulate venation. These arose from the tip of the rachis giving a fan-like or palmate leaf. The female reproductive axis bore short pinnae each of which ended in a round hollow structure or *cupule containing the ovules. The pinnae of the male reproductive organs bore synangia longitudinally divided into four pollen sacs, which appear to have been very similar to the anthers of flowering plants.

cell The fundamental unit of a living organism and the basis of its structure and physiology. A living plant cell comprises *cell wall and *protoplast. Bacterial cells and those of the blue-green algae are *prokaryotic, i.e. they have no nucleus. The cells of all other organisms are *eukaryotic and have a well defined nucleus containing the chromosomes enclosed in a nuclear membrane. In many plant tissues, e.g. sclerenchyma, cells are in a nonliving state, consisting of cell wall only or cell wall thickened with additional nonliving material such as lignin.

cell biology *See* cytology.

cell cycle The sequence of events in a cell resulting in the formation of two complete daughter cells. In addition to the observable changes during *mitosis, biochemical processes occur in the nucleus and cytoplasm. *Cytokinesis is followed by a growth phase (*G_1 phase*) during which there is a high rate of RNA formation and protein synthesis. A lag phase of variable duration is followed by a period of DNA replication and histone synthesis (*S phase*) as the chromosomes divide into chromatids. Finally there is a short growth period (*G_2 phase*) before prophase of mitosis begins. The length of each period varies considerably. In actively dividing cells in culture the full cycle takes twenty four hours while in living tissue the time taken to double the number of cells can be as short as eight hours.

cell division The division of the nucleus (karyokinesis) followed by the division of the cytoplasm (cytokinesis). Two daughter cells are formed, each with a nucleus containing the same number of chromosomes as the mother cell (*see* mitosis). After reduction division (*see* meiosis) the nuclei of the daughter cells contain half the number of chromosomes of the mother cell.

cell extension The increase in cell length and volume that occurs a certain distance behind the apex following cell division. The process involves both the uptake of large quantities of water into the cell vacuole by osmosis (*vacuolation*) and the synthesis of new cytoplasmic and cell wall materials. Cell extension accounts for a far greater proportion of overall plant growth than cell division.

cell fractionation The separation of cell components by differential centrifugation. The tissue is broken down in a homogenizer and the homogenate is spun in the cell of an ultracentrifuge. Protoplasmic particles settle out in order of size and weight, separation also depending on the speed and duration of centrifugation. The rate of sedimentation is expressed as the *sedimentation coefficient, *s*. After sedimentation is complete the fractions can be identified by their biochemical activity and by electron microscopy.

cell lineage The sequence of cells, produced by cell division and cytological changes, that results in the formation of differentiated cells with a specific function. For example, the germ cells in flowering plants are produced by the sequence of changes, commencing with undifferentiated cells in the anthers and nucellus respectively, that produce the microspores (pollen grains) and megaspores (embryo sacs) and eventually the male and female gametes.

cell membrane (membrane) Any membrane surrounding either the protoplast or any of the organelles within the protoplasm. The term is frequently used as a synonym for the plasma membrane. All cell membranes have essentially the same structure (*see* plasma membrane) and their selective permeability controls the passage of molecules and ions within the cell, particularly into and out of organelles. This ensures that metabolic processes are segregated without preventing links between biochemical pathways.

cellobiose A reducing disaccharide composed of two glucose units linked by a $\beta(1-4)$ glycosidic bond. It is formed by the incomplete hydrolysis of cellulose by the enzyme cellulase. Complete hydrolysis to glucose units is achieved by the enzyme cellobiase (β-glycosidase).

cell plate An opaque colloidal layer established by membrane-bound cavities in the equatorial region of the *spindle at the *telophase stage of mitosis. It separates the cytoplasm into two daughter protoplasts, each with a nucleus. Pectic substances within the vesicles contribute to the formation of a *middle lamella between the newly formed cells.

cellular slime moulds *See* Acrasiomycetes.

cellulase A hydrolytic enzyme found in bacteria and some seedlings that attacks alternate $\beta(1-4)$ linkages in *cellulose, degrading it to the disaccharide cellobiose. Cellulase is one of the very few enzymes capable of attacking the $\beta(1-4)$

glycosidic bond; cellulose is not degraded by other glycolytic enzymes such as amylase. Cellulase, produced by bacteria living in the gut of ruminant animals, enables such animals to derive nutrients from the fibrous component of their diet.

cellulose A polysaccharide composed solely of glucose units linked by $\beta(1-4)$ glycosidic bonds. It is the most abundant cell wall and structural polysaccharide in the plant kingdom and probably the most abundant of all compounds found in living organisms. In cell walls it is present in highly organized *microfibrils, the formation and organization of which is not fully understood at present.

Synthesis of cellulose is thought to take place in the golgi apparatus. Enzymes capable of cellulose formation from primer molecules and nucleotide diphosphate derivatives of glucose (e.g. UDP- and GDP-glucose) have been found in particulate cell fractions arising from the golgi apparatus. Study of cellulose synthetic enzymes has proved difficult because the enzymes are tightly membrane bound.

The enzyme *cellulase is responsible for the breakdown of cellulose to cellobiose, which is hydrolyzed to its constituent glucose units by the enzyme cellobiase.

cell wall The external covering of a plant cell composed mainly of cellulose molecules organized into *microfibrils. Small quantities of proteins have also been identified, e.g. enzymes such as invertase, phosphatase, and ATPase, whose functions are related to the uptake of nutrients and their passage from cell to cell. Enzymes responsible for cellulose synthesis are produced in the *golgi apparatus and reach the cell surface, where they become functional, within the membranes of vesicles. *Microtubules beneath the plasma membrane are thought to have a role in the organization of cellulose into microfibrils. Following cell division the *primary wall* is laid down by the deposition of microfibrils on the *middle lamella. The orientation of the microfibrils differs depending on whether the cells are destined to be parenchymatous or more specialized tissue. The *secondary wall* normally consists of three layers laid down after cell extension is complete. The microfibrils are closely packed and aligned in the same direction in each layer but in different directions in successive layers. In maturing fibres, the microfibrils are mainly parallel to the long axis, while in developing xylem vessels they are laid down in rings or in helical strands. The hydrated nature of the microfibrils renders the wall elastic and permeable to water and solutes, including the soluble respiratory gases. The pressure of cell contents against the walls causes stretching and confers turgidity to plant tissues. This is a major factor in the provision of mechanical support to the nonwoody tissues of plants. In the cells of sclerenchyma and xylem tissues, lignin is deposited within the layers of the secondary walls and the protoplasts of these cells eventually disintegrate. Lignin confers considerable strength and these tissues form the wood of plants. Other substances, including suberin and callose, may also be deposited. The walls of spores are often impregnated with sporopollenin.

The cell walls of fungi differ in that they usually contain chitin, while those of bacteria consist mainly of mucopeptide substances. Silica is an important constituent of certain algal cell walls (e.g. diatoms) and is also found in some grasses and sedges.

Cenozoic (Cainozoic, Caenozoic) The present era of geological time beginning about 65 million years ago. It has seen four periods of extensive land elevation separated by periods when the oceans expanded. The Andes, Rockies, Himalayas, and Alps were formed in the Cenozoic. Traditionally it has been divided into the *Tertiary and *Quaternary periods. Certain authorities now recognize the Palaeogene, equivalent to the earlier epochs (Palaeocene, Eocene, and Oligocene) of the Tertiary, and the Neogene for the later Tertiary epochs (Miocene and Pliocene) and the Quaternary. *See* geological time scale.

centre of diversity *See* gene centre.

centric bundle *See* concentric bundle.

centrifugal Developing from the centre outwards so the oldest structures are in the middle and the youngest at the perimeter. An example is the differentiation of secondary xylem. *Compare* centripetal. *See also* endarch.

centrifuge An apparatus designed for rapid separation of solids in suspension by rotating them in a container at very high speeds. The organelles of ruptured cells can be separated in this way. An *ultracentrifuge* rotates at much higher speeds (up to a million times per second) and is used to separate colloidal particles such as large protein molecules. As the sedimentation rate (*see* sedimentation coefficient) depends on the particle size, the size of particles can be estimated. The ultracentrifuge operates under refrigeration in a vacuum. *See also* density-gradient centrifugation.

centriole A cylindrical structure some 500 nm long and 150 nm in diameter composed of nine longitudinal fibrils. Centrioles are not found in the cells of higher plants, but are commonly found in pairs in animal cells and have been identified in the cells of some primitive plants, e.g. brown algae. At the commencement of cell division fibrils appear, radiating from the centrioles and forming *asters. These form the poles from which the *spindle fibres arise. The basal bodies (*see* kinetosome) of *flagella and cilia arise from centrioles and have a similar structure.

centripetal Developing from the outside inwards, so the oldest structures are at the edge and the youngest are in the middle. An example is the differentiation of xylem in club mosses (*Lycopodium*). *Compare* centrifugal. *See also* exarch.

centromere The region within a chromosome where DNA is present, as part of the structure, but has no genetic significance. When chromosomes are contracted in cell division, the centromere appears as a narrowed region because the DNA is not coiled. Its position on a chromosome is a distinctive feature enabling homologues to be identified. Centromere positions are described as median, terminal, subterminal, etc. (*see also* acrocentric, metacentric, telocentric) and are useful taxonomic characters. *Kinetochores are situated in the centromere region and these attach chromosomes to the equatorial region of the *spindle during cell division.

centrosome A *centriole with *aster fibres.

Centrospermae *See* Caryophyllidae.

chalaza The basal region of the ovule where the nucellus and integuments fuse. It may or may not coincide with the position of the funiculus depending on the mode of ovule orientation. *See also* chalazogamy.

chalazogamy A method of fertilization in seed plants in which the pollen tube does not reach the nucellus through the micropyle (*porogamy*) but grows or digests its way through the placenta and *chalaza instead. This may be observed in certain trees and shrubs, such as beech, elm, and hazel.

chalcones One of the two groups of *anthochlor flower pigments. Chalcones are less variable than other types of *flavonoid pigments with only some 20 kinds being known. In the synthesis of the various flavonoids from acetates and cinnamic acid the chalcones are the first compounds to be formed along the biosynthetic pathway. This has led some to consider the possession of chalcones a primitive characteristic as compared to the possession of *aurones, which are oxidized derivatives of chalcones. The yellow colours of gorse, dahlia, and carnation flowers are due to chalcones.

chalk grassland *See* heath.

chamaephyte A low-growing plant whose perennating buds are borne either at or near (within 0.25 m of) the soil surface. Chamaephytes include small bushes and herbaceous perennials and are commonly found in cold or semiarid climates. The buds are often protected by snow in cold climates. *See also* Raunkiaer system of classification.

chaparral *See* maquis.

characters *See* taxonomic characters.

Charophyta (stoneworts) A division of *algae containing one order, the Charales. These are macroscopic plants often found growing in freshwater ponds attached to the bottom by rhizoids. They have a distinct main axis with whorls of branches arising from the nodes and superficially resemble certain higher plants. The thallus often becomes encrusted with calcium carbonate, hence the vernacular name. Sexual reproduction is oogamous and very complex. The zygote germinates to give a filamentous protonema, a situation only found among algae in the Charales. This and the disclike form of the plastids suggests the Charales are more similar to the higher plants than the Chlorophyta.
The Charales are separated from the Chlorophyta because of their complex vegetative organization and reproduction, but many classifications retain them in the Chlorophyta on the basis of similar metabolism and pigmentation.

chasmogamy The production of flowers that open to expose the reproductive organs. This allows cross pollination but does not preclude self pollination. *Compare* cleistogamy.

chemical fossil Any of various chemicals, such as alkanes and porphyrins, found in certain rocks and thought to indicate the former presence of life. Such compounds are often the only evidence for the existence of living organisms in rocks of the Precambrian era.

chemoautotroph An organism that synthesizes organic molecules using energy derived from the oxidation of inorganic compounds present in the environment. Examples are the bacteria *Nitrosomonas*, which obtains energy by oxidizing ammonia to nitrite, and *Nitrobacter*, which oxidizes nitrite to nitrate. *Compare* chemoheterotroph. *See also* autotroph, chemotroph.

chemoheterotroph An organism that obtains the energy required for the synthesis of organic molecules from the oxidation of organic compounds. All *heterotrophs, with the exception of the *photoheterotrophs, are chemoheterotrophs.

chemosterilant Any chemical used to sterilize pests and hence reduce the number of offspring in the subsequent generation. A common technique is to regularly release large numbers of laboratory-reared sterile males into the wild population, which compete with normal males. The method has been used successfully to control various fruit flies.

chemosynthesis The elaboration of organic materials by organisms (*chemotrophs) using chemical energy derived from chemical reactions. *Compare* photosynthesis.

chemotaxis A *taxis in response to a chemical stimulus. For example, bacteria commonly show *positive chemotaxis*, swimming towards regions containing higher concentrations of food substances such as peptone and lactose. *Negative chemotaxis* may be shown in response to toxic substances.

chemotaxonomy (biochemical taxonomy) The application of the principles and procedures of chemical analysis and the results obtained to the classification of plants. Although the science has diverse and ancient origins (e.g. the medicinal and ethnic interest in the chemical constituents of plants) it has only made significant growth in recent decades. Plant product analysis has been simplified by the development of chromatography and electrophoresis, which with other new techniques, make routine analysis of a representative selection of material a possibility that was not once available. Many different compounds are potentially of taxonomic value and three main groups are recognized; primary and secondary metabolites and *semantides. Primary metabolites are those involved in the essential biochemical pathways of the plant. As such they are virtually ubiquitous through the plant kingdom and are rarely useful taxonomically. Secondary metabolites are those substances accumulated by plants that have traditionally been regarded as waste products, e.g. terpenoids, alkaloids, phenolics, etc. Often distinct discontinuities are found in the distribution of secondary metabolites, which have been used to delimit taxa.

chemotroph An organism that derives the energy necessary for the synthesis of organic compounds from the oxidation of either organic or inorganic compounds (as in *chemoheterotrophs and *chemoautotrophs respectively). Light energy plays no part in the process. *Compare* phototroph.

chemotropism A *tropism in response to a chemical stimulus. Thus, when a pollen tube grows through an agar culture towards a piece of implanted pistil, it exhibits two forms of chemotropism, namely *negative aerotropism* in growing away from oxygen and *positive chemotropism* in growing towards those chemicals exuding from the pistil.

chernozem (black earth) A type of zonal alkaline *soil (a *pedocal) found in regions with light summer rainfall, as in the pampas of South America and the prairies of North America. The top of the A horizon is very thick, rich in humus, and black, and the brown lime-rich layer beneath merges into, and is often indistinct from, the B horizon. The slight leaching due to the spring and summer rains is counteracted by surface evaporation in the hot summer resulting in the accumulation of lime in the A horizon. Such soils are very fertile over a long period of time.

chestnut-brown soil A type of zonal alkaline *soil (a *pedocal) that forms in areas of dry grassland receiving about 340–360 mm of precipitation per annum. The A horizon is much thinner than in a *chernozem soil because the supply of organic matter is limited.

chiasma (*pl.* chiasmata) A point at which *chromatids from different homologues in a bivalent are seen to be in close contact as separation of the paired chromosomes commences during the first meiotic *prophase. Genetic evidence shows that at these points the chromatids break and rejoin so portions are exchanged, resulting in the recombination of maternal and paternal genes on the *homologous chromosomes. Chiasma formation is also important for the distribution of chromosomes into the nuclei of gametes. If chiasmata fail to form, homologues do not pair effectively at first prophase and they are unable to move to the poles of the spindle. *See also* crossing over.

chiasma interference The effect a crossover occurring at one point has on the chances of a second crossover occurring in the same pair of chromatids. If the term is used without qualification, it normally implies the chances of a second crossover are reduced (*positive interference*). *Negative interference*, where the chances of a second crossover are increased, is rarer.

chimaera A plant or plant part that consists of two or more genetically different types of cell. In a *graft hybrid* the distinct cell types are derived from two different species. Such chimaeras are formed when a bud grows from the region of a graft union and contains tissues from both scion and stock (*see* periclinal chimaera). Chimaeras can also arise by a mutation in a meristematic region. This may give rise either to a sector of different cells (*sectorial chimaera*) or to a complete layer of different cells, as might occur if the mutation is in the top layer of the tunica, which subsequently gives rise to the epidermis. Chimaeras consisting of cells of different chromosome number may occur in cytologically unstable tissues. This may result from wounding. It is also seen in trisomic plants in which some cells lose the extra chromosome to revert to the normal chromosome number. In such cases both normal shoots and shoots showing the morphological changes associated with trisomy can occur.

chi-squared test (χ^2 test) A test usually used to find how well observed frequencies of results correspond with expected frequencies. It is used when the observations fall into discrete classes, such as red, pink, or white flowered plants. χ^2 is found by calculating

$$(\text{observed}-\text{expected})^2/\text{expected}$$

for each class and adding these together. The sum is then checked in a table of χ^2 values for different degrees of freedom at various probability levels.

chitin A polymer containing chains of acetylglucosamine units, having the empirical formula $C_{15}H_{26}O_{10}N_2$. Microfibrils of chitin make up the hyphal wall of most fungi.

Chlamydobacteriales (filamentous iron bacteria) An order of colourless rod-shaped gram-negative bacteria that aggregate into filaments, which sometimes become branched. The filaments are surrounded by a sheath of organic material impregnated with ferric and manganese oxides. Such bacteria derive energy from oxidizing organic substrates. However they also oxidize ferrous compounds to ferric oxides and probably derive additional energy from these reactions.

Chlamydomonadales *See* Volvocales.

chlamydospore A thick-walled resting spore formed by certain fungi. The term refers both to thickened segments of hyphae (*gemmae*), as found in the Saprolegniaceae and Mucoraceae, and to the dikaryotic *brand spores* of the Ustilaginales that develop in sori on infected plants.

chlorazol black A temporary acid stain that colours walls grey or black and is especially suitable for demonstrating pitting. It is also used for chromosome counts, in which case the tissue is first fixed in acetic acid. After staining, a drop of acetocarmine is added and the chromosomes stain a deep reddish-black colour. The stain gives a clear-cut picture and is therefore useful for photomicrography.

chlorenchyma Photosynthetic tissue containing numerous chloroplasts and often having relatively large intercellular spaces. The *mesophyll of leaves is composed of chlorenchyma.

chlorinated hydrocarbon Any hydrocarbon with any or all of the hydrogen atoms replaced by chlorine atoms. Such compounds are often used as insecticides, examples being aldrin, dieldrin, and DDT.

Chlorococcales An order of the *Chlorophyceae containing nonflagellate unicellular (e.g. *Chlorella*), colonial (e.g. *Scenedesmus*), coenobial (e.g. *Pediastrum*), and coenocytic coenobial (e.g. *Hydrodictyon*) algae. Motility is restricted to the gametes and zoospores. Sexual reproduction is isogamous or anisogamous but never, so far as is known, oogamous. Asexual reproduction is by motile zoospores or nonmotile aplanospores (autospores).

Chlorophyceae A class of the *Chlorophyta. In some classifications it contains all the orders of the Chlorophyta except the Charales (*see* Charophyta). In other classifications the orders are divided between the Chlorophyceae, *Oedogoniophyceae, *Bryopsidophyceae, and *Zygnemaphyceae. In the latter system the Chlorophyceae contains the orders *Volvocales, Tetrasporales, *Chlorococcales, *Ulotrichales, *Ulvales, Prasiolales, Microsporales, and Chaetophorales.

chlorophylls The main class of photosynthetic pigments. They absorb red and blue-violet light and thus reflect green

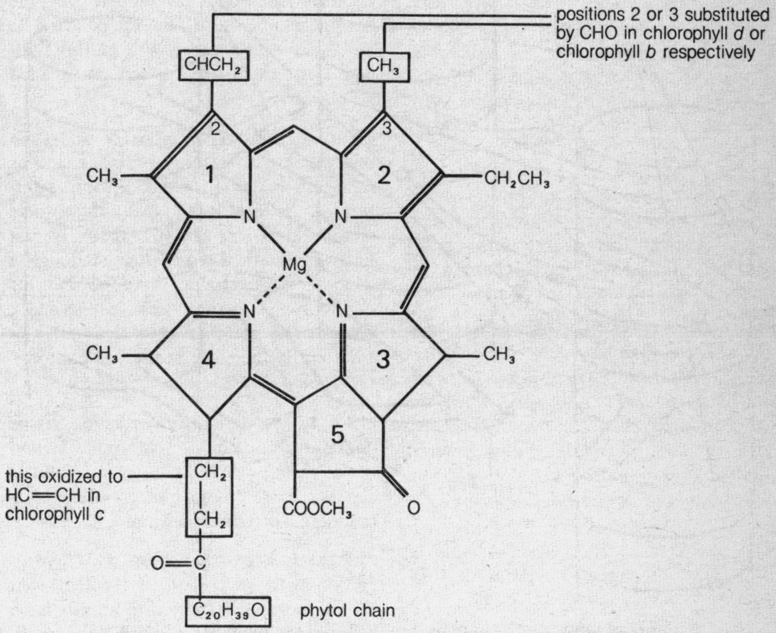

The structure of chlorophyll *a*.

light, so giving plants their characteristic green colour. Chlorophylls are involved in the light reactions of photosynthesis and are located in the chloroplast. The molecule consists of a planar porphyrin 'head' and a long phytol 'tail' (*see* diagram). A molecule of magnesium is present in the centre of the porphyrin ring. In vivo the chlorophyll molecule is bound to membrane proteins via the phytol chain. Its function in photosynthesis is to absorb light energy and initiate electron transport. There are four groups of chlorophylls: *a*, *b*, *c*, and *d*. Chlorophyll *a* is found in all autotrophic plants, chlorophyll *b* in the Chlorophyta, Charophyta, and land plants, and chlorophylls *c* and *d* only in certain algae. *See also* bacteriochlorophyll.

Chlorophyta (Isokontae, green algae) A large division containing the green-pigmented *algae, commonly found in freshwater or moist terrestrial habitats. The green algae show a wide range of form and of sexual and asexual reproductive methods. They contain chlorophylls *a* and *b*, carotenes, and xanthophylls, store food reserves as starch, and always have cellulose cell walls. In these

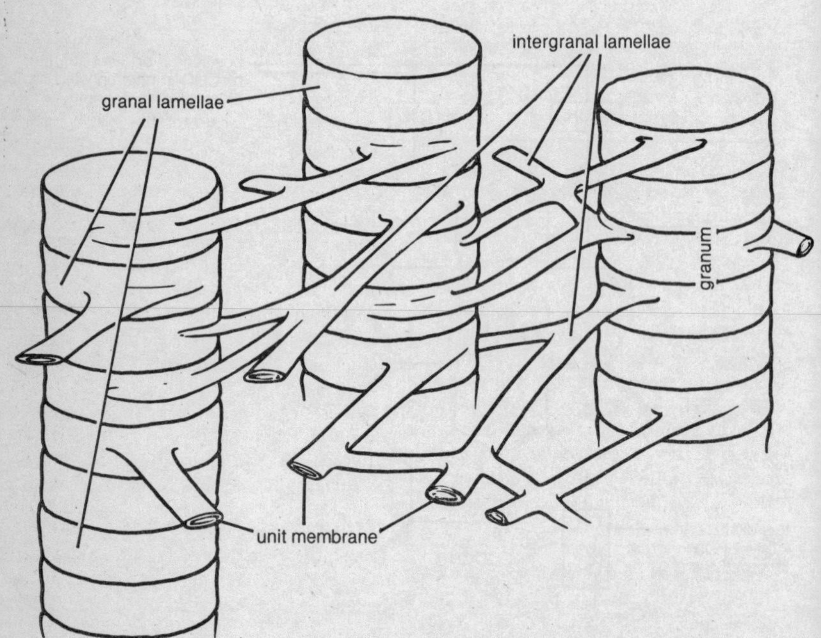

The arrangement of granal and intergranal lamellae in a chloroplast.

respects and in their ultrastructure they resemble the bryophytes and tracheophytes more than any other algal division (except possibly the Charophyta, when these are given divisional rank).
Some classifications separate the division into the two classes, Chlorophyceae and Charophyceae. Another system elevates the Charophyceae to divisional rank, the *Charophyta, and divides the remainder into four classes: the *Chlorophyceae, *Oedogoniophyceae, *Bryopsidophyceae, and *Zygnemaphyceae.

chloroplast A green *plastid with an internal membrane system incorporating the pigment molecules that are essential to photosynthesis. Chloroplasts are present in most of the cells of autotrophic plants that are exposed to light. In algae they are frequently large and may have a complex shape, e.g. the spiral chloroplasts of *Spirogyra*. Many algal cells contain only one chloroplast. In bryophytes and all higher plants they are generally lens shaped and present in large numbers. They vary in size, average dimensions being 5 μm by 2.3 μm, and probably develop from the minute *proplastids of meristematic cells.
The chloroplast is surrounded by a double membrane with no pores. The matrix or *stroma* of the chloroplast is a complex hydrophilic proteinaceous sol, containing particles that vary in size and distribution. Some are the temporarily stored products of photosynthesis, e.g. starch grains, plastoglobuli, and various crystals. Others are aggregates of ribulose bisphosphate carboxylase, ribosomes, and polysomes, and circular fibrils of DNA. Chloroplast ribosomes are smaller than cytoplasmic ribosomes. In this respect, and in their reactions to antibiotics, chloroplast ribosomes resemble those of prokaryotes (*see* serial endosymbiotic theory).
The internal chloroplast membranes form a complex system of granal and intergranal lamellae (*see* diagram). *Grana are formed from two or three up to approximately a hundred disclike flattened vesicles (thylakoids) stacked on top of each other. They are orientated in a variety of directions relative to the long axis of the chloroplast and their number and size varies with different species. The chloroplasts in the bundle sheath cells of C_4 plants do not contain grana. The intergranal lamellae are flexible interconnecting channels, continuous with and linking together the channels of individual grana. The precise arrangement of pigment molecules within the membranes is not clear but the thylakoid membranes have the densest concentration. Spherical structures (*quantasomes*), some 17.5 nm and others 11.0 nm in diameter have been identified, which are believed to be embedded in the lipid layers with their edges protruding from the membrane surfaces. These particles have been shown to be essential to the light reactions of photosynthesis. The pigment in the smaller particles is the P_{700} form of chlorophyll *a* associated with photosystem I, and that of the larger particles is the P_{680} form associated with photosystem II. Superficial particles, loosely attached to the granal membranes, have also been identified. These consist of molecules of a calcium-dependent ATPase.
The light reactions of photosynthesis thus occur within the granal layers. Water molecules are split, oxygen is evolved, and ATP and NADPH are formed. The dark reactions occur in the stroma, phosphoglyceric acid being reduced to glyceraldehyde 3-phosphate by NADPH. This is the basis for the biochemical reactions that ultimately produce all the organic molecules of the plant.

chlorosis A condition of green plants in which they become unhealthy and pale or yellow in colour. The yellowing is due to a fall in chlorophyll levels, which may be caused by a number of factors. Deficiency of iron, magnesium, or copper is a common cause (*see* deficiency diseases). Poor light conditions, and certain plant diseases, notably virus infections, also cause chlorosis. *See also* yellows.

chlor-zinc-iodide *See* Schultze's solution.

chromatid A thread formed by the longitudinal division of a *chromosome. During the S phase of the *cell cycle, the DNA of the chromosomes replicates and the protein components of *chromatin are synthesized. As chromatin becomes condensed in *mitosis, the chromosomes can be seen to be divided along their length, with the exception of the centromere, into two chromatids. In *meiosis chromatids do not become apparent until the diplotene stage of the first prophase. Each is structurally a complete chromosome. During anaphase of mitosis and the second division of meiosis, when the centromeres divide, chromatids are separated to become the chromosomes of the daughter nuclei.

chromatin The complex of proteins, DNA, and small amounts of RNA of which *chromosomes are composed. *Histones are the principal nuclear proteins, although nonhistone acidic proteins are also present. *See also* euchromatin, heterochromatin.

chromatin loop A region of relatively unwound DNA, about 200 nm long, extending outwards in a loop from the chromosome. It is thought that in this uncoiled condition transcription can occur. If so, chromatin loops correspond to genes actively involved in protein synthesis.

chromatography A technique used to separate and identify the components of mixtures of similar compounds, such as different amino acids or chlorophyll pigments, by selective adsorption. The mixture, dissolved in a liquid or gas *mobile phase* is passed through another solid or liquid medium the *stationary phase*. This may be, for example, a column of charged resin (*see* ion-exchange chromatography) or a piece of filter paper (*see* paper chromatography). Some of the compounds in the mixture will pass more slowly through the stationary phase than others, as they are more strongly adsorbed by the solid, or more soluble in the liquid. Consequently, the mixture will separate out. The column, strip of paper, or other material containing the separated components is termed a *chromatogram*. In some techniques, the compounds are collected as they emerge from the column, or their progress along the column is measured against a known standard. Colourless components of the mixture can be identified by electronic detection, ninhydrin developer, or radioactive labelling. *See also* gas–liquid chromatography, thin-layer chromatography.

chromatophore A pigmented lamellar or vesicular structure that can be isolated from disrupted photosynthetic bacteria or blue-green algae. These organisms have no chloroplasts. Their plasma membrane may be projected in folds into the cytoplasm (*see* lomasome) forming lamellae that have, therefore, double unit-membrane structure. The pigments and most of the enzymes required for the light-induced electron transport and phosphorylation processes of photosynthesis, are located in the plasma membrane and lamellae. The principle photosynthetic pigment in the blue-green algae is chlorophyll *a*; in purple bacteria, bacteriochlorophyll *a* or *d*; and in green bacteria, bacteriochlorophyll *b* or *c*.

chromomere A particulate portion of a *chromosome that becomes apparent as contraction and condensation of *chromatin occurs in *meiosis. Chromomeres vary in size and shape, homologous chromosomes having identical patterns along their lengths. An average chromomere has been found to contain sufficient DNA to code for approximately 30–35 polypeptides of average length. Genetic evidence, however, suggests that chromomeres contain only a single functional structural gene.

chromoneme The genetic material of a bacterium.

chromoplast A coloured *plastid. The *chloroplast is a specialized form of chromoplast. Other chromoplasts, in which carotenoid pigments predominate, give the characteristic colours to red, yellow, and orange flower petals, stamens, and fruits.

chromosome A threadlike structure, several of which can be seen in the nuclei of eukaryotic cells during cell division. They are composed of *chromatin and hence carry the genetic information. The chromatin becomes condensed prior to cell division, which renders the chromosomes visible under the light microscope. The chromosome is divided along its length into two identical strands, the *chromatids, joined at some point by the *centromere. The chromosome thus has four arms, the length of these depending upon the position of the centromere. The number, size, and shape of the chromosomes as seen at prophase are generally characteristic of a species (*see* karyotype).

During interphase the chromosomes uncoil into long narrow threads of *DNA about 2 nm in diameter. These bear beadlike structures, the *nucleosomes, which are highly organized aggregations of histones and DNA. The complex coiling and recoiling of the DNA and histones that occurs prior to cell division considerably reduces the length of the chromosome and increases the diameter to between 10 and 30 nm.

Bacterial and viral 'chromosomes' are much simpler. There is only one per cell and they consist solely of a single or double strand of DNA or RNA without any associated histones. They do not become condensed and hence are only visible using an electron microscope.

See also homologous chromosomes, B-chromosome.

chromosome aberration *See* chromosome mutation.

chromosome inversion *See* inversion.

chromosome mapping The assigning of genes to chromosomes and the calculation of their relative sequence and distance from each other on particular chromosomes. In initially assigning a gene to a particular chromosome, the inheritance of particular chromosome mutations may be studied, coupled with microscopic techniques. Linkage studies will indicate whether genes are on the same or different chromosomes and, if the former, their relative order and distance. Absolute distances cannot be determined by linkage studies. Chromosome mapping indicates that genes are arranged in linear order along chromosomes. In contrast to this, circular genetic maps are often obtained in prokaryotic (nonchromosomal) organisms and viruses.

chromosome mutation (chromosome aberration) An alteration in the number, order, or arrangement of genes caused by structural changes to a chromosome. Sometimes chromosome mutation may involve a large number of genes and the aberration may be visible by microscopy. There are four classes of chromosome mutations; *deletions, *duplications, *inversions, and *translocations.

chromosome number The number of chromosomes possessed by a species, which may be given as either the haploid (n) or diploid ($2n$) number. Chromosome number is important taxonomically since it is usually constant within a species. There are however many exceptions especially in species exhibiting reproductive abnormalities, e.g. apomixis. Among flowering plants the average haploid chromosome number is about 16 though this varies from $n = 2$ in the composite *Haplopappus gracilis* to extremely high numbers in certain polyploid series. For example, in the allopolyploid *Spartina townsendii* (saltmarsh grass) $n = 60$. It is thought that the original haploid chromosome number in angiosperms was in the region 7–9 and that numbers higher than $n = 14$ are polyploid in origin.

chromosome puff (Balbiani ring) A swollen region of a polytene chromosome, corresponding to a site where messenger RNA synthesis (transcription) is actively occurring. It is probably the large size of polytene chromosomes that makes visualization of the process possible. Although polytene chromosomes are found only in dipterous insects, it is thought that the puffing phenomenon has wider genetic implications.

chronistics The study of the time taken for evolutionary events to occur. If phylogenetic trees are drawn with the vertical axis representing the time scale then a diagram is obtained showing the chronistic relationship between organisms. Organisms that, from fossil evidence, appear to have evolved rapidly are termed *tachytelic* while those that appear to have changed slowly are termed *bradytelic*. *Horotelic* is used of organisms showing moderate rates of evolution.

Chroococcales An order of the *Cyanophyta containing the unicellular and colonial blue-green algae. The colonial (or *palmelloid*) forms arise when daughter cells become bound together in a mucilaginous sheath after cell division. Heterocysts and hormogonia are lacking.

Chrysophyta (golden-brown algae) A division made up predominantly of flagellate unicellular *algae but also containing some colonial and filamentous forms. These algae usually have two golden-brown plastids in each cell and food is stored as leucosin or fat.

In some classifications the Chrysophyta also includes the yellow-green algae, the diatoms, and the Haptophyceae on the basis that these three groups share a number of features with the golden-browns. These include accumulating food reserves as the polysaccharide leucosin, having siliceous or calcareous cell walls more or less conspicuously divided into two parts, and having a flimmer flagellum.

Chytridiales An order of the *Chytridiomycetes containing unicellular or coenocytic fungi that are considered the most primitive members of the Eumycota. Chytrids number about 550 species in some 93 genera and are usually aquatic although some species, such as *Synchytrium endobioticum* (which causes potato wart disease), are found in damp soil. They produce motile cells, which may behave asexually as zoospores or sexually as gametes, fusing in pairs.

Chytridiomycetes A class of the *Mastigomycotina containing simple fungi that typically produce motile cells with a posterior whiplash flagellum. It contains 110 genera and some 685 species and comprises the four orders *Chytridiales, Harpochytridiales, Blastocladiales, and Monoblepharidales.

ciliary body The basal body of a cilium. *See* flagellum, kinetosome.

ciliate Describing a structure, such as a leaf margin, fringed with fine hairs.

cilium *See* flagellum.

cincinnus A monochasium in which successive branches are borne on opposite sides of the stem. The inflorescence is often bent to one side. An example is that of *Strelitzia reginae* (the bird of paradise flower).

circadian rhythm (diurnal rhythm) An *endogenous rhythm in which physiological responses occur at 24-hourly intervals. For example, some plants exhibit a characteristic change in leaf position on a daily cycle. The opening and closing of stomata and flowers and the frequency of cell divisions also exhibit a circadian rhythm. Such rhythms will persist for several days if the plant is kept in continual darkness. It also appears that, in some species, the photoperiodic response is governed by circadian rhythms (*see* photophile).

circinate vernation A form of vernation in which the leaf primordia are rolled in on themselves from the apex to the base, so that the apex is in the middle of the coil (*see illustration at* vernation). It is seen in most ferns (except the Ophioglossales) and in certain of the cycads and extinct seed ferns.

circumnutation (nutation) The circular or elliptical movement of the stem tips of many plants. Due to a shift in the region of most active cell division in the stem, the growing tip swings in a characteristic manner, either clockwise or anticlockwise depending on the species. If during these movements (which seem to be dictated both by internal stimuli and by gravity) contact is made with a solid object, the plant begins to twine around it. The importance of gravity to the response can be demonstrated by inverting a plant that has already entwined a support. The tip and the last two or three coils straighten out and become erect before once again coiling upwards around the support in the characteristic direction.

***cis* arrangement** (coupling) The situation in which an individual is heterozygous for two linked *genes (either two units of function or two mutations in the same functional unit), and one chromosome carries both the recessive (or mutant) alleles. The homologous chromosome thus carries both the dominant (or normal) alleles and will mask the effects of the recessives in the phenotype. *Compare trans* arrangement. *See also cis-trans* test.

cisternae Interconnecting intracellular compartments consisting of tubular or flattened channels and vesicles that extend throughout the cytoplasm. The *endoplasmic reticulum and *golgi apparatus are made up of membranes formed into cisternae.

***cis-trans* test** A method for deciding whether two mutations having similar effects occur in the same or different genes (defined as units of function). If the mutations occur in the same gene, only heterozygotes with the **cis* arrangement will show the normal phenotype because the gene in the homologous chromosome is dominant wild type and codes for normal protein. In the **trans* arrangement both copies of the gene contain a mutation and thus there is no wild-type copy of the gene and hence no normal protein. If the mutations occur in different genes, specifying different polypeptides, all heterozygotes will be normal in appearance because the homologous chromosomes carry wild-type copies of the genes.

cistron A gene defined as a unit of function, i.e. a length of DNA that codes for a single polypeptide. The end of one cistron and the beginning of the next can be determined by the **cis-trans* test.

citric acid A six-carbon tricarboxylic acid with the formula $HOOCCH_2C(OH)(COOH)CH_2COOH$. Citric acid is the first intermediate in the TCA cycle, being formed by condensation of acetyl CoA and oxaloacetate. The reaction is the rate-limiting step in the TCA cycle. Citrate is itself an allosteric modulator for several reactions, e.g. the formation of fructose bisphosphate in glycolysis is inhibited by citrate. High concentrations of citrate are found in the fruits of some plants, notably lemons.

citric acid cycle *See* TCA cycle.

cladistic Describing genetic relationships due to recentness of common ancestry. In describing cladistic relationships, as opposed to patristic relationships, more emphasis is placed on the actual pathways of genetic divergence.

cladistics A method of classification in which the relationships between organisms are represented by a diagram, or *cladogram, based on selected shared characteristics. It is basically a phylogenetic system since the selected characteristics are those that are assumed to have been derived from a common ancestor. However some practitioners of cladistics, the 'transformed cladists' believe that cladograms are simply summaries of patterns of shared characteristics that provide a logical basis for classifications but do not have to be linked with and explained by the evolutionary process.

Cladistics differs from traditional phylogenetic approaches in the methods used to determine evolutionary relationships and in its definition of a true natural group. Cladistics assumes the closeness of relationship depends on the recentness of common ancestry, which itself is decided by the number and distribution of shared 'derived' character states (*synapomorphies*) in the organisms in question. Synapomorphies are those character states that can be traced back to the same character states in a recent common ancestor. They do not include shared derived characters that have arisen by convergent evolution. 'Primitive' character states (*symplesiomorphies*), inherited from more remote common ancestors, are ignored as misleading because some may have been inherited by one descendant but not by another, closely related, descendant. The same character can be both derived and primitive, depending at what level in the taxonomic hierarchy one is working. Thus a derived character at the class level will probably be a primitive character at the species level. In constructing a cladogram, branches are based upon the simplest explanation (often called the most parsimonious) of the distribution of shared derived characters.

Cladistics also differs from some traditional phylogenetic methods in regarding the only true natural groups as those that contain *all* the descendants of a common ancestor (*see* monophyletic, paraphyletic). Cladists also believe that those groups that are each others closest relative – sister groups – should be given equal taxonomic rank.

cladode (phylloclade) A specialized stem structure resembling and performing the functions of a leaf. Examples are the stem joints or pads in *Opuntia* (prickly pears) and the 'leaves' of *Ruscus aculeatus* (butcher's broom). In these genera the true leaves are reduced to spines or scales respectively. The formation of cladodes is an adaptation to arid conditions. A cladode may be distinguished from a leaf by the presence of buds on its surface. *Compare* phyllode.

cladogenesis The branchings of a *cladogram.

cladogram A type of branching diagram (dendrogram) representing the relationships between organisms as determined by cladistic methods. A cladogram may be regarded as showing the phylogenetic relationships of organisms, in which case it is effectively an evolutionary tree, or it may be seen as simply representing a pattern based on shared characteristics. In the first case, the points of branching are thought of as actual or hypothetical ancestral species, while in the second case (the position held by 'transformed' cladists), the points simply represent shared characteristics.

Cladophorales An order of the *Bryopsidophyceae containing both freshwater and marine algae. The thallus of such algae is divided into a number of multinucleate cells.

Cladoxylales *See* Filicinae.

clamp connection A characteristic form of growth seen in dividing dikaryotic hyphae of basidiomycete fungi (*see* illustration). Before cell division occurs the two nuclei divide simultaneously and at the same time a backward-pointing side branch develops. One of the daughter nuclei moves into this while another daughter nucleus of different genetic constitution moves down the main filament away from the hyphal tip. Two cross walls are then formed, one across the main filament, so separating off the other two nuclei in the hyphal tip, and the other across the side branch. The wall between the branch and the subapical cell then dissolves so the nucleus in the side branch can pass through to join the remaining daughter nucleus. This process has apparently developed to avoid the random movement of daughter nuclei, which could lead to the production of homokaryotic cells. *See also* crozier.

classical genetics (Mendelian genetics) The branch of genetics concerned with the study of inheritance through obser-

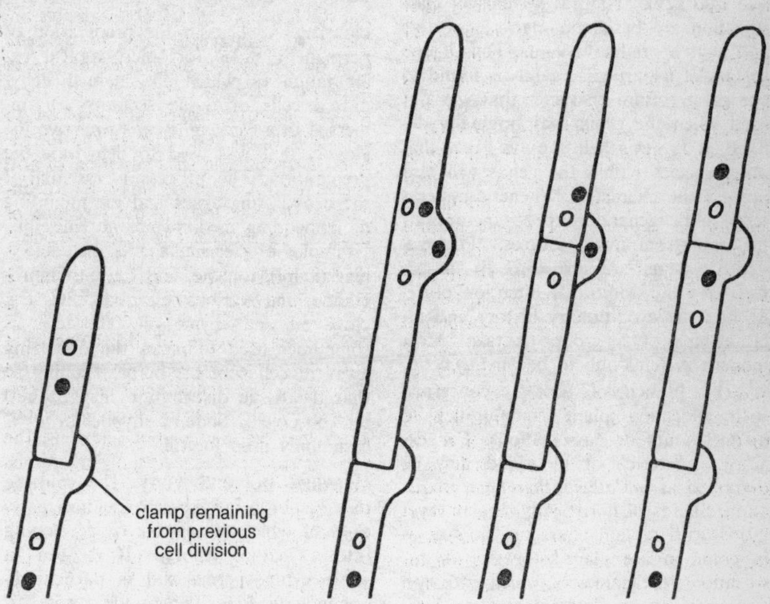

Stages in the formation of a clamp connection.

vation of the progeny obtained in breeding experiments. *See* Mendel's laws, Mendelism.

classification **1.** The process of establishing and delimiting taxa within the hierarchy of classes. The resulting system of classification should either help in the rapid identification of organisms or should express their natural interrelationships (preferably both). Four kinds of classification can be recognized: artificial, natural, phenetic, and phylogenetic. Artificial systems are based on a few predetermined characters usually selected to aid identification and consequently may not reflect the true relationships of the plants. The sexual system of Linnaeus based on the numbers of stamens and carpels is an example (*see also* key). *Natural systems of classification are based on many characters and have a predictive value. Thus if one species of a particular genus is found to possess a certain character that was not used when the genus was originally defined, it is nevertheless quite likely that other species within the genus will also possess the character. *Phenetic systems tend to be general purpose, incorporating data from many sources. *Phylogenetic systems are also broadly based, but have the additional intention of reflecting the evolutionary history and relationships of a group of taxa. Most modern systems aim to be phylogenetic. **2.** The branch of ecology concerned with allocating plant communities to distinct units or *associations. For example, a stretch of moorland may be described as a *Calluna/Pteridium* association. Stands of floristically similar vegetation with certain characteristic species in common are placed together in associations, the names of which end with the suffix -etum. Thus associations dominated by pine or oak are grouped as Pinetum or Quercetum respectively. The method provides a useful summary of the vegetation types found in a region. However it has the disadvantage that vegetation intermediate between two recognizable associations may be ignored for convenience' sake. It is a very subjective method of assessing vegetation. *Compare* ordination.

Clavicipitales An order of the *Pyrenomycetes whose members, which number about 175 species in 7 genera, produce small perithecia contained in the rounded heads of *stromata. Each ascus contains eight threadlike asci. The order includes the ergot fungus, *Claviceps purpurea.*

clay Mineral particles with a diameter below 0.002 mm. They consist of aluminium and silica arranged to form a layered crystal structure. *See* soil texture.

clearing A stage in the preparation of permanent slides for microscopical examination in which the stained dehydrated cells or tissue sections are immersed in a *clearing agent* to remove the alcohol that was used to dehydrate the preparation. The process is essential if the dehydrating agent and the mounting or embedding medium are not miscible. Examples of clearing agents include xylene (xylol), toluene, and 1,1,1,-trichloroethane and various essential oils, e.g. clove oil and thyme oil. The term is sometimes used to mean the dissolving away of cell contents before staining so that the tissue distribution may be better observed. Sodium hypochlorite is commonly used for this.

cleavage polyembryony The splitting of an embryo into several identical parts each of which is capable of developing into a mature embryo. It is seen in some gymnosperms and is particularly common in *Pinus*, where the zygote divides to form four nuclei each of which divides to give a chain of cells or suspensor. At the tip of each suspensor is a group of embryo initials. Usually, however, only one embryo develops to maturity.

cleistocarp *See* cleistothecium.

cleistogamy The production of flowers that do not open to expose the reproductive organs, so preventing cross pollination. Some species are obligately cleistogamous but others may be cleistogamous only under certain climatic conditions, as in *Viola* where flowers may not open at low temperatures. In some grasses only the lower florets are cleistogamous. The evolutionary origin of this mechanism is not known. *Compare* chasmogamy. *See also* autogamy.

cleistothecium (cleistocarp) The type of *ascocarp characteristic of fungi in the Plectomycetes. It is a globose structure with no specialized opening to the exterior.

climacteric The rise in respiration rate observed in certain fruits during ripening. It can be artificially induced by exposure to low concentrations of *ethylene, a gas evolved by ripening fruit. Ripening can therefore be stimulated by placing unripe fruit alongside ripe fruit. Conditions that prevent the onset of the climacteric (low oxygen and high carbon dioxide concentrations) can be employed during storage to maintain the quality of stored fruit.

climatic factors Those aspects of the weather, such as temperature, rainfall, light, humidity, and air movement, that influence the life of organisms. *See also* biotic factors, edaphic factors.

climax The final stable community that results after a series of changes (*succession) in the vegetation and animal life in a particular area. However, there are often factors that halt or slow down succession and prevent the establishment of a climax community. These factors include changes in climate, natural hazards, disease, animal interference, fire, and human activity. Often intensive grazing prevents the growth of shrubs and trees. The equilibrium achieved in this way is termed a *biotic climax*, *disclimax*, *plagioclimax*, or *subclimax vegetation*. *See also* clisere.

climbing root Any of the short adventitious roots that develop from the stems of certain climbing plants, e.g. ivy (*Hedera helix*), and serve to attach the plant to its support. Climbing roots are negatively phototropic and thus grow into darkened tissues in bark or crevices in walls. Their function is enhanced by the secretion of mucilaginous substances from the root tip.

cline A gradual change in a character over the range of a species. The term may also be applied to a change in the relative proportions of two or more distinct forms. For example, the leaves of lords and ladies (*Arum maculatum*) may or may not be spotted. The frequency of the spotted form increases towards the southern part of its range in Britain. *See also* ecocline, ecotype.

clinostat (klinostat) An apparatus used to investigate plant growth in the absence of directional stimuli (*tropisms). It consists of an electrically driven motor that slowly rotates a platform to which a seedling is attached in some manner. The rotation of the platform prevents any one side of the plant receiving an uninterrupted stimulus. Using a clinostat with a vertical rotating platform it can be demonstrated that a normally geotropic root will grow horizontally when not subject to the influence of gravity.

clisere A succession of different climax communities in a particular area brought about by changes in the climate.

clone A population of genetically identical cells or individuals. Such a population is obtained by mitotic division or by asexual reproduction. The term *strain* is sometimes used synonymously. *See also* pure line.

club fungi *See* Aphyllophorales.

club mosses *See* Lycopodiales.

club root (finger-and-toe disease) A disease of crucifers caused by the fungus *Plasmodiophora brassicae* in which the roots become swollen and malformed. Above ground the symptoms, which may not appear for a while, are wilting and later stunting and yellowing of infected plants. Infection occurs via the root hairs and once the cambium is infected club roots develop rapidly. Spores can remain viable in the soil for several years making control difficult. Liming of the soil is a traditional control method.

CoA *See* coenzyme A.

coacervate *See* protobiont.

coating The application of a thin film of heavy metal, commonly gold, to the surface of a specimen during preparation for scanning electron microscopy. Coating is particularly useful in preparation of biological specimens because it results in the production of many more secondary electrons and the improved conductivity helps prevent a charge building up in the specimen and causing image deterioration. Coating units are of several types; some use fine filaments as the source of coating metal while others, known as sputter coaters, have a large solid block of metal that is ionized by a beam of ions.

coccolith *See* Haptophyta.

coccolithophorids *See* Haptophyta.

coccus Any spherical or ellipsoidal bacterium. Cocci are rarely flagellate. *Compare* bacillus, spirillum, vibrio.

Codiales *See* Caulerpales.

codominance The phenomenon in which both alleles in a heterozygote are expressed in the phenotype. Thus in certain plants that possess a series of alleles governing incompatibility (*see S* alleles), an individual plant that has two different alleles both relatively high in the dominance series may be incompatible with any plant possessing either of these dominant alleles (i.e. both alleles are exhibited by the phenotype). In other cases the one allele may be completely dominant to the other and such a plant would be able to cross with another plant possessing the same recessive allele. This illustrates that codominance relationships do not necessarily exist for all the alleles of a gene. *Compare* incomplete dominance.

codon A sequence of three nucleotides (*triplet) that codes for one amino acid in a polypeptide. A sequence of codons will thus determine the type and sequence of amino acids in a polypeptide. There are about 20 different kinds of amino acids in proteins. Most amino acids are coded for by more than one codon: the *genetic code is thus said to be *degenerate*. The term codon is given both to the sequence of triplets on DNA and also to the sequence on messenger RNA. However these are not identical but complementary copies of each other.

Coelomycetes A class of the *Deuteromycotina containing mycelial imperfect fungi in which the conidia are borne in pycnidia (as in the *Sphaeropsidales) or acervuli (as in the *Melanconiales). It contains over 7000 species in some 870 genera.

coenobium A colony of cells that behaves as a single integrated unit, the number of component cells remaining constant. A coenobium often gives rise to internal daughter colonies. An example is the colonial alga *Volvox*.

coenocarpium A *multiple fruit that incorporates the ovaries, floral parts, and receptacles of many flowers. It has a fleshy axis. The pineapple (*Ananas comosus*) is an example. *Compare* sorosis.

coenocytic Describing a cell containing many nuclei or a thallus that lacks separating walls between nuclei. Such a structure is often threadlike. Examples are the hyphae of many fungi and the thalli of certain algae in the Xanthophyta and Siphonales. *See* acellular.

Coenopteridales *See* Filicinae.

coenosorus A continuous line of sporangia. It is seen in certain ferns, e.g. *Pteridium*, where discrete sori are not produced.

coenzyme An organic molecule that acts as an enzyme *cofactor. Coenzymes usually function as intermediate carrier molecules, transferring or removing functional groups, atoms, or electrons. The coenzyme may be tightly bound to the enzyme, in which case it is known as a *prosthetic group, or it may be only loosely bound and act like a second substrate of the enzyme. Examples of important coenzymes are NAD and NADP, coenzyme A, and biotin.

coenzyme I *See* NAD.

coenzyme II *See* NADP.

coenzyme A A heat-stable coenzyme that acts as a carrier of acyl groups in many metabolic processes, such as fatty-acid and pyruvate oxidation. Coenzyme A is a sulphur-containing molecule formed from ATP, pantothenic acid, and cysteine. The –SH group of coenzyme A can react with acetate to form acetyl CoA in the reversible reaction;

$$\text{CoA–SH} + CH_3COOH \rightleftharpoons \text{CoA–}SCOCH_3$$

Acetyl CoA can then react with an acyl acceptor to form an acylated product and release free coenzyme A. An example is the reaction of acetyl CoA and oxaloacetate to form citrate.

coenzyme Q (ubiquinone) A lipid-soluble electron-carrying coenzyme that participates in the transport of electrons from NAD to oxygen in the *respiratory chain. It is a reversibly reducible quinone. Plastoquinone, a closely related compound, performs a similar function but in photosynthetic electron transport.

cofactor A nonprotein component of some enzymes that is necessary for catalytic activity. Cofactors may be either metal ions or coenzymes. They are usually heat stable, unlike most enzyme proteins, which are denatured by heat. Many enzymes require metal ions as cofactors. Examples are ferredoxin, which requires ferrous, Fe(II), and ferric, Fe(III), ions, and alcohol dehydrogenase, which requires zinc ions. In some enzymes the metal ion is the primary catalytic centre, while in others it may act to bind enzyme and substrate together, or to stabilize the enzyme in its catalytically active conformation. *See also* prosthetic group, apoenzyme, holoenzyme.

cohesion theory The currently accepted theory explaining the ascent of sap in the xylem. It postulates that the reduction in water potential in the leaf (due to an increase in the rate of evaporation) causes water to flow through the plant to replace the lost water. The column of water in the xylem can withstand considerable tension before breaking because of the very strong cohesive forces between the water molecules. *See also* cavitation.

colchicine An alkaloid drug, extracted from the autumn crocus (*Colchicum autumnale*), that reacts reversibly with the protein tubulin and prevents its polymerization into microtubules. Consequently microtubule-assisted processes, such as the movement of organelles, are inhibited. Colchicine is often applied to dividing cells when making a chromosome preparation. It inhibits spindle formation and thus the condensed chromosomes are scattered rather than aligned on the spindle. This greatly facilitates their examination. If colchicine is applied to cells undergoing meiosis then the failure of the homologous chromosomes to separate results in the formation of diploid gametes. Colchicine is thus an important tool for the artificial induction of polyploidy and in other experimental studies, such as investigations of pollen development. It may also be used to render sterile hybrids fertile by doubling their chromosome number so the chromosomes have a homologue to pair with at meiosis.

coleoptile The sheathlike structure around the *epicotyl in the seeds of grasses. It protects the erect plumule during its growth to the soil surface. In most grasses the coleoptile is very sensitive to light and has thus been used to investigate light-directed tropisms. *Compare* coleorhiza.

coleorhiza The protective sheathlike structure around the *radicle in the seeds of grasses. *Compare* coleoptile.

coliphage Any DNA bacteriophage that infects the bacterium *Escherichia coli*. Examples are the intensively studied T series of phages (T1 through to T7), which have been developed on a particular strain of *E. coli*.

collateral bundle A *vascular bundle in which the phloem occurs on only one side of the xylem. In the stem the xylem is usually internal to the phloem. This type of bundle is common in both angiosperms and gymnosperms. *Compare* bicollateral bundle, concentric bundle.

collenchyma A tissue composed of relatively elongated cells with thickened nonlignified primary cell walls. Collenchyma is sometimes considered to be *parenchyma with exceptionally thick cellulose cell walls, specialized as a supporting and strengthening tissue in organs lacking (or almost lacking) secondary growth. Examples are young herbaceous stems and leaves. The thickened cell walls of collenchyma cells are sometimes used as a supply of cellulose for the other tissues in times of shortage. The wall thickenings may be mainly on the tangential walls (*lamellar collenchyma*), in the corners of the cells (*angular collenchyma*), or adjacent to the intercellular spaces (*lacunar collenchyma*). *Compare* sclerenchyma.

colonial Describing a type of growth form in which many cells of similar form and function are more or less loosely grouped together. Colonies of cells are characteristic of many algal genera, e.g. *Volvox* and *Gonium*. The colony is often surrounded by mucilaginous substances. *See also* coenobium.

colpus An oblong to elliptic germinal aperture in a pollen grain, at least twice as long as it is broad. Pollen with such apertures is termed *colpate*. *Compare* pore.

columella A small column, especially:
1. The central portion of the root cap of certain roots in which the cells are arranged in longitudinal lines.
2. The sterile tissue in the centre of the sporangium in mosses, liverworts, and fungi.
3. A rodlike radial element in a pollen or spore wall.

Commelinidae A subclass of the monocotyledons containing mainly herbaceous plants, most of which have bisexual flowers with a more or less reduced perianth. It contains the following orders: Commelinales, which comprises four families including the Commelinaceae (spiderwort family); Eriocaulales; Restionales; Poales, which includes the one family *Gramineae or Poaceae; Juncales, which consists of two families including the Juncaceae (rushes); Cyperales, made up of the one family *Cyperaceae (reeds and sedges); Bromeliales, including the one family Bromeliaceae (bromeliads), which contains the pineapple (*Ananas comosus*); and the Zingiberales, which includes the families Musaceae (banana family), Strelitziaceae (containing bird of paradise flowers, *Strelitzia*), and Zingiberaceae (containing various aromatic plants, e.g. ginger, *Zingiber officinale*). The Zingiberales is sometimes placed in the Liliidae. In some classifications the Typhales is included in the Commelinidae, while in others it is placed in the *Arecidae.

commensalism A relationship between two living organisms in which one of the participants, (the *commensal*) benefits but the other neither benefits nor loses. For example, an epiphyte, by germinating on the upper branches of a tree, benefits from the increased supply of light but the tree simply acts as a support and is not affected either beneficially or adversely. *See also* amensalism, mutualism.

community A group of plants and animals living together and interacting with one another in the same environmental conditions. Communities range in size from, for example, small woodland communities to large expanses of temperate deciduous forest or temperate grassland.

companion cells A specialized *parenchyma cell, characterized by a dense cytoplasm and conspicuous nucleus, associated with an adjacent *sieve tube element in the phloem of angiosperms. Companion cells are analogous to the *albuminous cells of gymnosperms. A sieve tube element and its associated companion cell arise from a single division of the same mother cell. The protoplast of the companion cell is connected with the enucleate protoplast of the sieve tube element by means of *plasmodesmata and is believed to regulate the rate of flow along the *sieve tube.

comparative biochemistry The recognition of discontinuities in biochemical variation and their subsequent use in the construction of classifications or correlation with existing classifications. *See also* chemotaxonomy.

compensation point The lowest steady carbon dioxide concentration achievable in a closed system containing a photosynthesizing plant. When this minimum level is reached, the photosynthetic uptake of carbon dioxide is exactly balanced by its respiratory release, indicating that the rate of synthesis of organic material is equal to the rate of breakdown by respiration. Low compensation points are indicative of photosynthetic efficiency as the plant is then using the maximum amount of available carbon dioxide (*see* C_3 plant, C_4 plant). Following a period of darkness a plant will take a certain amount of time, termed the *compensation period*, to reach its compensation point. Shade plants often have shorter compensation periods than sun plants as they can make better use of dim light.

competence A characteristic exhibited by embryonic cells whereby they can go on to develop into any one of several different types of cell. Competence is therefore a feature that they possess before their ultimate cellular type is determined. *See also* totipotency.

competition The situation that arises when two or more organisms of the same or different species need the same limited resources. Organisms compete with one another for light, water, nutrients, etc. Competition may be resolved in various ways, namely: certain organisms or species being eliminated completely; migration of the less successful competitors; use of the resources at different times; physiological changes occurring; or the organisms involved living together, but not as successfully as they would at lower densities (i.e. the reproductive rate falls off). When members of different species compete for most or all of the same resources, one species flourishes and the other declines. This is known as the *competitive exclusion principle* and means that two different species rarely occupy the same ecological niche.

competitive inhibition A form of enzyme inhibition in which the inhibitor competes with the substrate for the enzyme active site. The inhibitor can bind to the active site but does not react to form products. An example is the inhibition of the enzyme succinate dehydrogenase by certain compounds resembling succinate, e.g. malonate and oxalate. Studies of competitive inhibitors have provided valuable information about the way in which substrates are bound to the active site.

complementary genes Nonallelic genes that cooperate in the expression of a single characteristic. The genes may be linked or unlinked. In some cases a standard Mendelian ratio is maintained, but there may be unexpected phenotypes. Thus in the inheritance of flower colour in sweet peas, purple-flowered offspring may be obtained when two white-flowered plants are crossed. This is because flower colour is affected by two genes and if either is a double recessive then no colour develops. In other cases, no unusual phenotypes appear but there may be some apparent modification of the Mendelian ratio in the F_2, such as 9:3:4 (= 9:3:[3+1]), or 9:7 (= 9:[3+3+1]).

complementation The phenomenon whereby two recessive mutant strains can supply each other's deficiencies so that growth of both can occur. Thus adenine-requiring (ad^-) strains or histidine-requiring (his^-) strains of the fungus *Neurospora* will grow in the presence of the other in minimal media (lacking both adenine and histidine) but not separately.

complex gene *See* super gene.

Compositae (Asteraceae) One of the largest dicotyledonous families containing about 25 000 species in about 1100 genera and commonly referred to as the daisy or sunflower family. Its members are recognized by their characteristic headlike inflorescence, the *capitulum, which superficially appears to be a single flower. The fruit (*see* cypsela) often possesses a ring of hairs, the *pappus, to aid dispersal. All composites have resin canals except most of those in the tribe Lactuceae, which have latex ducts. Commercially important composites include the sunflower (*Helianthus annuus*), grown for the oil in its seeds, and lettuce (*Lactuca sativa*), and numerous ornamental species, e.g. *Dahlia*, *Chrysanthemum*, and *Aster*.

composite fruit *See* multiple fruit.

compost Rotted plant material, made from garden rubbish and kitchen vegetable waste and used as an organic fertilizer. The decomposition of compost can be hastened by adding a compost activator, such as nitrochalk.

compression wood *See* reaction wood.

concentric bundle (centric bundle) A *vascular bundle in which one vascular tissue surrounds the other so they appear to be arranged concentrically when viewed in transverse section. *Compare* bicollateral bundle, collateral bundle, meristele. *See also* amphicribal, amphivasal.

conceptacle A flask-shaped cavity in the thallus of some brown algae, e.g. *Fucus*, in which gametangia are formed. A female conceptacle is lined with unbranched sterile hairs or *paraphyses, and the *oogonia develop from short stalks arising from the chamber wall. The male conceptacle is filled with branched hairs that bear *antheridia. The conceptacle opens via an osteole.

conditioned media *See* nurse tissue.

conducting tissue 1. *See* vascular system.
2. *See* transmitting tissue.

conduplicate Folded, as certain grass leaves are. The term also refers to a form of vernation in which each leaf is folded in a U-shape around the next youngest leaf (*see illustration at* vernation).

cone *See* strobilus.

Congo red A scarlet benzidine dye often used to stain fungal hyphae.

conidium (conidiospore) An asexual spore produced by many fungi, especially ascomycetes, on specialized erect hyphae termed *conidiophores*. The conidia are usually formed in chains, which radiate from the globose or branched tip of the conidiophore. *See also* phialospore.

Coniferales (Pinales) The largest and most widely distributed order of gymnosperms, containing about 49 genera with about 570 species most of which are evergreen trees. These are divided among the six families Pinaceae, Taxodiaceae, Cupressaceae, Podocarpaceae, Cephalotaxaceae, and Araucariaceae. The Cupressaceae and Taxodiaceae are sometimes included as subfamilies in the Pinaceae.
Conifers are particularly abundant in the high latitudes of the northern hemisphere, where they form the climax vegetation. Typically conifers show a pyramidal growth pattern and bear simple leaves, often needles or scales. Most conifers are monoecious but bear the male and female reproductive organs in separate compact cones on different parts of the tree. Conifers are more advanced than the cycads and ginkgo in that they produce pollen grains; thus antherozoids are replaced by the two sperm nuclei, which are delivered to the egg by the pollen tube. It is thought this feature may have contributed to the success of the conifers as compared with other gymnosperms.
Conifers are commercially important as a source of timber for the paper-making, building, and furniture industries. They are generally faster growing and develop a less dense wood than angiosperm trees and are thus commonly called softwoods. In cross section distinct pores are normally absent in the wood (consequently termed nonporous wood) in contrast to the *ring-porous or *diffuse-porous wood of angiosperms. *See also* Taxales.

Conjugales (Conjugatophyceae) In older classifications, an order (or class) of the Chlorophyta the members of which undergo the characteristic reproductive process of *conjugation. More recently the order has been renamed *Zygnemaphyceae.

conjugated protein A protein that contains other components besides polypeptide chains. Conjugated proteins are classified according to the nature of the nonpeptide portion of the molecule. Glycoproteins are found in plant and bacterial cell walls, while cell and organelle membranes contain lipoproteins. Many enzymes contain metal ions and some dehydrogenases are *flavoproteins. Ribosomes and viruses are nucleoproteins.

conjugation A form of *sexual reproduction observed in green algae of the class Zygnemaphyceae, certain fungi (e.g. *Rhizopus*), and various bacteria. Isogametes fuse by forming a bridgelike outgrowth from one organism to the other, through which the contents of one cell migrate to fuse with the contents of the other, so forming a zygote (*see* diagram).

connate Describing similar organs that are joined together, such as petals fused to form a tube. The term is often used of paired leaves at a node, the bases of which have become fused to completely encircle the stem. *Compare* adnate.

connective The parenchymatous tissue that joins the pollen sacs in the *anther. It is a continuation of the filament and contains the vascular strand supplying each lobe.

consociation A comparatively small climax community named after one dominant species, e.g. oakwood, beechwood, etc. Consociations can be grouped together to form an *association. For example, in England the deciduous-forest association consists of consociations, i.e. smaller woods, of oak, ash, and beech.

constitutive enzyme An enzyme that is always present in nearly constant amounts in a given cell. A constitutive

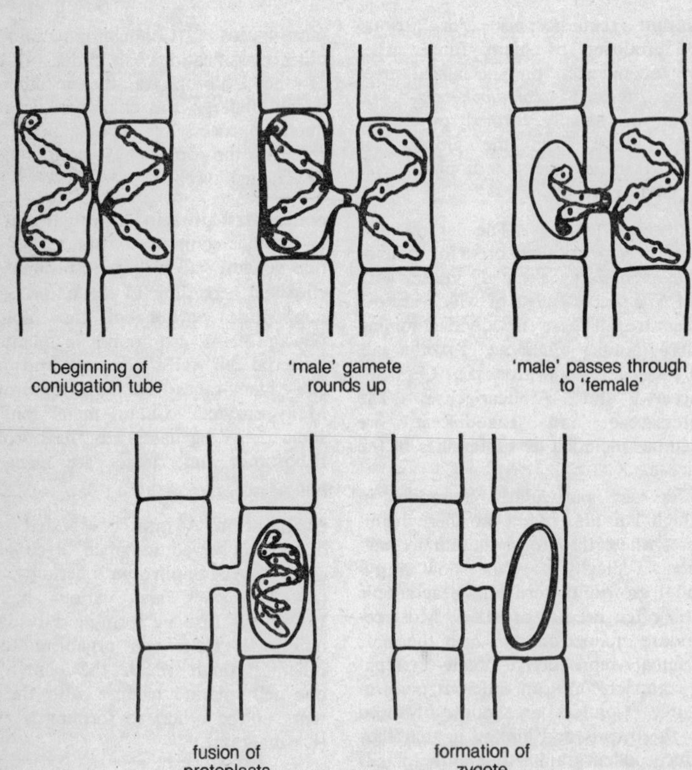

Conjugation stages in *Spirogyra.*

enzyme is part of the permanent and basic machinery of a cell. The enzymes of glycolysis are examples of constitutive enzymes in nearly all organisms. *Compare* induced enzyme.

consumer An organism that feeds on other living organisms, i.e. animals and parasitic and insectivorous plants. In a *food chain, herbivores that eat green plants are *primary consumers*, and carnivores that eat herbivores are *secondary consumers*. *Compare* decomposer, producer.

contact Describing an insecticide or herbicide that kills an insect or plant on contact rather than relying on ingestion or absorption. Examples of contact insecticides are DDT and pyrethrin. Contact herbicides include paraquat and diquat. *Compare* stomach insecticide, systemic.

continental drift The theory that the existing continents moved to their present position following the break up of an ancient landmass. The idea was first proposed by A. Wegener in 1912 but was discarded by the geophysicists of the time. However evidence from biogeographic and oceanographic studies and the emergence of plate tectonics theory has established its validity. It is thought that the original landmass, *Pangaea*, broke up in the Mesozoic into *Gondwana* (or Gondwanaland) and *Laurasia*. The subsequent break-up of Gondwana resulted in the formation of the landmasses of the southern hemisphere, i.e. Africa, India, South America, Australia, Antarctica, and New Zealand, while the break-up of Laurasia gave rise to North America and Eurasia. The theory helps to explain the present disjunct distributions of certain plant and animal groups. *See also* vicariance.

continuous variation *See* quantitative variation.

contractile root A specialized thickened root that serves to pull down a corm, bulb, rhizome, etc. to an appropriate level in the soil. Once the tips of the contractile roots are firmly anchored in the soil the upper region contracts, due to changes in the shape of the cortical cells. Prominent contractile roots are seen, for example, in *Crocus* and *Allium* species.

contractile vacuole A cavity in the cytoplasm of many freshwater unicellular algae that enlarges as it fills with water and contracts as water is expelled from it. The external medium is hypotonic with respect to the algal cell contents. To help prevent osmotic lysis, energy from ATP is used to pump water into the contractile vacuoles, the pulsations being visible with a light microscope. Water is released through the cell membrane or, in the case of *Euglena*, into the gullet. Such vacuoles are particularly important in those algae that do not have a rigid cell wall to help limit water intake.

convergent evolution (convergence) The development of *analogous structures in unrelated groups of organisms as a result of living in similar habitats or sharing other features, e.g. similar pollinators. An example is the development of the insectivorous habit in the families Sarraceniaceae, Nepenthaceae, and Droseraceae. *See also* parallel evolution.

coordinated growth *See* symplast.

copper Symbol: Cu. A metal element, atomic number 29, atomic weight 63.55, essential in trace amounts for growth. It is found in the enzyme cytochrome *c* oxidase, which is thought to catalyse the transfer of electrons from cytochrome a_3 to oxygen in respiration. In this reaction the two copper atoms of the molecule undergo transition from the cupric, Cu (II), to the cuprous, Cu(I) state. Copper also plays a role in photosynthetic elec-

tron transport, being found in the protein *plastocyanin.
Copper deficiency leads to chlorosis, which suggests the element may also play a role in the synthesis of chlorophyll.

copper fungicide An inorganic *fungicide containing copper. Some of the earliest known fungicides were copper compounds including the *Bordeaux and *Burgundy mixtures. Other copper-containing fungicides are often based on copper oxychloride or copper oxide and may be formulated with other types of fungicide.

coppicing A form of woodland management in which trees are cut back to ground level regularly (every 10–15 years) to encourage growth of numerous adventitious shoots from the base. The resulting thicket is termed a copse or coppice. The young shoots are harvested for firewood, charcoal burning, fencing, etc. Many British woodlands have been managed in this way. Often selected trees may be left to mature, giving a coppice with standards.

coprophilous Describing organisms that live on or in dung as, for example, fungi of the Pilobolaceae.

Cordaitales An extinct order of gymnosperms, thought to have originated some time before the Carboniferous period and to have persisted into the Triassic. They were arborescent and in some of their vegetative structures resembled certain modern conifers, notably the monkey-puzzle (*Araucaria araucana*).

cordate Heart shaped, such as the leaves of the sweet violet (*Viola odorata*) or the lemmas of quaking grasses (*Briza*).

coriaceous Having a leathery texture.

cork *See* phellem.

cork cambium *See* phellogen.

corm A short swollen underground stem that serves as an organ of perennation and of vegetative propagation in such genera as *Crocus* and *Gladiolus*. The foliage leaves and flowers form from one or more axillary buds and grow at the expense of the food reserves in the corm. At the end of the season the corm is reduced to a withered mass and one or more new corms form above it at the base of each flowering stem. The protective brown scales surrounding the corm are the remains of the previous season's leaf bases. *Compare* bulb, rhizome.

corolla The collective term for the *petals, constituting the inner whorl of the *perianth. A *corolla tube* is formed when the margins of the individual petals are completely or partially fused. *Compare* calyx.

corona **1.** A crownlike outgrowth from a corolla tube. It is especially prominent in daffodils (*Narcissus*).
2. A crownlike structure at the tip of the oogonium in certain algae of the Charophyta.

corpus **1.** *See* tunica–corpus theory.
2. The main body of a bladdered pollen grain, as in many conifers, e.g. *Pinus*.

correlative inhibition The suppression of the growth of certain plant parts by a compound, such as a food substance or growth substance, produced in another area of the plant. An example is the production of a substance by the root system that suppresses the development of roots on the stem. If the influence of the root is removed a detached petiole or stem may subsequently produce a profusion of adventitious roots.

cortex The tissue (including the endodermis) between the epidermis and the *stele in the stem or root. Although consisting mainly of ground parenchyma, the cortex often contains some collenchyma and sclerenchyma cells.

Sometimes part of the cortex is photosynthetic chlorenchyma, especially in xerophytes.

corymb A simple racemose inflorescence in which the flowers are formed on lateral stalks of different lengths, the longest at the base, resulting in a flat-topped cluster of flowers (*see illustration at* racemose inflorescence). This form of floral arrangement, found frequently in the Cruciferae, may aid insect pollination by giving easy access to individual flowers and in providing a flat landing platform, as in candytufts (*Iberis*).

cosmopolitan Describing species that have a worldwide distribution and are not restricted to specific areas. *Compare* endemic.

costa 1. A longitudinal rib, as seen in many cacti.

2. nerve.

cotyledon (seed leaf) The first leaf or leaves of the embryo in seed plants. In angiosperms the number of cotyledons is an important taxonomic character used to separate the two subclasses Monocotyledonae and Dicotyledonae. In endospermic seeds the cotyledons may be thin and membranous, as in the castor oil seed, but in nonendospermic seeds, e.g. pea, they take over the food-storing functions of the endosperm. Depending on the pattern of germination the cotyledons may remain in the seed (*see* hypogeal) or emerge to become the first photosynthetic organs (*see* epigeal). Gymnosperms may have several cotyledons.

coumarin A toxic white crystalline substance with the formula $C_9H_6O_2$, found in certain plants, particularly in the testa or fruit where it inhibits germination. Only when it has been removed or destroyed, for example by exposure to light, will the seed germinate. It has a characteristic smell of new-mown hay and has been manufactured artificially for use in the perfume industry.

counterstaining *See* staining.

coupling *See cis* arrangement.

covered smuts *See* smuts.

craspedromous Describing a form of leaf venation in which there is a single primary vein running down the centre of the leaf with secondary veins branching off this in an essentially parallel fashion. In traditional terminology such venation is described as *penni-parallel.* An example is the leaf of the sweet chestnut (*Castanea sativa*). *See illustration at* venation.

crassula (bar of Sanio) One of a pair of barlike thickenings made up of primary wall and intercellular material found adjacent to the *bordered pits of most gymnosperm tracheids.

crassulacean acid metabolism (CAM) A form of photosynthesis first described in the family Crassulaceae and since found in many other succulent plants. CAM plants keep their stomata closed during the day to reduce water loss by transpiration. Carbon dioxide can therefore only enter at night when, instead of combining with ribulose bisphosphate (as in conventional *C_3 plants) it combines with the three-carbon compound phosphoenol pyruvate (PEP) to give the four-carbon oxaloacetic acid. This is then converted to malic acid, which can be stored in the cell vacuoles until daylight, when it is transferred to the cytoplasm. Here it is broken down to release carbon dioxide, which is then fixed in the normal manner. This adaptation allows such plants to flourish in arid habitats but their growth rate is slow. CAM can be induced in certain C_3 plants by water shortage. *See also* Hatch–Slack pathway.

creationism (special creation) A view that opposes evolutionary theory and envisages the vast variety of living organisms, both existing and fossilized, as having been specially designed by a Creator. The attempted construction of phylogenetic pathways based on the idea that living forms have evolved from ancestral forms is interpreted by creationists as evidence of a 'Great Design'. Creationism is difficult to disprove by experiment. Some creationists believe in the theory of *catastrophism* in which it is thought that there have been a number of creations at different times, each having been destroyed by some kind of natural catastrophe, such as a flood.

cremocarp A type of *schizocarp, derived from two fused carpels, that divides into two one-seeded units at maturity. It is typical of the Umbelliferae.

crenate Describing a leaf margin that has rounded projections. A leaf margin with very small such projections is termed *crenulate*.

Cretaceous The final period of the Mesozoic era between about 136 and 65 million years ago during which much of the present-day land surface was covered by shallow seas. Cretaceous rocks largely consist of chalk formed from fossilized calcareous plates (coccoliths) of marine plankton. It is thought that forms similar to present-day members of the Filicales probably evolved towards the end of the period. The Bennettitales and Caytoniales (both gymnosperm orders) died out and the gymnosperms generally declined in importance, although forms similar to modern species, such as pines, yews, firs, and giant redwoods, arose. Most significant was the emergence of the angiosperms, which became the dominant vegetation, forming large areas of broad-leaved forest.

crista A shelflike structure in a *mitochondrion formed by infolding of the inner membrane. The cristae greatly increase the surface area of the inner mitochondrial membrane so providing a large area for the reactions of the electron transfer system to occur.

critical day length The period of daylight, specific in length for any given species, that appears to initiate flowering in *long-day plants or inhibit flowering in *short-day plants. In actual fact long-day plants will not flower if the dark period exceeds a certain maximum and conversely short-day plants will not flower unless the dark period exceeds a certain minimum. These periods are termed *critical dark periods* and must be continuous to have effect (*see* night-break effect).

critical-point drying A technique for removing the water from biological specimens without causing shrinkage or collapse. The specimen is dehydrated by suspending it in acetone or alcohol and driving off the solvent under controlled conditions of temperature and pressure. *Compare* freeze drying.

cross fertilization *See* allogamy.

cross field In the wood of conifers, the area where the walls of ray cells abut onto those of axial tracheids, as seen in radial longitudinal section. This region is usually pitted and the form of these cross-field pits is often characteristic of a genus or group of genera.

crossing over The exchange of corresponding segments between chromatids of homologous chromosomes. The process is inferred genetically from the recombination of linked genes and cytologically from the formation of *chiasmata. Normally crossing over is reciprocal (equal) but unusual segregation products in ordered tetrads indicate that it can very occasionally be nonreciprocal (unequal). Work with imperfect (asexual) fungi indicates that crossing over can also occur during mitosis. Chromosome reassortment and crossing over together constitute the main source of genetic variation in sexually reproducing organisms. Crossing over is suppressed by chromosome inversions. *See also* independent assortment, parasexual recombination, tetrad analysis.

crossover suppressor *See* inversion.

cross pollination The transfer of pollen from the anthers of one individual to the stigma of another individual of the same species. This is achieved by intermediary agents such as insects (*see* entomophily), wind (*see* anemophily), or water (*see* hydrophily). Birds (*see* ornithophily) and bats often act as pollinators in the tropics. Cross pollination does not necessarily preclude *self pollination, although some mechanisms, such as the closing of the stigma lobes after pollination with compatible pollen (as in *Mimulus*), may reduce the possibility of it occurring.

cross protection The inoculation of a young plant with a mild strain of a virus to protect it from other virulent strains. This procedure has been used in the control of tobacco mosaic virus and citrus tristeza virus.

crown gall A *gall disease caused by the soil-borne bacterium *Agrobacterium tumefaciens*. Galls are produced above and below ground on a wide range of host plants – particularly fruit trees. It is not generally a serious disease and even large galls on fruit trees seem to have little economic effect. The disease is more serious on nursery stock. Careful inspection of fruit-tree nursery stock and the use of resistant root stocks are helpful control measures.
The galls are of interest to physiologists because they display *habituation and are self-sufficient for auxin when grown in culture. *See also* genetic engineering, tumour-inducing principle.

crozier The hooked dikaryotic tip of an *ascogenous hypha. Its two nuclei divide simultaneously and the resulting four daughter nuclei are partitioned by two septa into three cells. The middle cell contains two genetically different nuclei and the apical and basal cells either side contain one nucleus each. The apical cell bends round to fuse with the basal cell, thus reestablishing the dikaryotic condition in both cells resulting from the cell division. The middle cell, now at the tip of the ascogenous hypha, develops into an ascus. The cell below

Stages in crozier formation.

may extend and go on to form another crozier. Similarities between this process and the *clamp connections of certain basidiomycetes have led to the suggestion that basidiomycetes evolved from ascomycetes.

Cruciferae (Brassicaceae) A large dicotyledonous family, commonly called the mustard family, members of which occur predominantly in north temperate latitudes. It contains about 3000 species in about 375 genera. Most cruciferous plants are herbs with alternate leaves and four-petalled flowers borne in a raceme or corymb. The petals are inserted in the shape of a cross (cruciform), hence the name of the family. The fruit is also characteristic (*see* siliqua, silicula, lomentum).

Many crucifers are important crop plants. Cultivars of *Brassica oleracea* include the cabbage, cauliflower, and Brussels sprout, while other species include the turnips, swedes, and oilseed rape. Ornamental crucifers include the wallflower (*Cheiranthus cheiri*), honesty (*Lunaria annua*), and stocks (*Matthiola*).

crustose (crustaceous) Describing lichens that have a crustlike thallus closely appressed to and virtually inseparable from the substratum. An example is *Lecanora*.

Cryptococcales *See* Blastomycetes.

cryptogam In early classifications, a plant whose method of reproduction is not immediately apparent, i.e. a plant in which the reproductive structures are not borne in conspicuous flowers or cones. Cryptogams thus included the algae, fungi, bryophytes, and most pteridophytes. *Compare* phanerogam.

Cryptophyta A division of unicellular flagellate *algae that are dorsoventrally flattened and have an oblique furrow. Cryptophytes contain green, blue, and red pigments and store food as starch in a pyrenoid outside the plastid. Reproduction is by longitudinal fission and sexual reproduction does not occur.

cryptophyte A plant with its perennating buds hidden either below ground (*see* geophyte, helophyte) or in water (*see* hydrophyte). Cryptophytes are more common in arctic and temperate regions than in tropical regions. *See also* Raunkiaer system of classification.

crystal A nonliving cytoplasmic inclusion, often of calcium oxalate or less frequently of calcium carbonate. Crystals occur in a variety of forms (*see* druse, raphide).

culm The jointed stem of members of the Gramineae and Cyperaceae. In grasses it is usually hollow or more rarely filled with pith. In sedges it is usually solid and triangular in cross section.

cultivar Any variety or strain produced by horticultural or agricultural techniques and not normally found in natural populations; a *culti*vated *var*iety.

culture medium (nutrient medium) A mixture of nutrients used to support the growth of microorganisms or of the cells, tissues, or organs of animals or plants. Such media may be solid, if mixed with a gelling agent (e.g. *agar), or liquid. Whenever possible media used in experimental work should be chemically defined, i.e. the identity and concentration of their constituents should be known. Sometimes however a culture can only be grown when complex undefined additional substances, e.g. coconut milk, are added. A culture medium must contain all the macro- and micronutrients essential to growth. A carbohydrate source is necessary and is usually supplied as sucrose. Carbohydrates are necessary even for plant tissue cultures since photosynthesis either does not occur or is insufficient in culture. Plant tissue cultures also require growth substances, normally an auxin or cytokinin or very often both in a particular ratio. Other factors, e.g. gibberellins or vitamins, are often necessary. Ideal media for plant tissue culture are usually also suitable for the growth of microorganisms. Culture must therefore be carried out under sterile conditions to avoid contamination.

cuneate Describing leaves that are wedge shaped with the point of the wedge forming the base of the lamina.

cup fungi *See* Pezizales, Helotiales.

cupule Any of various cup-shaped structures, especially: **1.** The structure that partially or completely encloses the fruits of trees in the Fagaceae, such as the cup encircling the bottom of an acorn and the spiny husk around sweet chestnuts (*Castanea sativa*). It is formed from fused extensions of the pedicel and is believed by some to provide a link between angiosperms and seed ferns.
2. The structure formed from fused glandular bracts that enclosed the seed in certain of the extinct seed ferns, e.g. *Lyginopteris*.
3. The hollow spherical structure that contained the ovules in members of the extinct gymnosperm order Caytoniales. Such cupules were borne at the tips of short pinnae on the female reproductive axis and had a pore at the base to enable pollen to enter.

cuspidate Suddenly narrowing to a point. The term is especially used of leaves.

cuticle The continuous layer of *cutin secreted by the *epidermis and covering the aerial parts of the plant body, broken only by stomata and lenticels. The cuticle affords some physical protection, but its main function is to prevent excessive water loss.

cutin A mixture of complex macromolecules forming the waxy cuticle that covers the aerial parts of most higher plants. Cutin contains long-chain fatty acids and polyhydroxy derivatives of these fatty acids (e.g. 10,16-dihydroxypalmitic acid). These molecules are cross-linked through ester bonds in a more or less random fashion to form a large interlinking matrix, relatively impervious to water and gases.

cutting A detached portion of a living plant, such as a bud or leaf, that can produce a new daughter plant if grown in soil or in a suitable culture medium. Taking cuttings is a common horticultural practice for plant propagation. Often the cut end of a stem cutting is dipped into a rooting powder, which contains a natural or synthetic auxin that stimulates the rapid development of adventitious roots and the establishment of a new daughter plant.

Cyanobacteria *See* Cyanophyta.

cyanocobalamin (vitamin B_{12}) A complex molecule consisting of a corrin ring (similar to a porphyrin ring) and a ribonucleotide. The corrin ring contains a cobalt atom, which is bonded to a cyanide group and to the ribonucleotide. Cyanocobalamin is not synthesized by plants or animals and commercial production is from bacterial cultures, e.g. of *Streptomyces olivaceus*. It is not needed by plants but animals need it for the normal development of erythrocytes.

cyanogenesis The production of cyanide *glycosides by certain plants, including cherry laurel (*Prunus laurocerasus*), birdsfoot trefoil (*Lotus corniculatus*), and white clover (*Trifolium repens*). The glycosides, found largely in water solution in the vacuole, yield a sugar and hydrogen cyanide on hydrolysis. Their function is uncertain but they may serve to detoxify harmful by-products of necessary metabolic reactions. They may also act as an emergency sugar store and, because of their poisonous nature, can play a role in deterring invasion or damage by disease and pest organisms.

cyanophycin A food reserve characteristic of algae belonging to the Cyanophyta.

Cyanophyta The prokaryotic division containing the blue-green *algae. It contains unicellular, colonial, and filamentous forms, many of which are surrounded by a mucilaginous sheath, hence the older name Myxophyceae. Sexual stages and flagellate forms are unknown. The predominant pigment is phycocyanin, which confers a wide range of different colours depending on species and environment. Blue-green algae are abundant in fresh water and the soil; they also occur in the marine littoral zone, in hot springs, as symbionts in lichens and root nodules, and as parasites in certain plants.

The nuclear and plastid material of blue-greens is not bound by membranes, chromosomes are lacking, and the pigments are dispersed in a primitive chromatophore. Some can also fix nitrogen. These characteristics have led many to classify them with the bacteria under various names, such as Cyanobacteria and Schizophyceae. However they differ from bacteria in their lack of motility and lack of fermentative activity.

The presence or absence of *heterocysts and *hormogonia and the tendency of cells to aggregate into distinct filaments are the main features by which orders are recognized. The *Chroococcales and the *Hormogonales are the most important orders.

cyathium A specialized cup-shaped type of inflorescence found in members of the Euphorbiaceae. It resembles a single flower and consists of a female flower in the centre surrounded by many simple male flowers enclosed within an involucre. The involucre also bears a number of petaloid glands along its rim.

Cycadales An order of tropical and subtropical gymnosperms containing one family, the Cycadaceae, divided into some 9 genera and about 100 species. Cycads have a thick, generally unbranched, stem bearing a terminal rosette of large palmlike leaves. Scale leaves are also present on the upper part of the stem and apex. Cycads are very slow-growing and vary considerably in height between species: some have a large proportion of the stem underground and reach a height of only a few metres, while others attain heights up to 15 m. Most of the mechanical support of the stem is provided by the sclerenchymatous leaf bases. The starchy nature of the stem (and of the seeds) has led to the use of certain cycads, the sago palms, as a food source.

Cycads are dioecious and bear exceptionally large female cones and ovules. The egg cell of the archegonium is larger than that found in any other group of plants. Cycads first appeared in the late Palaeozoic and are thought to be most closely related to the Pteridospermales.

Cycadeoidales *See* Bennettitales.

Cycadicae *See* Pinophyta.

Cycadofilicales *See* Pteridospermales.

cyclosis (cytoplasmic streaming) The movement of *cytoplasm within cells. The viscosity of cytoplasm can vary as a result of changes in the nature of protein molecules, particularly actin. In the fluid form it flows freely and can be observed carrying organelles and other inclusions in the cell. Cyclosis is very variable in plant cells. For example, in leaves of *Elodea* (waterweeds) it is more active in cells in the midrib region than in cells near the margin. The rate of flow increases with rise in temperature but is normally less than 0.1 mm per second. It is probable that contractile forces generated by F-actin microfilaments in the cytoplasm contribute to the movement. The streaming of cytoplasm probably assists the distribution of metabolites between cells via *plasmodesmata, augmenting diffusion, which is a slower process. It is also believed to contribute to the passage of material through sieve tubes and is more pronounced in young phloem sieve tubes than in the older elements. Cyclosis is not seen in prokaryotic cells.

cymose inflorescence (cyme, definite inflorescence) An *inflorescence in which the apical tissues of the main stem and laterals lose their meristematic capacity and differentiate into flowers. New growth arises from continued cell division in the axillary meristems. Older flowers are usually found near the stem apex. *Compare* racemose inflorescence. *See also* dichasium, monochasium.

Cyperaceae A large family of monocotyledonous plants, the reeds and sedges, numbering some 4000 species in about 90 genera. It is cosmopolitan in distribution but members are concentrated in the wetter regions of temperate and cold latitudes. Reeds and sedges superficially resemble grasses but may be distinguished by their solid three-angled stems, by the absence of ligules, and by the leaf sheaths, which are usually closed as compared to the open sheathing bases of grasses. The plants arise from swollen underground stems, which in some species, e.g. Chinese water chestnut (*Eleocharis tuberosus*), are edible. The flowers are small, having a reduced perianth, and arranged in spikes, each flower being subtended by a single glume. The fruit is an achene.

A few species are of limited commercial importance being used for making mats, baskets, hats, and paper. Others are used as ornamentals in water gardens.

cypsela A fruit similar to an *achene except that it develops from an inferior ovary and thus also includes noncarpellary tissue. It is typical of the Compositae, where the fruit is surrounded by hairs derived from the calyx (the *pappus). This type of fruit is traditionally considered a *pseudocarp.

cyst Any thick-walled resting spore. *See also* akinete, zygospore.

cysteine A polar sulphur-containing amino acid with the formula $HSCH_2CH(NH_2)COOH$ (*see illustration at* amino acid). It is synthesized from the amino acids methionine and serine. Breakdown is to pyruvate and may occur by several routes.

Cysteine is involved in the synthesis of the cofactor biotin. Sulphur-substituted derivatives of cysteine are found in the storage organs of many plants and the sulphide thus stored can be utilized in periods of rapid growth. *See also* cystine.

cystine A derivative of the amino acid *cysteine, formed by the oxidative linkage of two cysteine molecules through a disulphide bridge. The formation of sulphide bridges is important in the maintenance of the tertiary structure of many globular proteins. Cysteine residues in different parts of a protein chain may cross-link so causing folding of the chain. Cystine is also necessary for the synthesis of thiamin pyrophosphate, a coenzyme in many decarboxylation reactions.

cystocarp A swollen urn-shaped fruiting body found in some red algae, e.g. *Polysiphonia*. It forms following fertilization of a *carpogonium and contains the developing carpospores. The surrounding pseudoparenchymatous tissues develop from cells of the filament below the carpogonium.

cystolith A large intracellular structure formed by the deposition of lime on an ingrowth of the cell wall. An enlarged epidermal cell containing such a structure is termed a *lithocyst*. Species in which cystoliths are formed include *Ficus elastica* (rubber plant) and *Urtica dioica* (stinging nettle).

cytochrome Any iron–porphyrin containing electron-carrying protein forming part of the electron transport chains of respiration and photosynthesis. The porphyrin prosthetic group of cytochromes contains a chelated Fe ion, which changes oxidation state from ferric, Fe (III), to ferrous, Fe(II), when the cytochrome is reduced. Most cytochromes are tightly membrane bound.

Cytochrome aa_3 (also called cytochrome *c* oxidase or the *respiratory enzyme*) is responsible for the final step in the respiratory chain in which water is formed from oxygen and hydrogen ions.

cytogenetics The application of techniques used in *cytology to the problems of inheritance, particularly the study of chromosome structure and its relation to the phenotype.

cytokinesis The process of cytoplasmic division as distinct from nuclear division or karyokinesis. As *telophase proceeds, a *cell plate is organized in the cytoplasm in the equatorial region of the spindle. The cell is thus divided into two daughter cells.

cytokinin (kinin) Any of a group of growth substances whose primary effect is to stimulate cell division. Cytokinins were discovered during work on tissue culture media when it was found that cells of tobacco pith explants could be stimulated to divide by adding the purine adenine to the medium. Subsequently various adenine derivatives, e.g. kinetin, were found to have even greater effects on cell division. However these effects are not seen in the absence of auxin. Moreover, by changing the proportion of cytokinin to auxin, different types of meristematic activity may be induced (*see* auxin). Cytokinins are probably active in most aspects of plant growth and development. Their most obvious effects include the delay of senescence, the induction of flowering in certain species, and the breaking of dormancy in axillary buds and some seeds. All natural cytokinins are derivatives of the base adenine. Some synthetic cytokinins are substituted phenylureas. It has been suggested they may act by regulating nucleic acid activity, particularly that of transfer RNA (*see* IPA). Endogenous cytokinins are found in the greatest concentrations in embryos and developing fruits, e.g. in the 'milk' of the coconut. Before fruit set, it is probable that most cytokinin synthesis is in the root. Abnormally high levels of cytokinin are associated with certain plant diseases, e.g. witches broom and crown gall. *See* kinetin, zeatin.

cytology (cell biology) The study of the structure, organization, and functioning of cellular material.

cytomixis The apparent migration of cytoplasm and nuclear material from one pollen mother cell to another through pores (cytomictic channels) in the cell walls. It occurs prior to and during the prophase stage of meiosis. The reason for this is unknown though it may be a response to the nutritional problems of developing reproductive cells.

cytopharynx *See* gullet.

cytoplasm (hyaloplasm) The part of the *protoplasm outside the nucleus in which the cisternal elements and membrane-bound *organelles lie. The soluble phase of cytoplasm contains the principal components of the cell's biochemical pathways, i.e. ions, dissolved gases, and most of the enzymes and substrate molecules for metabolic processes. The cytoplasmic matrix also contains storage products, e.g. lipid droplets, aleurone grains, and starch grains. Complex networks and linear arrangements of fine microfilaments and *microtubules are commonly present. The microfilaments, 0.5–7.0 nm in length, are composed of helical chains of F-actin, and when supplied with ATP they contract. Those at the periphery of the cell are involved in movements of the components of the plasma membrane, e.g. in *endocytosis. Cytoplasm shows transitions between viscous (*plasmagel*) and fluid (*plasmasol*) phases. This is displayed in *cyclosis and in amoeboid movement shown by some algal unicells. The F-actin molecules of the microfilaments, particularly at the periphery of cells, are unstable, readily breaking up into soluble components so that the cytoplasm becomes fluid.

cytoplasmic inheritance (extrachromosomal inheritance) The determination of a characteristic by genes in the cytoplasm (plasmagenes) rather than in the chromosomes. The expression of a cytoplasmically determined character is therefore not related to the behaviour or movement of chromosomes, and consequently such characters fail to segregate in Mendelian ratios. They are normally transmitted through the female gamete, which contributes most of the cytoplasm to the zygote, the male gametes only contributing a nucleus. Genes inherited in this manner are found on DNA present in small quantities in the chloroplasts, mitochondria, and sometimes the cytosol itself. An example of a characteristic inherited in this fashion is a form of male sterility in maize (*Zea mays*).

cytoplasmic streaming *See* cyclosis.

cytosine A nitrogenous base, more correctly described as 4-amino-2-oxypyrimidine, derived from amino acids and sugars and found in all living organisms. Cytosine is found in one of the *nucleotides present in *DNA and *RNA and pairs specifically with the purine base guanine. Substantial amounts of cytosine in DNA and RNA are found in the modified forms of 5-methylcytosine and 5-hydroxymethylcytosine. The affinity of these substances for guanine is similar to that of cytosine but the significance of their occurrence is uncertain. Recent evidence suggests that the regions of DNA with a large number of methylated cytosines are inactive. Thus gene activation may be brought about, at least in part, by selective demethylation.

cytosome *See* gullet.

cytotaxonomy The use of chromosome studies (i.e. number, structure, and behaviour) in taxonomic work. Chromosome number is probably the most frequently quoted feature of the *karyotype used by taxonomists. Chromosome counts are usually made on sporophytic tissue at mitosis and are therefore the diploid number ($2n$). When dealing with a polyploid series, the base number (i.e. the number of chromosomes present in the original haploid *genome) may be given. The position of the centromere is a reliable feature of chromosome structure and consequently makes a good *taxonomic character. More detailed studies of meiotic behaviour can reveal, for example, the heterozygosity of some *inversions. Such a feature may be consistent for a particular taxon, thus providing additional taxonomic evidence. Cytological data is sometimes thought to be of special significance, and may be ascribed more weight (*see* weighting) than other taxonomic evidence.

CZI *See* Schultze's solution.

D

2,4-D (2,4-dichlorophenoxyacetic acid) A synthetic auxin of the *phenoxyacetic acid type with two substituted chlorine atoms. It is the auxin most widely used commercially as a selective weedkiller. *See also* 2,4,5-T, MCPA.

damping-off A disease of seedlings encouraged by cold wet soil and crowded conditions. In pre-emergence damping-off the germinating seed rots before the plumule breaks through the soil surface. In post-emergence damping-off the seedlings rot at soil level and then collapse and die. Damping-off is caused by numerous soil-inhabiting fungi including *Fusarium*, *Phytophthora*, *Pythium*, and *Rhizoctonia* (*Corticium*). Correct planting conditions, seed dressings, and the use of sterile soil in seed trays are the usual preventative methods. *See also* seedling blight.

dark-ground illumination A microscopical technique used to examine living cells or microorganisms. These are not suitable for examining under a normal light microscope (bright-field illumination) as they are usually transparent. Thus little detail can be seen against a light background. In dark-ground illumination direct light is prevented from forming an image by illuminating the specimen from the side so only diffracted light from the specimen passes through into the objective. The image of the specimen then appears luminous against a completely dark background. This technique exploits the differences in refractive index between organelles and the surrounded protoplasm that causes the organelle boundaries to reflect more light.

dark reactions The sequence of light-independent reactions that utilize the energy (in the form of ATP) and reducing power (in the form of NADPH), produced during the *light reactions of *photosynthesis, to reduce carbon dioxide. In eukaryotes, this process usually occurs in the chloroplast stroma and it can take one of two forms, depending on whether the subject is a *C_3 or a *C_4 plant. The details of the fixation of carbon dioxide differ in the two types of plants but the end result in both cases is the production of carbohydrates via the *Calvin cycle.

Darwinism The theory, put forward by Charles Darwin in 1858, that species evolve by *natural selection. Darwin noted that although organisms produce more than enough offspring to replace themselves, the numbers of a species tend to remain constant. He concluded that there is a struggle for existence and organisms compete with one another for food, space, etc. He also noted the considerable variation exhibited by members of the same species and concluded that only those best adapted to the environment survive. As many of the variations are hereditary, those that survive because of favourable variations will transmit these to their offspring. Those with unsuitable characteristics will not survive and such characteristics will be lost. This process of selective birth and selective death or 'survival of the fittest' as it came to be called Darwin termed natural selection. Finally, because the environment changes, natural selection continuously acting on a large number of variations could result in the formation of new species, the origin of species. Darwin could not explain how variation arose and made no distinction between quantitative and qualitative variation. He was later driven to accept Lamarck's theory of the inheritance of acquired characteristics (*see* pangenesis). The origin and maintenance of variation was later explained by work in genetics (*see* Neo-Darwinism). *See also* Weismannism.

dating The determination of the age of rocks, minerals, and organic materials. Organic materials may be dated by *radiocarbon dating and *dendrochronology. However these methods can only date specimens from the later part of the Quaternary period. *Varve dating is also limited to this period. Age determinations of older rocks rely on *radiometric dating.

day length *See* critical day length.

day-neutral plant A plant that is capable of flowering regardless of the amount of light it receives each day. The onset of flowering is therefore controlled by other factors. *Compare* long-day plant, short-day plant.

deamination The removal of an amino group from a compound. Breakdown of amino acids is by oxidative deamination. Glutamate, which is formed from many amino acids by transamination, is deaminated by the enzyme glutamate dehydrogenase, yielding α-ketoglutarate, ammonium ions, and reduced NAD or NADP.

decarboxylase Any enzyme that catalyses the removal of carbon dioxide from a substrate. An example is the enzyme pyruvate decarboxylase, which catalyses the formation of acetaldehyde from pyruvate, a reaction in the alcoholic fermentation of glucose. Decarboxylases have a tightly bound coenzyme, often biotin or thiamin pyrophosphate, which acts as an intermediate carboxyl carrier. *See also* carboxylase.

deciduous Describing woody *perennial plants that shed their leaves before the winter or dry season. Leaf fall is an adaptation that reduces water loss by transpiration when little or no water is available to the plant roots (*see* physiological drought). Various environmental factors, such as daylength, temperature, and light intensity, influence the onset of leaf *abscission. *Compare* evergreen.

decomposer An organism that feeds on dead organic material, breaking it down into simpler substances and bringing about decay. In this way, organic material is recycled as the products of decomposition can be used by plants (*producers). Examples of decomposers are some bacteria and fungi. *Compare* consumer. *See also* food chain.

decumbent Describing a stem that lies along the ground. Decumbent stems often have an upturned tip.

decurrent Describing leaf bases that extend down the stem beyond their point of insertion, forming a wing on the stem. The leaves of the common mullein (*Verbascum thapsus*) are examples. The term is also used of the gills of basidiomycete fungi that run down the stem, as seen, for example, in the genus *Clitocybe*.

decussate *See* opposite.

dedifferentiation The loss of the specialized features and reversal to a meristematic state of a *differentiated cell. Dedifferentiation often occurs immediately prior to further reorganizational changes as in the production of a secondary meristem. The process may also be brought about by wounding or stimulation by growth substances. It is thought to involve the removal of inhibitors that normally prevent the expression of the complete genetic complement. *See* totipotency. *See also* callus.

deficiency *See* deletion.

deficiency disease A disease caused by the lack of an essential nutrient element. Mineral deficiencies can often be recognized by characteristic symptoms. For example, nitrogen deficiency results in stunting of the plant and chlorosis of older leaves. Iron deficiency results in interveinal chlorosis, particularly of younger leaves, giving a striped appearance in cereal and grass leaves. Boron deficiency causes the disease of heart rot in sugar beet and mangolds (*Beta vulgaris*). *See also* yellows.

definite growth A form of growth where a maximum size is reached beyond which there is no further increase, i.e. growth is limited. It is seen in annual and biennial plants and in many plant organs, e.g. internodes, which do not continue to lengthen throughout the life of the plant. The term is also used of cymose inflorescences, in which growth is terminated by the production of a terminal flower bud. *Compare* indefinite growth.

definite inflorescence *See* cymose inflorescence.

definitive nucleus The diploid nucleus found in the centre of the *embryo sac after the fusion of the two haploid *polar nuclei. A male gamete released from the pollen tube may fuse with this to form the triploid primary endosperm nucleus from which the endosperm develops. *See also* double fertilization.

deflexed Describing a structure that is bent sharply downwards.

degree of freedom In an analysis of experimental results, a comparison between data made independently of any other comparisons. The *number of degrees of freedom* (N) in an analysis is therefore one less than the number of observations.

dehiscence The splitting open along predetermined lines of certain plant organs, such as anthers, spore capsules, and fruits, to release their contents. Dehiscence is often caused by the gradual drying out of the enclosing walls. In addition, variations in the degree of wall thickening may create tensions that result in a violent opening of the structure. *See also* annulus.

dehydration A stage in the preparation of permanent slides for microscopical examination, in which water is removed from the specimen. The material is dehydrated slowly by being placed in a series of successively more concentrated ethanol solutions culminating with absolute alcohol. Plant cells must be dehydrated far more gradually than animal cells as there is a tendency for the protoplasm to shrink from the cell wall. Dehydration is followed by *clearing.

dehydrogenase Any *oxidoreductase enzyme that catalyses the transfer of electrons from a reduced substrate or electron donor to an electron-accepting coenzyme; the transfer usually involves hydrogen. The equation for a dehydrogenase reaction is generally: reduced substrate + oxidized coenzyme $\rightleftharpoons$ oxidized product + reduced coenzyme. The reduced coenzyme is usually reoxidized in the *respiratory chain. Dehydrogenases are classified according to the nature of the coenzyme, for example, those requiring the pyridine dinucleotide coenzymes NAD and NADP are termed pyridine-linked dehydrogenases, while those requiring FAD or FMN are termed flavin-linked dehydrogenases or flavoproteins.

deletion (deficiency) A chromosome mutation involving the loss of part of a chromosome. Since a number of genes are lost as a result, deletions are usually harmful and often lethal if the corresponding segment of the homologous chromosome contains any recessive deleterious mutations. *Compare* aneuploidy.

deliquescence A gradual dissolving of tissue by *autolysis. It is seen, for example, in *Coprinus* (ink-caps), in which the gills change into an inky black fluid after the basidiospores have been discharged.

denaturation Change in the structure of a globular protein or a DNA molecule following exposure to temperatures above 60–70°C, to pH outside the normal physiological range, or to certain chemicals. Denaturation of proteins involves the uncoiling of the polypeptide chain so that its tertiary structure is lost, resulting in loss of biological activity. The process is sometimes reversible, the protein regaining its characteristic coiled structure (native form) and its activity. This is termed *renaturation*. Denaturation of DNA involves the uncoiling of the double helix, forming single-stranded DNA. The double helix molecule will quickly reform if denaturation has not proceeded too far (i.e. if a small part of the molecule has retained the double helical form). Renaturation (or annealing) is slow if the strands have completely separated.

dendrochronology The use of the annual rings of trees to date historical events. The dating of archaeological sites depends largely on *cross dating.* This involves taking small cores from old living trees and comparing them with the rings of timbers at the site. The year the timbers were felled is determined by finding the point at which ring patterns of the living tree correspond to those of the archaeological specimens. Bristlecone pines (*Pinus longaeva*) are often used as they can live for over 4000 years. If sufficiently aged living trees cannot be found in a particular region then chronologies may be built up successively further back in time by matching the ring patterns of a number of wood samples whose ages overlap.

Annual rings also provide a record of past environmental conditions. Thus ring width is normally positively correlated with water abundance. The presence and level of certain pollutants, e.g. lead, may also be recorded in the rings. It is necessary in such work to ensure that only those tree species that produce one growth ring a year are used. Certain trees, e.g. many juniper species, produce multiple rings each year, while in others it may not be possible to distinguish one season's growth from the next.

dendrogram A branching usually 'rooted' diagram that reflects the relationships of a group of taxa. The *taxo-

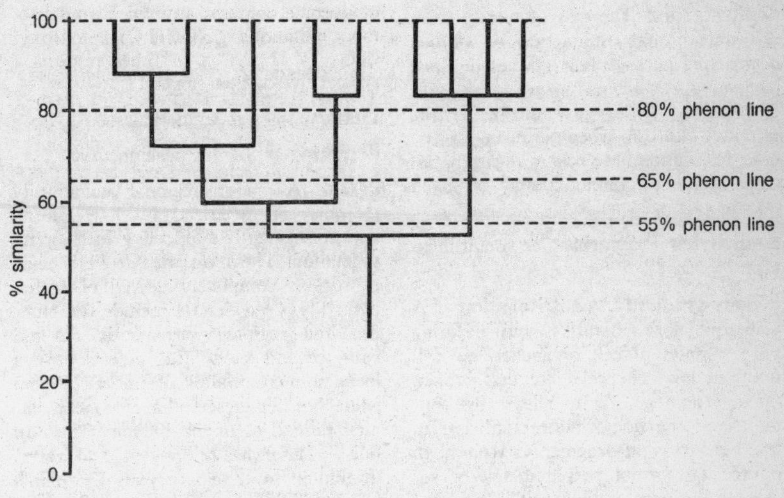

A dendrogram showing phenetic similarity. The phenon lines delimit groups of the same rank.

nomic hierarchy can be represented in a dendrogram, with kingdom at the base and subform as the terminal branches. Specific types of dendrogram can be recognized. *Phenograms* represent the degree of phenetic similarity and are based solely on phenetic data. *Phenon lines* drawn at right angles to the phenogram (*see* diagram), show values of similarity as percentages (the greater the value the higher the level of similarity). Clusters of similar groups are termed phenons and phenons may be assigned rank according to the level on the phenogram at which they branch off. Thus in the diagram the four clusters above the 80% phenon line could be assigned the rank of genus while the two clusters at the 55% similarity level might be assigned the rank of subfamily. Phylogenetic trees are also dendrograms, but unlike phenograms, the vertical axes represent time (though not necessarily to scale) or relative advancement. *See also* cladogram.

denitrification The loss of nitrate from the soil through the action of various *denitrifying bacteria* (e.g. species of *Clostridium*, *Pseudomonas*, *Micrococcus*, and *Thiobacillus*), which use nitrate as the terminal electron acceptor during anaerobic respiration. Molecular nitrogen, nitrous oxide, or ammonia may be given off. *See* nitrogen cycle.

denitrifying bacteria *See* denitrification.

density-gradient centrifugation A technique for separating and isolating pure samples of cell organelles, i.e. cell fractionation. The cells are first broken up in a homogenizer to release the contents and the homogenate is filtered to remove cell wall fragments. It is then poured on top of prepared salt or sucrose solutions of different concentrations that have been layered according to their density in a glass tube. Provided an appropriate centrifugal force is applied, the cell inclusions will band together at the regions in the density gradient that correspond to their own density and can then be separated off by various methods. Another method of isolating cell organelles, *differential centrifugation*, involves the filtered homogenate being centrifuged and decanted several times in sequence, increasing the time and speed of centrifugation at each successive stage. Nuclei, membranes, chloroplasts, mitochondria, vesicles, endoplasmic reticulum, and ribosomes, are collected in that order.

dentate Describing a leaf margin that is toothed, with outward-pointing notches. Leaf margins finely toothed in this way are termed *denticulate*.

deoxyribonucleic acid *See* DNA.

deoxyribose A pentose sugar and a component of the nucleotides present in DNA, i.e. adenosine, cytosine, guanidine, and thymidine phosphates (AMP, CMP, GMP, and TMP). Although not conforming to the general formula of monosaccharides ($C_x(H_2O)_y$), deoxyribose ($C_5H_{10}O_4$) is in other respects a typical five-carbon sugar.

Derbesiales *See* Caulerpales.

dermatogen *See* histogen theory.

desert A major regional community (*biome) characterized by low rainfall and consequently supporting little or no vegetation. The term *true desert* is used of regions completely devoid of higher plant life. Cold deserts include the *tundra and regions permanently covered with ice and snow. Hot deserts have a mean annual rainfall of under 250 mm while hot semideserts have a mean annual rainfall of under 400 mm. The rain falls in brief heavy showers and varies in amount from year to year. The plants are few in number and are usually short and sparsely distributed. The permanent perennial plants consist of succulent or xerophytic trees, shrubs, and herbs.

There are also *ephemeral plants whose seeds lie dormant until a brief rainstorm prompts them to germinate, flower, and set seed in a short space of time.

desmids *See* Zygnemaphyceae.

Deuteromycotina (Fungi Imperfecti) An artificial subdivision of the *Eumycota containing all those fungi in which the sexual stage has either not been found or appears to have been replaced by other mechanisms, e.g. by *parasexual recombination. It contains over 15 000 species in some 1825 genera. Most of the imperfect fungi show affinities with the ascomycetes. Fungi in which the sexual (perfect) stage is found may be given a different generic name from closely related species only showing the imperfect stage, e.g. the name *Aspergillus* is given to certain moulds that show no sexual stages but similar species that form ascocarps are called *Eurotium*. Some imperfect fungi are classified with their perfect forms. The group is subdivided into the classes *Blastomycetes, *Hyphomycetes, and *Coelomycetes.

developmental spiral *See* genetic spiral.

deviation Symbol: *d*. The departure of a value or observation from what was expected. *See also* mean deviation, standard deviation.

Devonian The first period of the Upper Palaeozoic era from about 395 to 345 million years ago in which the first extensive invasion of the land by plants and animals took place. A possible bryophyte fossil, *Sporognites*, with a well developed capsule has been found in the Lower Devonian, but most fossil bryophytes appear later. Many well preserved fossils of the Psilophytales have been found in the *Rhynie chert of the Devonian in Scotland. Lycopsids and sphenopsids are also found. The ferns (Filicinae) arose in Devonian times possibly from a form resembling *Protopteridium*, an intermediate between the psilophytes and the Filicinae. The gymnosperms probably arose in the late Devonian, being represented by the extinct Pteridospermales, which include the earliest seed plants *Archaeopteris* and *Archaeosperma*, and also the extinct Cordaitales. *See* geological time scale.

dextrose The former name for *glucose.

diadelphous Describing stamen filaments that are fused into two groups.

diagnosis The statement, in Latin, that describes how a new taxon differs from its closest relatives.

diakinesis *See* prophase.

diallel cross A breeding experiment in which each of a number of males is mated to each of a number of females.

dialysis A process in which large molecules, such as proteins and starch, and small molecules, such as amino acids, glucose, and salts, are separated by selective diffusion through a semipermeable membrane. For example, a mixture of starch and glucose molecules in solution can be separated by placing them in a semipermeable dialysis tube made, for example, of cellophane, which is immersed in a container of distilled water. The large starch molecules remain in the dialysis tube and the small glucose molecules pass out through the membrane and can be detected in the water in the container.

diaminopimelic acid An amino acid found in bacteria and blue-green algae. It is involved in the biosynthesis of lysine in such organisms. It is also often found in the peptide side chains of the acetylmuramic acid molecules that partly make up the mucopeptides of bacterial cell walls.

diastase The name originally given to the active preparation from malt extract that breaks down starch to sugar. It includes the enzyme β-amylase. *See* amylase.

diatoms *See* Bacillariophyta.

diatropism A *tropism in which the plant part is aligned at right angles to the direction of the stimulus acting on it. A diatropic response to gravity (*diageotropism*) is seen in the horizontal growth of rhizomes and stolons. A diatropic response to light (*diaphototropism*) is often seen in the positioning of leaf blades at right angles to the incident light. *Compare* orthotropism, plagiotropism.

dicaryotic *See* dikaryotic.

dichasium (dichasial cyme) A *cymose inflorescence in which the apices develop into flowers and two lateral branches develop below each apex from axillary buds at a common node. This gives a symmetrical appearance to the inflorescence, as seen in stitchwort (*Stellaria holostea*). *Compare* monochasium.

2,4-dichlorophenoxyacetic acid *See* 2,4-D.

dichogamy The maturation of anthers and stigmas at different times in the same flower so that pollen reception and pollen presentation do not coincide. This reduces the chances of self fertilization and enhances outcrossing. *Compare* homogamy. *See also* protandry, protogyny.

dichotomous Describing the system of branching where each division is into two equal parts. It occurs in many lower plants, for example the branching of the thallus in *Fucus* and the branching of the stem in *Lycopodium* and *Selaginella*.

dicliny The separation of male and female reproductive parts into different flowers. Diclinous plants may either have female and male unisexual flowers on the same individual (monoecy) or on different individuals (dioecy). The term diclinous may also refer to the flowers themselves, in which case it is equivalent to the term unisexual. *Compare* hermaphrodite.

Dicotyledonae A subclass of the *Angiospermae containing all the flowering plants having embryos with two cotyledons. Its members usually possess a cambium and may thus be either woody or herbaceous. Other general features that can often be used to distinguish the Dicotyledonae from the *Monocotyledonae include: having broad leaves with branching veins; having flower parts inserted in fours or fives; having a persistent primary root that develops into a taproot; and having the vascular bundles in a ring (*see* eustele).

The Dicotyledonae contains about 250 families, ranging from the primitive Magnoliaceae to the advanced Compositae. In some classifications, dicotyledons are placed in the class Magnoliopsida. This is divided into some six or seven subclasses in recent classifications (*see* Magnoliidae, Hamamelidae, Caryophyllidae, Dilleniidae, Rosidae, Asteridae). In certain other classifications the Dilleniidae, the Rosidae, and sometimes also the Asteridae are grouped together.

dictyosome *See* golgi apparatus.

dictyostele A type of dissected *siphonostele in which the vascular tissue (as viewed in transverse section) is divided into a number of amphicribal vascular bundles called *meristeles. These are separated by parenchymatous areas, which are usually associated with large closely placed leaf gaps. Dictyosteles are typical of many ferns, e.g. *Dryopteris*, and are essentially similar to *solenosteles, except that the amphiphloic cylinder of the solenostele is reduced to a meshwork by the occurrence of numerous overlapping leaf gaps. *See also* eustele.

didynamous Having two long stamens and two short stamens, as do plants in the family Labiatae.

differential centrifugation *See* density-gradient centrifugation.

differentiation 1. The process by which amorphous cells, produced by meristematic cell division, undergo various physiological and morphological changes during maturation in order to become specialized for a particular function. Differentiation thus produces the specific cell types that make up the separate tissues composing a plant body. At a higher level, differentiation results in the formation of the various plant organs and the division into shoot and root and into vegetative and reproductive structures. *Compare* growth. *See also* dedifferentiation.
2. *See* staining.

diffuse-porous wood Secondary xylem in which there is little or no seasonal variation in tracheary-element diameter as, for example, in the wood of yellow birch (*Betula lutea*). *Growth rings are therefore difficult to discern. *Compare* ring-porous wood.

diffusion The movement of molecules or ions in a fluid from areas of high concentration to areas of low concentration. In a closed system this will continue until the solution or gas mixture is evenly mixed. Diffusion is responsible for such processes as *transpiration and the uptake of carbon dioxide. The rate of diffusion depends on concentration differences and on the nature of the pathway the molecules have to take. Thus transpiration is greater when the air is dry and when the stomata are fully open.

diffusion pressure deficit (DPD, suction pressure) In older terminology, the net force or pressure that causes water to enter a plant cell. Its magnitude is the difference between the *osmotic pressure (Π) and the *turgor pressure (TP), i.e. DPD = Π – TP. *See* water potential.

digitate *See* palmate.

dihybrid An organism that is heterozygous for two particular genes, as in AaBb. Dihybrids are most conveniently produced from crossing parents that are homozygous for the genes in question, e.g. AABB × aabb or AAbb × aaBB. A *test cross normally yields four kinds of offspring: A_B_ (both dominant characters expressed); A_bb, and aaB_ (only one dominant character expressed); and aabb (no dominant characters expressed). If the genes A/a and B/b are not linked, equal numbers of all kinds of offspring will be produced. If they are linked, two classes, the parentals, will be more common, and two classes, the recombinants (cross overs), will be rarer. Selfing a dihybrid normally yields a 9:3:3:1 ratio (*see* dihybrid ratio) in the case of independent assortment but a much more complex result in the case of linkage. *See also* monohybrid, trihybrid.

dihybrid ratio A 9:3:3:1 ratio of phenotypes among the offspring of a single cross. Such a ratio can only be obtained providing that all the following conditions are met: two genes at different loci are involved, e.g. gene A/a and gene B/b; the immediate parents of these offspring are heterozygous for both genes, i.e. AaBb; these heterozygotes are selfed or crossed with one another, i.e. AaBb × AaBb; the genes are unlinked, i.e. independent assortment; there is no interaction between the genes, e.g. no epistasis; A is dominant to a, and B is dominant to b, i.e. no codominance, etc.; fertilization is random, and all the gametes, zygotes, and offspring have an equal chance of survival.
If there is reason to expect a 9:3:3:1 ratio in a particular cross and this result is not obtained, one or more of the

above conditions is not being fulfilled. For example, the ratio may be modified to 9:3:4 (= 9:3:[3+1]) in some cases of gene interaction (epistasis).
A dihybrid ratio is an example of a simple *Mendelian ratio*, being an expansion of the monohybrid 3:1 ratio, viz: $(3:1)^2$.

dikaryotic (dicaryotic) Describing mycelia in which each segment contains two genetically distinct nuclei. Dikaryosis is common in basidiomycete fungi and results when two monokaryotic mycelia of different mating types fuse. *See* heterokaryosis.

Dilleniidae A subclass of the dicotyledons containing both woody and herbaceous plants. The flowers may be unisexual or bisexual and the stamens develop in a centrifugal manner. Some 12–17 orders are recognized including: the Dilleniales; the Theales, including the Theaceae (e.g. camellia, tea) and Guttiferae; the Malvales, including the Tiliaceae (e.g. limes) and Malvaceae (e.g. mallows); the Violales, including the Violaceae (e.g. violets), Flacourtiaceae, and Begoniaceae (begonias); the Salicales, containing the Salicaceae (e.g. willows); the Capparales, including the Capparaceae (e.g. capers), and *Cruciferae (e.g. mustards); the Ericales, including the *Ericaceae (heath family); the Diapensiales; the Ebenales, including the Sapotaceae (e.g. sapodillas) and Ebenaceae (e.g. ebonies); and the Primulales, including the Primulaceae (primrose family). Other orders sometimes recognized include: the Paeoniales (often placed with the Dilleniales); the Passiflorales, Cucurbitales, and Tamarales (often placed with the Violales); the Euphorbiales (often placed in the subclass Rosidae); and the Thymeleales (often placed either with the family Flacourtiaceae or in the order Myrtales of the subclass Rosidae). The Urticales, including the Ulmaceae (e.g. elms), Moraceae (e.g. mulberries, figs), and Urticaceae (e.g. nettles), is sometimes included in the Dilleniidae, and sometimes in the Hamamelidae.

dimerous (2-merous) Describing flowers in which the parts of each whorl are inserted in twos, as in the enchanter's nightshades (*Circaea*).

dimorphism The existence of two distinct forms. The term may be applied to organelles (e.g. bundle sheath chloroplasts and mesophyll chloroplasts), appendages (e.g. sun and shade leaves or juvenile and adult leaves), stages of a life cycle (gametophyte and sporophyte), individuals (e.g. males and females, i.e. *sexual dimorphism*), etc. *See also* polymorphism.

dinitro compounds A group of chemicals including dinocap, which is effective in controlling powdery mildews, and dinitro-orthocresol (DNOC), used as a fungicide, herbicide, and insecticide.

2,4-dinitrophenol *See* uncoupling agent.

dinoflagellates *See* Dinophyta.

Dinophyta (Pyrrophyta) A division consisting predominantly of flagellate unicellular marine *algae, often dark brown in colour due to the presence of the xanthophyll pigment peridinin. It is divided into two classes, the Desmophyceae and the Dinophyceae, or *dinoflagellates*. The Dinophyceae are abundant in the phytoplankton. They characteristically have one longitudinal and one transverse groove in the cell wall and two flagella, one situated in each groove. The Desmophyceae either lack a cell wall or have a wall divided into two semicircular halves by a transverse groove. Certain tropical members of the Desmophyceae exhibit bioluminescence.

dinucleotide *See* nucleotide.

dioecious Describing plants in which the female and male reproductive organs are separated on different individuals.

Dioecy makes cross fertilization obligatory and ensures genetic variability in the population but this is done at the cost of lower seed-setting efficiency and also prevents isolated individuals reproducing. The sex of a dioecious plant is thought to be determined by *sex chromosomes, as in red campion (*Silene dioica*). *Compare* hermaphrodite, monoecious. *See also* dicliny.

diphosphopyridine nucleotide *See* NAD.

diplanetic Describing organisms that successively produce two different types of zoospore (planospore), as do certain species of *Saprolegnia*. *Compare* monoplanetic.

diplobiontic Describing a life cycle in which two types of vegetative plant are formed, one haploid and the other diploid (i.e. a life cycle in which there is an *alternation of generations). The life cycles of ferns are diplobiontic. *Compare* haplobiontic.

diploid A nucleus or individual having twice the haploid number of chromosomes in the nuclei of its somatic cells. In many lower plants and about 50% of flowering plants, diploidy is established at fertilization by the union of two haploid gametes, each containing a single set of chromosomes from its respective parent. The symbol $2n$ is used to denote the diploid number. Diploid plants usually have between 12 and 26 chromosomes in their nuclei. If they contain more than this then it is likely they are *polyploid in origin.

diplontic Describing a life cycle in which the diploid phase predominates and where the haploid stage is limited to the gametes. Such life cycles are seen in the diatoms and members of the Fucales. *Compare* haplontic. *See illustration at* life cycle.

diplospory A form of *apomixis in which the embryo forms directly from the megaspore mother cell. It is found, for example, in mountain everlastings (*Antennaria*).

diplostemonous Describing *stamens that are inserted in two whorls with the outer opposite the sepals and the inner opposite the petals. *Compare* obdiplostemonous.

diplotene *See* prophase.

disaccharide A carbohydrate in which two *monosaccharide units are joined by a glycosidic bond. The commonest disaccharide in plants is *sucrose. Other common disaccharides include *maltose and *cellobiose. Less common disaccharides include the nonreducing trehalose, composed of two glucose units linked by an $\alpha(1-1)$ glycosidic bond. This is the main free sugar in *Selaginella* species.
Some disaccharides have specific functions, such as the translocation function of sucrose, while others (e.g. cellobiose) are breakdown products of larger molecules.

disc floret *See* capitulum.

disclimax *See* climax.

Discomycetes A class of the *Ascomycotina containing those ascomycetes that produce an *apothecium (the exception being the Tuberales, where the hymenium is enclosed). It contains over 3000 species in some 425 genera, which are distributed among the orders *Phacidiales, *Helotiales, Ostropales, *Pezizales, and *Tuberales. The majority of fungi that form lichens are Discomycetes (*see* Lecanorales).

discontinuous variation *See* qualitative variation.

disease resistance *See* resistance.

disjunct species *See* endemic.

dissepiment *See* septum.

distal Denoting the region of an organ that is furthest away from the organ's point of attachment. *Compare* proximal.

distely A special case of *polystely in which the vascular system is divided into two separate *steles. This condition is exhibited in certain species of *Selaginella*, e.g. *S. kraussiana*.

distichous Describing a form of alternate leaf arrangement in which successive leaves arise on opposite sides of the stem so that two vertical rows of leaves result. It is seen, for example, in grasses. The term also describes a form of opposite leaf arrangement in which the pairs of leaves all arise in the same plane, again giving two rows of leaves.

dithiocarbamate fungicide Any fungicide based on the organic sulphur compound dithiocarbamic acid. Such compounds include the commonly used fungicides ziram, thiram, zineb, and maneb.

diurnal rhythm *See* circadian rhythm.

divergent evolution (divergence) The development of different forms from a single basic structure as a result of different selective pressures acting on that structure. These changes may be associated with a new function that is added to or replaces the original function. For example, the general form of the angiosperm stem is aerial and elongated. Its function is to support and space out the leaves and flowers as well as conducting food and water. However certain xerophytic plants with reduced leaves have expanded leaflike stems to increase photosynthetic capacity. Other plants have short swollen stems that serve as underground perennating organs. *See* adaptive radiation, homologous.

division The second highest category in the taxonomic hierarchy. Divisions are composed of classes. The number of divisions within the plant kingdom varies with the system of classification. The Latin names of divisions terminate in -phyta, e.g. Tracheophyta. Several botanists now adopt the term *phylum in place of division.

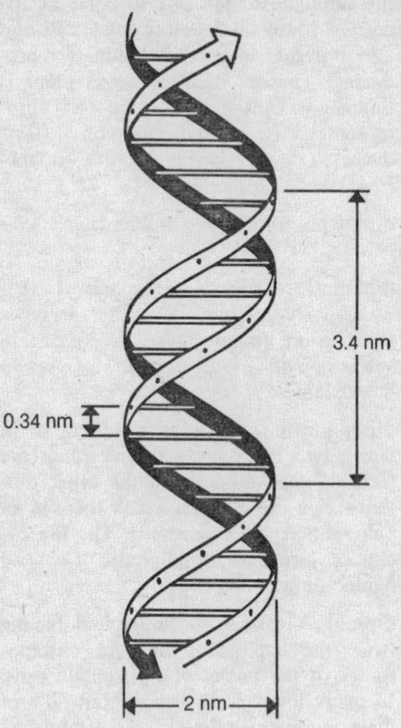

Dimensions of the DNA double helix.

DNA (deoxyribonucleic acid) The substance of which genes are made and that determines the inherited characteristics. In eukaryotes, it is mostly confined to the chromosomes, where it is found in association with basic proteins called histones. Bacterial, mitochondrial, and chloroplast DNA is naked and forms circles or loops. Typically a DNA molecule consists of two polynucleotide chains, forming a right-handed helix with one coil positioned more or less

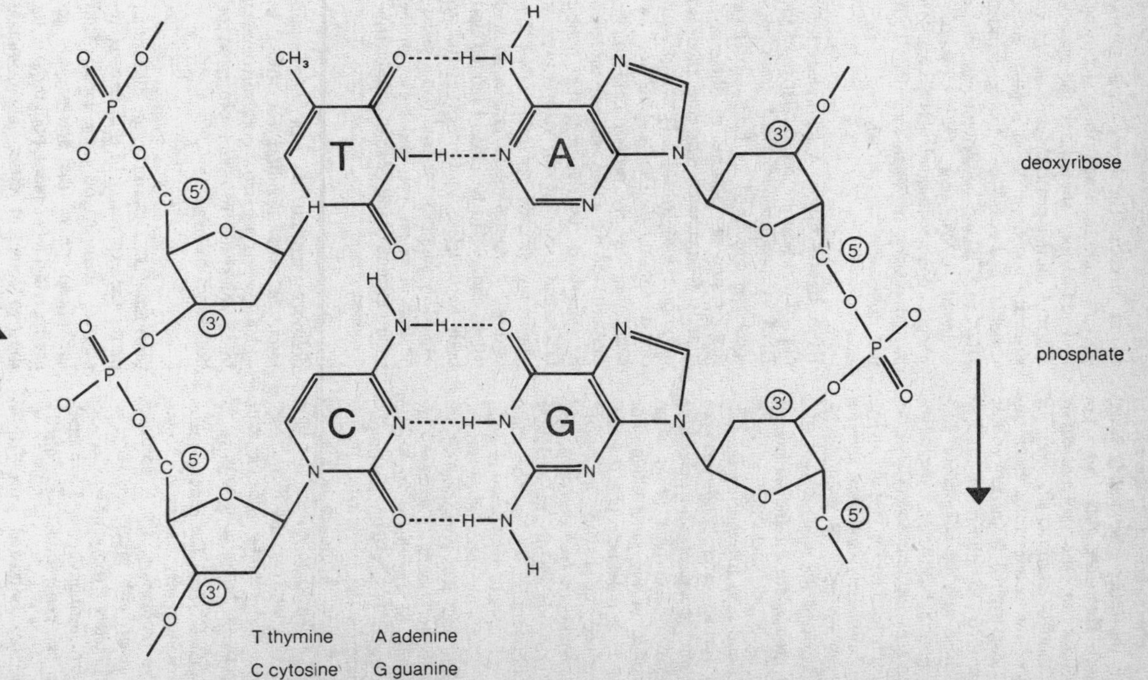

Segment of the DNA molecule showing the sugar-phosphate backbones running in opposite directions and hydrogen bonding between complementary base pairs.

underneath the other (the double helix, *see* diagram). There are only four types of nucleotides in the polynucleotide chains, although each chain can be thousands of nucleotides long. Each nucleotide is constructed from deoxyribose sugar, esterified to phosphate and a base. The base may be adenine or guanine (the purine bases), or cytosine or thymine (the pyrimidine bases). The sequence of nucleotides in a polynucleotide chain ultimately specifies the sequence of amino acids in proteins, i.e. forms the basis of the genetic code. The phosphate groups are on the outside of the helix. Each polynucleotide chain is established and held together by the formation of further ester bonds between the phosphate group of one nucleotide and the deoxyribose sugar of an adjacent nucleotide in the same chain. For this reason, each polynucleotide is sometimes described as having a 'sugar–phosphate' backbone (*see* diagram). The two polynucleotide chains in a double helix are antiparallel, i.e. in one chain the sugar–phosphate backbone is composed of 3′–5′ phosphodiester bonds, while in the other, it is composed of 5′–3′ phosphodiester bonds. The bases are on the inside of the helix, and each forms hydrogen bonds with a complementary base on the other helix. By the formation of these hydrogen bonds, the two helices are held together. The base pairing is highly specific, adenine:thymine and guanine:cytosine being the only permissible combinations.

Evidence that DNA is the genetic material is substantial. By means of *semi-conservative replication, DNA produces copies of itself with great accuracy, so that the genes (polynucleotide sequences) are normally passed on unaltered from one generation to the next.

DNA may be contrasted with *RNA, the other main type of nucleic acid found in cells. The latter plays an important role in transcribing (copying) and translating (decoding) the genetic message during protein synthesis. In viruses the genetic material is much more variable and can be double-stranded DNA (e.g. T-even bacteriophage), single-stranded DNA (e.g. Ø174 bacteriophage), double-stranded RNA (e.g. Ø6 bacteriophage), or single-stranded RNA (e.g. tobacco mosaic virus).

DNA hybridization A laboratory technique for measuring the similarity of the DNA of two species. DNA from each species is denatured by heat into single strands and a mixture of the DNA strands is incubated allowing them to recombine into hybrid DNA, a process called annealing. The amount of annealing is directly proportional to the similarity of the DNA strands and can be measured by heating the hybrid DNA and recording the temperature at which the strands separate. DNA hybridization studies in the grasses showed wheat and rye to be closely related and both to be close to barley but not to oats. This supports the traditional classification of wheat, rye, and barley in the tribe Triticeae and oats in the tribe Aveneae.

dolipore septum A type of septum found in many basidiomycete fungi, in which there is a narrow pore surrounded by a thickened rim and protected on both sides by caps.

dominant 1. Describing an allele that masks the expression of a different (*recessive) allele at the same locus. Thus the phenotype of the heterozygote resembles that of a plant homozygous for the two dominant alleles. The dominant allele is usually the normal and more common form in natural populations, i.e. the wild-type allele. Not all characteristics are governed by simple dominant/recessive relationships. *See also* codominance, incomplete dominance.

2. Describing the most abundant plant species in a community, e.g. in the UK *Quercus robur* (pedunculate oak) is dominant in most deciduous woodlands on nonacid clays and loams.

3. In forestry, a tree whose crown is more than half exposed to full illumination.

Domin scale *See* Braun-Blanquet scale.

dormancy An inactive phase often exhibited by seeds, spores, and buds, during which growth and developmental processes are deferred. It may be a means of surviving adverse environmental conditions, as in the formation of dormant perennating buds and the dormant resting spores of certain bacteria, fungi, and algae. However seeds often exhibit dormancy despite prevailing favourable conditions, in which case it may serve to give the seed time to mature fully. Seed dormancy may be broken in various ways, including the gradual degradation of an impervious *testa or the progressive destruction of growth inhibiting substances such as coumarin, sometimes found in the seed coat. Seed or bud dormancy may also be broken by a specific light treatment. *See also* after-ripening.

dormin *See* abscisic acid.

dorsal 1. In thallose plants, the upper surface away from the substrate. *Compare* ventral.

2. In lateral organs, *abaxial.

dorsifixed Describing an anther that is joined to the filament for some distance along its dorsal edge. *Compare* basifixed, versatile.

double fertilization The process found in most flowering plants where two male gametes enter the embryo sac and both participate in fertilization. One male gamete fuses with the female gamete or egg nucleus to form a *zygote, which develops into the embryo. The other male gamete fuses with either the *polar nuclei or the *definitive nucleus to form a triploid primary endosperm nucleus, which will give rise to the endosperm.

A form of double fertilization is also seen in some species of the gymnosperm genus *Ephedra*, in which one male gamete fuses with the egg cell and the other fuses with the ventral canal cell. However the product of the second fertilization does not undergo further development.

double flower A flower having more than the usual number of petals usually due to the transformation by a mutation of stamens into petals. Extremely double flowers, in which the carpels have also become petals, are termed *flore pleno*. An example is *Ranunculus aconitifolius* 'flore pleno' (fair maids of France). The transformation of reproductive organs into petals is called *petalody*.

double helix A molecule composed of two similar polymeric chains coiled in the same direction about the same axis, as in DNA. Actin, a fibrous protein found in most eukaryotic cells also takes the form of a double helix. Double helices are, however, relatively rare configurations. Much more common is the single helix (α-helix) characteristic of globular protein.

double recessive A diploid individual homozygous for (containing two copies of) the same recessive allele of a gene, as indicated by the expression of the recessive allele in the phenotype.

double staining *See* staining.

downy mildew A plant disease caused by fungi of the family Peronosporaceae in the order *Peronosporales. The pathogen penetrates the tissues more deeply than *powdery mildews and produces a white or grey mealy growth on the leaves and stems of the host in humid conditions. This surface growth is composed of sporangiophores, which grow out from the stomata. Examples include downy mildew of vine (*Plasmopara viticola*), downy mildew of onion (*Peronospora destructor*), and downy mildew of lettuce (*Bremia lactucae*).

DPD *See* diffusion pressure deficit.

DPN *See* NAD.

drupe A fleshy indehiscent fruit in which the seed or seeds are surrounded by a hardened schlerenchymatous endocarp, as in wild cherry (*Prunus avium*) and holly (*Ilex aquifolium*). The endocarp may replace the testa in its protective role and may also play a part in the dormancy mechanism. *See illustration at* fruit.

druse (sphaeroraphide) A globular mass of needle-like crystals found either attached to the cell wall or free in the cytoplasm.

dry rot A plant disease in which there is disintegration of the tissues and the affected cells crumble into a powdery mass. Dry rot of stored potatoes is caused by the fungus *Fusarium solani* var. *caeruleum* and that of timber by *Serpula lacrimans*.

duplication 1. The occurrence of two or more copies of the same gene on a chromosome.
2. The occurrence of two or more copies of the same segment of chromosome sequentially on a chromosome.

duramen *See* heartwood.

dwarfism Stunted growth, often due to a fault in genetic composition. The condition often involves the alteration of the proportions of the various body parts. In plants, dwarfism is frequently caused by a *gibberellin deficiency, as in the case of the pea. The application of exogenous gibberellin usually causes internode elongation and the production of a plant of normal size.

dynein *See* flagellum.

E

earth stars *See* Lycoperdales.

ecesis The establishment of the first stage or *sere in a *succession.

ecocline A *cline that has been shown to be due to a specific environmental factor. For example, a gradual increase in heavy-metal tolerance can be correlated with increasing metal concentration in the soil.

ecological niche (niche) The place and role occupied by an organism in a community, determined by its nutritional requirements, habit, etc. Different species may occupy a similar niche in different areas, for example, the grass species of the Australian grasslands, though different from those of the North American grasslands, occupy the same niche. Also, the same species may occupy a different niche in different areas. Generally the breadth of the niche varies depending on the adaptability of the species, the more adaptable species occupying a wider niche than the less adaptable or more specialized species.

ecological system *See* ecosystem.

ecology The study of the relationships between living organisms and the living (biotic) and nonliving (abiotic) factors in the environment.

ecospecies *See* ecotype.

ecosphere *See* biosphere.

ecosystem (ecological system) A unit comprising a *community of living organisms and their environment. There is a continuous flow of energy and matter through the system. Ecosystems may be small or large, and simple or complex in structure. They range from small freshwater ponds or pools to the earth itself.

ecotone A transition zone or region separating two *biomes. For example, in central Asia an area of wooded steppe, the grove belt, separates the coniferous forest from the temperate grassland.

ecotype A distinct population of organisms within a species that has adapted genetically to its local habitat. For example, some organisms may be able to tolerate different conditions of temperature or light intensity from other members of the same species. This may result in changes in their morphology or physiology. However, they are able to reproduce with other ecotypes of the same species and produce fertile offspring. These ecotypes may be sufficiently distinct to be given subspecific names, in which case they may be termed *ecospecies*. *Compare* cline.

ectocarp *See* exocarp.

ectophloic siphonostele *See* siphonostele.

ectotrophic mycorrhiza (ectophytic mycorrhiza) A form of mycorrhizal association in which there is a well developed mycelium forming a mantle on the outside of the root. Such associations are found on most trees and often a tree will not grow properly unless the appropriate fungus is present. It is thus a common practice when planting young trees to inoculate the soil with mycorrhizal fungi. These are frequently members of the *Agaricales. Infected roots commonly show a characteristic *coralloid* form of growth in which the lateral roots fail to elongate and instead branch repeatedly forming a swollen mass. Initially some of the hyphae penetrate the cortex and form an intercellular meshwork termed the *Hartig net*. The outer mantle, which replaces the piliferous layer of the root, subsequently arises from this net. It has been demonstrated that mycorrhizal roots take up nutrients better than uninfected roots, while the fungi obtain carbohydrates and possibly B-group vitamins from the tree. *See also* endotrophic mycorrhiza, vesicular-arbuscular mycorrhiza.

edaphic factors (soil factors) The physical, chemical, and biological properties of *soil that influence the life of organisms. The main edaphic factors include water content, organic content, texture, and pH. *See also* biotic factors, climatic factors.

EDTA (ethylenediamine tetra-acetic acid) A compound that reversibly binds iron, magnesium, and other positive ions, i.e. that acts as a chelating agent. Used in culture media complexed with iron it slowly releases iron into the medium as required. It may also act as a noncompetitive inhibitor of certain enzymes that require metal ions as cofactors. *See also* sequestrol.

egg A large nonmotile female gamete, such as an *oosphere or a *megaspore.

ektexine *See* exine.

elaioplast A *leucoplast in which oil is stored.

elater Any of the numerous elongate helically thickened cells intermixed with the spores in the capsule of most liverwort sporophytes. As the mature capsule dries out, tension builds up in the thickening of the elater. This tension is released when the capsule dehisces, dispersing the spores over a distance. *See also* hapteron.

electron density The contrast between parts of a specimen observed by transmission electron microscopy. Electron dense structures deflect most of the electron beam and appear dark whereas electron lucent features do not deflect many electrons and appear bright. Such differences are often accentuated by the use of electron stains that render parts of the specimen electron dense that would otherwise appear electron lucent.

electron micrograph *See* micrograph.

electron microscope An instrument that uses electromagnetic lenses to focus a parallel beam of electrons and produce an image by differential electron scattering. The *resolving power of an electron microscope is far greater than that of a *light microscope because the wavelength of an electron beam is 0.005 nm as compared to about 600 nm for yellow light. In theory this should mean that objects only 0.0025 nm apart could be distinguished. However in practice aberrations in the lenses (electromagnetic fields), deterioration of the specimen during observation, and other technical difficulties reduce resolution to about 1 nm. This is still some 300 times better than that using light.
Two principal types of electron microscope are used. In the *transmission electron microscope* (TEM) the electron beam, which is usually produced from a tungsten filament, passes through a condenser lens system and is focused onto the specimen, which is normally an ultrathin section held on a fine copper grid. The image of the specimen, focused by the objective lens and magnified by the projector lens, is then projected onto a fluorescent screen or photographic plate. The entire microscope column is under high vacuum to prevent the beam of electrons from being scattered by the molecules in air. The casing of the microscope must therefore be completely dry and the specimen dehydrated. The magnification is altered and the focusing adjusted by varying the current through the electromagnetic lenses.
In the more recently developed *scanning electron microscope* (SEM), commercially available since the mid 1960s, the surface of the specimen is observed and a three-dimensional appearance is obtained. The beam of incident electrons ejects secondary electrons and back-scattered electrons from the specimen, which is usually treated by *coating. These reflected electrons are collected by a scintillator and an image is built up on a high-resolution cathode-ray tube as the electron beam passes over the specimen in a sequence of scanning movements. The resolution obtained in the scanning electron microscope is lower than in the transmission electron microscope but is constantly being improved upon and instruments capable of resolving 2 nm are now available.
Some electron microscopes are capable of operating in several different modes, either as scanning or transmission electron microscopes or in a scanning transmission electron microscope (STEM) mode. Microscopes combining optical and transmission electron microscopy are also available for making high and low magnification observations of the same specimen.
Since the electron beam produces x-rays from the specimen, attachments are also available that permit the composition of the specimen, or part of it, to be determined by comparison of the radiation produced.
One of the main limitations of the electron microscope apart from its expense is that living material cannot be viewed. The specimen has to be fixed, dehydrated, etc., before observation. Hydrated specimens can, however, be examined by *low-temperature scanning electron microscopy and one direction of current research is aimed at producing a special specimen chamber that could be operated at atmospheric pressure and isolated from the high vacuum

system of the microscope. *See also* coating, staining, freeze drying, freeze fracturing, replica plating, shadowing.

electron stain *See* staining.

electron transport chain A series of membrane-linked oxidation–reduction reactions in which electrons are transferred from an initial electron donor through a series of intermediates to a final electron acceptor (usually oxygen). The electron transport systems of respiration and photosynthesis are the major chemical energy sources in aerobic organisms.

A number of other electron transport systems are known besides those of respiration and photosynthesis. Heterotrophic bacteria have a *respiratory chain similar to that in eukaryotes and a microsomal electron transport system is responsible for desaturation of fatty acids. *See also* photosystems I and II.

electrophoresis A technique used to separate mixtures of solute molecules or colloidal particles in solution by placing them in an electric field. Molecules with a net positive charge (cations) move towards the cathode (negative electrode) and those with a net negative charge (anions) move towards the anode (positive electrode). The speed of movement of the molecules depends on their net charge, which in turn depends on the pH of the medium. The size of the molecule and the strength of the voltage applied also affect the speed of movement. Often the solvent is a gel of, for example, starch, agar, or polyacrylamide. These prevent the passage of small molecules, which become caught within the molecules of the gel (*see* gel filtration), so enabling clearer separation of the larger molecules. Electrophoresis is often used to separate the components of protein mixtures and may be used in conjunction with paper chromatography. *See also* immunoelectrophoresis.

elfin forest (Krummholz) The dwarfed and deformed trees that are found in the zone between the *timber line and the *tree line. They tend to grow along the ground and are thus better able to withstand the strong winds in such regions. The zone of elfin forest is termed the *Kampfzone.*

emarginate Describing a leaf, petal, or sepal that is indented at its tip. The notched petals of many cranesbills (e.g. *Geranium versicolor* and *G. pyrenacium*) are examples.

emasculation The removal of the anthers of a flower to prevent either self-pollination or the pollination of surrounding plants.

Embden–Meyerhof–Parnas pathway *See* glycolysis.

embedding A process in microscopical preparation in which delicate tissue is impregnated by, and embedded in, a solid supporting medium, such as paraffin wax, to enable thin sections of the material to be cut with a microtome. Tougher embedding materials (e.g. epoxy resins) are used in preparations for the electron microscope.

embryo The young plant individual after fertilization or parthenogenesis when the *proembryo has differentiated into embryo and suspensor. Embryo cells are typically thin walled with dense cytoplasm and maintain mitotic activity. Cells exhibit polarity from early on, and in higher plants a plumule, radicle, and cotyledons may be identified. In seed-bearing plants the embryo is protected by integuments, which later form the seed coat. Further embryo development in the seed is usually controlled until environmental conditions are appropriate. *Compare* zygote. *See* embryogeny.

Embryobionta *See* Embryophyta.

embryo culture The growth of isolated plant embryos on suitable media in vitro. The technique is useful in plant breeding as it enables certain hybrids to be raised that would abort if left on the plant, either because of endosperm breakdown or endosperm incompatibility. *See also* ovule culture.

embryogeny (embryogenesis, embryony) The development of an embryo, normally from a fertilized egg cell. The first divisions of the zygote are at right angles to the long axis of the archegonium or embryo sac. In angiosperms and in some gymnosperms and pteridophytes these initial divisions give rise to a chain of cells, the *suspensor. The embryo proper develops from a large cell at the tip of the suspensor at the end furthest from the micropyle. The pattern of embryo development differs markedly between groups of plants but normally globular, heart-shaped, and torpedo-shaped stages can be recognized. Embryogeny often lags behind the development of other parts of the seed so the embryo may grow at the expense of the previously formed endosperm. Certain cells of the embryo may be polyploid if DNA synthesis outstrips nuclear division.

The term embryogenesis is also applied to the development of embryos from diploid somatic cells in suspension cultures. This phenomenon has been described in a number of angiosperm species and demonstrates the *totipotency of cells when removed from the inhibiting influences of the plant body. Such nonzygotic embryos are often termed *embryoids* to distinguish them from zygotic embryos. However they closely resemble zygotic embryos in their development and if transferred to suitable culture conditions may give rise to normal plants. Embryoids can also be induced to form from pollen grains and subsequently may develop into haploid plants. *See also* adventive embryony, endoscopic embryogeny, exoscopic embryogeny.

embryoid *See* embryogenesis.

Embryophyta (Embryobionta) In certain classifications, one of two subkingdoms of plants (the other being the *Thallophyta). It contains the bryophytes and vascular plants. These plants are grouped together because they all have an embryonic stage that is dependent on the parent plant for a greater or lesser period. This is generally not seen in the Thallophyta, though a situation approaching it is found in some red algae.

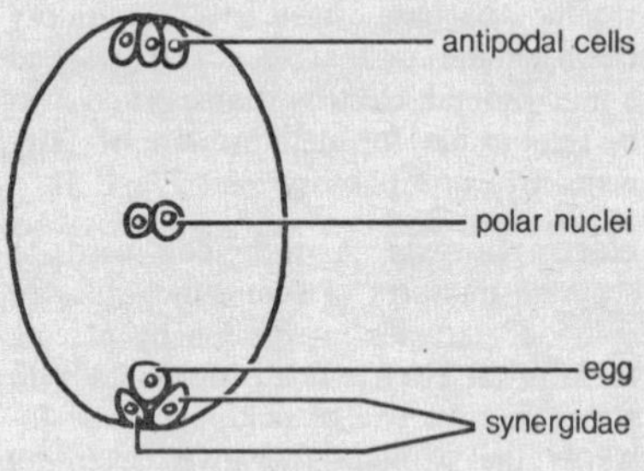

A typical embryo sac.

embryo sac A large oval cell in the nucellus of the ovule of flowering plants within which fertilization occurs. Initially it is the megaspore mother cell, which divides meiotically, normally forming four megaspores. Typically three of these abort and the remaining one divides mitotically to give the haploid cells of the embryo sac. This form of development is termed *monosporic.* Usually there are three mitotic divisions to give an eight-nucleate embryo sac. The nuclei are arranged as shown in the illustration. However in some plants, e.g. *Oenothera*, there are only two mitotic di-

visions giving a four-nucleate embryo sac (lacking the three antipodal cells and one of the polar nuclei).
In *bisporic* development, two megaspores contribute to the formation of the mature embryo sac. For example, in *Allium* the cells of the embryo sac develop from one of the two cells formed after the first division of meiosis (the other product of the reduction division aborting at this stage). Three divisions of this cell give rise firstly to two megaspores (a dyad) and then to an eight-nucleate embryo sac.
In some plants all four megaspores may continue development (*tetrasporic* development). In such cases there may be one mitotic division to give an eight-nucleate embryo sac, e.g. *Adoxa*, or two mitotic divisions, giving a sixteen-nucleate embryo sac, e.g. *Drusa*. The arrangement of cells in embryo sacs derived by tetrasporic development may show a variety of forms.
The mature embryo sac represents the female gametophyte, the egg cell being the gamete. *See also* double fertilization.

Emerson effect (enhancement effect) The observation (made by Robert Emerson in 1957) that photosynthesis, which proceeds very slowly using light of 700 nm wavelength, can be greatly increased when chloroplasts are also illuminated with light of shorter wavelength (650 nm). This was a surprising observation as it was then thought that light absorbed by the chlorophylls and other pigments was all passed on to a small percentage of chlorophyll *a* molecules (the energy trap) absorbing at 700 nm. This and later work indicated a second energy trap absorbing at 680 nm. *See* photosystems I and II.

enation 1. An outgrowth, usually from a leaf. Enations may occur in response to virus infection, e.g. cotton leaf curl virus.
2. A small lateral outgrowth of the stem that is produced by certain primitive vascular plants and may be an early stage in *microphyll evolution.

endarch Describing xylem maturation in which the older cells (protoxylem) are nearer the centre of the axis than the younger cells. Development is thus centrifugal. *Compare* exarch, mesarch.

endemic Describing a plant species that grows in a specific area and has a restricted distribution. Some species (*broad endemics*) are restricted to a particular large region. Other species (*narrow endemics*) are confined to a much smaller area, such as a few square kilometres, and tend to be very specialized. Examples of broad endemics are the sugar maple (*Acer saccharum*) in the eastern United States and the cacao tree (*Theobroma cacao*) in the Amazon basin. An example of a narrow endemic is *Darcycarpus viellardii* in the island of New Caledonia. Some endemic species (palaeoendemics) represent the relicts of once widespread species. For example, *Lyonothamnus floribundus*, which today is only found on the California Islands, grew throughout California in the Tertiary. Species found in two or more widely separated regions are termed *discontinuous* or *disjunct*, e.g. magnolias grow in southeast Asia, eastern North America, and Central America, but nowhere in between. *Compare* cosmopolitan.

endexine *See* exine.

endocarp The innermost layer of the *pericarp of an angiosperm fruit, internal to the mesocarp and exocarp and external to the seed(s). Sometimes the endocarp consists of a stony layer, as in the fruit of the peach (*Prunus persica*).

endocytosis The entry of particles or fluid into cells by methods other than diffusion or active transport across the plasma membrane. In *phagocytosis*, exhibited by some unicellular holozoic algae, protrusions of the outer regions of the cell, formed by flowing movements of the cytoplasm, surround food particles, enclosing them in a membrane-bound food vacuole. Lysosomes become associated with the food vacuoles, the intervening membranes break down, and hydrolytic enzymes are released to digest the particles.

Pinocytosis is a process by which submicroscopic particles (macromolecules or molecular aggregates) or droplets of extracellular fluid enter cells. The particles adhere to the plasma membrane (absorptive endocytosis) and an invagination forms. This becomes pinched off so taking the particles into the cytoplasm in a membrane-bound *vesicle – a *pinocytotic* or *endocytotic vesicle*. Alternatively extracellular fluid enter a pitlike invagination that becomes pinched off and enters the cytoplasm (fluid endocytosis). Water is absorbed from the vesicles as they move through the cytoplasm. Eventually the membrane breaks down, releasing the contents into the cytoplasm.

endodermis A layer of cells at the boundary of the *cortex and stele, usually regarded as the innermost layer of the cortex. It is generally clearly seen in all roots and in pteridophyte and certain dicotyledon stems. A *Casparian strip is usually present. When the endodermis contains numerous starch grains, it is known as a *starch sheath*.

endogamy *See* autogamy.

endogenous Describing any process, substance, or organ that arises from within an organism. Examples are *endogenous rhythms, the origin of lateral roots from the pericycle, and the capacity of certain tissue cultures to manufacture their own growth substances. *Compare* exogenous.

endogenous rhythms Sequential physical or biochemical processes that occur in a plant or plant part in response to internal stimuli. An example is the apparently *autonomic component of the *nyctinastic movements of certain plants, the leaflets of which continue to follow a periodic folding and unfolding sequence even if kept in continuous darkness. *See also* circadian rhythm, biological clock.

endogenous root *See* root.

endomitosis A sequence of changes in the nucleus resulting in division of the chromosomes, as in *mitosis, but no separation of the chromatids into daughter nuclei. The resulting nucleus is therefore *polyploid. The process can be induced in isolated tissues by treatment with colchicine, which prevents spindle formation so the *centromeres of the daughter chromosomes are unable to move apart into separate nuclei. It may occur as an error in part of a plant, producing, for example, a tetraploid branch on a diploid plant. It occurs as a normal feature in some tissues of higher plants, e.g. the phloem cells of some leguminous plants are polyploid. This type of polyploidy, where some of the cells of a plant have more than the normal complement of chromosomes for the species, is known as *endopolyploidy*. If endomitosis occurs in cells in the germ line or during the second division of meiosis then unreduced gametes may result.

Endomycetales An order of the *Hemiascomycetes containing about 200 species in some 50 genera. It includes the family Saccharomycetaceae, the yeasts, which contains the commercially important species *Saccharomyces cerevisiae* (brewer's yeast), used in bread and beer making, and *S. ellipsoideus*, used in wine making. The yeasts reproduce asexually by budding or by dividing into two equal cells as in *Schizosaccharomyces* (fission yeasts). Sexual reproduction is by ascus formation.

endophytic mycorrhiza *See* endotrophic mycorrhiza.

endoplasmic reticulum A complex interconnecting system of flattened membrane-surrounded channels and vesicles (cisternae) that spreads throughout the cytoplasm of eukaryotic cells. The membranes appear to be continuous with the *plasma membrane at the outer surface and with the *tonoplast and *nuclear membrane, and have the same general structure. Some cisternae have a granular appearance in electron micrographs due to ribosomes that are attached to the cytoplasm side of the enclosing membranes. They are thus called *rough endoplasmic reticulum*. Others are devoid of ribosomes and are called *smooth endoplasmic reticulum*. Polypeptides can be identified in the channels of the rough endoplasmic reticulum. The function of the endoplasmic reticulum is the segregation of newly synthesized products and their further processing and intracellular transport. *See also* golgi apparatus.

endopolyploidy *See* endomitosis.

endoscopic embryogeny *Embryogeny in which the embryo develops from the inner of the two cells that result from the first division of the zygote. The outer cell often forms the suspensor. Endoscopic embryogeny is seen in the Lycopsida, most of the Filicinae, and all the Gymnospermae and Angiospermae. *Compare* exoscopic embryogeny.

endosperm The storage tissue in the seeds of most angiosperms, derived from the fusion of one male gamete with two female polar nuclei. The endosperm is a compact triploid tissue, lacking intercellular spaces and storing starch, hemicelluloses, proteins, oils, and fats. Seeds containing an endosperm or *perisperm at maturity are termed *albuminous*, whereas those lacking such tissue are termed *exalbuminous. *See also* aleurone layer.

endospore **1.** An extremely resistant resting spore formed by certain bacteria, especially those of the Bacillaceae, e.g. *Bacillus* and *Clostridium*. Only one endospore is formed in any given cell. Sporulating cells are termed sporangia and the size, shape, and position of the endospore is characteristic of the species. Endospores are resistant to heat, desiccation, radiation, and chemicals, including organic solvents. They retain the capacity to germinate for a considerable number of years.
2. A naked cell, many of which are formed by division of the protoplast in certain, usually unicellular, blue-green algae. On release they form a cell wall and develop into new individuals.

endosporic gametophyte A gametophyte that develops within a spore. For example, the female gametophyte of *Selaginella* is contained within the megaspore and at maturity only a portion, bearing the archegonia, is exposed through the spore wall. Such gametophytes are better able to withstand dry conditions and this pattern of development may be seen as a step in the evolution of the seed habit.

endothecium **1.** (fibrous layer) The subepidermal layer of the *anther wall in angiosperms. The cells of the endothecium often develop thickenings on the anticlinal and inner tangential walls as the anther matures, which are believed to aid in anther dehiscence.
2. The inner layer of cells in the young sporophyte of bryophytes, giving rise to the columella and (except in *Sphagnum*) sporogenous tissue. *Compare* amphithecium.

endotoxin *See* toxin.

endotrophic mycorrhiza (endophytic mycorrhiza) A form of mycorrhizal association in which the fungus lives between and within the cells of the cortex and growth on the outside of the root is limited. Such associations are found on many herbaceous species, especially orchids and heathers, and on certain woody plants, such as rhododendrons. The fungi involved are usually species of *Rhizoctonia* on orchids or *Phoma* on heathers. The hyphae form tightly coiled masses, termed *pelotons*, within the cells of the outer cortex. Plants that commonly develop such associations cannot grow normally without the appropriate fungus. *See also* ectotrophic mycorrhiza, vesicular-arbuscular mycorrhiza.

enhancement effect *See* Emerson effect.

entire Describing a leaf, petal, or sepal margin that has a smooth undivided outline.

entomophily (insect pollination) *Pollination by pollen carried on insects. Insect-pollinated plants need to be able to attract the pollinating agent and provide a suitable landing space for it and then deposit pollen onto and collect pollen off the visitor. Insects may be attracted to a flower through the provision of food, by pseudomating signals (for example, some orchids mimic the shape, colour, and odour of the female insect), or by provision of brood or shelter sites. Pollen itself may be provided as the food source as it is protein rich, but it needs to be produced in quantities sufficient to offset the loss. More often nectar is offered as a high-energy food. This is usually secreted from a nectary so placed relative to the reproductive parts that pollen collection and deposition is ensured. Fats, oils, and water are also used to attract pollinating insects.

Adequate landing sites may be provided by increase in individual petal size, by increase in size of all the petals, or by the clustering of flowers into a compact inflorescence, such as an umbel. Recognition of the plant by the pollinating insect is achieved through secondary attractants, such as brightly coloured petals or insect-like movements.

Special arrangements of the reproductive parts can be seen in many insect-pollinated species. Flowers that can be pollinated by a number of different insects are termed *allophilic* or promiscuous, while those that can only be pollinated by one specific agent are termed *euphilic* (*see* mutualism). Some flowers may deposit pollen all over the agent but more specialized flowers deposit pollen only on certain areas of the vector. The pollen grains tend to be heavily sculptured and sticky in order to adhere to the agent's body.

Compare anemophily, hydrophily.

Entomophthorales *See* Zygomycetes.

enumeration data Data in which the material under observation falls into discrete classes.

environment The conditions in which an organism lives. The external environment includes the nonliving physical and chemical factors such as light, nutrients, etc., together with the effects of other living organisms. Genetically similar plants growing in different environments may appear different (*see* plasticity). Such intraspecific variation is termed environmental variation and is not inherited. An organism also has an internal environment, which is the result of its own metabolism.

enzyme A protein molecule specialized to catalyse biological reactions. The extremely high specificity and activity of enzymes enables the living cell to function at physiological temperatures and pH values. Without enzymes, nearly all metabolic processes would require high temperatures and extreme pH values, or would produce excessive amounts of heat.

The catalytic activity of an enzyme is due to its possession of an *active site. The three-dimensional conformation and the charge distribution of the active site are critical; in some enzymes these are entirely maintained by the tertiary structure of the protein, but in other enzymes *cofactors or *coenzymes are required for catalytic activity.

Enzymes are classified according to their function by an internationally recognized system. There are six classes of enzyme, the *oxidoreductases, the *transferases, the *hydrolases, the *lyases, the *isomerases, and the *ligases. Each class has a code number, and each is subdivided into subclasses and subsubclasses. Thus any enzyme has a common or recommended name, a systematic name indicating the reaction that it catalyses, and a four-digit code number. *See also* substrate.

enzyme technology The use of isolated purified enzymes as catalysts in industrial processes. The enzymes used are normally extracellular enzymes with no requirements for complex cofactors. Examples are proteases, amylases, cellulases, and lipases extracted from bacterial or yeast cultures. Cell-free enzymes have certain distinct advantages over the use of microorganisms, a major one being that only the desired enzymatic reaction will occur and substrate is not wasted in the formation of unwanted by-products or microbial biomass.

Eocene *See* Tertiary.

eosin An acid stain that colours cytoplasm pink and cellulose red. It is often used with a counterstain, such as methylene blue or haematoxylin. Wright's stain, Leishman's stain, and Giemsa stains are all mixtures of eosin and methylene blue mixed with alcohol.

Ephedrales *See* Gnetales.

ephemeral A plant with a short life cycle that may be completed many times in one growing season. Examples are groundsel (*Senecio vulgaris*) and shepherd's purse (*Capsella bursa-pastoris*). Some desert plants also have short life cycles that can be completed in a short period following rain. *Compare* annual, biennial, perennial.

epibasal cell The outer of the two cells formed by the first division of the zygote. It gives rise to the embryo in plants showing exoscopic embryogeny and to the foot in those showing endoscopic embryogeny. *Compare* hypobasal cell.

epiblem *See* epidermis.

epicalyx A calyx-like extra whorl of floral appendages, positioned below the calyx. The individual segments resemble sepals and are termed *episepals*. An epicalyx is found in the strawberry (*Fragaria vesca*) flower and in the tree mallow (*Lavatera arborea*). *Compare* involucre.

epicarp *See* exocarp.

epicotyl The apical end of the axis of an embryo above the cotyledon(s), which gives rise to the stem and associated organs. *Compare* hypocotyl, plumule.

epidemiology The study of factors causing and influencing widespread outbreaks of disease.

epidermis The outermost cells of the primary plant body, usually consisting of a single layer but sometimes several layers thick (*multiple* or *multiseriate epidermis*). In either case the cells differentiate from the *protoderm. In stems and roots exhibiting secondary growth, the epidermis is usually replaced by the *periderm. The term *epiblem* (*rhizodermis*) is sometimes used in place of epidermis for the outermost layer of cells in the root. The main function of the epidermis is to protect the underlying tissues from excessive water loss and, to some extent, from physical injury and attack from pathogens.

epigeal Describing seed germination in which there is considerable elongation of the hypocotyl so the cotyledons are raised above the surface of the ground to form the first leaves (seed leaves) of the plant. Epigeal germination is seen in the sycamore (*Acer pseudoplatanus*). In some species, e.g. onion, the cotyledon is raised above ground level by the rapid growth of the cotyledon itself, rather than by extension of the hypocotyl. *Compare* hypogeal.

epigenesis The theory that an organism develops from a fertilized egg by a gradual series of interdependent physiological and physical changes, brought about by the genes, that result in an increase in the complexity of its separate parts. It refutes the earlier *preformation* theory, which stated that either the male or female gamete contained a miniature version of the adult and that only an increase in size followed thereafter.

epigyny An arrangement of floral parts in which the stamens, sepals, and petals are inserted above the ovary, giving an *inferior ovary*. Epigynous flowers are seen in many members of the Rosaceae, e.g. quinces (*Chaenomeles*). The receptacle is urn shaped and completely encloses the ovary, the other floral parts arising from the top of the receptacle. A type of pseudocarp, the *pome, commonly forms from such flowers. *Compare* hypogyny, perigyny.

epimatium A specialized type of *ovuliferous scale that bears and completely encloses a single inverted ovule. It is found in the monkey-puzzle tree (*Araucaria araucana*) and in certain of the Podocarpaceae.

epimerase An enzyme that catalyses the transfer of a hydroxyl group from one position to another within a molecule. Epimerases are a form of *isomerase and are important biologically in the interconversion of sugars.

epinasty A *nastic movement in which the resultant bending of the plant part is downwards, due to increased growth on the upper side of an organ. This occurs in the opening of many flowers when the bracts and perianth parts curl downwards to expose the sexual organs. Epinasty can also occur in leaves where the petiole bends so that the leaf points to the ground rather than upwards. *Compare* hyponasty.

epipetalous Describing stamens that arise from the petals, as occur in many flowers with tubular corollas.

epiphyte (air plant) A plant that has no roots in the soil and lives above the ground surface, supported by another plant or object. It obtains its nutrients from the air, rain water, and from organic debris on its support. Many orchids are epiphytes and numerous species are found growing in the canopies of tropical rain forests. Their aerial roots form a tangled network that catches falling leaves and other organic material, which provide a source of mineral salts for the plant. In addition, the mesh of roots and organic debris acts as a sponge to collect and hold water. Mosses and lichens, growing on the bark of trees, are examples of epiphytes in temperate regions and bromeliads are examples of xerophytic epiphytes. Epiphytes growing on the leaves of another plant are termed *epiphyllous* while those growing on rock outcrops are called *lithophytes. Epiphytes are also a group of plants in the Raunkiaer system of classification.

epiphytotic A widespread outbreak of disease among plants.

episepal *See* epicalyx.

episome *See* plasmid.

epistasis A form of gene interaction in which one gene affects the expression of a second gene. It usually arises because the two genes affect sequential steps in the same biochemical pathway. It is most easily detected as a modification of the Mendelian ratio in the F_2. The phenomenon was first reported in the inheritance of flower colour. For example, suppose there are two enzymes, A′ and B′, produced by the normal alleles of two unlinked genes, A and B, but not by their recessive alleles, a and b. The substrate of A′ is a white pigment and the product of B′ is a purple pigment. In addition the product of A′, a red pigment, is also the substrate of B′. Thus the sequence of reactions white pigment to red pigment to purple pigment can only take place if there is at least one normal allele at both gene loci. The genotype aa_ _ would give a white phenotype (no enzyme A′; gene B is immaterial), genotype A_bb would give a red phenotype (enzyme A′ but not B′), and genotype A_B_ would give a purple phenotype (both A′ and B′). Instead of the F_2 forming a 9:3:3:1 ratio, a 9:3:4 ratio (= 9:3:[3 + 1]) would be obtained, since aaB_ would be indistinguishable from aabb. In this case, gene A is the *epistatic gene*, and gene B the *hypostatic gene*. Other variants are 9:7, 13:3, and 12:3:1 ratios. *See also* complementary genes.

epithelium The lining of either a resin canal in gymnosperms or, more rarely, of a gum duct in dicotyledons. When present the epithelium usually consists of the cells pulled apart during the often *schizogenous formation of the canal. These epithelial cells often have a secretory function.

Equisetales *See* Sphenopsida.

ergastic matter Nonprotoplasmic substances that are produced as by-products of protoplasmic activity. It includes crystals, starch grains, tannins, and oil droplets. Many ergastic substances are regarded as waste matter, some are storage products, while the functions of other materials are unknown.

Ericaceae A large family of dicotyledonous usually shrubby plants, commonly called the heath or heather family, numbering about 3000 species in some 100 genera. It is virtually cosmopolitan in distribution though members are poorly represented in Australia. Ericaceous plants have simple often evergreen leaves that lack stipules and may be reduced to needles, e.g. bell heather (*Erica cinerea*), or rolled, e.g. cowberry (*Vaccinium vitis-idaea*). The flowers are usually actinomorphic and bisexual and may be borne singly or in racemes. The petals, normally four or five, are usually fused at the base. Over half the described species are in the two genera *Rhododendron* (about 1200 species) and *Erica* (about 500 species). Both show unusual distributions, species of *Erica* being concentrated in South Africa (the Cape heaths number some 450 species) while *Rhododendron* species are concentrated in the Himalayan foothills and New Guinea. Over half the *Rhododendron* species are cultivated as ornamentals (the azaleas are included in this genus). Many heaths (*Erica*) and heathers (*Calluna*) are also grown as ornamentals. The berries of certain species, notably bilberry (*Vaccinium myrtilus*) and cranberry (*Vaccinium oxycoccus*), are valued as food in many areas.

Erysiphales An order of the *Plectomycetes containing about 150 species (20 genera) of obligate parasites that cause *powdery mildew diseases of plants. The diseases are so named because the superficial hyphae produce masses of powdery conidiospores.

erythrose A four-carbon aldose sugar (*see illustration at* aldose). In the phosphorylated form, erythrose 4-phosphate, it is an intermediate in the Calvin cycle. In the shikimic acid pathway, the first reaction step is the joining of erythrose 4-phosphate with phosphoenolpyruvate to form a seven-carbon keto sugar acid.

essential element *See* macronutrient.

essential fatty acid *See* fatty acid.

essential oil (ethereal oil) Any of the volatile oils secreted by aromatic plants that give them their characteristic taste or odour. Most essential oils are *terpenoids, while some are derivatives of benzene. Many xerophytic plants produce essential oils as a means of reducing their transpiration rates. In hot conditions the evaporating oils contribute to the density of the *boundary layer at the leaf surface. The higher the density of the boundary layer, the greater resistance it offers to diffusion of water vapour. Some essential oils repel insects while others, e.g. those of *Myoporum deserti*, are toxic to grazing animals. Many allelopathic substances are essential oils.

etaerio *See* aggregate fruit.

ethanal *See* acetaldehyde.

ethanoic acid *See* acetic acid.

ethene *See* ethylene.

ethereal oil *See* essential oil.

ethylene (ethene) A gaseous hydrocarbon, formula C_2H_4, that is produced in small quantities by many plants and that can be considered to act as a plant growth substance. The production of ethylene is frequently stimulated by *auxins and the ethylene so produced acts, by a feedback mechanism, to inhibit auxin synthesis. Ethylene production also increases after wounding and exposure to disease. Ethylene may enhance an auxin response but it normally inhibits longitudinal growth and causes radial expansion of tissues. It can promote flowering in certain species, e.g. pineapple (*Ananas comosus*), and speeds up the ripening of fruit, an effect that has been put to commercial use in the citrus industry. Other effects mediated by ethylene include: the induction of epinasty; the induction of root hairs; the stimulation of seed germination in certain species; the promotion of leaf abscission; and the inhibition of auxin transport.
The route of ethylene synthesis in plants is uncertain though the amino acid methionine has been suggested as a precursor.

ethylene chlorohydrin (ethylene chlorhydrin) A chlorinated derivative of ethylene glycol, formula CH_2ClCH_2OH, used to break bud dormancy.

ethylenediamine tetra-acetic acid *See* EDTA.

etiolation The condition, seen when plants are growing with an inadequate light supply, in which there is abnormal chlorophyll development. It causes the plant to look pale and there is excessive elongation of the internodes and poor development of the leaves. It is an adaptation to hasten the plant's location of a light source. When plants are growing from underground stems, etiolation is manifested by extreme elongation of the petioles. This phenomenon is exploited in the growth of certain vegetables, such as rhubarb and celery, which may be kept in darkness to either blanch (whiten) the leaf stalks or to force them on early.

etiology The study of the causal agents of a disease.

etioplast A partially differentiated *plastid, 1.5–2.0 μm in diameter, found in the cells of plants that have been kept in darkness (etiolated plants). A three-dimensional lattice, the *prolamellar body, occupies almost the whole of the central region. Etioplasts contain protochlorophyll, which is converted to chlorophyll on exposure to light. *Compare* proplastid.

Eubacteriales An order containing easily stained rod-shaped or spherical bacteria with, in the motile forms, numerous flagella (*see* peritrichous). The rod-shaped organisms are assigned to the following families: Azotobacteraceae (which contains free-living nitrogen-fixing soil bacteria, e.g. *Azotobacter*), Achromobacteraceae, Rhizobiaceae (which contains symbiotic nitrogen fixers, e.g. *Rhizobium*, and some gall-forming bacteria, e.g. *Agrobacterium*), Enterobacteriaceae (which contains some plant parasites, e.g. *Erwinia*), Brucellaceae, and Bacteroidaceae. All these families are gram-negative. Gram-positive rods are assigned to the families Brevibacteriaceae, Lactobacillaceae (*see* lactic acid bacteria), Propionibacteriaceae, Corynebacteriaceae (containing *Corynebacterium*, which causes certain plant rots), and Bacillaceae. The spherical organisms are divided into the gram-negative Neisseriaceae and the gram-positive Micrococcaceae.

eucamptodromous *See* camptodromous.

eucaryotic *See* eukaryotic.

euchromatin An expanded region of *chromatin in the interphase nucleus, which stains lightly with basic dyes. Metabolically active nuclei show large amounts of euchromatin. Autoradiographic investigations support the view that messenger RNA is transcribed in the euchromatic regions, but the amount of euchromatin is related to the number of genes being transcribed rather than the amount of mRNA produced. *Compare* heterochromatin.

Euglenophyta A division of very variable, predominantly unicellular, flagellate *algae. It contains both pigmented (autotrophic or facultatively heterotrophic) and unpigmented (obligately heterotrophic) genera. When pigments are present they are similar to those found in the Chlorophyta. A cell wall is absent, enabling certain of the unpigmented forms to ingest food by phagocytosis. Such forms are probably better classified as Protozoa even though they most likely evolved from photosynthetic forms. Food is stored as paramylum and fat.

eukaryotic (eucaryotic) Describing cells that have a nucleus, or organisms made up of such cells (*compare* prokaryotic). The genetic material of eukaryotes consists of *chromatin and is divided into a number of chromosomes, which are located in the nucleus. Eukaryotic cells usually divide either by mitosis or meiosis while division in prokaryotes is amitotic. In certain classifications all eukaryotic organisms are placed in the kingdom Eukaryota, to emphasize the difference between these and the prokaryotes, or Prokaryota.

Eumycota (Mycobionta) The division containing the true fungi, which differ from the *Myxomycota in having a mycelial (or occasionally unicellular) thallus. It contains the subdivisions *Mastigomycotina, *Zygomycotina, *Ascomycotina, *Basidiomycotina, and *Deuteromycotina.

Euphorbiaceae A large family of dicotyledonous plants, mainly tropical in distribution, containing over 5000 species in some 300 genera. It is commonly known as the spurge family. Most members have simple alternately arranged leaves and unisexual flowers. In the tribe Euphorbieae the small flowers are grouped into an inflorescence, which, due to the presence of various petal-like structures (e.g. highly coloured bracts and glandular appendages), resembles a single flower. The 'flowers' of poinsettia (*Euphorbia pulcherrima*) are examples. The specialized inflorescence of *Euphorbia* species is termed a *cyathium. The fruit is often a regma and in many genera the seeds are carunculate.

The family contains many commercially important species including *Hevea brasiliensis*, the main source of natural rubber, *Manihot esculenta* (cassava), and *Ricinus communis* (castor oil).

euploidy The condition in which the number of chromosomes in a nucleus is an exact multiple of the haploid number. This is the most common and stable condition. *Compare* aneuploidy.

Eurotiales (Plectascales, Aspergillales) An order of the *Plectomycetes containing about 150 species (80 genera) of various moulds (e.g. *Eurotium*) that show a perfect stage. Their imperfect stages resemble those of certain members of the *Hyphomycetales. Asci are produced by *crozier formation.

eusporangium A type of *sporangium that is derived from several initials. There are several layers of cells in its wall and it usually produces a large number of spores. The eusporangium is typical of ferns in the Marattiales and Ophioglossales. *Compare* leptosporangium.

eustele A type of dissected *siphonostele, typical of dicotyledons, in which the vascular cylinder appears (in transverse section) as a ring of *collateral or *bicollateral vascular bundles separated by medullary rays. *Compare* atactostele. *See also* dictyostele.

eutrophic Rich in minerals and bases. The term is usually used of nutrient-rich lakes and ponds but it may be applied in other contexts, e.g. to fen peats that contain relatively high proportions of minerals and bases. Eutrophic lakes and ponds support a prolific growth of aquatic plants. When these plants die they accumulate as a thick layer on the lake bottom. The organisms decomposing this material diminish the dissolved oxygen in the water so fish are scarce or absent. Eutrophication may unwittingly be accelerated by man when, for example, sewage effluents enter a lake, or water rich in dissolved fertilizers seeps in from surrounding fields. In such instances the amount of oxygen used to decompose the plant remains may be so great that the fish die from oxygen deprivation. *Compare* oligotrophic.

evaporation The conversion of liquids to the vapour state by the expenditure of heat energy. Losses of water due to evaporation at plant surfaces can be serious and many species use various means, such as waxy or thick cuticles, to minimize the problem. However evaporation does play an important role in keeping the leaf cool. Evaporation of water from the stomata gives rise to the *transpiration stream.

evergreen A woody perennial plant that retains its foliage throughout the year by continuously shedding and replacing a few leaves at a time. Many evergreens are tropical or equatorial plants that are not subject to periods of winter cold or dry seasons. Evergreens growing in temperate and cold latitudes show various adaptations to overcome water loss by transpiration. For example, many coniferous trees have needle-like or scalelike leaves to reduce surface area. *Compare* deciduous.

evolution The process by which genetic changes have taken place in populations of animals and plants over millions of years in response to environmental changes (*see* natural selection). Evolution has resulted in the formation of new species and, usually, an increase in complexity. Evidence for evolution comes from palaeontology, biogeography, genetics, and comparative anatomy and physiology. *Compare* creationism. *See* Darwinism.

exalbuminous seed A seed that lacks an *endosperm at maturity. In such seeds the cotyledons absorb the food reserves from the endosperm and act as storage organs.

exarch Describing *xylem maturation in which the older cells (protoxylem) are further from the centre of the axis than the younger cells. Development is thus centripetal. *Compare* endarch, mesarch.

exine The outer part of the wall of a pollen grain external to the *intine. Several alternative systems exist for recognizing different layers in the exine (*see* diagram). In some morphological systems it is divided into an outer *sexine* and an inner *nexine*. Other systems based on chemical differences recognize an outer *ektexine* and an inner *endexine*.

Various projections of the exine surface may confer a characteristic sculpturing to the pollen grain. This sculpturing, together with information on pollen size and shape and the positioning of the germinal apertures, can be so distinctive that individual genera and even species can be recognized.

The exine is composed of a carotenoid polymer called sporopollenin, which is highly resistant to decay. This has enabled palynologists to study fossil pollen and deduce the composition of earlier floras. Such studies have also thrown light on the evolution of plants.

exocarp (epicarp, ectocarp) The outermost layer of the *pericarp of an angiosperm fruit, external to the mesocarp. It is usually only a thin layer, as in the plum, where it forms the outer skin.

exodermis A layer of cells, often lying immediately beneath the epidermis or velamen of the root, that is essentially a specialized *hypodermis. Exodermal cells often resemble those of the *endodermis and usually possess a suberized lamella on the inside of the primary cell wall.

exogamy *See* allogamy.

exogenous Describing any process or substance arising from outside the organism, or organs arising in the peripheral layers of the body. Examples are the external supplies of nutrients and growth substances needed to maintain growth in cultures, and the origin of branches from the stem. *Compare* endogenous.

exon A polynucleotide sequence in a structural gene that codes for a protein, i.e. a section between two introns. This terminology is not applicable to bacteria, which do not have introns.

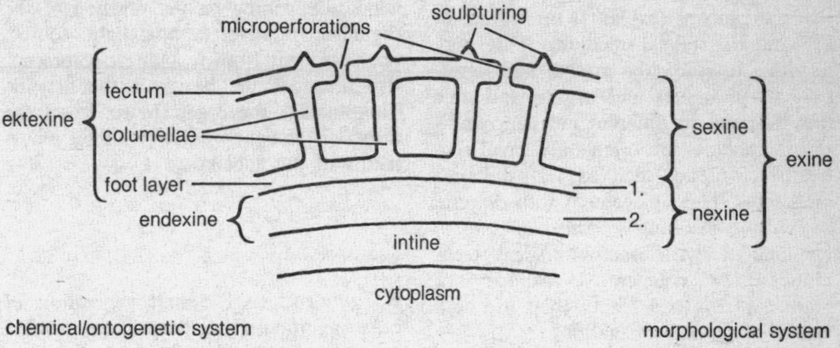

Section through a pollen grain wall showing details of the exine.

exoscopic embryogeny *Embryogeny in which, of the two cells resulting from the first division of the egg cell, the outer gives rise to the embryo and the inner to the foot. It is seen in the Bryophyta, Psilopsida, and Sphenopsida. *Compare* endoscopic embryogeny.

exoskeleton A hard supportive external covering, such as the deposition of calcium carbonate on certain algae in the Charophyta and Rhodophyta. Such encrustations are involved in the formation of coral reefs.

exotic (alien) Describing an organism that has originated from another region and is not native to the area in question. The term is used especially of tropical plants grown in greenhouses or indoors in temperate regions. An exotic plant that has become adapted to its new environment and can grow successfully without help from man is termed *naturalized.* An example in the UK is the monkey flower (*Mimulus guttatus*), which originated from North America. Many naturalized plants are garden escapes.

exotoxin *See* toxin.

explantation The removal of parts of living organisms (explants) for culture in a suitable artificial medium. This may be done to study the growth and development of tissues and organs and how they respond to different nutrient conditions. Explants are often only small segments of tissue, such as leaf discs or meristems. Their placement into or onto the culture medium is termed *inoculation* by analogy with microbiological techniques. The regenerative capacity of plant explants had led to their use as a means of plant propagation.

exponential growth A type of population growth in which the rate of increase in the number of members is proportional to the number present. Exponential growth involves an increasing *rate* of growth: the more individuals there are, the faster the population increases. A graph of number against time in such growth has a J-shaped curve. In practice, exponential growth does not continue indefinitely. Other factors (e.g. shortage of food) become important and slow down the growth rate. The resulting curve is S-shaped (sigmoid). *See also* logarithmic phase.

extracellular Taking place or located outside the cell.

extrachromosomal inheritance *See* cytoplasmic inheritance.

extrafloral nectary *See* nectary.

extrorse Describing anthers that release their pollen to the outside of the flower so promoting cross pollination. *Compare* introrse.

eyespot (stigma) A pigmented structure in motile unicellular algae and in flagellated reproductive bodies of some multicellular algae. It consists of one or more concentrically arranged curved plates composed of orange-red carotenoid granules. The eyespot is situated close to the bases of the flagella. It functions as a light absorbing shield, which, depending on the position of the organism relative to the light source, prevents light from reaching photoreceptive areas at the bases of the flagella. This enables the organism to detect the direction of light and respond by movements of the flagella.

F

F_1 generation The first generation of offspring (first *F*ilial generation) from a cross between two individuals homozygous for contrasting alleles. In this case the F_1 must necessarily be heterozygous for the gene under investigation and the phenotypes are consequently similar.

The F_1 generation thus shows far less variation than subsequent generations (*see* F_2 generation) in which segregation becomes apparent. The term F_1 is also commonly, though incorrectly, used to mean the offspring produced from unspecified parents. *See also* F_1 hybrid.

F_2 generation The offspring (the second *F*ilial generation) produced by selfing the *F_1 generation, or allowing them to breed among themselves. Provided that a characteristic is governed by simple Mendelian genes, it is in the F_2 that Mendelian ratios, such as 3:1, 9:3:3:1, become apparent (*see* monohybrid, dihybrid).

F_1 hybrid A crop variety that has been produced by crossing two selected parental pure lines. F_1 hybrids are favoured because they can combine the qualities of the parental lines and because they usually show hybrid vigour. However they do not breed true and seed must consequently be produced each year. In such seed production selfing within the parental lines, which would give nonhybrid seed, must be prevented. This may be achieved by emasculation and hand pollination but this is only commercially feasible for particularly valuable crops. Selfing can be prevented by ensuring that the parental lines both contain (different) reliable incompatibility alleles, i.e. *S* alleles high in the dominance series. This method is employed in producing F_1 hybrid varieties of certain brassica vegetables. Alternatively one parental line may be male sterile; however use of male sterility halves the yield of hybrid seed since seed set is only possible on one parent line. F_1 hybrid varieties of maize may be produced this way.

FAA *See* formalin-acetic-alcohol.

Fabaceae *See* Leguminosae.

factor (Mendelian factor) The inherited component that is responsible for the determination of a characteristic. In modern terminology, chromosomal *gene or cistron may be used instead.

factorial experiment An experiment in which the material is divided into a number of groups, such that each combination of treatments can be tried separately on at least one group and there are enough groups for every combination to be applied. *See also* randomized block, Latin square.

facultative Possessing the ability to utilize certain circumstances or environmental conditions but not being dependent upon them. For example, a facultative parasite can grow either parasitically or saprobically. *Compare* obligate.

FAD (flavin adenine dinucleotide) A riboflavin-derived coenzyme that acts as a prosthetic group to several dehydrogenase enzymes, accepting electrons from the substrate and thus being reduced to $FADH_2$. The most important FAD-linked enzymes are succinate dehydrogenase, which oxidizes succinate to fumarate, and dihydrolipoyl dehydrogenase, a component of the pyruvate dehydrogenase complex, which oxidizes pyruvate to acetyl CoA. *See also* flavoprotein.

Fagaceae A family of hardwood trees containing about 1000 species but only eight genera. However these include *Quercus* (oaks), *Fagus* (beeches), *Castanea* (chestnuts), and *Nothofagus* (southern beeches), which together make up a large proportion of the broadleaved forests of the world. It is probable that, in terms of biomass, this is the most abundant family of flowering plants. Representatives are not, however, found in the equatorial rain forests of Africa and South America.
The flowers are unisexual and usually borne in catkins or spikes. The fruit is a nut, which is partly or completely enclosed within a *cupule. Certain species of chestnut, especially sweet chestnut (*C.*

sativa), are grown for their fruits. However timber, obviously, is the most economically important product of this family and, on a far smaller scale, certain species of oak yield cork (*Q. suber*) and tannins. Many species are grown as ornamentals.

falcate Sickle-shaped. The term is usually used of leaves.

fall Any of the three large drooping outer petals of an *Iris* flower. In some species (e.g. *I. germanica*) they are bearded. *See also* standard.

false fruit *See* pseudocarp.

family A major category in the taxonomic hierarchy, comprising groups of similar genera. Families are thought by some to represent the highest natural grouping. The Latin names of families usually terminate in -aceae, e.g. Ranunculaceae, Papaveraceae, etc. However, there are eight exceptions, (Compositae, Cruciferae, Gramineae, Guttiferae, Labiatae, Leguminosae, Palmae, and Umbelliferae) whose names have been conserved by the International Code of Botanical Nomenclature. Alternative names have been proposed and are now often used in preference in many works. These are Asteraceae, Brassicaceae, Poaceae, Hypericaceae, Lamiaceae, Fabiaceae, Arecaceae, and Apiaceae, respectively. Groups of similar families are placed in *orders. Large families may be split into *tribes. *See also* natural order.

far-red light Electromagnetic radiation of approximately 740 nm wavelength. It has been shown that far-red light inhibits many physiological processes in plants, e.g. the germination of lettuce seeds. It also counteracts the effect of *red light. For example, if an etiolated plant is exposed to short periods of red light this results in more normal growth. However if the periods of red light are followed by periods of far-red light the plant remains etiolated. It is believed that far-red light is absorbed by a pigment, *phytochrome, which mediates such responses.

fasciation An abnormal form of growth in which a shoot becomes enlarged and flattened, giving the appearance of several shoots fused together. It is often seen in dandelions (*Taraxacum*). It may be caused by mechanical injury, by fungi or mite attack, or by infection with a bacterium of the genus *Phytomonas*, especially *P. fasciens.* In some plants, e.g. cock's comb (*Celosia argentea cristata*), it is due to a mutation.

fascicular cambium (intrafascicular cambium) The part of the *vascular cambium that originates within the *vascular bundles (fascicles) between the xylem and phloem. *Compare* interfascicular cambium.

fast green A permanent stain used to stain cellulose and cytoplasm. It is often used as a counterstain with safranin and, when mixed with acid fuchsin, stains pollen tubes.

fastigiate Describing a tree in which the branches grow almost vertically, such as the Lombardy poplar (*Populus nigra italica*).

fat A triacylglycerol that is solid at room temperature. The predominant fatty acid in such compounds is palmitic acid. The term is often used in a wider sense in much the same way as the term *lipid.

fatty acid A long-chain aliphatic carboxylic acid. Fatty acids are to a large extent responsible for the physical properties of complex *lipids. They may be saturated, monounsaturated (monoenoic), or polyunsaturated (polyenoic). Some of the less common ones may also contain polar or cyclic groups. The nature of the hydrocarbon chain greatly alters the solubility and reactivity of the fatty acid.

The majority of fatty acids have an even number of carbon atoms, although some odd-carbon fatty acids are found. The commonest fatty acids in plants are *oleic acid (C_{18}, monounsaturated) and *palmitic acid (C_{16}, saturated), but in specialized tissues, such as the chloroplast, *linolenic acid (C_{18}, triunsaturated) is predominant. Other important fatty acids are *stearic, *linoleic, myristic, lauric, and palmitoleic acids. Linoleic and linolenic acid can be synthesized by plants but not by animals. As precursors of prostaglandins they are essential in animal diets and are termed *essential fatty acids*. *See also* triacylglycerol.

fatty acid metabolism The breakdown or synthesis of *fatty acids. In plants fatty acids can be broken down to yield carbon dioxide, water, and energy but in many lipid-metabolizing tissues, especially germinating oil-rich seeds, fatty acids are metabolized to sugars. The major route for fatty acid breakdown is the *β-oxidation cycle, which degrades fatty acids to acetyl CoA. The acetyl CoA can then either undergo further oxidation or can be converted into sugar via the *glyoxylate cycle.

Fatty acid synthesis is a membrane-associated process. The enzymes responsible for fatty acid synthesis aggregate in synthetase complexes, which are found associated with the various membrane-bound organelles of the cell. Three different types of synthetase complex have been identified, each with a different specificity. *See also* malonyl ACP.

feedback inhibition A type of inhibition in which the end product of a multienzyme sequence inhibits the activity of an enzyme at or near the beginning of the pathway. An example is the inhibition by isoleucine of the enzyme threonine dehydratase, the first enzyme in the biosynthetic pathway from threonine to isoleucine. The opposite of feedback inhibition, *feedforward stimulation*, occurs when the first substrate of a reaction sequence stimulates subsequent reactions in the sequence.

Variations of simple feedback inhibition are found in branching pathways.

Fehling's solution A mixture of two solutions used to detect the presence of reducing sugars and aldehydes in solution. Fehling's A, which is a copper(II) sulphate solution, is added to the test solution followed by an equal amount of Fehling's B (a mixture of sodium potassium tartrate and sodium hydroxide solutions). On boiling, a brick red precipitate of copper(I) oxide indicates a positive result.

female Describing either the reproductive parts or a whole organism that bears the megaspore-producing apparatus. After fertilization the female may nurture the developing embryo. *Compare* male. *See also* carpel, archegonium, egg.

fen A flat region of land that has developed from open stretches of base-rich water that have gradually silted up and passed through the *hydrosere or *halosere stage of vegetation. There is a build up of basic peat but the persistence of wet marshy conditions means that large trees cannot grow, resulting in a subclimax community. If the level of peat is raised sufficiently then certain woody plants that can tolerate wet conditions, such as alders (*Alnus*), willows (*Salix*), and guelder rose (*Viburnum opulus*), may become established. These fen woodlands are known as *carr*. Drained fenland, as for example that of the eastern English counties of Lincolnshire, Cambridgeshire, and Norfolk, is agriculturally very productive.

fermentation The anaerobic breakdown of glucose and other organic fuels to obtain energy. Fermentations are thought to be the oldest energy-yielding catabolic pathways.

Various types of fermentation are known, which differ in their substrate and their final product or products. The best known fermentation is *glycolysis, in which glucose is broken down to pyruvic acid. *Alcoholic fermentation in yeast yields ethanol and carbon dioxide as end products. Other fermentations in microorganisms and bacteria lead to such end products as butyric acid and acetone.

fermenter An apparatus that is designed to contain, control, and monitor a biological process and collect the product of that process. For example, alcoholic fermentation, in which alcohol is produced by yeast cells, can be carried out on an industrial scale in a suitably designed fermenter. Industrial fermenters are used in brewing, effluent disposal, yeast production for bakeries, and in the production of important organic substances such as antibiotics, amino acids, single-cell protein, etc.

ferns *See* Filicinae.

ferredoxin One of the components of the electron transport chain in chloroplasts. Ferredoxins are nonhaem iron–sulphur proteins, i.e. they utilize an iron-containing reaction centre but the structure is not that of a haem, as in cytochromes, but iron and acid-labile sulphur are present instead. In chloroplasts ferredoxin mediates the transfer of electrons between photosystem I and the ultimate electron acceptor, NADP. It accepts an electron, via a ferredoxin reducing substance, from the excited P700 chlorophyll *a* molecule, and transfers it, through a reaction catalysed by ferredoxin-NADP reductase, to NADP. Bacteria and animals also contain iron–sulphur proteins, some free in solution and some membrane bound.

fertilization (syngamy) An essential part of sexual reproduction involving the fusion of haploid male and female gametes to form a diploid zygote, from which a new individual develops. *External fertilization* occurs in lower plants, where gametes are released and exist independently of the parent before fusion occurs. In this process the gametes are totally dependent on the availability of water. *Internal fertilization* occurs in higher plants, where fusion takes place within the tissues of the female parts, which serve to protect and nurture the developing zygote. Internal fertilization, in its more advanced forms, removes the dependence on water and has enabled such plants to exploit drier habitats.

fertilizer Any substance that is applied to land to raise soil fertility and increase plant growth. Fertilizers include natural products such as farmyard manure, guano, and bonemeal, as well as proprietary chemicals. Organic fertilizers tend to have longer lasting effects than inorganic fertilizers but are generally more expensive and slower acting. However they have the advantage of improving soil structure. The three main constituents of fertilizers are usually nitrates, phosphates, and potash (potassium oxide). The relative amounts of these in a compound fertilizer are expressed by the N:P:K ratio. *See also* compost, macronutrients, micronutrients.

Feulgen's test A test used to detect DNA in nuclei, especially during cell division. A section of tissue is first placed in dilute hydrochloric acid for ten minutes at 60°C to hydrolyse DNA, removing the purine bases and exposing the aldehyde groups of deoxyribose. When the tissue is soaked in *Schiff's reagent the location of the DNA is shown by the development of a magenta colour.

F factor *See* plasmid.

fibre A relatively long *sclerenchyma cell, often with inconspicuous simple pits, that is usually differentiated directly from meristematic cells. Fibres that occur in the xylem are called *xylary fibres* and include the *fibre-tracheids.

The inner layers of the secondary wall of certain xylem fibres, notably in tension wood, may be gelatinous and swell with the uptake of water. Of the *extraxylary fibres*, those occurring in the phloem are often referred to as *bast fibres*.
Fibres are of great economic importance in the textile industry and for the making of rope and baskets. Commercially, fibres are divided into 'soft' and 'hard' categories. The 'hard' fibres come from monocotyledons; examples are those of the leaves of *Musa textilis* (abaca) and *Agave sisalana* (sisal). Examples of soft fibres are the pappus of *Gossypium* (cotton) and the bast fibres of *Linum usitatissimum* (flax), *Corchorus capsularis* (jute), and *Cannabis sativa* (hemp).

fibre-tracheid A xylary *fibre that resembles a *tracheid in its possession of bordered pits, although these are usually less conspicuous than those of tracheids. Fibre-tracheids appear to exhibit characters intermediate between those of fibres and tracheids and are believed to represent evidence for a phylogenetic relationship between the two.

fibrous layer *See* endothecium.

fibrous root system The type of root system that develops when the radicle or primary root does not persist as the main root but is either replaced by adventitious (seminal) roots, as occurs in grasses, or becomes highly branched. Fibrous root systems are typical of monocotyledons.

field capacity The amount of water held in the soil by capillary action after excess or gravitational water has run off or percolated down to the water table. At field capacity the water potential of the soil is high. The percentage of water held at field capacity varies with soil texture, clay having a high field capacity and sand a low field capacity. *See also* permanent wilting point.

field resistance *See* resistance.

field theory *See* repulsion theory.

filament **1.** The stalk of a *stamen, bearing the anther at its apex.
2. A long strand composed of numerous similar cells joined end to end. The thalli of many algae, e.g. *Spirogyra*, are filamentous. Certain higher plants, e.g. the Musci and Filicales have a short-lived phase of filamentous growth following germination of the spore (*see* protonema). The nonphotosynthetic filaments forming the mycelia of fungi are termed *hyphae.

filamentous iron bacteria *See* Chlamydobacteriales.

filamentous sulphur bacteria *See* Beggiatoales.

Filicales (Polypodiales) The largest order of the Filicinae containing about 9000 species. Its members differ from those of the Marattiales and Ophioglossales in the development of the sporangium, which originates from one initial cell, while in the other orders it develops from a group of initials. Although secondary vascular tissue is not clearly developed in any of the Filicinae some members of the Filicales, the tree ferns, attain heights of 10 m. In these species bands of schlerenchyma strengthen the stem. The lamina of the fern frond usually shows differentiation into palisade and spongy mesophyll but in the filmy ferns the leaf is only one cell thick.
The larger homosporous families of the Filicales include (according to one of many classifications): the Osmundaceae (royals ferns) with some 20 species in 3 genera; the Gleicheniaceae (staghorn ferns) with some 150 species in 5 genera; the Polypodiaceae with about 1050 species in some 55 genera; the Schizaeaceae, including the climbing ferns, with about 160 species in 6 genera; the Adiantaceae, including the maidenhair ferns, with about 850 species

in some 20 genera; the Cyatheaceae, including the brackens, filmy ferns, and tree ferns, with about 1500 species; and the Aspleniaceae, including the spleenworts and lady ferns, with about 1900 species in some 75 genera. The heterosporous families all contain aquatic ferns and are often classified into separate orders, the Marsileales, comprising the Marsileaceae (about 70 species in 3 genera), and the Salviniales, comprising the Salviniaceae (10 species in the genus *Salvinia*) and the Azollaceae (5 species in the genus *Azolla*).

Filicinae (Polypodiopsida) A class of the *Pteropsida or *Pteridophyta containing about 10 000 species of chiefly tropical herbaceous plants, the ferns, most of which may be distinguished by their characteristic *fronds. All ferns, except the heterosporous water ferns, are homosporous and normally bear the spores on the abaxial surface of the frond in distinct groups of sporangia called sori (*see* sorus). The free-living gametophyte requires humid conditions to survive and is a vulnerable part of the life cycle. If often resembles a heart-shaped liverwort, but filamentous and subterranean gametophytes also occur. It is thought the Filicinae evolved from plants similar to *Psilophyton* (*see* Psilopsida). The Filicinae contains two subclasses, the Leptosporangiatae, in which the sporangium develops from a single cell, and the Eusporangiatae, where the sporangium develops from a number of cells. The *Filicales belong to the Leptosporangiatae and the *Ophioglossales and *Marattiales to the Eusporangiatae.

The Filicinae have a rich fossil record going back to the Devonian and made up a large part of the Carboniferous flora. Many forms became extinct during the Palaeozoic and two extinct orders are recognized, the Cladoxylales and the Coenopteridales. Both of these appear to have been predominantly homosporous.

filiform Having a threadlike form.

fimbria *See* pilus.

fimbriate Having a margin that is fringed, usually with hairs.

fine structure *See* ultrastructure.

finger-and-toe disease *See* club root.

fireblight A serious disease of pears and other trees of the family Rosaceae caused by the bacterium *Erwinia amylovora.* The disease attacks blossoms and leaves in the spring and, as the disease progresses, individual branches or whole trees look as if they have been scorched by fire. The host forms *cankers in response to infection. The bacteria overwinter in the cankers and are spread by insects and rainsplash to infect flowers in the spring. Hawthorn (*Crataegus*), an *alternate host, can also be a source of inoculum.

fission A method of *asexual reproduction in which the nucleus and then the cytoplasm splits into equal parts to form new individuals genetically identical to the parent. It is seen in many microorganisms, such as bacteria and the fission yeasts (*Schizosaccharomyces*). The new cells may often sporulate in adverse conditions. The parent usually divides into two equal parts (*binary fission*). *Compare* budding, fragmentation.

fission fungi *See* Schizomycetes.

fission yeasts *See* Endomycetales.

fitness The relative ability of an organism to produce large numbers of viable offspring that survive to a reproductive age and themselves contribute to the gene pool of the next generation. Fitness in this sense is not synonymous with the common usage of the word, i.e. healthy, as it does not always follow that the physically fittest are the fittest genetically.

fixation 1. An initial stage in the preparation of cells or tissues for microscopical examination, in which the material is killed and preserved to prevent distortion, decay, and self-digestion (autolysis). Various chemicals (*fixatives*) can be used to penetrate the cells and preserve their structure either by denaturing the protein (e.g. picric acid, ethanol) or by tanning (e.g. acetic acid, formaldehyde). Plant specimens are often prepared for electron microscopy by fixing with Luft's fixative, a 1% solution of potassium permanganate. Common botanical fixatives include Navashin's solution (chromic-acetic-formaldehyde) and Carnoy's solution, an alcoholic fixative that rapidly penetrates the tissues and is especially suitable for hard materials such as seeds.
2. The attainment of a frequency of 100% by an allele in a population due to the complete loss, either by chance or by natural selection, of all other allelic forms of that gene. The likelihood of fixation occurring is greater in small populations.

fixative *See* fixation.

flaccid Describing the wilted condition that results when a plant is deprived of water. The cells have lost their *turgor and the plant therefore lacks rigidity.

flagellin *See* flagellum.

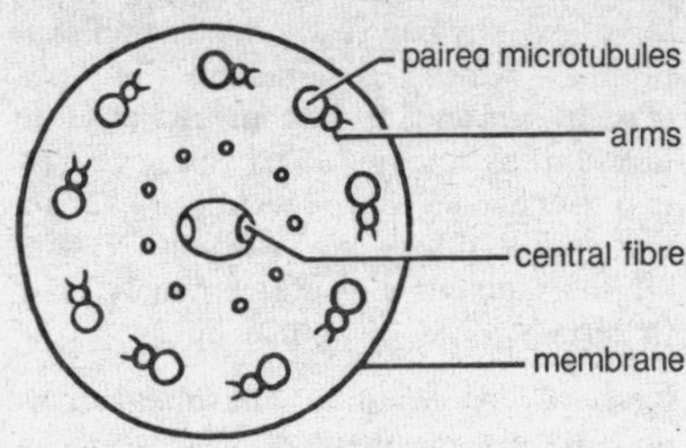

Cross section of a eukaryotic flagellum.

flagellum A threadlike projection arising from the surface of motile unicellular algae, bacteria, and fungi and from spores and gametes. Flagella and *cilia* (*see* undulipodia) of eukaryotic cells have the same structure, cilia (5–10 μm) being shorter than flagella (up to 150 μm). Cilia are usually numerous and their movements create currents that carry extracellular materials over the cell surface; flagella occur singly or in pairs and their activity moves the cell. Both arise from a *kinetosome in the cytoplasm. Internally they consist of eleven fibres running lengthwise and constituting the *axoneme*, surrounded by a membrane continuous with the plasma membrane. Nine of the fibres form a peripheral outer cylinder and the remaining two are located in the centre, surrounded by a central sheath (the so-called 9+2 arrangement, *see* diagram). Each central fibre is a single *microtubule, but the outer ones are paired microtubules, one of each pair being slightly wider than the other. At regular intervals along the length of the narrower tubule, short paired projections or arms arise, each positioned in a clockwise direction around the axoneme when viewed from the base. These consist of *dynein*, a protein with ATPase activity. The production of force in flagella depends on sliding movements of one peripheral fibre against its neighbour, energy being derived from dynein activity. Localized sliding of this kind, occurring in sequence around and along the axoneme produces linear forces that slightly distort the shape of the flagellum and induce bending to bring about the typical waves from base to tip. Flagella of prokaryotic cells have no membrane or axoneme. They are composed of three intertwined strands of the protein *flagellin*, which resembles the protein myosin in contractile muscle fibres.

flavanones A group of *flavonoid compounds the members of which lack a double bond between carbons two and three of the C_3 group in the centre of the *flavonoid nucleus. An example is naringenin, which is responsible for the bitterness of lemons.

flavin Any of a group of yellow plant pigments with an absorption peak around 370 nm that are active in certain plant growth responses to light. Flavins have been implicated in phototropism and are also thought to transfer light energy from blue to red wavelengths so influencing the function of phytochrome. *Riboflavin is an example. *See also* FAD, FMN.

flavin adenine dinucleotide *See* FAD.

flavin mononucleotide *See* FMN.

flavones A group of flavonoid pigments the members of which contain an unaltered *flavonoid nucleus. It includes flavone, found in certain *Primula* species, and the more widespread apigenin and luteolin. *Isoflavones* are isomers of flavones in which the B group of the flavonoid nucleus is attached to the third rather than second carbon of the central C_3 group. Isoflavones are particularly common in the Leguminosae. Some isoflavones have oestrogen-like properties, which has led to sterility in sheep grazing on certain legume fodder crops, notably subterranean clover (*Trifolium subterraneum*).

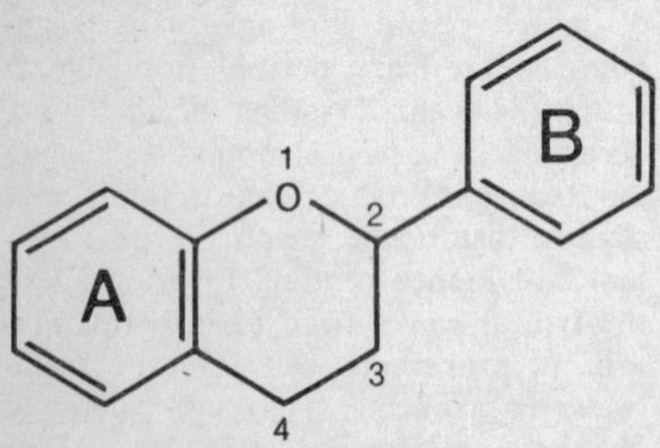

The flavonoid nucleus (2-phenylbenzopyran).

flavonoids A group of plant compounds all of which contain a 2-phenylbenzopyran nucleus (*see* diagram). They include the *flavones, *anthocyanins, *flavanones, *chalcones and *aurones, and *flavonols. Flavonoids usually occur in combination with sugars as glycosides. They have been isolated from bryophytes and vascular plants but not from algae, fungi, or bacteria. Certain types of flavonoid may be characteristic of a group of plants, the best example being the widespread occurrence of *biflavonyls* (dimers of the flavone apigenin) in the gymnosperms. Biflavonyls are scarce or absent among other plant groups. Flavonoids have been more often used than any other group of plant substances in chemotaxonomic studies.

Flavonoids have traditionally been thought of as waste products and their functions largely remain obscure. Many act as flower pigments while some have been implicated in the control of IAA activity. For example, flavonols, such as quercitin, with two hydroxyls in the B ring (forming a catechol group) inhibit IAA oxidase while flavonols with one hydroxyl (e.g. kaempferol) promote the enzyme's activity. Other flavonoids, especially certain of the isoflavonoids, serve as phytoalexins. An example is pisatin, which accumulates in pea (*Pisum sativum*) tissues in response to invasion by many fungal species. Certain isoflavones have oestrogen-like activity while others, e.g. rotenone, resemble the saponins and are extremely toxic to certain forms of animal life.

flavonols A group of *flavonoid compounds, the members of which have a hydroxyl group on the third carbon of the C_3 group in the centre of the flavonoid nucleus. They are formed by the addition of a hydroxyl ion to a flavanone. Common flavonols include quercitin and kaempferol.

flavoprotein An enzyme containing a flavin molecule (e.g. *FMN, *FAD) as a prosthetic group. Flavoproteins are usually dehydrogenases, e.g. succinate dehydrogenase, which is a TCA cycle enzyme. The general equation for a flavoprotein reaction is: flavoprotein + reduced substrate $\rightleftharpoons$ reduced flavoprotein + oxidized substrate.
Like NAD, reduced flavoproteins can be reoxidized in the *respiratory chain. Flavoproteins are also components of the electron transport system of photosynthesis and the microsomal electron transport system.

flimmer flagellum (tinsel flagellum, pantonematic flagellum) A threadlike projection arising from motile algal and fungal cells that has longitudinal rows of fine hairs (mastigonemata) along its outer membrane. Internally the fibrillar structure is typical of a *flagellum.

flocculation *See* liming.

floral diagram A stylized representation of the structure of a flower in which the whorls of floral parts are shown as a series of concentric circles. All floral segments arising at the same level are placed, in their correct relative positions, in the same circle. When appropriate, fusion of parts is also indicated. The ovary is represented in cross section in the centre of the whorls. The bottom of the diagram represents the anterior side of the flower, i.e. the side furthest from the axis of the inflorescence. Any bracts subtending the flower are inserted at the base of the diagram, while the inflorescence axis is shown as a circle at the top (posterior) side of the diagram. Floral diagrams are usually presented in conjunction with a longitudinal section through the centre of the flower (*see* illustration).

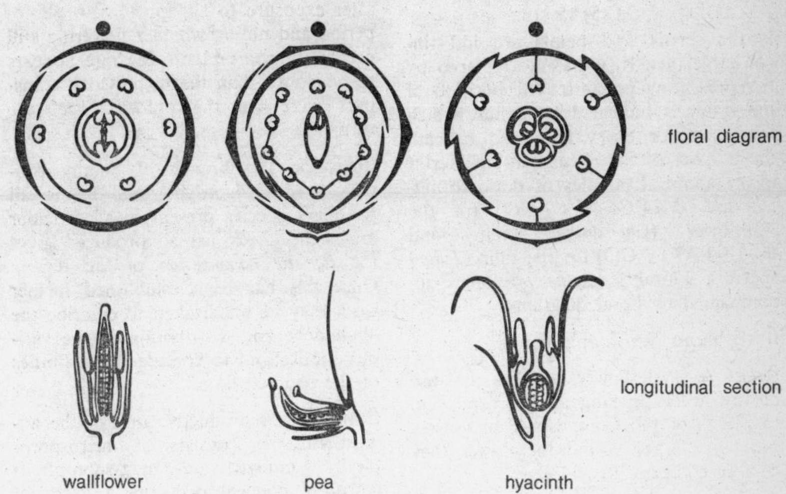

floral formula A method of recording floral structure by a series of symbols, letters, and numbers. The formula begins either with the sign ⊕, representing actinomorphy, or with ·|· or ↑, representing zygomorphy. This is followed by a series of letters in the order K (for calyx), C (for corolla), A (for androecium) and G (for gynoecium). The number of parts (sepals, petals, stamens, and carpels respectively) in each whorl is given by a number following the appropriate letter. If this number exceeds 12 the symbol ∞ (infinity) is used. If the parts are fused then the number is bracketed. If the parts are in distinct whorls or groups this is indicated by splitting the number appropriately. Thus in the garden pea (*Pisum sativum*) there are ten stamens, nine of which are fused and the tenth separate. This is represented A(9)+1. If two whorls are united then the representative symbols are joined by a single bracket. Thus in the primrose (*Primula vulgaris*) the stamens are inserted on the corolla tube. This is represented

$$\overset{\frown}{C(5)A5}$$

If the sepals and petals are indistinguishable then K and C are replaced by P representing perianth. The position of the ovary is indicated by a line, which in a superior ovary is placed beneath the number of carpels and in an inferior ovary above. Examples of floral formulae are ⊕K4 C4 A4+2 G($\underline{2}$) for the wallflower (*Cheiranthus cheiri*) and ⊕P3+3 A3+3 G($\underline{3}$) for the tulip (*Tulipa* species). Floral formulae are often accompanied by floral diagrams.

flore pleno *See* double flower.

floret A small flower, such as the disc or ray florets making up the capitulum in plants of the Compositae. In grasses the lemma, palea, and the flower they enclose comprise the floret.

Florideae (Florideophycidae) *See* Rhodophyta.

floridean starch The assimilation product of algae in the Rhodophyta. It has a structure similar to amylopectin.

florigen A hypothetical plant growth substance that is said to cause flower initiation in many species. Florigen has yet to be isolated from plants although there is much circumstantial evidence for its existence. Thus although flowering often depends on the photoperiod experienced by the plant it is the leaves and not the growing apex that perceive photoperiod. Thus the stimulus must be transported from the leaves to the apex. The flowering stimulus can be transferred, by grafting, from one plant to another. If leaves from short-day plants grown under short days are grafted onto long-day plants grown under short days then the stimulus from the grafted leaves will promote flowering in the long-day plant. Such work shows that florigen is similar in long- and short-day plants. The rate of florigen movement has been measured by removing the leaves from a plant at varying intervals after exposure to an appropriate photoperiod and noting whether flowering still occurs. In some plants the rate appears to be slower than that expected if transport were via the phloem. *See also* vernalin.

floristics The branch of botany concerned with identifying and listing all the plant species present in a particular area. The check list so produced gives the *floristic composition* of the region. Once this has been established further work may be undertaken to describe the abundance and distribution of the various species and to consider the affinities of the regional flora.

flower The reproductive unit of the angiosperms. It consists of microsporophylls (stamens) and megasporophylls (carpels) concentrated on a terminal apex of limited growth. The unit comprises a central axis or *receptacle, and usually nonessential floral parts (the *sepals and *petals) surrounding the essential floral parts, the *stamens and *carpels. The arrangement of floral parts within the flower is usually related to the method of pollination and seed dispersal. The essential floral parts are involved in seed production, and the female parts persist after fertilization to form the fruit. The nonessential parts may be incorporated into the fruit after fertilization but more usually senesce and disintegrate. *See also* inflorescence.

flowering glume *See* lemma.

flowering plants *See* Angiospermae.

fluid-mosaic model *See* plasma membrane.

fluorescence microscopy A type of microscopy that uses ultraviolet light as the source of illumination and involves the treatment of the specimen (prior to viewing) with a fluorescent dye (*fluorochroming*). The short wavelengths of ultraviolet light passing through the specimen cause the dye to fluoresce, i.e. emit long wavelengths of visible light. A fluorescent image is thus produced against a dark background. The technique is used widely. In bacteriology, *fluorescent antibody reactions* involve the addition of a fluorescent dye to either the antibody or antigen to locate the site of antibody-antigen reactions.

flushing The process by which soluble substances in the lower layers of the soil are washed upwards and deposited on or near the soil surface. It occurs, for example, in water-meadows, fenland, and where there are springs. *Compare* leaching.

FMN (flavin mononucleotide) A riboflavin-derived coenzyme, similar to *FAD, that acts as a prosthetic group to several dehydrogenases. NADH dehydrogenase, the enzyme that catalyses transfer of electrons from NADH to coenzyme Q in the respiratory chain, is an FMN-containing enzyme. *See also* flavoprotein.

folic acid A member of the B group of *vitamins first isolated from spinach leaves. In its activated form tetrahydrofolic acid it acts as a coenzyme in reactions involving the transfer of hydroxymethyl, formyl, and methyl groups. Such transfers are important in the synthesis of the purines and the pyrimidine thymine.

foliose Describing organisms that bear leaflike structures, used especially of leafy lichens (e.g. *Peltigera*) and sometimes of leafy liverworts (e.g. *Diplophyllum*).

follicle A dry dehiscent many-seeded fruit derived from one carpel, which on ripening splits down one side only, usually the ventral suture, to expose the seeds, as in *Delphinium*.

food chain A series of organisms, each successive group of which feeds on the group immediately previous in the chain, and is in turn eaten by the succeeding group. Green plants (producers) are normally the first step in a food chain, herbivores (primary consumers) the second step, with carnivores (secondary consumers) making up the remaining two or three stages. The different stages are termed *trophic levels* and organisms that are removed from the beginning of the chain by the same number of steps are said to occupy the same trophic level. The chain is effectively a series of energy transfers, with considerable amounts of energy being lost at each transfer (*see* bioenergetics). A food chain consequently rarely has more than four or five stages since there is not enough energy in the system to maintain more. A consequence of this is that, for a given consumer level, considerably more individuals can be maintained if an earlier consumer stage is omitted. For example, a given amount of grain will support many more people if it is eaten as grain, rather than being first converted to meat by feeding to livestock.

Food chains rarely exist in isolation but are interconnected to form a *food web*. Thus any of the links in one food chain may, through the activity of decomposers, saprophytes, or parasites, lead into a number of different food chains. In addition, most plants and herbivores have a number of different predators.

foot The basal portion of an embryo, sporophyte, or spore-producing body, which is embedded in the parental tissue. It serves as an anchor and to absorb nutrients.

foot rot A plant disease in which there is rotting of the root system and the bottom part of the stem. Foot rot of tomato is caused by *Phytophthora cryptogea* and *P. parasitica*. Stem tissues around soil level appear brown or blackish and the root system is found to be rotten.

forest Any major community (*biome) in which the dominant plants are trees. There are three main types of forest; cold forests, temperate forests, and tropical forests. The cold forests include the *northern coniferous forest* (*boreal forest* or *taiga*), where there are relatively few types of tree, mainly fir, pine, and spruce with some deciduous birches and larches. The latter are found mainly on the borders of the tundra where it is too cold and exposed for the conifers to grow. However conifers are well adapted to survive long cold winters and can take advantage of the short growing season as they retain their leaves. Boreal forest is found in North America and northern Eurasia. The *mountain coniferous forest* and *pine forest* of the southeast United States are also boreal forests.

Temperate forests have abundant and fairly evenly distributed rainfall with moderate temperatures, but there is a marked seasonal change. The vegetation consists of broad-leaved deciduous trees, an adaptation to the lack of available water during winter (*see* physiological drought). There are usually several different species of trees and the term *mixed deciduous forest* is often used. *Warm temperate* and *cool temperate rain forests* also exist, e.g. in south China and New Zealand respectively. *Broad-leaved evergreen forest* is characteristic of a Mediterranean-type climate with hot dry summers and mild wet winters.

Tropical forests include tropical rain forest, monsoon forest, and thorn forest. *Rain forests* have regular heavy rain, constantly high temperatures, and therefore prolific plant growth. Rain forest regions, such as the Amazon basin and the Malay-Indonesian region, contain thousands of different species. Essentially similar forests, known as *riverine forests*, occur along river banks in drier regions. *Monsoon forests* have a period of drought, which may last for several months. They do not have such a great variety of species as tropical rain forest and have a more open canopy and very dense undergrowth. Many of the trees shed their leaves in the dry season. They are found in India, Burma, and Indo-China. *Thorn forest* is a transitional type of vegetation with some similarities to *savanna and semidesert vegetation. It grades into tropical forests and is found in Central and South America, Australia, and Africa.

Generally trees need more light and moisture than other plant forms, although some trees are adapted to survive long periods of drought and cold.

form (forma) The lowest rank normally used by practising taxonomists for sporadic distinct variants that sometimes occur in populations. These may be relatively minor genetic variants (i.e. based on one or a few linked characters) but their effects can be conspicuous. The most obviously recognizable forms are those in which flower colour is modified (e.g. albino individuals occurring in a population of purple-flowered plants). The category form may relate directly to species but it is more commonly used to distinguish variants of subspecies and varieties.

formalin-acetic-alcohol (FAA) A fixative containing 17 parts of 70% alcohol to 2 parts 40% formaldehyde to 1 part glacial acetic acid.

forma specialis (*pl.* formae speciales) *See* physiological race.

form genus 1. Any genus in which imperfect states of fungi are placed. Similarly a *form species* is a species of imperfect fungus. When the connection between an imperfect state and a perfect state is established then the name of the perfect state takes precedence, although both names are still valid.
2. A group of fossil species whose affinities are unknown.

fossils The remains of organisms from past geological ages preserved in sedimentary rocks either as actual structures or as impressions, casts, or moulds. Plant fossils are rarely as well preserved as animal fossils because their tissues do not normally contain calcified structures. They are usually therefore completely decomposed before the processes of fossilization, including carbonization and petrification, act to preserve them. The remains of macroscopic structures, such as branches, leaves, fruits, and seeds, are termed *megafossils* while those of pollen and spores are called *microfossils*. If a fossil cannot be assigned to any genera containing extant species then its genus is termed an organ genus. Similarly if it cannot be assigned to a family it is placed in a form genus.
The study of fossils has helped in the construction of phylogenetic classification schemes and has also thrown light on how some of the complex structures of extant plants have evolved. *See also* palaeontology.

founder principle The proposition that a small pioneer community that is establishing itself in genetic isolation from the main population will only possess a small and possibly nonrandom selection of genes from the parental gene pool. Consequently, the community that develops may take a different evolutionary path from the parental population. Relative uniformity among the community's members and the frequent occurrence of distinct or unusual characteristics is seen as evidence in support of the principle.

fraction 1 protein *See* ribulose bisphosphate carboxylase.

fragmentation A form of *asexual reproduction in which the parent splits into several pieces, each of which may develop into a new individual. It is seen in many filamentous algae, e.g. *Ulothrix* and *Spirogyra*. It is also a feature of some aquatic plants, e.g. waterweeds (*Elodea*), and of the gametophytes of certain bryophytes. *Compare* budding, fission.

free-central placentation A form of *placentation in which the placentae develop on a central dome or column of tissue. It is seen in unilocular compound ovaries, such as those of the Primulaceae and Caryophyllaceae.

free space The part of the protoplasm where there is relatively free exchange of inorganic ions between the protoplasm and extracellular fluids. This occurs because the outer membrane of the protoplasm is permeable to most inorganic ions. The membrane surrounding the cell vacuole is not so permeable and though ions may enter by *active transport they cannot pass out again; certain salts thus tend to accumulate in the vacuole. The composition of the free space and the vacuole may therefore be quite different.

freeze drying A technique for the removal of water from a specimen by sublimation of ice under vacuum. It is used as a method of drying in cases where the action of heat would cause damage, e.g. the concentration of labile solutes such as enzymes and the drying of fragile specimens for electron microscopy and storage (e.g. herbarium specimens). It is also used when the presence of liquid water would cause distortion of tissue by surface tension. Rapid freezing prevents the formation of ice crystals that would otherwise cause disruption.

freeze etching *See* freeze fracturing.

freeze fracturing A technique used in electron microscopy in which the specimen is frozen quickly at very low temperatures and then fractured along lines of weakness. The new surface so formed is coated with a thin layer of carbon or platinum, which is then floated off to form a replica of the surface. *Freeze etching* is a modification of the technique in which the water vapour is allowed to evaporate from the fracture surface before the replica is made. This takes longer but allows higher resolution of the specimen.

frond A large leaf or leaflike structure. The term is most frequently used of the dissected leaves of ferns, cycads, and palms. More rarely it is applied to the thalli of foliose lichens and certain macroscopic algae.

frost resistance The ability, possessed by many temperate and arctic plants, to withstand subzero temperatures. The resistance is technically *frost tolerance* as the plant cannot avoid the cold conditions and resultant freezing of water in the intercellular spaces. This causes a lowering of the water potential in the intercellular spaces so water moves out of the protoplast to regions of ice crystallization. Frost damage is thus often a result of severe cellular dehydration. However this dehydration increases the solute concentration of the cell and thus makes ice formation within the cell less likely. Dormant organs and seeds tend to show more frost resistance than actively growing tissue. The mechanism of frost resistance is uncertain but membrane elasticity and reduction in cell size appear to play a part. *See also* hardening.

fructification Any seed- or spore-bearing structure, especially one that is easily visible as, for example, the aerial fruiting bodies of many higher fungi.

fructose (fruit sugar, laevulose) A ketohexose sugar (*see illustration at* ketose). In solution fructose forms a *furanose ring to give fructofuranose. Fructose is the commonest keto sugar and is a component of the disaccharide *sucrose. Fructose is, with glucose, one of the few monosaccharides to occur free in plant tissues, e.g. fruit juices. Polymers of fructose are termed *fructosans*, the most important of which is inulin.

fructose 1,6-bisphosphate A phosphorylated derivative of fructose, situated at the main control point of glycolysis. In glycolysis fructose 1,6-bisphosphate is formed from fructose 6-phosphate by the allosteric enzyme phosphofructokinase, which is inhibited by ATP and citrate, and stimulated by ADP and AMP. This reaction is the rate-limiting step of glycolysis. The reverse reaction, which is a step in glucose synthesis, also involves an allosteric enzyme, fructose bisphosphatase. Fructose 1,6-bisphosphate is also an intermediate in the pentose phosphate pathway.

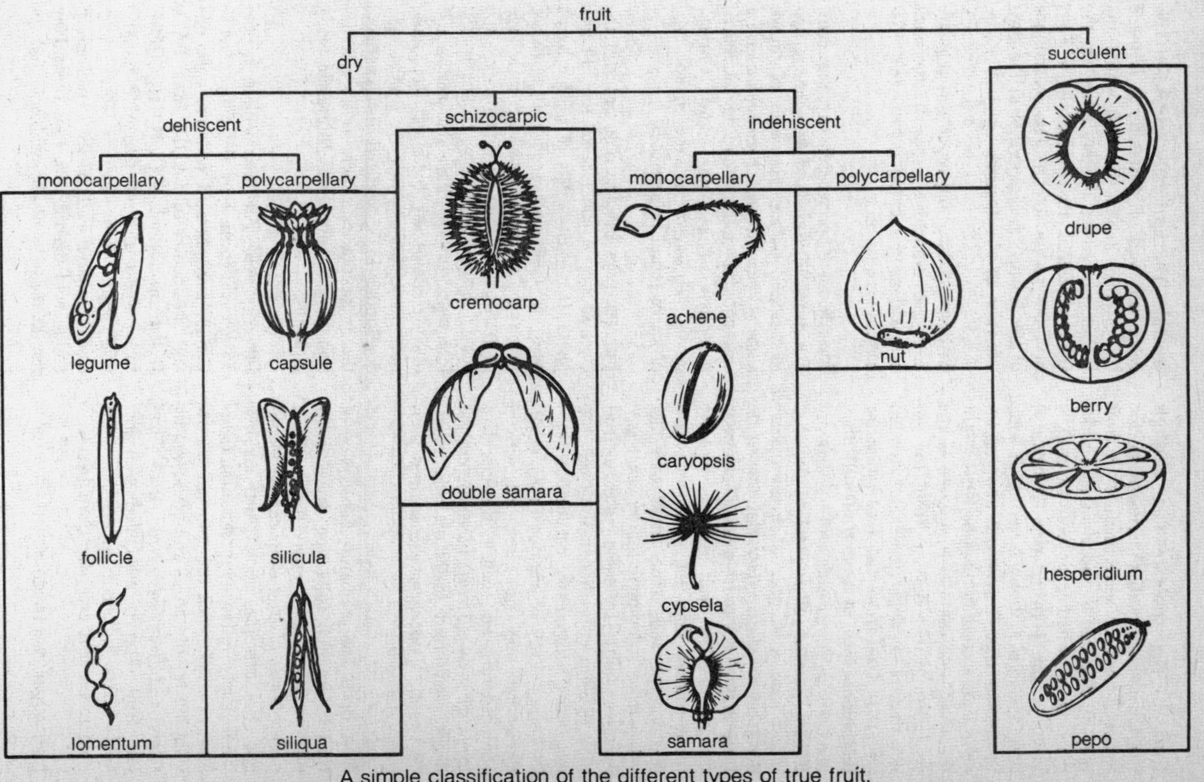

A simple classification of the different types of true fruit.

fruit **1.** The structure that develops from the ovary wall (*pericarp) as the enclosed seed or seeds mature. A fruit may be classified as succulent or dry depending on whether or not the middle layer of the pericarp (the *mesocarp) develops into a fleshy covering. It may be further classified as dehiscent or indehiscent according to whether or not the fruit wall splits open to release the seed. Fruits that develop from the gynoecium of a single flower are termed simple or true fruits (*see* illustration for the main types of true fruits). If they are derived from a single ovary they are termed *monocarpellary* while those that incorporate a number of fused ovaries are termed *polycarpellary*. An *aggregate fruit may develop from an apocarpous gynoecium. The fruit may incorporate tissues other than the gynoecium (*see* pseudocarp) and some fruits may develop from a complete inflorescence (*see* multiple fruit). In some cases a fruit may develop even though the ovule has not been fertilized (*see* parthenocarpy).
2. Loosely, any of various fleshy structures that may be associated with a gymnosperm seed, such as the succulent *aril of yew (*Taxus baccata*) or the fleshy ovuliferous scales of some members of the Podocarpaceae, such as junipers (*Juniperus*).

fruit drop The premature *abscission of fruit before it is fully ripe. It is a normal process and in many fruits certain peak periods of fruit drop can be identified. For example, apple fruits are lost immediately following pollination (*post-blossom drop*), when the embryos are developing rapidly (*June drop*), and during ripening (*preharvest drop*). Like *leaf fall, fruit drop is associated with low auxin levels and auxin sprays have been used to prevent excessive fruit drop.

fruit sugar *See* fructose.

frustule The intricately sculptured siliceous cell wall of a diatom. It is made up of two halves, one of which fits within the edges of the other in the same way that the bottom of a Petri dish fits within its lid. The patternings of the frustule are important in the classification of the diatoms.

fruticose Describing organisms that are erect and branching, especially the erect or pendent branching lichens (e.g. *Usnea*).

fuc-, fuco- A prefix denoting seaweeds. For example, *fucin* and *fucoidin* are colloids found only in the walls of brown algae (Phaeophyta). *Fucoxanthin* is a carotenoid pigment found in the Bacillariophyta, Phaeophyta, and Chrysophyta. *Fucosterol* is the main type of sterol found in seaweeds.

Fucales (wracks) The only order of the Cyclosporae, a subclass of the *Phaeophyta. It consists solely of marine forms with the commoner species (e.g. *Fucus*, *Pelvetia*, *Ascophyllum*) making up a large proportion of the vegetation of the coastal littoral zone. The thallus is differentiated into holdfast and lamina and often contains air bladders. Reproduction is oogamous with the gametangia developing in specialized invaginations, the *conceptacles. There is no alternation of generations, the *life cycle being diplontic.

fumaric acid An intermediate in the TCA cycle formed by the dehydrogenation of succinic acid with the concomitant production of $FADH_2$ from FAD. It has the formula $COOH(CH)_2COOH$. In the next step of the cycle fumaric acid reacts with water to form malic acid. Fumaric acid is also involved in amino acid metabolism. It is formed by the degradation of phenylalanine and tyrosine. In amino acid synthesis it is thought ammonia may react with fumaric acid to form aspartic acid.

fumigant A volatile chemical used to kill pathogens in the soil or in glasshouses. Fumigants are usually applied to the soil by injection. Chemicals used as fumigants include methyl bromide, chloropicrin, formaldehyde, metham, ethylene, dibromide, and carbon disulphide.

fungi A group of saprobic, symbiotic, or parasitic eukaryotic organisms containing some 50 000 recognized species divided among some 5100 genera. However it has been estimated that the actual number of species may be between 100 000 and 250 000. The name fungi has been used as a general term, lacking any systematic meaning, but now such organisms are often placed in a separate kingdom Fungi (Mycota), distinct from the green plants. They differ from green plants in not possessing chlorophyll and are thus heterotrophic, usually obtaining food by absorption, though some lower fungi, possibly more closely allied to the Protozoa, take in food by ingestion. Fungal cell walls are characteristically chitinized or composed of fungal cellulose. The kingdom Fungi contains two divisions, the *Myxomycota and the *Eumycota.

fungicide A chemical that kills fungi. Fungicides are used to prevent or control fungal diseases. They are applied to seeds as a dust or slurry and to standing crops usually as a spray. For control of postharvest diseases fruits may be dipped or sprayed, or packed in fungicide-impregnated materials. Fungicides can also be applied to the soil for control of damping-off and root diseases.

Fungicides, like other pesticides, usually have three names – the chemical name of the active ingredient, the approved common name, and the trade name. For example, methyl N-[1-(butylcarbamoyl)-2-benzimidazole] carbamate is the chemical name of benomyl, which has the trade name Benlate.

Fungicides fall into two major categories: the inorganic fungicides including *copper fungicides and *sulphur dust; and the organic fungicides, a group comprising the more modern fungicides, including *systemic fungicides. *See also* quinone fungicide, organomercurial fungicide, dithiocarbamate fungicide, dinitro compounds, fumigant, seed dressing.

Fungi Imperfecti *See* Deuteromycotina.

funiculus (funicle) The stalk attaching the ovule, and later the seed, to the *placenta or ovary wall in angiosperms. It serves as an anchor and provides a vascular supply to the ovule and seed. *See also* raphe.

furanose ring A ring containing four carbon atoms and an oxygen atom. Such rings are formed by ketose sugars with five or more carbon atoms. When such sugars are in solution the ketone group on the second carbon of the sugar reacts with the hydroxyl group on the fifth carbon atom. Fructose existing in this form is termed fructofuranose. Aldose sugars can also form furanose rings. However aldoses also form *pyranose rings, which are far more stable and so predominate in solutions of aldose sugars.

6-furfurylaminopurine *See* kinetin.

fusiform Elongated and tapering at each end, as does a fusiform initial.

fusiform initials More or less elongate *initials in the vascular cambium that give rise to the components of the axial (longitudinal) system of the secondary xylem and secondary phloem (e.g. vessel elements, tracheids, and sieve tube elements). They also give rise to the *ray initials.

G

G_1 phase, G_2 phase *See* cell cycle.

galactan A polysaccharide in which the major monosaccharide subunit is galactose, though other sugars, notably arabinose, are often also present. Galactans are structural polysaccharides, often found among the pectic substances of cell walls.

galactose An aldohexose sugar commonly found in plants (*see illustration at* aldose). It normally does not exist in the free state but as polymers (*see* galactan). It is also a constituent of the oligosaccharides raffinose and stachyose. Oxidation of galactose yields galacturonic acid, which on polymerization forms pectic acid.

galacturonic acid *See* uronic acids.

gall An abnormal localized swelling or outgrowth produced by a plant as a result of attack by a parasite. Galls are caused by bacteria, fungi, nematodes, insects, or mites or by a combination of these agents. *Crown gall is a common gall produced as a result of bacterial infection. The root knots produced by nematodes of the genus *Meloidogyne* are another example. Galls caused by insects are very diverse and include the oak apple caused by the gall wasp *Neuroterus lenticularis* and the pincushion galls on roses caused by the gall wasp *Rhodites rosae.*

gametangiophore An upright structure that bears the female gametes (*see* archegoniophore) or male gametes (*see* antheridiophore) in certain liverworts. Gametangiophores are extensions of the thallus and are most elaborate in the genus *Marchantia.*

gametangium A cell or organ in which gametes are formed. The term is mostly used of the single-celled gametangia of certain algae and fungi in which there is no differentiation into male and female. However antheridia, oogonia, archegonia, etc. may also be described as gametangia.

gamete A cell or nucleus that may participate in sexual fusion to form a *zygote. It is normally haploid and thus on fusion of two gametes a diploid zygote is formed. In virtually all plants (exceptions are those with a *diplontic life cycle) meiosis is separated from *gametogenesis by the development of a somatic gametophyte generation (*see* alternation of generations). In the primitive algae and fungi the gametes are often naked and *isogamous. In more advanced forms there is a trend through *anisogamy to *oogamy and specialization of the gametes, so that they become better protected and less dependent on water for survival and dispersal. *See also* sexual reproduction.

gametogenesis 1. The formation of gametes. Gametes are usually formed mitotically by the *gametophyte generation though in some groups (*see* diplontic) gametogenesis corresponds with meiosis. Gametes may form in any cell of the gametophyte, as occurs in the green alga *Ulva*, or they may be confined to specialized organs, such as antheridia and archegonia. *Compare* sporogenesis.
2. The growth and development of the gametophyte generation.

gametophyte The gamete-producing, usually haploid generation in the life cycle of a plant. The gametophyte arises from haploid spores produced as a result of meiosis in the *sporophyte, or diploid, generation. In the bryophytes the gametophyte constitutes a major part of the life cycle and the sporophyte is either partially or totally dependent on it for anchorage and nutrition. In the vascular plants the gametophyte becomes progressively less prominent so that in the angiosperms it is represented by the pollen tube and its nuclei and the embryo sac. In lower plants the gametophyte is unprotected and tends to be highly susceptible to dehydration. It therefore tends to be confined to damp habitats. *See also* alternation of generations.

gamone Any substance released by a gamete that serves to attract another gamete. For example, in certain ferns malic acid is believed to be released from the archegonia and to attract the male gametes.

gamopetalous (sympetalous) Having petals that are fused along their margins to the base, forming a corolla tube. Most plants possessing such flowers are grouped together in the Sympetalae (*see* Asteridae).

gamosepalous Having sepals that are fused along their margins to form a tubular calyx.

garigue (garrigue) A type of scrub woodland characteristic of limestone areas with low rainfall and thin poor dry soils. The low-growing vegetation consists of aromatic, often spiny, species, such as sage (*Salvia officinalis*), thymes (*Thymus*), and lavender (*Lavandula vera*). Many bulbous plants grow between the shrubs and there are numerous species that are also present in *maquis. This type of vegetation is widespread in Mediterranean countries.

gaseous exchange The *diffusion of gases into and out of the cells of an organism. For exchange to occur, the gases usually have to pass through a liquid boundary layer. The rate of diffusion of gases dissolved in water is much slower than diffusion through air, an important consideration in calculations of gaseous exchange. During photosynthesis carbon dioxide is taken in and oxygen given off; the reverse occurs in respiration. The passage of water vapour is usually in one direction only – out of the plant.

gas–liquid chromatography (GLC) A chromatographic technique in which the mixture of substances to be analysed is vaporized and carried along a column containing a liquid (the stationary phase) by an inert carrier gas such as nitrogen (the mobile phase). The components of the mixture separate out at different times according to their gas–liquid partition coefficients (*see* partition chromatography). Those that are most soluble in the gas emerge from the column first and those most soluble in the liquid emerge last. The nature and amount of the different gases emerging from the column is registered using a gas detector. Fatty acids and other substances, e.g. sterols and hydrocarbons, that are volatile at reasonably low temperatures are separated in this way. *See also* chromatography.

Gasteromycetes A class of the *Basidiomycotina containing the most advanced basidiomycetes. These have a well developed basidiocarp and basidiospores that are not discharged forcibly. It contains about 700 species in some 150 genera and includes the orders Hymenogastrales (false truffles), *Lycoperdales, *Nidulariales, and *Phallales.

gas vacuole A structure made up of numerous gas-filled vesicles, many of which may be found in the cells of certain aquatic bacteria and blue-green algae. The density of the cell depends on how much gas is contained within the vesicles. The position of the organism in a column of water can thus be regulated by these structures. The vesicles are unusual in being bound by a protein rather than phosphoglyceride membrane.

gelatinous fungi *See* Tremellales.

gel filtration A form of *chromatography in which the mixture to be separated is washed through a column packed with beads of an inert gel. The polymer forming the gel contains internal pores, into which molecules below a certain size can penetrate. In general, large molecules, which cannot penetrate the pores, move quickly down the column; smaller molecules, which enter the gel, travel more slowly. The technique exploits differences in size (i.e. molecular weight) and measurements of the rate of travel can give estimates of molecular weight. It is used particularly to separate protein mixtures, but can also be applied to cell nuclei, viruses, and other large molecules. It is sometimes referred to as *molecular-exclusion chromatography*.

gemma **1.** A specialized multicellular unit of vegetative reproduction found in certain mosses and liverworts, and in *Psilotum*. It may take various forms, from being disc- or platelike to being filamentous or heart shaped. When separated from the parent it develops into a new individual identical to the parent. *Gemma cups* may be formed where several gemmae are produced on a protective receptacle, as seen in the moss *Tetraphis pellucida*. Clusters of gemmae may also be seen at leaf tips or in the leaf axils, on specialized stalks or pseudopodia, and on rhizoids. Reproduction by gemmation occurs more frequently than spore formation and in some species it is the only form of reproduction seen.
2. *See* chlamydospore.

gene The unit of inheritance. Defined no more precisely than this, the term 'gene' corresponds with Mendel's 'factors' responsible for the manifestation of a particular characteristic.
A particular gene may be further defined as being located at a specific point on one of the chromosomes and capable of existing in alternative forms or alleles. Thus there would be a gene for 'seed colour' in peas, with alternative alleles of 'green' and 'yellow'. In a diploid cell, alleles of a gene are paired, one on each of a pair of homologous chromosomes. If contrasting alleles are present in the same organism, e.g. one for 'green' and one for 'yellow', then the phenotype of the organism will depend upon the dominance relationship between the alleles.
However, in addition to determining characteristics, inherited information also mutates and recombines. A gene defined as a unit of function (*cistron) may be a piece of *DNA several hundred nucleotides long, whereas a gene defined as a unit of mutation (*muton*) may be a single nucleotide. The unit of recombination (*recon*) lies between these two extremes. In modern usage the term gene is most commonly applied to functional units of DNA. Thus *structural genes determine protein structure and regulator, promoter, and operator genes control the regulation of protein production. Genes are extremely stable and so spontaneous *mutation occurs very rarely. However it is the ultimate source of all genetic variation, i.e. the reason why the 'seed colour' gene exists in different forms, such as green and yellow. Once mutation has created alternative alleles, a further source of genetic variation is by genetic *recombination, i.e. the rearrangement of whole chromosomes or parts of chromosomes during meiosis.
Though the majority of DNA, and hence the genes, is confined to the chromosomes, some DNA is present in such organelles as mitochondria and chloroplasts. The pattern of inheritance shown by cytoplasmic genes differs from that of chromosomal (Mendelian) genes and they are referred to as plasmagenes in order to distinguish them from the latter.

gene bank An institution where plant material, in danger either of becoming extinct in the wild or of being lost from cultivation, is stored in a viable condition. In many such centres the emphasis is on maintaining collections of plants of demonstrated or potential use to man, e.g. crop varieties containing useful genes that have been superseded by improved cultivars. Normally seed material is dried down to about 4% moisture and stored at around 0°C preferably in hermetically sealed containers. For many seeds such conditions will maintain viability for 10–20 years and longer. The seeds of some species, including many tropical plants, cannot be dried without killing them and yet have a limited lifespan in the hydrated state. Such material must be maintained in the growing condition. The problems of space and maintenance that this entails may in some instances be overcome by tissue-culture methods. All material must periodically be sown to check viability and multiply seedstock. Pollen may also be stored in gene banks though generally the longevity of pollen is less than that of seeds. *See also* genetic erosion, genetic resources.

gene centre (centre of diversity) An area that shows considerable genetic diversity of certain crop plants and their relatives. Some gene centres, the primary centres, are believed to correspond to the region where a particular crop originated. For example, numerous forms of wheat exist in the Middle East and, since many of the wild relatives of wheat are also found in this area, it is believed the crop was first domesticated there. Other centres, the secondary centres, do not contain wild relatives and are believed simply to be areas where the crops have been cultivated for a considerable period. Ethiopia is an example of a secondary centre.

Gene centres are typically found in mountainous regions. This may be because human communities are more isolated and climatic conditions more variable in such areas – both conditions that would lead to more rapid divergence of crop populations. The recognition of gene centres is important in conservation and plant breeding work as such areas are important reservoirs of natural genetic variability. *See also* genetic erosion.

genecology The study of variations in gene frequency within a species in relation to changes in the environment.

gene flow The movement of genes between populations of the same species by interbreeding. *Compare* introgression.

gene frequency The proportion of an allele in a population compared with other alleles of that gene. Thus if a gene A has two alleles, a and A, and the frequency of a = 0.2 (20%) then the frequency of A must be 0.8 (80%) since A + a = 1 (100%). Although the term 'gene frequency' is commonly used by population geneticists, it would be more consistent with standard definitions of the word 'gene' to use the term 'allele frequency'. *See also* Hardy–Weinberg law.

gene mutation An alteration to a single gene resulting from a change in the number, type, or sequence of bases specifying the amino acids in a polypeptide. Such mutations are not visible by microscopy. Gene mutations result in new allelic forms of a gene. *See also* chromosome mutation, mutation, mutagen.

gene pool The sum total and variety of all the genes and their alleles present in a breeding population or species at one time. *See also* gene centre, genetic erosion.

generation time The interval between the commencement of consecutive cell divisions. In meristematic tissue or in a colony of unicells a period of time elapses after cell division while the daughter cells enlarge and establish their complete cellular organization (*see* cell cycle) before they divide.

generative cell One of the cells found in the pollen tube of seed plants (*see also* tube cell, vegetative nucleus). In gymnosperms it gives rise to the *body cell and stalk cell whereas in angiosperms it gives rise directly to the two male gametes or *generative nuclei.

generative nuclei In angiosperms, the two male gametes that are formed by division of the generative cell. They migrate down the pollen tube behind the vegetative nucleus. When the pollen tube enters the embryo sac the tip of the tube breaks down to release the generative nuclei. On entering the nucellus and embryo sac they participate in the fertilization process. One fuses with the egg nucleus to form the zygote and the other usually fuses with the polar nuclei or definitive nucleus to form the primary endosperm nucleus.

genetic code The sequence of nucleotides along the DNA of an organism within which is incorporated the information necessary for protein synthesis. By controlling the type and amount of protein manufactured the genetic code controls the growth, development, and characteristics of an organism. The observation that DNA could specify the incorporation of twenty different types of amino acids into protein but itself consisted of only four bases implied that more than one base must code for a single amino acid. The shortest sequence of any four bases that could specify all the types of amino acids found in protein is three, since $4^3 = 64$ possible combinations. (Pairs of any four bases could only code for $4^2 = 16$ amino acids, which is not enough.) Thus arose the *base triplet* or *triplet code hypothesis*. It was further proposed that the *triplets were sequential, not overlapping. This would account for the almost infinite variation in the sequence of amino acids in a protein. If there was a tendency for some amino acids to follow others in a polypeptide, then this might indicate an overlapping code, i.e. one where the last one or two bases of one triplet acted as the first one or two in the next triplet. However, this is not the case.

The theory that the genetic code consists of base triplets received considerable support from *frame-shift experiments*. These demonstrated that mutations resulting from the addition or loss of three base pairs close together produced protein resembling that of normal individuals. However, mutations resulting from the addition or loss of one, two, or four pairs produced significantly different protein. When three bases were added or lost only a short segment of protein was affected, whereas if other numbers were added or lost, the code for the whole protein was thrown out of sequence.

genetic drift (Sewall Wright effect) A change in *gene frequency that is due purely to chance, as opposed to natural selection. Genetic drift is more likely to occur in very small populations, since in such populations mating is likely to be nonrandom compared to matings in larger populations. *See also* fixation, founder principle.

genetic engineering (recombinant-DNA technology) The isolation of useful genes from a donor organism or tissue and their incorporation into an organism that does not normally possess them. For example, genes from a donor may be incorporated into the DNA of a microorganism, which, by its own replication, will produce many copies of the foreign gene and hence substantial quantities of the gene's products. These products, which may be enzymes, hormones, antibodies, etc., may have commercial or medical value. Genetically engineered microorganisms are already proving commercially successful, e.g. in the production of human insulin. However the manipulation of eukaryotic organisms is considerably more difficult.

A possible method of introducing foreign genes into higher plant cells involves using the tumour-inducing plasmid that is associated with the bacterium *Agrobacterium tumefaciens*, the causative agent of crown gall disease. Once introduced into a plant cell by the bacterium part of this plasmid becomes incorporated into the plant genome and disrupts the control of cell division. Often cell differentiation is also disrupted leading to the development of disorganized tumorous tissue. However certain strains of *A. tumefaciens* contain plasmids that do not cause disorganized growth. If new DNA sequences are incorporated into such plasmids and the plasmid introduced into a culture containing totipotent plant cells then a genetically transformed plant could be obtained. All dicotyledons are susceptible to *A. tumefaciens* but monocotyledons are not naturally infected.

Other natural vectors that have been tried in plant genetic engineering are the caulimoviruses, which include the cauliflower mosaic virus. There are considerable difficulties to be overcome in other methods not employing natural vectors because, if the introduced genes do not also cause an identifiable disease, there is the added problem of introducing mutant marker genes so transformed tissues may be distinguished.

genetic erosion The loss of genetic variation and the consequent narrowing of the genetic base of cultivated plants through the introduction of new improved varieties of crops that can be grown over wide areas. The continued existence of locally adapted primitive varieties is threatened by their large-scale replacement with adaptable high-yielding varieties. The situation is particularly serious in *gene centres, where valuable genes for such characters as disease resistance may be lost. *See also* genetic resources.

genetic load The proportion of disadvantageous genes that are present in and sustained by a population. Genetic load has three components; *mutational load*, *segregational load*, and *substitutional load*. The first arises because most mutations are harmful. The second may occur in cases of heterozygous advantage, since selfing a fit heterozygote Aa generates deleterious AA and aa offspring. The third, also called *environmental load* or *frequency-dependent load*, occurs during transient *polymorphism.

genetic resources The diversity and availability of alleles in both natural and artificially maintained stocks of organisms. For a number of years concern has been expressed about the need to maintain a diverse range of genotypes in domesticated plants and animals and their close relatives in the wild. A broad genetic base increases the capacity for variation and successful adaptation, such as resistance to pathogens or hostile environments. Both national and international programmes exist to maintain genetic resources of agriculturally important organisms. This had led to the establishment of a number of *gene banks or seed banks where potentially valuable seeds are evaluated, multiplied, and stored under ideal conditions. *See also* genetic erosion.

genetics The study of inheritance. The subject is now so large that it impinges on all branches of biology. It also has wide applications in agriculture (e.g. plant and animal breeding), medicine (e.g. blood typing), and industry (e.g. synthesis of antibiotics, hormones, etc. by genetic engineering). *See also* biochemical genetics, classical genetics, cytogenetics, population genetics.

genetic spiral (developmental spiral) A hypothetical spiral formed by drawing a line through and joining the centres of successive leaf primordia at a shoot apex. The form of the spiral depends on the angle of divergence and the distance between successive primordia. The first

geological time scale

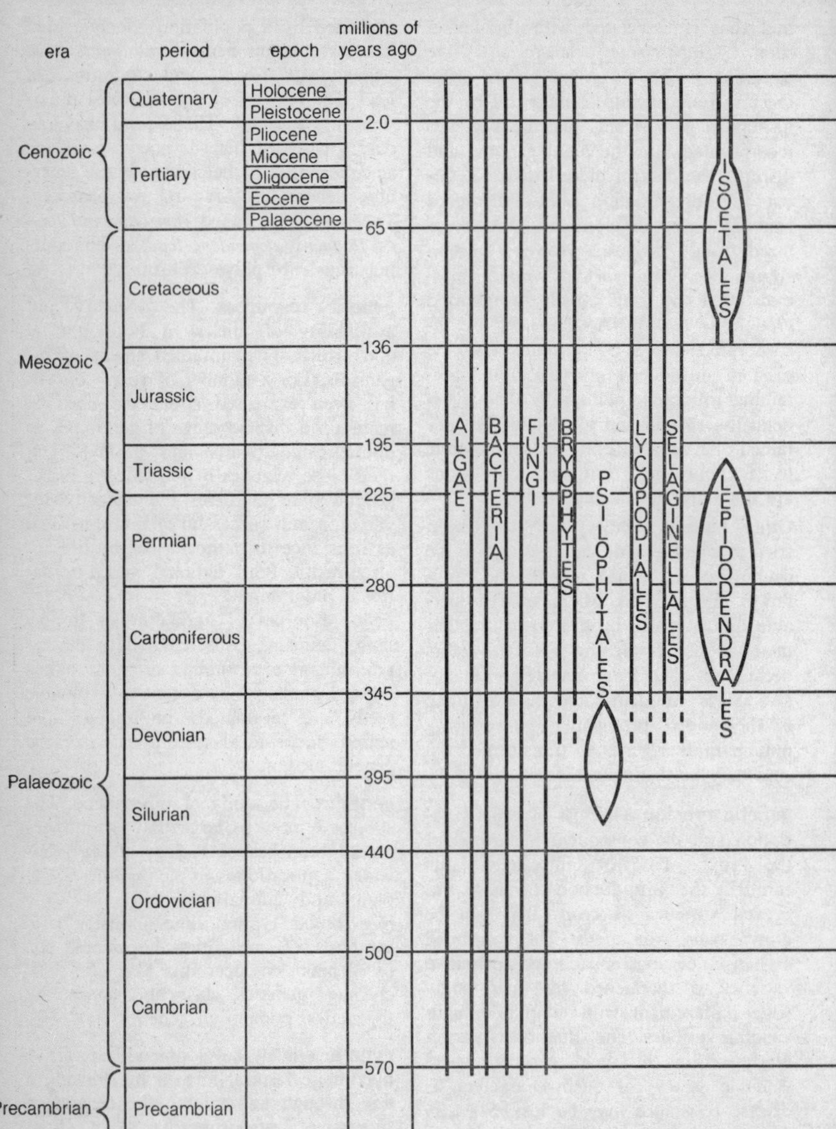

The geological time scale showing the approximate time of

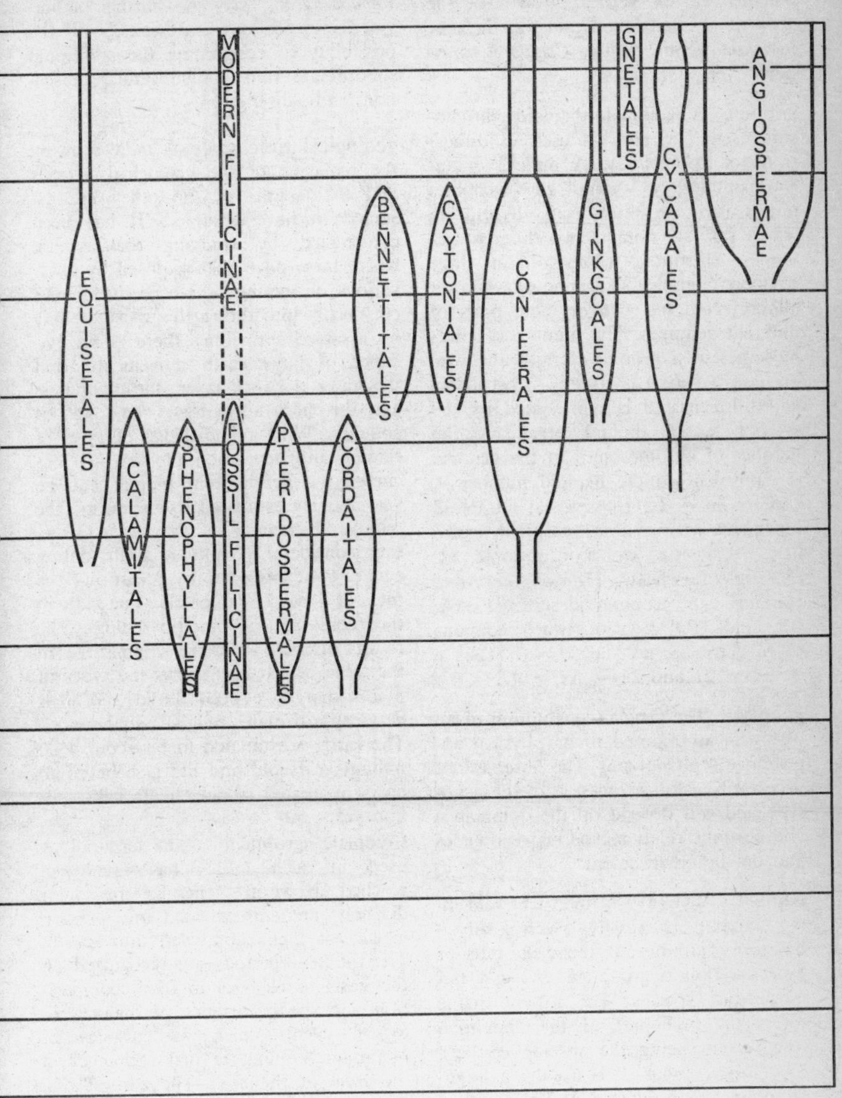

origin and the relative abundancies of different plant groups.

is often 137.5° and the second depends on the rate of vertical growth of the apex. It is often found that this distance increases geometrically. *Compare* parastichy. *See* phyllotaxis.

genome A complete haploid chromosome set. The term is used in discussions of *polyploidy. A diploid organism contains two homologous genomes unless it is an interspecific hybrid, in which case it contains two different genomes. An autopolyploid contains three or more homologous genomes while an allopolyploid has two or more pairs of different genomes. The number of chromosomes in a genome (termed the *base* or *basic number*) is usually between six and thirteen and is represented by the symbol x. In diploid organisms the number of chromosomes in the genome is equivalent to the haploid number n, and so $2n = 2x$. However in tetraploid organisms $2n = 4x$, in hexaploid organisms $2n = 6x$, etc. For example, allohexaploid wheat, *Triticum aestivum*, contains six genomes designated AA, BB, and DD each of which contains seven chromosomes, i.e. $x = 7$. Thus $n = 3x = 21$ and $2n = 6x = 42$.

genotype The genetic constitution of an organism as opposed to its physical appearance (*phenotype). The latter is not necessarily a full expression of the genotype and will depend on the dominance and epistatic relationships between genes and on the environment.

genus (*pl.* genera) An important rank in the taxonomic hierarchy, which is subordinate to *family, but above the rank of *species. It is a group of obviously homogeneous species. The generic name forms the first part of the *binomial (the second being the specific epithet), e.g. *Quercus robur*. It is usually a singular noun and is written in Latin with a capital initial letter, but lacks a uniform ending. Collections of similar genera are grouped into families. Large genera, e.g. *Rhododendron*, may be further subdivided into sections and series, with the possibility of designating the additional subordinate ranks of subgenus, subsection, and subseries.

geological time scale A table showing the sequence of the geological periods and the lengths of time they are assumed to have occupied. It has been constructed by studying rock strata, where these have been exposed by excavations or mining or where rivers have cut deeply into the earth's crust. It may be assumed, providing there is no evidence of large earth movements, that the lower the rock layer, the older it is and the more ancient are the fossils it contains. With a knowledge of rates of erosion and deposition the intervals occupied by the different periods can be very roughly estimated by measuring the relative thicknesses of the strata. However considerably more accurate dating can now be provided by measuring the rates of decay of radioactive materials in the rocks (*see* radiometric dating). The results of such work are summarized in the table, which also shows the types of plants that have been found and their very approximate relative abundances. The earth is estimated to be about 4600 million years old and life is believed to have originated about 3000–3500 million years ago.

Complete agreement on the terminology used in such tables has not been reached. However, generally the major divisions are *eras*, divided into *periods*, which are then subdivided into *epochs*. The different periods are recognized on the basis of changes in fossil composition and the occurrence of major geological events, such as episodes of mountain building or major changes in the level of the seas. The earliest era, the *Precambrian, has few fossils. The succeeding *Palaeozoic, *Mesozoic, and *Cenozoic eras have abundant fossils.

geophyte A plant with its perennating buds situated below ground on a rhizome, tuber, bulb, or corm. *See* cryptophyte.

geotropism A *tropism exhibited in response to gravity. Thus, when placed in a horizontal position with all-round illumination, shoots grow upwards and roots downwards, the former being negatively and the latter positively geotropic. A clinostat, consisting of a drum to which plant seedlings can be attached, revolves so as to cancel out the effects of gravity. Using it, one can demonstrate the lack of root and shoot curvature when no net gravitational forces are operating. Like *phototropism, the removal of the shoot tip destroys the geotropic response. Certain parts, such as tertiary roots, are naturally ageotropic. *See also* starch–statolith hypothesis.

germination The physiological and physical changes undergone by a reproductive body, such as a seed, pollen grain, spore, or zygote, immediately prior to and including the first visible indications of growth. The process will not occur unless both internal and external conditions are favourable. In seeds the various factors causing *dormancy must be overcome and dehydrated seeds must have a supply of water. Temperature and light can also affect germination. Following imbibition of water by a dry seed, the respiration rate increases markedly as food reserves are broken down and protein synthesis commences. The radicle is normally the first organ to emerge through the testa, followed by the plumule. *See also* epigeal, hypogeal.

germ plasm Genetic material, especially that contained within the reproductive (germ) cells. *See* Weismannism.

germ pore A thin-walled area on a spore or pollen-grain wall through which the germ tube may emerge on germination. In the pollen grain the germ pore is an area where the exine is either thinner or absent. Pollen grains may be identified to some extent by the shape and number of the germ pores. *See* colpus, pore.

germ tube The filament that emerges on germination of a spore. The structure that emerges from a pollen grain is termed a *pollen tube.

giant ferns *See* Marattiales.

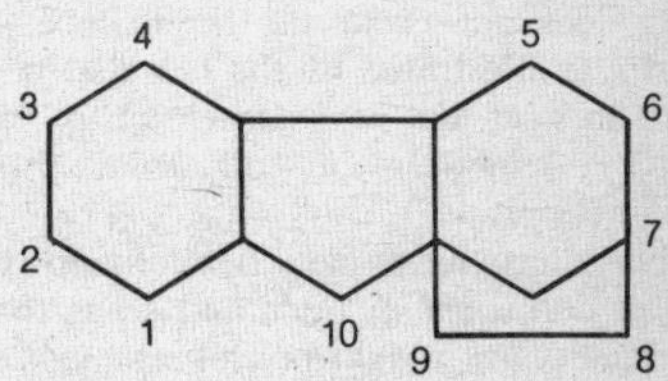

The gibbane skeleton.

gibberellic acid (GA_3) The first of the *gibberellins to be isolated and characterized. It was obtained from the fungus *Gibberella fujikuroi*, which causes a disease of rice seedlings characterized by excessive elongation of shoots and leaves. The basic structure of gibberellic acid and most other compounds with gibberellin-like activity is the gibbane carbon skeleton (*see* diagram). Gibberellic acid is a terpenoid and is synthesized from mevalonic acid. Other gibberellins are synthesized by the same pathway and there is probably much interconversion in the plant between the different gibberellins. Gibberellic acid is the gibberellin most widely used in experiments.

gibberellin Any of a group of plant growth substances first discovered through their ability to cause greatly increased stem elongation in intact plants. Gibberellins have subsequently been

shown to affect numerous aspects of plant growth and development. Most are chemically related to *gibberellic acid. A few compounds with gibberellin-like activity, e.g. helminthosporol from the fungus *Helminthosporium sativum*, have a quite different chemical structure. Gibberellin can overcome certain forms of genetic dwarfism and dwarf varieties are often used in bioassays for gibberellin. Gibberellin-induced stem extension is due to the effect of gibberellin on cell expansion, which it increases by influencing cell-wall expansibility. This however does not occur in the absence of auxin, as illustrated by the fact that gibberellin does not cause extension of excised internodes, cut off from their auxin supply.
Gibberellins have also been found to break dormancy in buds and seeds that normally have a light or chilling requirement. They can also partly or completely replace the photoperiod or cold requirement necessary to some species for flowering. Gibberellin levels are high in young leaves and, if applied to ageing leaves, can delay senescence. Levels are also high in developing seeds and fruits. Gibberellin applications can induce parthenocarpy and this has been put to commercial use in the production of seedless varieties of fruit. In barley seeds gibberellin has been found to stimulate the synthesis of the enzyme α-amylase. It does this by making available the messenger RNA responsible for α-amylase synthesis. This effect has proved of use in the brewing industry where α-amylase activity is essential for the production of malt. Gibberellin can even promote the enzyme's activity in inviable seed.
Gibberellins interact with other growth substances in various ways. There is evidence that *abscisic acid reduces gibberellin levels, hence the antagonistic effects of these substances. Gibberellin is believed to interact with auxin in the control of sex expression in dioecious plants. Gynoecious plants usually have low levels of gibberellin and high levels of auxin. Application of gibberellin can induce the formation of male flowers. This effect has been put to use in the cucumber industry where certain hybrid varieties are naturally gynoecious. Since pollen production is necessary for fruit set some of the plants are sprayed with gibberellin to produce the necessary male flowers.

gibbous Swollen or bulging on one side. The calyx of the narrow-leaved vetch (*Vicia angustifolia*) is an example of a gibbous structure.

gills The radially arranged lamellae that hang from the undersurface of the cap in certain fungi of the Basidiomycotina. The gills, which are covered by the *hymenium, are usually wedge shaped in longitudinal section and attached at the wider end of the wedge. This is believed to help prevent the spores becoming caught in the gills if the fruiting body is tilted slightly from the vertical. All gill-bearing fungi are traditionally placed in the family Agaricaceae, but this family is now sometimes split into a number of more homogeneous families.

Ginkgoales An order of gymnosperms containing only one living species, the maidenhair tree (*Ginkgo biloba*), which is known only in cultivation. It is a tall deciduous tree and bears fan-shaped leaves, usually with a distinct notch opposite the petiole. The leaf venation is exceptional among extant gymnosperms in being dichotomously branching. The Ginkgoales were more abundant in the Mesozoic and are possibly an offshoot of the Cordaitales.

glabrous Describing a surface that is devoid of hairs or other projections.

gland A group of one or more cells whose main function is to secrete a specific chemical substance or substances. A gland may be on or near the plant surface, discharging the secretion externally, or internal, releasing it into a canal or reservoir. Examples of the former are glandular hairs, nectaries, and hydathodes; the resin canals of *Pinus* are an example of the latter. *See also* laticifer, oil gland, trichome.

glaucous Describing surfaces having a waxy greyish-blue bloom, such as the leaves of rape or swede (*Brassica napus*).

GLC *See* gas–liquid chromatography.

gleba The mass of spore-producing tissue that forms in the fruiting bodies of such fungi as truffles, puffballs, earthstars, and stinkhorns. It is enclosed by the *peridium.

gley (glei) A waterlogged (hydromorphic) intrazonal *soil lacking in oxygen. In such conditions there is little decomposition of organic matter by bacteria, resulting in the accumulation of mor or raw humus. Beneath it is a *gley horizon* of blue-grey clay including ferrous compounds. Localized areas of oxidized rust-coloured ferric compounds may occur giving the soil a mottled appearance. *Gleyzation* may be seen in bog, meadow, and tundra soils.

gliadin A storage protein found in the caryopses of wheat. *See* prolamine.

gliding bacteria *See* Beggiatoales.

glochid A short barbed hair, many of which arise in tufts from the areoles of some cacti. It is a type of *glochidium. The presence of glochids is one character used to separate the subfamily Opuntioideae, whose members possess glochids, from the other two subfamilies of the Cactaceae, Pereskioideae and Cactoideae, in which glochids are absent.

glochidium Any hairlike projection with a hooked tip. In certain water ferns, e.g. *Azolla*, glochidia arise from the surface of the *massulae and serve to attach the *microsporangia to the *megasporangium prior to fertilization. *See also* glochid.

glucan Any polysaccharide made up exclusively of glucose subunits. Examples of glucans are starch and cellulose.

gluconeogenesis The formation of glucose from various precursors, such as pyruvate, certain amino acids (glucogenic amino acids), and intermediates of the TCA cycle. Most of the stages in the pathway are reversals of the reactions involved in glycolysis, with two exceptions. Phosphoenolpyruvate cannot be formed from pyruvate by reversal of the pyruvate kinase reaction of glycolysis. An alternative sequence of reactions involving the formation of oxaloacetate and malate and the input of a molecule each of ATP and GTP is performed to bypass this step. Similarly fructose 1,6-bisphosphate cannot be converted to fructose 6-phosphate by reversal of the glycolytic reaction catalysed by phosphofructokinase. The reaction is instead catalysed by the enzyme fructose bisphosphatase.

In plants glucose is formed predominantly by photosynthesis. Amino acids may be converted to glucose by the TCA and *glyoxylate cycles though, unlike animals, no distinction is made between glucogenic and nonglucogenic amino acids, all being suitable for gluconeogenesis by this pathway. Plants can also bring about the net synthesis of glucose from fatty acids via the succinate produced by the glyoxylate cycle. This process is utilized by germinating fatty seeds.

glucose (dextrose, grape sugar) An aldohexose sugar (*see illustration at* aldose). In solution it forms a six-membered *pyranose ring. Glucose is the major fuel source of nearly all organisms and the basic substrate from which starch, cellulose, sucrose, and other carbohydrates are synthesized. Glucose is synthesized by several routes; de novo synthesis from carbon dioxide is via the *Calvin cycle, but synthesis from fatty acids via acetyl CoA and the *glyoxylate cycle is also a major route in some cells (e.g. germinating oil-bearing seeds). It can be oxidized through *glycolysis and the *TCA cycle. Synthetic pathways from glucose involve activation, by phosphorylation or by formation of a nucleotide diphosphate sugar.

glucose 1-phosphate A phosphorylated derivative of glucose, formed by isomerization from glucose 6-phosphate. Glucose 1-phosphate is an intermediate in the synthesis of UDP-glucose, from which starch and cellulose are formed. It is also the predominant breakdown product of starch and cellulose.

glucose 6-phosphate A phosphorylated derivative of glucose. It can be formed by the action of hexokinase on glucose or by isomerization from glucose 1-phosphate. Glucose 6-phosphate is an important activated form of glucose. In glycolysis and in the pentose phosphate pathway the first step is activation of glucose to glucose 6-phosphate. It is also an intermediate in sucrose formation.

glucoside *See* glycoside.

glucuronic acid *See* uronic acids.

glume 1. One of a pair of *bracts subtending each spikelet in the inflorescence of grasses. *Compare* palea, lemma.
2. The bract subtending a flower in the inflorescence of reeds and sedges.

glutamic acid An acidic amino acid with the formula $HOOC(CH_2)_2CH(NH_2)COOH$ (*see illustration at* amino acid). Glutamic acid plays a central role in the cell's nitrogen metabolism. Along with glutamine it is the primary product of nitrogen assimilation, being formed from α-ketoglutaric acid by reductive addition of ammonia. Some amino acids are formed directly from glutamate (e.g. proline, ornithine), while others receive their amino group from glutamate in a *transamination reaction. Glutamate is also a precursor in the synthesis of purines, pyrimidines, and porphyrins.
In amino acid catabolism, glutamate is formed in transamination reactions with other amino acids. The carbon skeleton of these amino acids can then be further broken down, while the glutamate can undergo oxidative phosphorylation to reform α-ketoglutarate.

glutamine A polar uncharged amino acid with the formula $NH_2CO(CH_2)_2CH(NH_2)COOH$ (*see illustration at* amino acid). Glutamine is formed by the ATP-assisted addition of ammonia to glutamic acid. Breakdown is by a reversal of this reaction but without concomitant formation of ATP. Like asparagine, glutamine is important in the neutralization and storage of free ammonia. It is also an amino group donor, particularly in purine and pyrimidine synthesis.

glutelin Any of a group of plant proteins that are soluble in dilute acids and alkalis but insoluble in neutral salt solutions and in water and alcohol. *Glutenin* from wheat and *oryzenin* from rice are examples. Glutenin is the binding agent in flour pastes and dough. *Compare* prolamine.

gluten The mixture of gliadin and glutenin that remains when the starch of wheat grains has been removed.

glutenin *See* glutelin.

glycan *See* polysaccharide.

glyceraldehyde The simplest of the monosaccharides containing an aldehyde group. It has the formula $CHOCH(OH)CH_2OH$ (*see illustration at* aldose). In its phosphorylated form, glyceraldehyde 3-phosphate (formula $CHOCH(OH)CH_2OPO_3^{2-}$), it is important in many metabolic pathways. In glycolysis glyceraldehyde 3-phosphate is formed, together with dihydroxyacetone phosphate, by the cleavage of fructose 1,6-bisphosphate. Glyceraldehyde 3-phosphate is also one of the compounds formed in the Calvin cycle.

glyceride *See* acylglycerol.

glycerol (glycerine) A trihydroxy alcohol with the formula $CH_2OHCHOHCH_2OH$. Glycerol is a basic component of nearly all complex *lipids. *Triacylglycerols are esters of glycerol and fatty acids, while *phosphoglycerides and *glycolipids are esters of glycerol derivatives and fatty acids. Phosphorylated derivates of glycerol are intermediates in glycolysis and other areas of metabolism. Glycerol is often used as a *mounting medium in microscopy. *See also* triose phosphate.

glycerophosphatide *See* phosphoglyceride.

glycine The simplest amino acid. It has the formula NH_2CH_2COOH (*see illustration at* amino acid). There are several pathways of glycine biosynthesis. It can be made directly from ammonia, carbon dioxide, and a methyl group donated by the methylating agent N^5,N^{10} methylenetetrahydrofolic acid (mTHFA). In another reaction involving mTHFA, glycine can be formed from serine. A third pathway, from phosphoglyceric acid, involves transamination of glyoxylate. Glycine can be broken down either by reversal of the first or second synthetic pathway, or by deamination to glyoxylate and hence into the TCA cycle via malic acid.

Glycine is one of the basic components of porphyrins, along with *succinic acid. It is also necessary in purine synthesis.

glycogen A storage polysaccharide, similar in structure to *starch, important in animals but not found in plants. It is however present in certain bacteria, blue-green algae, and fungi. It is a branched glucan, like amylopectin, but the branching is more frequent, occurring every 8–12 glucose residues. The enzyme that catalyses glycogen synthesis in bacteria is similar to starch synthetase in that it requires ADP-glucose as a substrate. Breakdown of glycogen, like starch breakdown, is catalysed by *amylase.

glycolic acid The substrate that is oxidized during *photorespiration in C_3 plants. It is formed by the oxidation of ribulose bisphosphate to phosphoglyceric acid and phosphoglycolic acid. Hydrolysis of the phosphoglycolic acid yields free glycolic acid. During photorespiration the enzyme glycolate oxidase catalyses the oxidation, by molecular oxygen, of glycolic acid to glyoxylic acid. This reaction, which also produces hydrogen peroxide, takes place in the peroxisomes. The glyoxylic acid then appears to be transformed into glycine. In the cytoplasm two molecules of glycine react together to form serine and carbon dioxide. It is this carbon dioxide that is released in photorespiration.

glycolipid (glycosyldiacylglycerol) Any of a group of *acylglycerols containing a carbohydrate group, commonly a mono-, di-, or trisaccharide or an amino sugar. Glycolipids are the major lipid constituents of chloroplasts. No triacylglycerols are found in chloroplasts and phosphoglycerides are present only in small quantities. The commonest glycolipids are monogalactosyl- and digalactosyldiacylglycerols. They are synthesized from diacylglycerols and UDP derivatives of galactose.

glycolysis (Embden–Meyerhof–Parnas pathway) The metabolic pathway by which glucose is anaerobically degraded to pyruvic acid. In the glycolysis of one molecule of glucose, two molecules of ATP are used in the phosphorylation reactions at the beginning of the pathway and four molecules of ATP are formed later giving a net yield of two ATP molecules. Glycolysis is an example of an

anaerobic fermentation. Under aerobic conditions, however, it is important principally not as an energy-supplying pathway but as a preparation of glucose for entry into the *TCA cycle.
The enzymes of glycolysis are located in the cytosol, the most important being phosphofructokinase, which is the major regulatory enzyme in glycolysis. It is inhibited by ATP and citrate and stimulated by ADP and AMP.

glycoprotein A macromolecule consisting of a protein backbone along which short oligosaccharide branches are attached at intervals. The most important plant glycoprotein is extensin, a structural component of cell walls. The protein backbone of extensin is unusual in that it contains a large number of *hydroxyproline residues (about 30% of the total amino acid residues). The hydroxyproline serves to provide attachment points for the carbohydrate chains – in extensin these side chains are tetrasaccharides of arabinose.

glycoside A compound formed by the reaction of a *pyranose sugar with a nonsugar molecule (an aliphatic or aromatic hydrocarbon) termed the *aglycone*. The aglycone replaces the hydrogen in the hydroxyl group of carbon atom one of the sugar ring. Glucose is the sugar

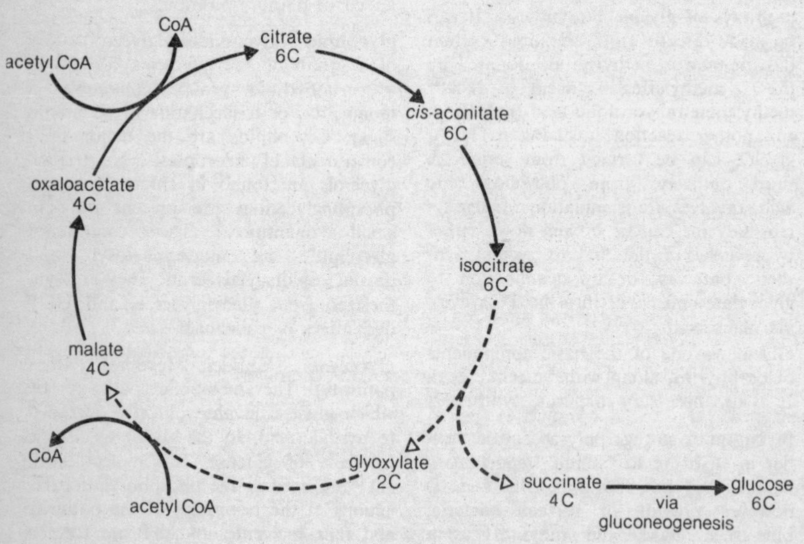

The glyoxylate cycle.

component of many glycosides, such compounds being called *glucosides*. Some rare sugars are only found in glycosides, e.g. *digitalose*, which has only been detected in certain *Digitalis* glycosides. Major classes of glycosides include the anthoxanthin glycosides, important as plant pigments, the steroid glycosides (*see* saponin, cardiac glycoside), and the cyanogenic glycosides, which release hydrogen cyanide on hydrolysis. An example of the last group is the glucoside amygdalin, which is obtained from certain members of the Rosaceae, e.g. almond (*Prunus amygdalus*) and peach (*Prunus persica*). Cyanogenic glycosides may act to deter grazing animals.

glyoxylate cycle (glyoxalate cycle) A cyclic series of reactions involving *TCA cycle intermediates in which one molecule of succinate is formed from two molecules of acetyl CoA. The reactions that differ from those of the TCA cycle are indicated by broken arrows in the diagram. These reactions are catalysed by isocitrate lyase and malate synthase respectively. The glyoxylate cycle avoids the carbon dioxide forming steps of the TCA cycle and thus allows net synthesis of carbohydrates from fatty acids via succinate (*see* gluconeogenesis). The enzymes of the glyoxylate cycle are active in germinating oil-bearing seeds and in other fat-metabolizing plant tissues. The cycle also takes place in microorganisms but not in higher animals.

glyoxysome A *microbody that contains the five enzymes of the glyoxylate cycle. Acetyl residues that are derived from the fatty acids of stored fats are converted to carbohydrates via succinic acid released in the glyoxylate cycle. Glyoxysomes are therefore found in those cells of higher plants where fats are utilized in this way, e.g. in the endosperm or cotyledons of fat-storing seeds. They also contain an enzyme that catalyses the decomposition of hydrogen peroxide to water and oxygen.

Gnetales A very distinct order of gymnosperms containing the three genera *Ephedra* (joint firs) with about 40 species, *Gnetum* with about 30 species, and *Welwitschia*, containing the one species *W. mirabilis*. Certain species of *Ephedra* are important as sources of the alkaloid drug ephedrine. Each genus shows certain advanced characteristics reminiscent of angiosperms. For example, their tracheids are arranged in columns and have highly perforated end walls similar to angiosperm vessels, while in *Gnetum* the sieve cells are closely associated with parenchymatous cells, reminiscent of companion cells. The leaves of *Gnetum* have a broad oval lamina with reticulate venation, very similar to certain angiosperm leaves. The Gnetales also have compound male and female strobili with superficially sepal- or petal-like structures, while the female gametophytes of *Gnetum* and *Welwitschia* lack archegonia. In some species of *Ephedra* there is a form of double fertilization, whereby one sperm nucleus fuses with the egg cell and the other with the ventral canal cell. However only the zygote undergoes further development. These angiosperm-like characteristics led some to consider the Gnetales as ancestral to the angiosperms but it is now generally considered that the group simply represents a specialized offshoot of the gymnosperms. The three genera of the Gnetales are considered so different from each other that in certain classifications they are raised to the status of orders (Ephedrales, Welwitschiales, and Gnetales) within the subdivision Gneticae (*see* Pinophyta).

Gneticae *See* Pinophyta.

golden-brown algae *See* Chrysophyta.

golgi apparatus (golgi body, dictyosome) An *organelle formed from flattened membrane-bound cisternae and a variable number of associated spherical vesicles. In most cells, between three and twelve saucer-like cisternae are stacked together with their concave surfaces directed towards the cell surface. The expanded 'rims' of the cisternae give rise to the surrounding spherical vesicles, which contain the products of biochemical activity within the golgi apparatus. If these vesicles move to the cell surface to release their contents, their membranes become incorporated into the *plasma membrane. The golgi apparatus is thus important in cell membrane synthesis.

Within the golgi cisternae, synthesis of glycoproteins, which commences on the polyribosomes of the rough *endoplasmic reticulum, is completed. These are moved within vesicles to the cell surface where they may be discharged from the cell or become incorporated into the cell membrane. Complex polysaccharides are also formed in the cisternae and are similarly distributed in vesicles to the cell surface. Cellulose synthesizing enzymes have been identified in the membranes of these vesicles, but they only become functional when incorporated in the plasma membrane. It is probable that hydrolytic enzymes in the golgi cisternae are organized into membrane-bound *lysosomes. In some unicellular algae (Haptophyta) scales that have been observed with the electron microscope on the cell surface originate in the golgi apparatus.

gonidium Any of the algal cells in a lichen.

graft hybrid *See* chimaera.

grafting A horticultural method of plant propagation in which a segment (the *scion) of the plant to be propagated is inserted onto another plant (the *stock) in such a way that their vascular tissues combine, so allowing growth of the grafted segment. The technique, which is mainly used on woody species, relies on the natural regenerative capacities of plants following wounding. In addition to the cambia of scion and stock growing together to form a continuous column, prolific development of callus around the graft area ensures a firm union. Grafting is only successful between closely related species. Thus apples are usually grafted on to different varieties of apple rootstock. Pears however are usually grafted onto quince rootstocks. *See also* chimaera.

grain *See* caryopsis.

Gramineae (Poaceae) A monocotyledonous family containing the grasses, which number about 9000 species in about 620 genera. Grasses generally have long narrow parallel-veined leaves inserted distichously on a round hollow stem. The inconspicuous flowers are usually borne in a terminal panicle, spike, or raceme consisting of a number of spikelets. Each flower is surrounded by two bracts. The fruit is a *caryopsis. Grasses are the dominant vegetation in savannas, prairies, and steppes (*see* grasslands). Economically they are the most important family of plants as they contain all the cereals, which are man's staple diet. Wheats (*Triticum*), maize (*Zea mays*), rice (*Oryza sativa*), barley (*Hordeum vulgare*), oats (*Avena sativa*), rye (*Secale cereale*), sugar cane (*Saccharum officinarum*), and sorghums (*Sorghum*) are all grasses. They are also widely planted for pasture and fodder.

gram-negative bacteria *See* Gram stain.

gram-positive bacteria *See* Gram stain.

Gram stain A stain used in bacteriology to distinguish between two physiologically distinct types of bacteria. The procedure involves staining the organisms with a basic dye, such as gentian violet or crystal violet, and adding a mordant, such as iodine or picric acid. In *gram-positive* bacteria these form a complex that cannot be removed by such decolorizing agents as acetone or alcohol. However *gram-negative* organisms lose the violet colour and on counterstaining with a red dye, such as carbol fuchsin or neutral red, will take up the red colour. Gram-positive organisms, on counterstaining, retain the original violet colour and can thus easily be distinguished from the red gram-negative bacteria. It is believed these differences may be related to cell wall structure. Thus gram-negative cell walls contain far more lipid than those of gram-positive cells, and a wider range of amino acids. However it is not clear whether it is the materials of the wall that act as the substrate for the differential staining reaction or whether they act by affecting the permeability of the wall.

Gram-positive and gram-negative bacteria differ in many other ways apart from their staining reactions. For example, gram-positive bacteria are often more exacting in their nutritional needs and more susceptible to antibiotics but more resistant to plasmolysis. Examples of gram-negative bacteria are the Pseudomonadales, Chlamydobacteriales, and various families of the Eubacteriales, e.g. Enterobacteriaceae and Brucellaceae. Gram-positive bacteria include those of the families Lactobacillaceae, Corynebacteriaceae, and Neisseriaceae.

grana Distinct stacks of *lamellae seen within *chloroplasts. They contain the pigments, electron transfer compounds, and enzymes essential to the light-dependent reactions of photosynthesis.

grape sugar *See* glucose.

grasses *See* Gramineae.

grassland A major regional community (*biome) in which grasses are the dominant vegetation. Usually the rainfall is insufficient for trees to grow (about 250–500 mm annually in temperate regions and 750–1500 mm in the tropics). Grasslands are common in continental interiors where the rain falls mainly in spring and summer. Most grasses are perennials but some are annuals or biennials. There are the low-growing closely packed turf-forming grasses of temperate regions and the tufted or tussock grasses, which grow in separate clumps and occur widely in temperate, tropical, and tundra regions.

Temperate grasslands do contain other herbaceous plants though trees are found only along rivers and streams. They include the North American *prairies*, the *steppes* of southwest Russia, the grasslands of Manchuria and Mongolia, and the South American *pampas*.

Tropical grasslands, known as *savanna*, have tall coarse tufted grasses and there are often scattered trees. In Africa baobab (*Adansonia digitata*) and *Acacia* and *Euphorbia* species are common. The climate alternates between cool dry winters and hot summers with heavy rains. Savanna is found in large areas of South America, East and South Africa, Southeast Asia, and northern Australia.

graticule *See* micrometer.

green algae *See* Chlorophyta.

green manure A fast-growing inexpensive crop that is sown towards the end of the growing season with the intention of ploughing or digging it in a short while later when it is still green. This increases the amount of organic matter in the soil. Legumes, such as clovers (*Trifolium*) and lupins (*Lupinus*), also increase the nitrogen content of the soil and are often used.

green sulphur bacteria *See* photosynthetic bacteria.

ground meristem The central part of the apical meristem, the derivatives of which give rise to the ground tissue (ground *parenchyma) and associated tissues. *Compare* procambium, protoderm.

ground tissue *See* parenchyma.

growth The sum total of the various physiological processes that combine to cause an increase in the dry weight of an organism and an irreversible increase in size. In most plants, growth is accomplished by the assimilation and fixation of inorganic substances from the surrounding environment. In contrast to animal growth, plant growth is usually confined to *meristems, where cell division occurs, and to the regions adjacent to these where *cell extension takes place.

growth correlation The relationship that exists between the different growth rates of various parts of a plant body. Growth rates depend on the balance of growth substances in the region and the competition between parts for nutrients. *See also* allometric growth.

growth factor Any substance that affects the growth rate of a plant or plant part. The term covers food reserves and minerals as well as the various *growth substances and *growth inhibitors found within plant systems.

growth inhibitor Any substance that retards the growth of a plant or plant part. Almost any substance will inhibit growth if present at high enough concentrations but two common hormonal inhibitors are *abscisic acid and *ethylene, which both have effect at very low concentrations, similar to the effective concentrations of auxins, gibberellins, and cytokinins. Other inhibitors, which must be present in much higher concentrations to take effect, include certain of the phenolics, quinones, terpenes, fatty acids, and amino acids.

growth ring The increment of secondary xylem produced in any one growing period in the stems and roots of many plants, especially those in temperate regions. In transverse section this is visible as a ring and is due to the difference in radial diameter of xylem elements formed at the beginning and end of the growing season. There may be one or several growth rings in any one year, i.e. an *annual ring may consist of more than one growth ring. The growth rings are then termed *false annual rings*. *See also* dendrochronology.

growth substance (hormone, phytohormone) Any substance that has a marked and specific effect on plant growth and that produces this effect when present in very low concentrations. Growth substances may stimulate plant growth by promoting cell division (*see* cytokinin) or cell elongation (*see* gibberellin) while others, such as *abscisic acid and *ethylene, inhibit certain developmental processes. The term includes substances produced within the plant and also artificial, often structurally related, chemicals that have similar effects. Endogenous growth substances are often produced in a particular region, for example *auxin is synthesized in shoot apices. They are then transported from these regions and take their effect at sites far removed from the point of production. Normal growth is only achieved when there is a correct balance of the various growth substances. Abnormal growth and even death may result from unusually high or low concentrations of a growth substance. This reaction is exploited in the use of certain weedkillers that contain synthetic auxins.

GTP (guanosine triphosphate) A nucleoside triphosphate containing the base guanine. It is formed from GDP (guanosine diphosphate) and phosphate during the deacylation of succinyl CoA to succinate in the TCA cycle. GTP plays a role in the activation of fatty acids, in gluconeogenesis, and in the initiation and elongation of polypeptide chains during protein synthesis.

guanine A nitrogenous base, more correctly described as 2-amino-6-oxypurine, derived from amino acids and sugars and found in all living organisms. Guanine combines with pentose sugar phosphates to form one of the nucleotides making up *RNA and *DNA. In addition, hydrolysis of the *nucleoside guanosine triphosphate (*GTP) releases energy to drive some energetically unfavourable reactions. *See also* nucleotide.

guanosine triphosphate *See* GTP.

guard cells The pair of specialized crescent-shaped epidermal cells immediately surrounding the stomatal pore and forming the *stoma. The opening and closing of the stoma is controlled by changes in turgidity of the guard cells, facilitated by the pronounced thickening of the cell walls adjacent to the stomatal pore. *Compare* subsidiary cell.

gullet An invagination at the anterior end of the cell of euglenoid algae. It has an expanded basal region or reservoir, a narrow passage rising to the surface, the *cytopharynx*, and an opening to the exterior, the *cytosome*. In the reservoir, in a region where there is no pellicle, particles from the surrounding medium can be taken into the cytoplasm by *endocytosis.

gum Any substance that swells in water to form gels or sticky solutions. Structurally they are mostly complex, highly branched polysaccharides, although a few gums with relatively simple structures are known.

Three main classes of gum are recognized. Acidic polysaccharides of glucuronic or galacturonic acids form glassy, hard gums; these are often produced by plants as a result of injury. Examples include gum arabic and tragacanth. In algae another class of acidic gums are found in which the acidity is due either to sulphate acid ester groups or to uronic acids. Examples are agar, alginic acid, and carrageenin. The third class contains gums that are obtained from seeds, examples being the seeds of the carob tree (*Ceratonia siliqua*). Natural gums are increasingly being replaced by synthetic substitutes in the manufacture of adhesives.

gum arabic A polysaccharide, found in the cell walls of certain plants, made up of D-galactose and D-glucuronic acid, and arabinose and rhamnose. It is obtained commercially from *Acacia* and is used in glues and pastes and as a mounting medium in microscopy.

gummosis A symptom of certain plant diseases in which there is an abundant formation of gum. Gummosis often occurs on trees but may occur on herbaceous plants, e.g. infection of young cucumber fruits with the cucumber scab fungus (*Cladosporium cucumerinum*) results in secretion of gum at the edge of the lesions.

guttation The exudation of liquid water onto a plant surface. It occurs under conditions of high humidity when the saturated atmosphere prevents *transpiration. The increase in *root pressure forces water out of special *hydathodes. The secreted water may contain calcium salts, which dry as a white crust at the leaf margins. Morning 'dew' on grass is often the product of guttation, as the lower temperatures at night provide ideal conditions for the process to occur.

Gymnospermae A class of the *Tracheophyta or *Spermatophyta containing about 700 species of chiefly arborescent seed-bearing plants. These are divided into some 67 genera and 12 families. Gymnosperms differ from the angiosperms in having naked seeds with no enclosing carpellary structure. Double fertilization does not occur and thus no true endosperm is developed. The gametophyte generation is not as reduced as in angiosperms, but neither is it autotrophic. The female gametophyte typically consists of at least 500 cells and distinct archegonia are produced in all genera except *Welwitschia* and *Gnetum*. Secondary vascular tissue develops in all members and consists of tracheids and sieve cells. Companion cells are lacking and vessels are absent except in *Gnetum*. The class contains the extinct orders *Pteridospermales, *Caytoniales, *Bennettitales, *Cordaitales, and *Pentoxylales and the extant orders *Cycadales, *Ginkgoales, *Coniferales, *Taxales, and *Gnetales. Gymnosperms have a particularly rich fossil record and were abundant in Carboniferous times, contributing largely to the formation of coal deposits. They originated during the Devonian period and were the dominant vegetation during the Jurassic and early Cretaceous. Towards the end of the Cretaceous their dominant position was taken over by the angiosperms. *See also* Pinophyta.

gynaecium *See* gynoecium.

gynandrous Describing stamens that are inserted on the gynoecium.

gynodioecious Describing plants that bear female and hermaphrodite flowers on separate individuals, as in thymes (*Thymus*) and oregano (*Origanum vulgare*). *Compare* androdioecious, gynomonoecious.

gynoecium (gynaecium) The female component of the angiosperm flower, which may be made up of one or more *carpels. The gynoecium usually occupies the most central part of the floral axis, although in more advanced forms it may not be symmetrically placed. If the gynoecium contains only one carpel it is termed *monocarpellary*, *unicarpellous*, or *stylodious*. If there are two or more separate carpels it is *apocarpous* and if the carpels are fused *syncarpous*. The gynoecium is represented by the letter G in the floral formula. *Compare* androecium. *See* ovary.

gynomonoecious Describing plants that bear female and hermaphrodite flowers on the same individual. For example, many members of the Compositae have female ray florets and hermaphrodite disc florets. *Compare* gynodioecious, andromonoecious.

gynophore An extension of the receptacle between the androecium and gynoecium that bears the ovary. Gynophores are found in many members of the Capparaceae, e.g. the spider flower, *Cleome spinosa*. *See also* androgynophore.

H

habitat The area in which an organism or group of organisms lives. It includes the climatic, topographic, *biotic, and, in the case of terrestrial habitats, *edaphic features of the area. There may be considerable variations in conditions within a habitat and also seasonal variations. Examples of habitats are seashore, woodland, pond, etc. Microhabitats are much smaller areas, for example, a bird's nest, a cow pat, or a log.

habituation The process of becoming accustomed to a new environment. The term is applied specifically to the ability of certain long-established callus cultures to synthesize auxin and hence become independent of exogenous supplies. Such habituated cultures are also called *anergized cultures*.

haematochrome A red pigment that accumulates in certain terrestrial green algae, e.g. *Trentepohlia*, under drought conditions.

haematoxylin A blue dye obtained from the leguminous tree logwood (*Haematoxylon campechianum*) that stains nuclei and cellulose cell walls blue. Haematoxylin itself is not a stain until it has been oxidized to haematin. It also usually requires a mordant, such as iron alum or an aluminium salt. Different haematoxylin solutions can be prepared, e.g. Delafield's haematoxylin, which can be used as a counterstain with safranin. *See also* staining.

hair A *trichome. The terms trichome and hair are often regarded as synonymous, but sometimes hair is restricted to relatively simple nonglandular trichomes. *See also* root hair.

halophyte A plant that is adapted to live in soil containing a high concentration of salt. Such plants are abundant in *salt marshes and mud flats. Halophytes must obtain water from soil water with a higher osmotic pressure than normal soil water. To achieve this the root cells of some halophytes have a very high concentration of salts and so are able to take up water by osmosis. *Succulent halophytes also store water for use when the salt concentration of the soil water rises further as a result of evaporation at low tide. An example of a succulent halophyte is sea rocket (*Cakile maritima*). There are also certain halophytic grasses that grow so abundantly they have played a major part in land reclamation. The most successful is the C_4 plant *Spartina townsendii*, which was deliberately introduced in the Netherlands in 1924 specifically for the purpose of land reclamation. *See also* halosere.

halosere A pioneer plant community that develops in a low-lying region of land originally beneath the sea and subsequently silted up with mud, silt, and shingle. The plants must be adapted to survive inundation by the tide twice a day. A major pioneer plant is glasswort (*Salicornia europaea*), which gradually gives way to salt-marsh grasses (*Puccinellia*) or sea aster (*Aster tripolium*). When the salinity decreases other species, such as thrift (*Armeria maritima*) and seablites (*Suaeda*), grow. The region becomes stabilized to form mudflats and *salt marshes. *See also* sere.

Hamamelidae (Amentiferae) A subclass of the dicotyledons containing mainly woody plants often with small unisexual apetalous flowers frequently borne in catkins. The number of orders recognized varies between about seven and fifteen. The Trochodendrales; the Hamamelidales, including the Platanaceae (e.g. plane trees) and Hamamelidaceae (e.g. witch hazels); the Eucommiales; the Leitneriales; the Myricales, including the Myricaceae (e.g. sweet gale); the Fagales, including the *Fagaceae (e.g. beeches, oaks, and chestnuts); and the Casuarinales, including the Casuarinaceae (e.g. she oaks) are usually recognized. Additional orders include the Cercidiphyllales and Eupteleales (both often included in the Trochodendrales); the Urticales (often included in the subclass *Dilleniidae); the Betulales, including the Betulaceae (e.g. alders, birches, hazels, and hornbeams), and the Balanopales (both often placed with the Fagales); the Juglandales (often included in the subclass Rosidae); and the Didymelales and Barbeyales.

hapaxanthic (monocarpic) Describing a plant that flowers only once during its life; after fruiting, the leaves senesce and the plant dies completely. It survives as a seed during the winter or dry season.

haplobiontic Describing a life cycle in which either the sporophyte or gametophyte generation is lacking. For example, in the green alga *Chlamydomonas* the plant is haploid and the diploid condition is represented only by the zygospore, which gives rise to haploid zoospores immediately on germination (*see* haplontic). In diatoms the plant is diploid and arises directly by the fusion of gametes formed by the sporophyte generation (*see* diplontic). *Compare* diplobiontic.

haplocheilic Describing a gymnosperm stomatal complex in which the subsidiary cells are not derived from the same initial as the guard cells, as occurs in cycads. *Compare* syndetocheilic. *See also* perigenous.

haploid A nucleus or individual containing only one representative of each chromosome of the chromosome complement. The haploid condition, denoted by the symbol *n*, is established by meiotic division of a diploid nucleus. In most plants (bryophytes, ferns, seed-plants, some algae) meiosis establishes a haploid generation, the gametophyte. Sooner or later the haploid gametophyte produces gametes by mitosis. In lower plants the situation is extremely variable. Some, such as *Fucus*, follow the pattern more characteristic of animals, where meiosis results directly in the formation of gametes. In other cases, such as *Spirogyra*, the only form of the plant is haploid, and the diploid stage is restricted to a single-celled zygote.

In flowering plants the haploid gametophyte generation is reduced to the pollen tube in the male and the embryo sac in the female. However haploid plants can be obtained by culturing pollen grains under suitable conditions. Haploid plants may also be obtained when a zygote formed from an interspecific cross sheds all the chromosomes of one parent as it undergoes development. This phenomenon has been demonstrated when barley (*Hordeum vulgare*) is fertilized with pollen from the wild barley *H. bulbosum*. Haploid plants have great potential in plant breeding as it is possible, by doubling the chromosomes of a haploid plant, to obtain a completely homozygous plant. This may be impossible by other means, especially with self-sterile plants.

haplontic Describing a life cycle in which the haploid phase predominates and the diploid stage is limited to the zygote. Such life cycles may be found in the filamentous Chlorophyta. *Compare* diplontic. *See also* haplobiontic.

haplostele A *protostele in which the central core of xylem is circular when seen in transverse section. This is believed to be the most primitive type of protostele and was present in the earliest vascular plants, e.g. *Rhynia*.

hapteron (*pl.* haptera) **1.** *See* holdfast.
2. The outer wall of the spore in species of *Equisetum* (horsetails). It differentiates into an elongated X-shaped structure, the arms of which are tightly coiled around the spore. Following release of the spores from the sporangium the haptera dry out and uncoil, this movement assisting in the dispersal of the spores. They are thus similar in function to elaters.

haptonasty A nongrowth *nastic movement in which a plant part moves in response to a tactile stimulus. An example is the rapid and progressive collapse of *Mimosa pudica* (sensitive plant) leaflets, often throughout the whole plant, after those of one particular branch are touched. *See also* pulvinus.

haptonema A threadlike structure that arises between the two flagella in algae belonging to the *Haptophyta. It is not involved in cell movement, but may be used for temporary anchorage. The length varies and in some cases it is coiled when the cell is motile. Internally it consists of three concentric membranes surrounding a ring of fibres and a central space.

Haptophyta A division comprising mainly unicellular biflagellate *algae most of which have a *haptonema between the two flagella. It contains a single class, the Haptophyceae with many genera that used to be included in the *Chrysophyta. The new division was recognized on the basis of evidence from electron-microscope studies, which revealed certain characteristic features, such as the presence of scales covering the cells. Certain marine forms, the *coccolithophorids*, have small intricate plates (*coccoliths*) of calcium carbonate completely covering the cell. Cretaceous chalks contain the remains of vast numbers of such algae.

haptotropism (thigmotropism) A *tropism in response to an external contact stimulus. The movements are most clearly exhibited by the tendrils of certain species. In passion flowers (*Passiflora*), the tendril is coiled spirally with the lower side facing outwards. The tendril later straightens and the sensitive tip undergoes *circumnutation. If it touches a solid body, it bends towards the stimulated side, a process occupying perhaps several minutes. The bending results in fresh parts of the tendril making contact with the stimulus and so the process is continuous, the tendril gradually encircling the support.

hardening The gradual exposure of a plant to increasingly lower temperatures to increase its resistance to frost. This is usually achieved by placing plants in a cold frame and gradually increasing the ventilation. Seedlings raised in greenhouses are often hardened off in this manner before planting out.

hard seed Any seed with a tough impervious outer coat that will not allow the entry of water. *Imbibition and germination therefore cannot occur until the seed coat is ruptured, either by scarification or microbial action. The sweet pea (*Lathyrus odoratus*) has hard seeds.

hardwood *See* secondary xylem.

Hardy–Weinberg law The law stating that, provided certain conditions are met, the gene (allele) frequencies in a population of organisms will remain constant and be distributed as p^2, $2pq$, and q^2 for the genotypes AA, Aa, and aa respectively where p = the frequency of the dominant allele and q = the frequency of the recessive allele, such that $p + q = 1$ (i.e. A and a are the only alleles). The law only holds providing that: the population is large (theoretically infinite); the population has been produced by random breeding; there is no natural selection for or against any particular genotype; there is no differential migration into or from the population; and there is no mutation.
Despite these conditions, Hardy and Weinberg's law is the basic theorem of population genetics. From it can be calculated the frequencies of A and a, even though a significant proportion of the a alleles are masked in heterozygotes. Thus the frequency of a (q) = $\sqrt{}$(frequency of homozygous recessives), and the frequency of A (p) = $1 - q$. If a population does not fit the distribution $p^2 + 2pq + q^2$ then one or more of the conditions stated above are not being fulfilled. The usual reason is natural selection against a particular phenotype. The theorem can be extended to enable the effects of natural selection on gene frequencies to be calculated. In effect, this provides a yardstick by which the rate of evolution can be measured and quantitatively defined.

Hartig net *See* ectotrophic mycorrhiza.

hastate Describing a leaf shaped like a spear, with three lobes, one pointing forwards and two pointing sideways either side of the petiole. The leaves of the copse buckwheat or bindweed (*Polygonum dumetorum*) are an example.

Hatch–Slack pathway An alternative form of carbon dioxide (CO_2) fixation found in *C_4 plants (*see* diagram). In such plants the first product of CO_2 fixation is not the three-carbon phosphoglyceric acid (*see* Calvin cycle) but the four-carbon oxaloacetate. This is formed by the carboxylation of phosphoenolpyruvate (PEP) by the enzyme PEP carboxylase. The oxaloacetate is then either reduced to malate or trans-

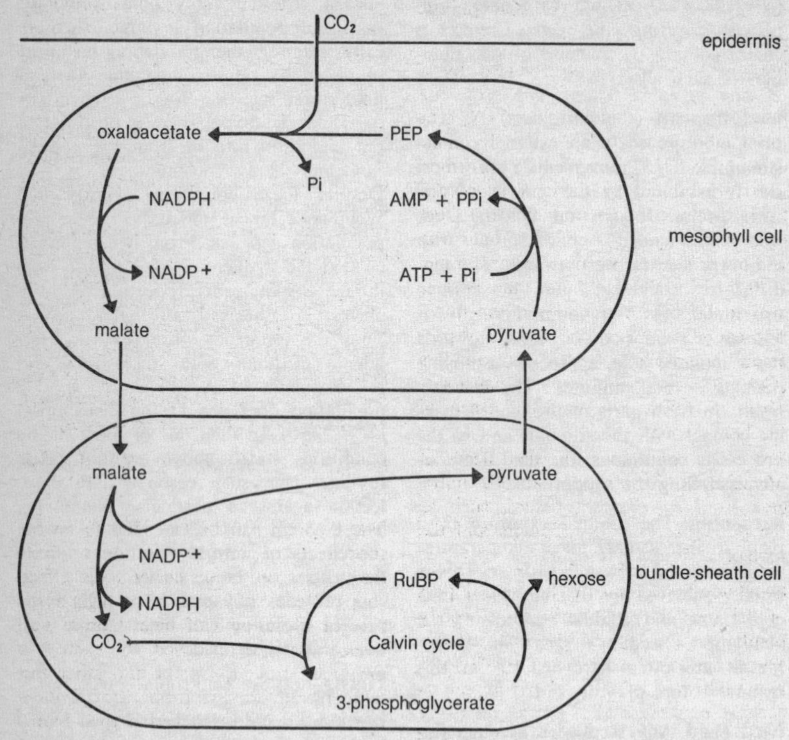

The Hatch-Slack pathway.

aminated to form aspartate. These reactions occur in the cells of the mesophyll. The malate or aspartate is then transported to bundle-sheath cells situated around the leaf veins (*see* Kranz structure) and decarboxylated to form CO_2 and pyruvate. The CO_2 so released reacts with ribulose 1,5-bisphosphate to form two molecules of phosphoglyceric acid. The normal Calvin sequence of reactions then commences. The pyruvate is returned to the mesophyll cells where it is converted to PEP with the concomitant formation of a molecule of AMP from ATP. This step, which uses up two high-energy phosphate bonds, is the reason why, overall, C_4 plants require 30 molecules of ATP for each molecule of glucose synthesized whereas C_3 plants only require 18.

PEP carboxylase has a far higher affinity for CO_2 than RUBP carboxylase and C_4

plants are consequently more efficient at fixing CO_2 than C_3 plants. This accounts for their lower *compensation points. *See also* photorespiration.

haustorium An organ produced by a parasite to absorb nutrients from its host. Most commonly the term refers to the hyphal extensions formed by obligate fungal parasites, which protrude into individual host cells. A successful fungal parasite does not kill the host cells and manages to insert haustoria without rupturing the plasma membrane. If the host cells are killed, the host is said to be hypersensitive and the infection cannot develop. The suckers of parasitic angiosperms, such as dodders (*Cuscuta*), are also called haustoria. These penetrate the vascular tissue of the host plants.

The term haustorial is also used of nonparasitic organs that absorb nutrients from surrounding tissues, such as the pollen tubes of seed plants and the foot of a sporophyte plant.

heartwood (duramen) The central part of the secondary xylem in some woody plants containing nonfunctional tracheary elements, often blocked by *tyloses and infiltrated with organic compounds such as resins, tannins, gums, and many aromatic substances and pigments. The heartwood is derived from the *sapwood as it deteriorates due to age and damage. Heartwood is more highly prized than sapwood for the making of furniture, due to its greater density, darker colour, and greater resistance to decay.

heath A region of land that is exposed to strong winds and has poor sandy well drained soil. These conditions are thought to be partly responsible for the absence of trees although heaths are surrounded by forest regions. However many heaths are subclimax communities and lack trees only because they have been regularly burnt to provide grazing land. Heathland is common near the Atlantic coast of Western Europe and on the North European Plain. Heath plants include bilberry (*Vaccinium myrtillus*), fine-leaved bell heather (*Erica cinerea*), gorses (*Ulex*), and broom (*Cytisus scoparius*). Despite the acidity, the peat layer is very thin.

Subclimax communities on calcareous soils as seen, for example, on the downs of southeast England, are sometimes called heaths but should correctly be called *chalk grassland*. These have a very different flora including such species as salad burnet (*Sanguisorba minor*), upright brome (*Zerna erecta*), and meadow oat grass (*Helictotrichon pratense*). If shrubs and trees can grow, as in ungrazed areas, these include yew (*Taxus baccata*), juniper (*Juniperus communis*), and hawthorn (*Crataegus monogyna*).

heavy-metal shadowing *See* shadowing.

heavy-metal tolerance The ability of certain plants to grow in areas where the concentration of heavy metals, e.g. copper, lead, and zinc, is so high as to prevent the growth of most plants. Tolerance may be achieved by excluding the metal from the plant altogether but more usually results from *detoxification*, when the metal is converted to a harmless form. Some *Agrostis* species exhibit copper tolerance, restricting the metal to the roots and thus protecting the more sensitive shoots.

helical thickening *See* spiral thickening.

heliotropism Another word for *phototropism, though implying movement in response to sunlight rather than artificial light.

helophyte A marsh plant with its perennating buds situated in the mud at the bottom of a lake or pond. Examples are bulrushes (*Typha*) and water plantains (*Alisma*). *See* cryptophyte.

Helotiales (cup fungi) An order of the *Discomycetes containing some 1750 species in about 225 genera. Its members are similar to the *Pezizales except that the asci lack an operculum. Most are saprobes but some are plant parasites, e.g. *Sclerotinia*, which causes brown rot of stored fruit. *See also* Lecanorales.

Hemiascomycetes A class of the *Ascomycotina containing ascomycetes that do not develop ascocarps. It contains about 330 species in some 60 genera. Such fungi are unicellular or produce a poorly developed mycelium. The Hemiascomycetes contains the orders *Endomycetales, *Taphrinales, and Protomycetales. In some classifications these fungi are considered a subclass, Hemiascomycetidae of the class Ascomycetes.

Hemibasidiomycetes *See* Teliomycetes.

hemicellulose Any of a variety of polysaccharides found in plant cell walls often in close association with cellulose. Hemicelluloses differ from cellulose in being composed of pentose sugars (arabinose or xylose) or hexose sugars other than glucose (e.g. mannose and galactose). They also differ in that, as well as having a structural function, like cellulose, in cell walls, they can be broken down by enzymes and so act as a nutrient reserve. Cellulose by contrast is metabolically inactive once it has been incorporated into the cell wall.

hemicryptophyte A plant with perennating buds situated at or just below the soil surface. Hemicryptophytes are usually herbaceous perennials and are commonly found in cold moist climates. *Rosette plants are hemicryptophytes. *See also* Raunkiaer system of classification.

hemizygous The situation in which the normal diploid nucleus contains only one copy of a particular gene (or chromosome). Thus the unpaired section of the X chromosome in the heterogametic sex (XY) is described as hemizygous.

hep *See* hip.

Hepaticae (Marchantiopsida) A class of the *Bryophyta containing the thallose and leafy liverworts, which number about 10 050 species in about 295 genera. The Hepaticae differ from the *Musci (mosses) in showing marked dorsiventrality in the gametophyte. The antheridia and archegonia may be borne on the surface of the thallus or on fleshy stalks (*see* gametangiophore). The capsule of the sporophyte, which contains sterile elaters as well as spores, matures before the seta lengthens, while in mosses the reverse occurs. The capsule does not contain a central pillar of sterile cells (columella) as is found in the Musci and Anthocerotae. The Hepaticae is divided into some five orders, of which the Jungermanniales (8000 species) and Marchantiales (2000 species) are the largest. A typical thallose liverwort is *Pellia*, which consists of a flattened dichotomously branching and frequently deeply lobed thallus. The ventral surface has numerous unicellular rhizoids growing from the area around the midrib. The leafy liverworts (suborder Jungermanninae) generally have three rows of leaves arising from a prostrate stem, although usually only the two dorsal rows are fully developed.

Herbaceae *See* Lignosae.

herbaceous perennial *See* perennial.

herbarium (hortus siccus) A collection of dried pressed plants, mounted on sheets of thin card, accompanied by data labels and stored in pest-proof wooden or metal cabinets. Smaller organs, including pollen grains, are perfectly preserved in this way, and many other features of the plant (e.g. anatomy, morphology, and chemistry) may be retained virtually unaltered. Details of floral structure can be observed by boiling or soaking in a wetting agent. The data labels, besides giving the plant's name, usually also include the data and place of collection and the collector's name. Notes on habitat, local names, and local uses of the plant (for food, medicine, etc.) may be invaluable in later searches for new sources of drugs, etc.

Specimens are normally arranged according to a particular taxonomic system (e.g. that of Bentham and Hooker in many UK herbaria) but occasionally material is filed in alphabetical order of family, genus, and species. Often specimens are further segregated into geographical regions within their family or generic groups. Herbaria may be small local collections, containing, for example, a county flora, or large international assemblages as at the Royal Botanic Gardens, Kew, where approximately five million specimens are housed. The larger herbaria are centres for taxonomic research and usually provide a plant identification service to other institutions and to the public.

herbicide (weedkiller) Any chemical that, when applied to a plant, will either destroy it or seriously inhibit its growth. Herbicides exist in many forms but may be subdivided into two basic categories: the *contact and the *systemic herbicides. *See also* selective.

heredity The phenomenon by which offspring resemble their parents and the laws that govern this.

heritability The proportion of the total variance of an observable characteristic that may be accounted for by genetic factors. Knowledge of the heritability of a trait is of particular value to agriculturalists. For example, if a variable characteristic like yield shows low heritability in a particular crop, this suggests that most of the variation is due to environmental factors and thus selection of high-yielding plants over a number of generations would not improve the yield.

hermaphrodite (monoclinous) Having both male and female reproductive parts in the same flower. This condition is thought to be derived from a primitive unisexual floral arrangement involving wind pollination and to have developed simultaneously with animal pollination. The hermaphrodite arrangement lends itself to self pollination and inbreeding but many species have intricate self incompatibility systems. *Compare* dicliny, dioecious, monoecious.

Heteroblastic leaf development in *Cannabis sativa.*

hesperidium A type of *berry that has a leathery epicarp, such as a citrus fruit. Fluid-filled trichomes fill the locule of each carpel to form the characteristic segments.

Heterobasidiomycetidae *See* Basidiomycotina.

heteroblastic development A progressive change in the form and size of successive organs. In many grasses, for example, successive leaf blades are progressively longer and only after a maximum length is achieved will the plant begin to flower. In many plants with compound leaves it may be seen that the younger leaves are simpler than the later leaves. For example, in hemp (*Cannabis sativa*) successive leaves have a greater number of lobes and the margins become more serrated (*see* diagram). This trend to greater complexity is often reversed after flowering.

heterochromatin A condensed region of *chromatin in the interphase nucleus that stains heavily with basic dyes. There are large amounts of heterochromatin in inactive nuclei. *Compare* euchromatin.

heterocyst Any of the large cells that occur at intervals in the filaments of certain species of blue-green algae. They do not contain chlorophyll but do possess large amounts of DNA. Narrow pores at one or both poles are also a distinguishing feature. It has been suggested that heterocysts are involved in nitrogen fixation. Some also consider they may be vestigial reproductive cells.

heteroecious Describing a rust fungus in which the various spore forms are developed on two different and usually unrelated hosts. An example is *Puccinia graminis*, which causes black stem-rust of grasses and cereals and also infects barberry (*Berberis*). The *urediospores, *teliospores, and *basidiospores develop on the grass host and the *pycnospores and *aeciospores develop on barberry. *Compare* autoecious.

heterogeneous RNA *See* messenger RNA.

heterokaryosis The occurrence of normally two nuclei of different genotypes in a fungal cell or mycelium. Heterokaryons tend to grow better than their constituent homokaryons. If one type of nucleus possesses a mutation such that it cannot synthesize a particular compound, then this deficiency is overcome by the other type of nucleus, a phenomenon termed *complementation. Heterokaryosis also allows recombination in imperfect and homothallic fungi (*see* parasexual recombination).

heterokont Describing an organism having two different types of flagella, as have the motile stages of algae in the Xanthophyta, hence the former name of the taxon, Heterokontae. *Compare* isokont.

Heterokontae *See* Xanthophyta.

heterolactic fermentation *See* mixed lactic fermentation.

heteromerous Describing lichens in which the algal cells are restricted to a layer situated between an upper plectenchymatous cortex and a lower loosely woven medulla of fungal tissue. It is the more usual tissue arrangement. *Compare* homoiomerous.

heteromorphic **1.** (antithetic) Describing a life cycle in which the alternating generations are morphologically and physiologically distinct, as in the bryophytes and pteridophytes. *Compare* isomorphic.
2. *See* dimorphism, polymorphism.

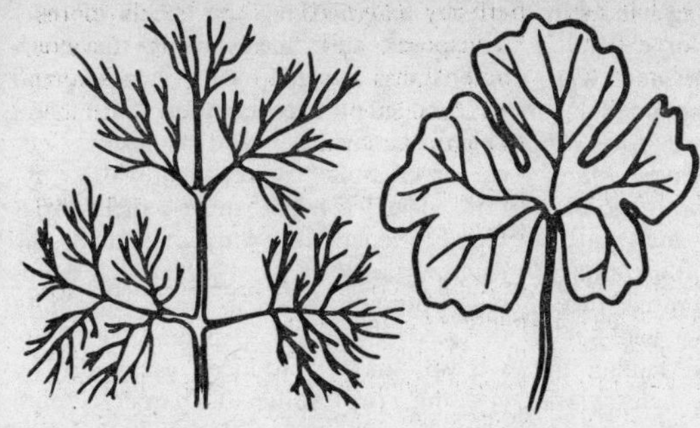

water buttercup
Ranunculus aquatilis

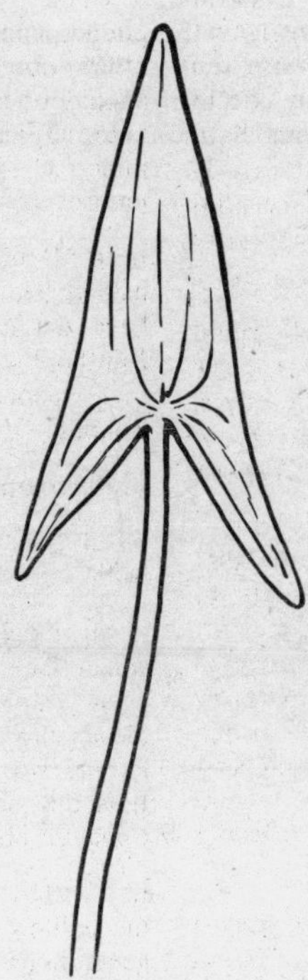

arrowhead
Sagittaria sagittifolia

submerged

aerial or floating

Heterophylly in two aquatic species.

heterophylly The possession of two or more leaf types, often differing widely in morphology and function. For example, certain species of *Lycopodium* and *Selaginella* have two rows of expanded lateral leaves and one or two rows of smaller adaxial or abaxial leaves. Many aquatic and semiaquatic plants have dissected submerged leaves and entire floating or aerial leaves (*see* illustration). The submerged leaves are adapted to reduce resistance to water flow while the floating leaves have a broad lamina to maintain buoyancy. *Compare* heteroblastic development.

heterosis (hybrid vigour) The exhibition by a hybrid of more vigorous growth, greater yield, or increased disease resistance than either parent. The effect is thought to be due to an accumulation of dominant alleles, each having additive effects, and masking the effects of deleterious recessive alleles. Thus, if a characteristic is affected by two genes X and Y, and the parents are XXyy and xxYY, the hybrid would be XxYy. Selfing or crossing between identical F_1 hybrids would produce an extremely variable F_2 generation with only half the heterozygosity of the F_1. Thus to maintain optimum heterosis and uniformity the seed of some crop plants (e.g. Brussels sprouts) is produced each year by crossing two different parental pure lines. Such crops are termed F_1 hybrids.

heterospory The production of more than one type of spore by a species. Usually large *megaspores containing food reserves and small *microspores

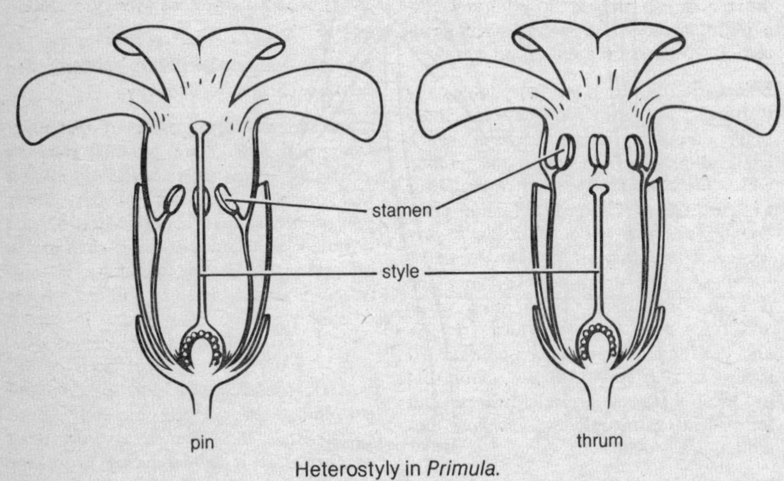

Heterostyly in *Primula*.

are formed. The megaspore eventually produces female gametes and the microspore male gametes. Heterospory is seen in certain lycopsids, e.g. *Selaginella*, in the water ferns, and in all the seed plants. The degree of heterospory is used as a diagnostic character in pteridophytes. *Compare* diplanetic.

heterostyly The existence of two or more different arrangements of the reproductive parts of a flower in one species. These differences are often related to the position assumed by the pollinating agent when visiting the flower. For example, in *Primula* two different arrangements are found (*see* illustration), one where the style is long and the stamens inserted at the base of the corolla on short filaments (*pin*), and the other where the style is short and the stamens are inserted in an almost sessile fashion on the neck of the corolla (*thrum*).
The different arrangements were thought to ensure cross pollination between the different types. Recent evidence does not support this hypothesis although the differences do appear to reinforce existing physiological barriers to self fertilization. *See* incompatibility, allogamy.

heterothallic Having morphologically identical but physiologically distinct thalli within a species between which fertilization can take place but among which fertilization does not occur. Thus gametes released from the same plant do not fuse and self fertilization is prevented. Heterothallism is thought to be the most primitive form of sexuality and is found in certain of the lower algae and fungi, e.g. the green alga *Ulothrix* and the fungus *Phytophthora infestans*. It is also seen in certain higher plants that produce morphologically identical but self sterile gametophytes. *Compare* homothallic.

heterotrichous Describing a filament made up of both erect and prostrate parts, as found in green algae of the order Chaetophorales and in brown algae of the Ectocarpales.

heterotroph An organism that depends for its nourishment on organic matter already produced by other organisms. All animals and fungi are heterotrophs. Parasitic plants and many bacteria also exhibit heterotrophism. *Compare* autotroph. *See also* photoheterotroph.

heterozygous Describing an individual that has been formed either from gametes possessing contrasting alleles of a single gene or from gametes differing in the arrangement of genes (*see* inversion). If only one of the alleles is expressed in the phenotype, then that allele is described as dominant, and the masked allele as recessive. Heterozygosity may be inferred from the failure of an individual to breed true for the characteristic under investigation. *Compare* homozygous.

hexose Any sugar with six carbon atoms. Most of the common monosaccharides are hexoses, e.g. glucose, fructose, galactose, and mannose. *See also* aldose, ketose.

hexose monophosphate shunt *See* pentose phosphate pathway.

hierarchical classification A system in which individuals are grouped into an ascending series of successively larger and broader categories, so that lower groups are always subordinate to, and included in, those that are higher in the hierarchy. Within the hierarchy, classification is based on the affinities of the component units. The order in which the 12 principal ranks are used is governed by the *International Code of Botanical Nomenclature and, although all available ranks in the hierarchy need not be used, their sequential order must not be altered. In ascending order the ranks are: form, variety, species, series, section, genus, tribe, family, order, class, division, and kingdom. Additional subcategories can also be designated (e.g. subseries and subclass) thus further extending the total number of ranks to 24.

high-energy phosphate bond A phosphate linkage with a negative standard free energy of hydrolysis, i.e. one that hydrolyses spontaneously, releasing energy. The last two phosphate bonds of ATP are examples.

Phosphate-bond energy is the major method of transfer of chemical energy from one enzymic reaction to another. For example, the high-energy phosphate bond of phosphoenolpyruvate can be broken to supply energy for the formation of ATP from ADP and phosphate. The ATP bond energy may then be used to drive an energy-requiring reaction, such as the formation of glutamine from glutamic acid.

Hill reaction The light-induced transport of electrons from water to nonphysiological electron acceptors, e.g. potassium ferricyanide, with the concomitant evolution of oxygen. The result is the reduction of the electron acceptor (*Hill reagent*) against the chemical gradient. Four moles of potassium ferricyanide are reduced for every mole of oxygen evolved. The reaction is named after Robin Hill, who discovered it in isolated chloroplasts in 1937.

hilum 1. The scar on the seed coat comprising a corky abscission layer where the seed or ovule was attached to the funiculus. In some species it may be large and highly coloured and may be used as a diagnostic character.

2. A point in a starch grain around which layers of carbohydrate are deposited. It can be detected when starch grains are stained with dilute iodine solution. The position of the hilum may be characteristic of a species. If the layering is uniform, the hilum is in the centre of the grain (e.g. maize starch) but if it is uneven, the hilum may be to one side (e.g. potato starch). In some cases (e.g. rice starch) there may be more than one hilum in each grain.

hip (hep) A type of *pseudocarp typical of members of the genus *Rosa* (roses). It consists of a large red cup-shaped hollow receptacle similar, except in colour and texture, to a *pome. The receptacle contains a collection of flask-shaped achenes each covered with small hooklike hairs.

hispid Covered with rough or stiff hairs as, for example, the leaves of the bristly hawk's-beard (*Crepis setosa*).

histidine A basic amino acid with the formula $C_3H_3N_2CH_2CH(NH_2)COOH$ (*see illustration at* amino acid). The occurrence of histidine in proteins is rare. Synthesis is via a complex pathway, starting from the condensation of ATP and 5-phosphoribosyl pyrophosphate, while breakdown is via glutamic acid. In bacteria, the synthesis of histidine has been found to be under genetic control, with histidine acting as its own repressor.

histochemistry The study of the distribution of certain chemical substances in cells and tissues by their colour reaction with specific chemicals. For example, cellulose will become bright blue when Schultze's solution is added. *See* staining.

histogen theory A concept of the organization and development of the apical meristem, in which the meristematic region is differentiated into three main zones, the dermatogen, periblem, and plerome. The *dermatogen* is supposed to give rise to the epidermis, the *periblem* to the cortex, and the *plerome* to all the primary tissues internal to the cortex. This theory has largely been replaced by the *tunica-corpus theory.

histogram A type of graph with vertical rectangular columns, the height of each column indicating the number of times that each class of result occurs in a particular sample or experiment. For example, the height of individual organisms in a given sample of a population can be plotted in this way.

histology The microscopical study of tissue structure.

histone Any of the basic proteins confined to the nuclei of eukaryotic cells. Their basic properties are due to a high proportion of the basic amino acids lysine and arginine. These properties enable them to interact very strongly with DNA. There are five types of histones: H1, H2b, H2a, H3, and H4. Two molecules of each type (except H1) aggregate together to form 7–10 nm nodules around which DNA winds. H1 appears to clamp DNA into position on the nodules. Each entire DNA-histone complex is called a *nucleosome. In forming nucleosomes, histones contribute to the packaging of relatively long DNA molecules into relatively minute chromosomes. Additionally, messenger RNA transcription only appears to be possible when H1 is released from the nucleosome. Histones may thus also play a role in the regulation of gene action. Histones are not associated with the genetic material of prokaryotes.

hnRNA *See* messenger RNA.

holdfast (hapteron) A structure found at the base of many algae in flowing or tidal water that serves to attach the plant to a support. It is often dissected into many finger-like processes or rhizoids.

Holocene *See* Quaternary.

holoenzyme The catalytically active complex, consisting of enzyme plus cofactor, that is formed by enzymes requiring a cofactor for activity. In some cases the holoenzyme may be extremely complex. For example, pyruvate dehydrogenase contains three different cofactors organized in a multienzyme complex. *See also* coenzyme, apoenzyme.

holophytic Designating organisms that obtain nourishment in the same manner as plants, i.e. that utilize the process of photosynthesis. The term is roughly equivalent to phototrophic. *Compare* holozoic.

holotype The sole specimen or other element either used by or designated by an author as the nomenclatural *type of a species when he first published the description. Whenever a new taxon is described it is now imperative both to designate a holotype and state where it is deposited. Duplicates are known as *isotypes.

holozoic Designating organisms that obtain nourishment in the same manner as animals, i.e. by the ingestion either of other organisms or their products, or other complex organic matter. The organic substances are digested, absorbed, and then assimilated. Holozoic organisms include, as well as animals, fungi and the insectivorous and parasitic green plants. The term holozoic is roughly equivalent to heterotrophic. *Compare* holophytic.

homeogenetic induction The influence of a differentiated cell on an adjacent undifferentiated cell such that it brings about similar differentiation in the adjacent cell. The process is seen when vascular tissues are cut. The severed ends of the vascular bundles induce differentiation in adjacent pith cells and thus re-establish a connection.

Homobasidiomycetidae *See* Basidiomycotina.

homoeology Similarity between chromosomes derived from different but related species. Depending on the degree of homoeology, pairing between homoeologous chromosomes may be seen at meiosis in segmental allopolyploids. Such polyploids display meiotic behaviour intermediate between autopolyploids and amphidiploids (*see* autopolyploidy, allopolyploidy).

homogamy The maturation of anthers and stigmas at the same time in the same flower so that the time of pollen presentation and reception coincides. Self pollination may be facilitated by this mechanism. *Compare* dichogamy.

homoiomerous Describing lichens in which the algal and fungal cells are evenly distributed, as in *Collema*. *Compare* heteromerous.

homologous **1.** Describing structures that have a common evolutionary or developmental origin. They may carry out similar or different functions. The stamens of angiosperms are generally said to be homologous with the microsporophylls of lower vascular plants. *Compare* analogous. *See also* divergent evolution.

2. *See* isomorphic.

homologous chromosomes (homologues) A pair of similar chromosomes, one of maternal and one of paternal origin. In the diploid cells of an organism, two sets of chromosomes are present, one set originating from the male gamete and one from the female gamete. When contraction and condensation of the chromatin of the chromosomes occurs in *prophase, their distinctive morphological features become apparent. The positions of *centromeres and *nucleolar organizers, and the *chromomere patterns are identical in homologous chromosomes and enable homologues to be identified. The genes in corresponding chromomeres influence the same characteristics and if one gene is dominant to the other, it will be expressed in the phenotype. Homologous chromosomes pair during prophase of the first meiotic division and subsequently separate to different daughter cells. *Compare* homoeology.

homonym A name identical in spelling to another, yet based on a different *type. The *International Code of Botanical Nomenclature rules that all homonyms except the earliest are illegitimate. For example, *Viburnum fragrans* Bunge, described in 1831, is a later homonym of *Viburnum fragrans* Loisel., which was described in 1824. Since this was the first published name for the taxon, it must be adopted.

homoplastic Describing *phenetic similarity resulting from convergent or parallel evolution rather than descent from a recent common ancestor. *Compare* patristic.

homospory The production of only one type of spore by a species. The term is commonly used of certain nonseed-bearing vascular plants, such as the psilopsids and most of the ferns, to distinguish them from other vascular plants that produce megaspores and microspores (*see* heterospory). *Compare* monoplanetic.

homothallic Describing species in which the thalli are morphologically and physiologically identical and fusion can occur between gametes produced on the same thallus, as in some lower algae and fungi. *Compare* heterothallic.

homozygous Describing an individual that has been formed either from gametes possessing identical alleles of a given gene or from gametes resembling each other in the arrangement of genes (i.e. not possessing any chromosome inversions or translocations). Homozygosity may be inferred from the ability of an organism to breed true for the characteristic under investigation. *Compare* heterozygous.

honey guides (nectar guides) Lines or dots on the petals that direct a pollinating insect to the nectaries. An example is the orange spot on the lower lip of the flower of the common toadflax (*Linaria vulgaris*).

hordein A storage protein found in the grains of barley. *See* prolamine.

horizon *See* soil profile.

hormocyst A resting stage formed by certain filamentous blue-green algae from side branches of the filament.

Hormogonales (Oscillatoriales) An order of the *Cyanophyta containing the filamentous blue-green algae. The filaments may be composed of roughly equal-sized cells, as in *Nostoc*, or cell size may diminish from base to apex, as in *Gloeotrichia*. Heterocysts and hormogonia are often present.

hormogonium A short filament of more or less spherical cells that may be formed on germination of an *akinete in certain filamentous blue-green algae. Although it does not possess flagella it is able to move, possibly by the streaming of mucilage along the surface. When it comes to rest on a suitable surface it gives rise to filaments.

hormone *See* growth substance.

hornworts (horned liverworts) *See* Anthocerotae.

horotelic *See* chronistics.

horsetails *See* Sphenopsida.

hortus siccus *See* herbarium.

hose-in-hose Describing a flower that has an extra set of petals within the normal corolla. Such flowers are seen in some azalea varieties, for example the evergreens 'Kirin' and 'Rosebud'.

host *See* parasitism.

hummock and hollow cycle A cyclic change in the vegetation that takes place on a developing raised bog. A pool of water on the peat is invaded by *Sphagnum* mosses and a low hummock is built up. Other plants, such as *Calluna vulgaris* (ling) and *Erica tetralix* (cross-leaved heath), then establish themselves. When the ling dies the hummock is easily eroded and reduced to a water-filled hollow surrounded by other hummocks in various stages. The cycle can then be repeated. Similar cycles occur in grassland, and in the tundra where frost hummocks of peat are formed by frost action and subsequently eroded.

humus The soft moist amorphous black or dark brown organic matter in soil derived from decaying plant and animal remains (especially leaf litter) and animal excrement. The complex chemical reactions involved in the decomposition by soil organisms result in the formation of end products that can be used by plants. Humus has a very complex and variable chemical composition. Its colloidal properties enable it to retain water, making it a useful addition to sandy soils. It also helps the formation of soil crumbs, which improve aeration and drainage in clay soil and can also provide insulation in soil by reducing temperature fluctuations. There are two main types of humus. *Mor* humus (raw humus) is formed in acid conditions (pH 3–4.5) where the plant litter is low in bases. It is thus acid and this, together with the extremes of moisture or temperature that are often found in such habitats, inhibits the activity of the soil fauna. The slow decay is thus brought about by saprophytic fungi. Bogs, heaths, and pine-forest soils have thick layers of mor humus. *Mull* humus (mild humus) is formed in less acid or alkaline soils, which are richer in calcium. Animals, e.g. earthworms, can consume it and the bacteria that bring about decay flourish. It may be found in warm or mild humid climates in such habitats as deciduous forest or grassland. An intermediate form of humus is termed *moder*. *See also* soil.

hyaline Thin and translucent.

hyaloplasm *See* cytoplasm.

hybrid An individual produced from genetically different parents. The term is often reserved by plant breeders for cases where the parents differ in several important respects. On the other hand, it may be used by geneticists for the F_1 of a monohybrid cross, where the only difference is in a pair of alleles at a single gene. Interspecific hybrids, resulting from a cross between different species, are often sterile (*hybrid sterility) if derived from haploid gametes, but this may be overcome by polyploidy (*see* allopolyploidy). Hybrids produced from different varieties of the same species (*F_1 hybrids) are often more vigorous than either parent, so although they cannot breed true, they may be favoured by agriculturalists. *See also* heterosis, graft hybrid.

hybridization Natural or artificial processes that lead to the formation of a *hybrid. The hybridization of normally self-pollinating species involves the removal of the anthers (emasculation) and the artificial transfer of pollen from another plant. A paintbrush is often used to place the foreign pollen on the stigma. Often hybridization techniques must overcome *incompatibility systems. The development of new horticultural and agricultural cultivars involves hybridization techniques. *See also* introgression.

hybrid sterility The reduced ability of some hybrids to produce viable gametes. This is caused by the absence of homologous pairs of chromosomes so that bivalents cannot form during meiosis. The resulting gametes are thus aneuploid. Hybrid sterility may be overcome by *polyploidy. Sterility does not necessarily correlate with poor growth or inviability. Indeed many hybrids show increased vigour (*see* heterosis).

hybrid swarm A continuous and variable series of hybrids between two original parental forms. It occurs where a barrier to reproduction between two distinct populations has broken down.

hybrid vigour *See* heterosis.

hydathode A secretory structure that removes water from the interior of a leaf and deposits it on the surface. This process is known as *guttation and often occurs during damp humid nights when water absorption is high but transpiration is minimal. In some plants, e.g. runner bean, the hydathodes are glandular hairs and water is exuded by active secretion, while in others they are water pores and exudation is passive, the water being forced out by hydrostatic pressure. Both types of hydathode are usually situated at vein endings. Water pores are generally incompletely differentiated stomata and in certain plants, e.g. sea lavenders (*Limonium*), function as salt glands. *See also* gland.

hydroid An elongated nonlignified water-conducting cell in the stems and leaves of certain bryophytes, such as *Polytrichum*, analogous to a *tracheary element in vascular plants. *Compare* leptoid.

hydrolase Any *enzyme that catalyses hydrolysis reactions. Examples are the phosphatases, glycosidases, and peptidases, which catalyse the hydrolysis of phosphate esters, glycosidic linkages, and peptide bonds respectively.

hydrophily *Pollination by pollen carried by water. The pollen may be transported on the water surface, as in tasselweeds (*Ruppia*) and water starworts (*Callitriche*), in which case it needs to be light enough to float and water repellent and the stigmas must be exposed at the water surface. Movement of such pollen grains may be enhanced by a natural outer coating of oil, which alters the surface tension of the water. In tape grass (*Vallisneria spiralis*) the whole male flower is released and attaches itself to the female flower at the surface of the water.

Pollen may also be transported through water. The mechanisms for this are very variable and are thought to be derived from entomophilic mechanisms. In naiads (*Najas*), pollen grains are heavy and sink due to gravity onto the female flowers underneath. In eel grasses (*Zostera*) pollen grains transform into structures that resemble a pollen tube, which wrap themselves around the stigma. *Compare* entomophily, anemophily.

hydrophyte (aquatic) A plant that is adapted to living either in waterlogged soil or partly or wholly submerged in water. Many hydrophytes absorb water and gases over the whole surface and have no stomata, e.g. the spiked water milfoil (*Myriophyllum spicata*), which is completely submerged in water. The mechanical and vascular tissue of many hydrophytes is reduced as water is plentiful and supports them. They often have large intercellular air spaces in their stems, roots, and leaves to overcome the difficulty of obtaining gases from the water. Hydrophytes that are partially submerged have floating leaves with stomata through which gases can be exchanged as in land plants. However, to prevent the leaves being flooded with water, the petioles may be very long or shaped like a corkscrew to adjust easily to changes in water level. In the giant water lily (*Victoria regia*), the enormous leaves have a vertical rim to prevent them from being flooded. Some species, e.g. water crowfoot (*Ranunculus aquatilis*), have both finely divided submerged leaves and floating leaves with stomata.

Hydrophytes are also a class in the *Raunkiaer system of classification, and are defined as having their perennating buds under water. The buds may become detached and sink to the bottom (*see* turion) or may be borne on submerged rhizomes as in water lilies (*Nuphar* and *Nymphaea*).

Compare mesophyte, xerophyte. *See also* hydrosere.

hydroponics (water culture) The growth of plants in a sterile medium, such as sand or vermiculite, to which nutrients are added in a balanced liquid fertilizer.

hydrosere A pioneer plant community that develops in water when the depth is decreased by silting. The species present depend on the chemical nature of the ground water, typical species being various water lilies and pondweeds. As the water becomes more shallow, reeds develop. The organic matter builds up to form peat and then the hydrosere gives way to swamp. Eventually *mesophytes, such as coarse grasses, develop followed by shrubs and trees. *See also* sere.

hydrotropism A *chemotropism in which water is the orientating factor. Thus most roots show positive hydrotropism, growing towards moister regions of the soil, while hypocotyls and the reproductive organs of certain fungi exhibit negative hydrotropism. If water is in short supply hydrotropism exerts a stronger influence on the direction of root growth than geotropism.

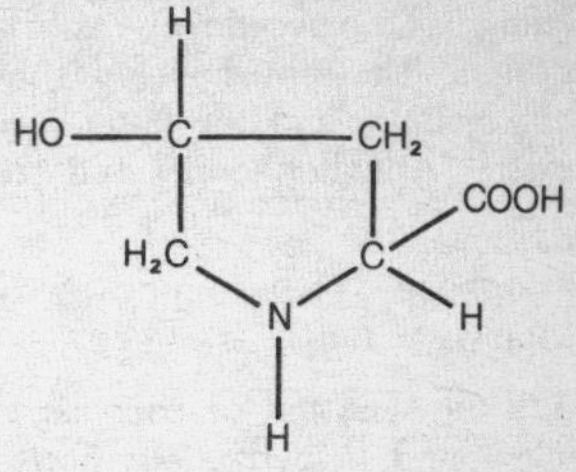

Hydroxyproline.

hydroxyproline A polar imino acid (*see* diagram) formed by the hydroxylation of proline. The enzyme that catalyses this reaction only acts on proline residues that have been incorporated into a polypeptide chain. Hydroxyproline is rare in most proteins but common in cell-wall proteins. Degradation of hydroxyproline does not follow that of proline; instead it is broken down to alanine and glycine.

hygroscopic Showing a readiness to take up moisture from the surroundings as, for example, imbibing seeds prior to germination.

hymenium In the higher fungi, a fertile layer consisting of asci or basidia. In the ascomycetes the hymenium lines the ascocarp while in basidiomycetes it lines the basidiocarp. The hymenium may be exposed at maturity, as in the Discomycetes and the Hymenomycetes, or enclosed, as in the Plectomycetes, Pyrenomycetes, and Gasteromycetes.

Hymenomycetes A class of the *Basidiomycotina containing those basidiomycetes with a well developed basidiocarp and basidiospores that are forcibly discharged. It contains over 5000 species in some 675 genera. It is divided into two subclasses: the Phragmobasidiomycetidae (or Heterobasidiomycetidae), which have septate basidia; and the Holobasidiomycetidae (or Homobasidiomycetidae), with aseptate basidia. The former includes the order *Tremellales and the latter includes the orders *Aphyllophorales and *Agaricales.

hymenophore The fruiting body of hymenomycete fungi.

hypanthium The flat or cup-shaped receptacle found in perigynous flowers. It is joined to the ovary when the ovary is inferior.

hyperplasia Abnormal overdevelopment as a result of an increase in the number of cells in response to a disease-producing agent. Common manifestations of hyperplasia are *witches' brooms, *cankers, *galls, *leaf curl, and *scab. *Compare* hypertrophy, hypoplasia.

hypersensitivity Increased sensitivity of a plant to attack by a particular pathogen so that the host tissue dies at the point of infection and thus prevents spread of the disease. The only symptoms are minute necrotic flecks. Plants showing this type of reaction to a particular pathogen are very resistant.

hypertonic Having a higher *osmotic pressure than an adjacent solution or a solution under comparison. If a hypertonic solution is placed within the confines of a permeable membrane and surrounded by a solution of lower osmotic pressure, then there is a net influx of solvent molecules, due to *diffusion, until the solute concentrations become equal. *Compare* hypotonic.

hypertrophy Abnormal overdevelopment due to an increase in cell size. It is seen, for example, in the roots of crucifers infected with club root (*Plasmodiophora brassicae*). *Compare* hyperplasia, hypoplasia.

hypha A branched filament, many of which together make up a fungal *mycelium. Hyphal walls are usually made up of chitin laid down in microfibrils. The hyphae of most higher fungi are septate, i.e. divided by cross walls (septa) into many uni- or multinucleate segments. Those of lower fungi are commonly coenocytic. Hyphae may aggregate and anastomose to form a tissue-like mass of *plectenchyma.

In certain fruiting bodies hyphae may be differentiated into thin-walled *generative hyphae*, thick-walled unbranched *skeletal hyphae*, and thick-walled branched *binding hyphae*. *Flexuous* describes the thin-walled branching hyphae found, for example, in pycnidia.

Hyphochytridiomycetes *See* Mastigomycotina.

Hyphomycetales The largest order of the *Hyphomycetes, containing those mycelial imperfect fungi in which the conidiophores are not compacted into specialized structures. It contains over 6000 species and some 635 genera including: *Penicillium* (common *moulds), important as the source of penicillin; *Aspergillus*, many species of which cause spoilage of stored food, in some cases (e.g. *A. flavus*) producing toxic substances (aflatoxins); *Botrytis* and *Cladosporium*, common causes of leaf spot diseases; and *Fusarium*, one of the damping-off fungi.

Hyphomycetes A class of the *Deuteromycotina containing mycelial imperfect fungi, which, if fertile, bear the conidia on hyphae. It contains over 7500 species in some 930 genera and is divided into the orders *Agonomycetales, *Hyphomycetales, Stilbellales, and Tuberculariales. In the latter two orders the conidiophores are compacted into specialized structures. *Compare* Coelomycetes.

hypobasal cell The inner of the two cells formed by the first division of the zygote. In plants showing exoscopic embryogeny it gives rise to the foot and in plants showing endoscopic embryogeny it gives rise to the embryo. *Compare* epibasal cell.

hypocotyl The region of the stem derived from that part of the embryo between the cotyledons and the radicle. The transition from the stelar arrangement of the stem to that of the root occurs in the hypocotyl. The 'root' of certain crop plants, e.g. turnip (*Brassica rapa*), is actually a swollen hypocotyl. *Compare* epicotyl.

hypodermis (hypoderm) One or more layers of cells lying immediately beneath the epidermis in the leaves and other organs of many plants, differing morphologically from the underlying tissues. A true hypodermis develops from the *ground meristem and therefore has an origin different from that of the epidermis as is evidenced by the noncoincidence of the anticlinal walls of the two tissues. However, the term is often used to refer to the layer or layers of cells immediately inside the outermost layer, regardless of their derivation. The hypodermis often contains large quantities of *sclerenchyma to enable it to act as a strengthening tissue, as in the leaves of many conifers. *See also* exodermis.

hypogeal Describing seed germination in which the cotyledons remain underground as there is no great lengthening of the hypocotyl. The cotyledons gradually give up their contents to feed the developing plumule and radicle. Hypogeal germination is seen in the broad bean (*Vicia faba*). The term is also used of fruits that develop underground, such as those of the peanut (*Arachis hypogea*) and of the Bambara groundnut (*Vigna subterranea*). *Compare* epigeal.

hypogyny The most commonly seen arrangement of floral parts in which the stamens, sepals, and petals are inserted below the ovary, giving a *superior ovary*. *Compare* epigyny, perigyny.

hyponasty A *nastic movement in which the plant responds by more rapid growth on the lower side of an organ, resulting in a curving upwards of that part. *Compare* epinasty.

hypoplasia Underdevelopment as a result of disease or nutrient deficiency, giving dwarfed or stunted plants. In some cases particular parts of the plant may develop abnormally or not at all. *Compare* hyperplasia, hypertrophy.

hypostasis The situation in which the expression of one gene (the hypostatic gene) is dependent on a second gene (the epistatic gene). The genes need not be on the same chromosome. *See also* epistasis.

hypotonic Having a lower *osmotic pressure than an adjacent solution or a solution under comparison. There is a net outflow of solvent molecules from a hypotonic medium into a more concentrated solution separated by a permeable membrane. *Compare* hypertonic.

I

IAA *See* indole acetic acid.

idioblast Any cell markedly differing from the cells of the surrounding tissue. The *brachysclereids of pear fruits are an example.

idiogram *See* karyogram.

i-gene *See* regulator gene.

illumination *See* dark-ground illumination, bright-field illumination.

imbibition The process by which a substance absorbs a liquid and, as a consequence, swells in volume but does not dissolve. Imbibition, which is usually reversible, is exhibited by many biological compounds, particularly by cell-wall constituents such as pectin, celluloses, and lignin. The testas of dehydrated seeds imbibe water just prior to germination and the process is also important in water uptake by roots. *See also* matric potential.

imbricate Overlapping. The term may be used to describe the form of aestivation in which the sepals and petals overlap in the bud. It is also used of mature structures, e.g. the overlapping spikelets of foxtails (*Alopecurus*). *Compare* valvate.

imino acid A molecule containing a carboxylic acid group and an imino (NH) group. *Proline and *hydroxyproline, often described as amino acids, are in fact imino acids. Other imino acids are probably formed during the degradation of amino acids.

immunity Complete *resistance to a particular disease. Plants do not have an immune system, as animals do, but instead rely on physical barriers to prevent entry of a pathogen, or on various physiological reactions, e.g. *hypersensitivity or production of *phytoalexins.

immunoelectrophoresis A serological technique used in chemotaxonomy to compare protein extracts from different plant species. An extract from each species (the antigens) is dispersed in a gel by *electrophoresis. Antisera, produced by injecting a laboratory animal with an extract from one particular species, is then allowed to diffuse through the gel in a direction at right angles to the electrophoretic separation. Precipitation arcs form where the concentrations of complementary antigen and antisera are greatest. The degree of similarity between the antigens in the gel and the antigens of the species used to produce the antisera may be estimated by observation of the number, shape, clarity, etc. of the precipitation arcs. *See also* serology.

imparipinnate Describing a pinnate leaf having a centrally placed unpaired terminal leaflet. Such leaves are seen in most milk vetches (*Astragalus*). *Compare* paripinnate.

imperfect fungi *See* Deuteromycotina.

inbreeding The production of offspring by the fusion of genetically closely related gametes. Self fertilization is the most intense form of inbreeding. It is widespread in the plant kingdom, being inevitable in cleistogamous flowers and during lateral conjugation (e.g. in *Spirogyra*). Inbreeding increases homozygosity and hence decreases genetic variability. Thus when a population heterozygous for a particular gene, e.g. Aa, is selfed, 50% of the next generation will be homozygous for that gene (i.e. either AA or aa). Since the frequency of homozygous recessive lethals and semilethals increases by inbreeding, this can lead to a decline in vigour in species that are normally outbreeding. This is called *inbreeding depression*. Some species have developed systems (outbreeding mechanisms) to inhibit excessive inbreeding. Nevertheless, some degree of inbreeding can be an advantage, for example to a rapidly speciating population where it prevents the population being 'swamped' by foreign genes. *Compare* outbreeding.

incipient plasmolysis *See* plasmolysis.

incompatibility The failure of gametes from genetically similar material to fuse due to physiological or morphological mechanisms. In some fungi, e.g. *Mucor*, sexual fusion cannot occur between similar races or strains. In seed plants complex mechanisms have developed involving the interactions of pollen and stigmatic tissue. *See* self incompatibility, allogamy.

incomplete dominance (partial dominance) The partial expression of both alleles in a heterozygote so that the phenotype is intermediate between those of the two homozygotes. Thus in *Antirrhinum*, the homozygotes ++ and ww are red and white flowered respectively, while the heterozygote +w is pink. Where one allele is expressed to a greater extent than another, the former may be described as *partially dominant*. There are all degrees of dominance relationships between complete dominance and incomplete dominance: the dividing lines between the various categories are to some extent arbitrary. *See also* codominance.

incubation period The phase following initial penetration of a pathogen when there are no visible symptoms of the disease. It may last several months in which case it is termed a *latent infection. When the first visible symptoms appear the disease is said to have entered the infection stage.

incubous Describing the leaf arrangement in leafy liverworts where the front edges of the leaves lie above the back edges of the leaves in front. *Compare* succubous.

indefinite growth The ability of certain plants or plant organs to show unlimited growth. For example, trees continue to grow each year and the leaves of monocotyledons grow throughout the life of the plant. The term is also used of racemose inflorescences, in which growth of the inflorescence is not terminated by the production of an apical flower bud but continues indefinitely. Meristems may also be said to exhibit indefinite growth. *Compare* definite growth. *See also* monopodial branching.

indefinite inflorescence *See* racemose inflorescence.

indehiscent Describing a fruit or fruiting body that does not open to disperse its contents. Indehiscent fruiting bodies are thought to have developed from suppression of the naturally occurring opening mechanisms found in dehiscent structures. The seeds or spores are released either when the surrounding wall decays or when it is eaten by an animal.

independent assortment The separation of a pair of alleles of one gene at meiosis independently of the separation of alleles of other genes. Thus, in a heterozygote AaBb, independent assortment would produce four kinds of gametes equally frequently: AB, Ab, aB, and ab. The Law of Independent Assortment was formulated by Mendel (Mendel's Second Law). However, it only holds true provided that genes are on separate chromosomes. If genes are on the same chromosome, then they tend to segregate together (*linkage). The term independent assortment can be applied to chromosomes as well as genes.

indicator A chemical substance that, by changing its colour, gives an indication of the pH of the solution to which it is added. Examples are litmus, phenolphthalein, sodium bicarbonate indicator, and universal indicator.

indicator species A species that by its presence or absence shows the type of environmental conditions that prevail. For example, the absence of lichen growth on trees is taken to indicate atmospheric pollution by sulphur dioxide. As some lichen genera, e.g. *Usnea*, are more sensitive to pollution than others, e.g. *Lecanora*, the degree of pollution can be ascertained by the relative abundances of the different lichens. Other examples of indicator species are *Gypsophila* species, which indicate alkaline conditions, and *Sphagnum* mosses and wavy hair grass (*Deschampsia flexuosa*), which indicate very acid conditions.

indigenous Describing an organism that is native to an area, i.e. that has not been introduced from another area. *Compare* exotic.

indole acetic acid (IAA) The principal *auxin of most plants. The other indole compounds that also occur in plants probably owe their auxin-like activity to conversion to IAA at, or near to, the site of action. Thus, indole pyruvic acid, indole acetaldehyde, and other compounds may act as IAA reserves. The route of IAA synthesis is from the amino acid tryptophan. Studies in vivo have shown that IAA is transported in greater quantity and inactivated or destroyed more quickly than synthetic auxins. This suggests that an important aspect in the control of auxin activity is the relative rates of synthesis, transport, and inactivation in different tissues. IAA is decomposed by light and by the enzyme IAA oxidase. It may also combine with other compounds to form an inert complex. The rapid oxidation of IAA by the plant's enzymes has limited its commercial use as a weedkiller, and synthetic auxins, e.g. 2,4-D, 2,4,5-T, and MCPA, are normally used.

indole alkaloids A group of alkaloids, the members of which contain an indole nucleus. They are derived either from tryptophan or phenylalanine. Examples are strychnine, from *Strychnos nux vomica*, and yohimbine from the bark of *Corynanthe johimbe*. Certain alkaloids containing quinoline are derived from indole compounds, an example being quinine, obtained from the bark of *Cinchona* species.

induced enzyme (adaptive enzyme) An enzyme that is synthesized in response to high concentrations of its substrate (*see* induction). When the substrate is absent the genes governing the enzyme's synthesis are repressed. *Compare* constitutive enzyme. *See also* repressor, operon.

inducer A substance that promotes the activity of a gene or block of genes. Inducers of enzymes in prokaryotes fall into two groups. In negative control systems (e.g. the lac operon of *Escherichia coli*) they may be substrates of the enzymes they induce or analogues, derivatives, or precursors of the enzyme substrates. In any event, they bind with a regulator protein and so prevent the latter complexing with the operator gene to inhibit mRNA synthesis (*see* operon). In positive control systems (e.g. the ara operon in *E. coli*) the inducer is a complex formed from the regulator-protein substrate. This complex must bind with the operator in order for mRNA synthesis to occur. In eukaryotes the nature of inducers is less well understood but appears more diverse. Steroid and other membrane-soluble growth substances, such as gibberellic acid, can act as inducers when complexed to cytoplasmic receptors.

induction A process in which the de novo synthesis of a particular enzyme or group of enzymes is stimulated by the presence of the substrate, especially when the substrate is the cell's only carbon source. The induction of enzymes has been studied for the most part in bacteria. The classical example of enzyme induction is that of β-galactosidase in *Escherichia coli*. This enzyme rapidly increases in activity when *E. coli* cells are given lactose as the sole carbon source. This induction effect is not seen, however, if glucose is present, even if lactose is present in high concentrations. In eukaryotic cells induction is not as common as in bacteria, nor is it so rapid. In primitive eukaryotes, such as *Saccharomyces* and *Neurospora*, induction of β-galactosidase can be observed, although at a much slower rate than in bacteria. In higher eukaryotes there are fewer proven examples of enzyme induction. One that has been studied in detail in white mustard (*Sinapis alba*) involves the synthesis of phenylalanine ammonia-lyase (PAL), which catalyses the formation of cinnamic acid (a precursor of the flavonoids) from phenylalanine. Transcription of the gene governing PAL synthesis is induced by the P_{fr} form of phytochrome. *See also* operon.

indumentum A dense or sparse covering, usually of hairs.

indusium A flap of tissue that partially or completely covers each sorus in certain ferns. It is an outgrowth of the placenta. Indusia are seen on the undersurface of the fronds of *Dryopteris* as lines of brown kidney-shaped outgrowths.

inferior ovary *See* ovary.

inflorescence Any flowering system consisting of more than one flower. It is usually separated from the vegetative parts by an extended internode, and normally comprises individual flowers, bracts and peduncles, and pedicels. The change from vegetative to reproductive growth may be triggered by photoperiod or temperature and is thought to be hormonally controlled (*see* florigen, phytochrome, vernalin). The arrangement of flowers into inflorescences is believed to have evolved from solitary flowers. The increase in flower number has been accompanied by a decrease in the size of the individual flowers, and specialization for different roles.

Development of inflorescences may terminate vegetative growth (*see* determinate growth) or allow it to continue (*see* indeterminate growth). The type of growth habit has considerable economic implications in such crops as the grain legumes where it directly affects total crop yield, and sequence and time of fruit ripening.

Inflorescences are classified morphologically by their branching behaviour into *cymose inflorescences, where the apical meristem differentiates into a flower and new growth is from the axils, and into *racemose inflorescences, where growth continues at the apices, and flowers are developed in the axils.

infrared gas analyser An instrument used for the quantitative measurement of the proportion of a particular gas in a mixture. Different gases absorb infrared at different characteristic frequencies. In the analyser a beam of infrared of the frequency known to be absorbed by the gas under investigation is passed through the gas sample. The instrument also contains a plate sensitized to detect infrared of this frequency. The amount of infrared detected by the plate will fall as the proportion of the gas in the sample increases. Infrared analysers are used for simple measurements, especially the continuous monitoring of air for a given pollutant. They are also used to measure rates of respiration or photosynthesis by recording changes in carbon dioxide level. More complicated analyses of gas mixtures can be made using a *spectrophotometer.

infraspecific variation Differences in appearance exhibited by different members of the same species or breeding population. The variation may be a result of environmental influences, such as nutrition, crowding, seasonal effects, etc. It may also be due to genetic factors, principally the effects of genetic recombination and mutation. Whatever the ultimate cause, natural selection plays a major role in governing the extent of the variation and types of individuals present. The term may be applied to characteristics only apparent by advanced laboratory techniques, as well as to those readily visible to the naked eye.

inhibitor A substance that inhibits the activity of an enzyme. It may do this by binding at the active site of the enzyme in place of the substrate (*see* competitive inhibition) or by binding with the enzyme–substrate complex. Alternatively the inhibitor may bind to some part of the enzyme other than the active site and in so doing change the three-dimensional configuration of the enzyme so it is not able to function normally. Certain enzymes containing an –SH group are inhibited by heavy metals in this manner. *See also* feedback inhibition.

initial (initiating cell, meristematic cell) Any actively dividing cell in a *meristem. Each time an initial divides, one of the daughter cells retains its meristematic properties and remains in the meristem as an initial while the other differentiates to form a cell in the plant body. The latter daughter cell may undergo further divisions but in this event all daughter cells differentiate further. Initials commonly possess a dense cytoplasm. *See also* vascular cambium.

inoculum 1. The material that initiates disease in a previously uninfected plant. Under artificial conditions this may be a suspension of spores, which is sprayed onto the plant. In nature the inoculum is carried to a healthy plant from its source, e.g. an infected plant, by an agent, such as wind or an insect.
2. In microbiology or tissue-culture work, the cells that are introduced into a sterile medium to initiate a culture.

inositol A sugar alcohol formed by the complete hydroxylation of cyclohexane. The *myo* isomer is the only isomer with biological significance. *Myo*-inositol is synthesized in a two-stage reaction from glucose 6-phosphate. Its hexaphosphate ester, *phytic acid*, is an important storage product in seeds; in germinating seeds and young seedlings it is thought that the phytic acid is converted to *myo*-inositol. This then undergoes oxidation to form glucuronic acid; thus phytic acid acts as a source of uronic acids for synthesis of cell-wall components.

insecticide Any chemical that kills insects. Loosely the term is also applied to chemicals that kill mites, nematodes, and other invertebrate pests. The substance often affects the nervous system and may be a stomach poison (*see* stomach insecticides). It may also act by *contact or as a *fumigant. *See also* botanical, chemosterilant, chlorinated hydrocarbon, dinitro compounds, organophosphorus insecticide, sulphur dust, systemic.

insectivorous plant (carnivorous plant) A plant that is adapted to obtain food by digesting small animals (particularly insects) in addition to feeding by photosynthesis. Such plants are found in regions where the soil is deficient in certain nutrients, particularly nitrates. Examples are butterworts (*Pinguicula*) and sundews (*Drosera*) growing on heathland and moorland, and tropical plants, such as the Venus' flytrap (*Dionaea muscipula*) and the various pitcher plants (families Nepenthaceae and Sarraceniaceae). The plants are adapted in a number of ingenious ways to attract, trap, and kill the insects, which are then digested by proteolytic enzymes secreted by the plant. Butterworts trap insects by having sticky infolded leaves. Venus' flytraps have folding leaves with spines along the edges that spring together and trap the insect. The leaves of pitcher plants form a long narrow container shaped like a pitcher, into which the insects fall and drown in the liquid at the bottom.

insect pollination *See* entomophily.

integument A protective structure that develops from the base of an ovule and encloses it almost entirely except for an opening, the micropyle, at the tip of the nucellus. In most angiosperms there are two integuments, which may or may not be fused. In gymnosperms and in most dicotyledons with fused petals there is only one integument. Occasionally a third integument is formed, which becomes conspicuous as an aril following fertilization. This is seen in the spindle tree (*Euonymous europaeus*). The integuments have a vascular supply continuous with the parent via the funiculus.

intercalary meristem A *meristem positioned (i.e. intercalated) between more or less differentiated tissues, some distance from the *apical meristem. Intercalary meristems may contain not only *initials, but also relatively mature cells. This prevents it from being a structurally weak part of the organ. The primary function of an intercalary meristem is to facilitate longitudinal growth of a plant organ, independent of activity of the apical meristem. Examples are the meristems in the leaf sheaths and internodes of grasses and horsetails.

intercellular Located or taking place between cells.

interchange *See* translocation.

interfascicular cambium The part of the *vascular cambium that forms between the vascular bundles (fascicles) and joins up with the *fascicular cambium to form a continuous meristematic ring.

interference *See* chiasma interference.

interference microscope An advanced development of the *phase contrast microscope in which the light from the condenser is split into two beams by a prism. The *object beam* passes through the specimen and the objective; the *reference beam* passes through a second matched objective without going through the specimen. The two beams are recombined before going through the eye piece. Interference between the beams produces a series of light and dark fringes in the field of view. These are the result of differences in refractive index of the specimen and allow detail of the specimen to be seen. The instrument is more sensitive than the phase contrast microscope and spurious effects (such as halos) are eliminated. With white light different parts of the specimen appear coloured as a result of interference.

International Code of Botanical Nomenclature (ICBN) A periodically revised publication outlining the procedures for the scientific naming of plants (i.e. vascular plants, bryophytes, algae, blue-green algae, fungi, and slime moulds).

The rules have been drawn up to sort out errors and ambiguities arising from past misunderstandings and misidentifications and to ensure correct naming of new taxa. Fundamental to the code are the principles that naming of families and lower ranks is by reference to nomenclatural *types, that the first valid name published is maintained (for vascular plants this is the first name published on or after 1 May, 1753, the date Linnaeus' *Species Plantarum* was published), and that once the circumscription, position, and rank of a group have been decided it can only have one correct name. However this does not pre-

clude a group of plants being treated in different ways (e.g. being given a different position) by different authors, with each author being able to assign a different nomenclaturally correct name. For example, watercress is thought by some authors to belong to the genus *Nasturtium*, while others believe it is better placed in *Rorippa*. Following the rules of the Code, the names *Nasturtium officinale* and *Rorippa nasturtium-aquaticum* are equally acceptable. Names cannot be changed simply because they are inappropriate. However *homonyms and superfluous names (names published for a taxon after that taxon already had a valid name) must be rejected. If two or more taxa are combined into one then the new taxon takes the name of the constituent that had the oldest valid name. If a taxon is split into two or more taxa, then one of the new taxa must be given the name of the old taxon.
The ICBN is only applicable to wild plants and because of the importance of many cultivated plants, a separate set of rules has been laid down in 57 Articles in the International Code of Nomenclature of Cultivated Plants or ICNCP.

internode A part of the stem lying between two adjacent *nodes.

interphase The period in the *cell cycle following cytokinesis and preceding the next nuclear division.

interpositional growth *See* intrusive growth.

intine The innermost layer of the pollen grain wall, composed mainly of cellulose. It is less resistant than the *exine and hence not preserved in geological deposits.

intracellular Taking place or situated within the cell.

intrafascicular cambium *See* fascicular cambium.

intraspecific selection *Natural selection among individuals of the same species.

intrazonal soil *See* soil.

introgression (introgressive hybridization) The incorporation of genes from one species or subspecies into another related species or subspecies. It arises as a result of successful hybridization and subsequent backcrossing of the hybrids with one of the parental populations. Introgression is believed to have been a major factor in the evolution of many plants, especially crop plants. For example, the present wide range of the sunflower (*Helianthus annuus*) over North America is thought to have been achieved fairly recently by the introgression of genes from a number of *Helianthus* species, each providing characteristics enabling adaptation to different environments. Introgression is the basis of some plant breeding techniques. *Compare* gene flow.

intron A noncoding polynucleotide sequence between two coding regions (exons) of DNA. Introns are present in eukaryotic cells, but not in bacteria. The length and number of introns in a gene varies, but in some genes introns can contain more DNA than exons. In other genes they are absent altogether. mRNA initially synthesized on intron-containing genes is called heterogeneous mRNA, or hnRNA. While in the nucleus, probably in the nucleolus, introns are removed and the ends of the mRNA are annealed to produce mature mRNA. The latter is then translated into protein. How introns can be removed with such precision, and what happens to them after their removal, is obscure. Several functions have been ascribed to them. For example, they may speed evolution by enhancing recombination between exons. It has also been suggested that they play a role in differentiation. Introns should not be confused with *repeated sequences, which occur between genes.

introrse Describing anthers that release their pollen to the inside of the flower so promoting self pollination. *Compare* extrorse.

intrusive growth (interpositional growth) The type of growth exhibited by certain cells when they intrude between other cells that are either not growing or growing at a slower rate. For example, certain members of the Liliaceae have vascular tracheids that elongate at both ends to achieve a length of between 15 and 40 times that of the original. These tracheids insinuate themselves amongst the surrounding cells, disrupting established plasmodesmata. *Compare* sliding growth, symplastic growth.

intussusception The insertion of cellulose into spaces within the wall of an elongating cell. Water passes into the

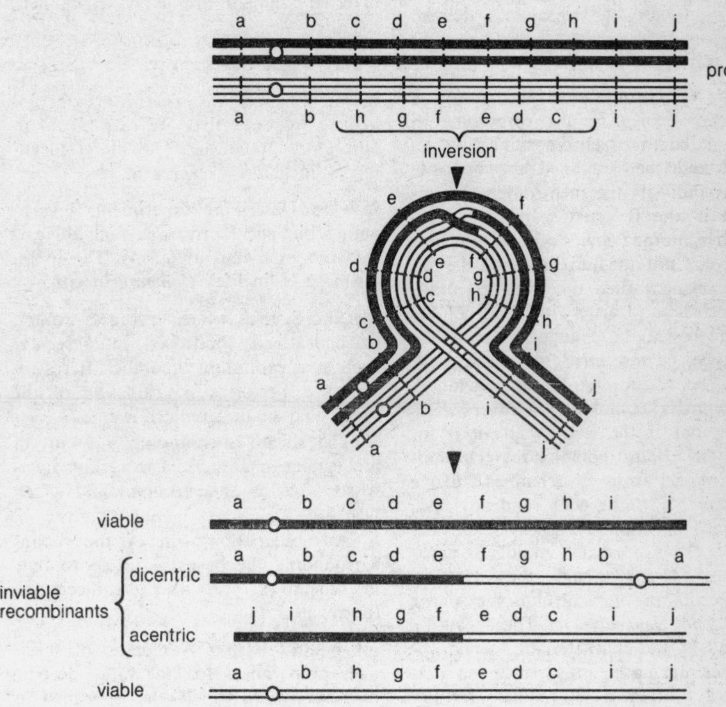

Consequences of crossing over in an inversion heterozygote.

vacuole of the cell, creating an outward pressure on the wall, which consequently stretches. The new cellulose is then incorporated, the result being an increase in wall area. The next step is usually the thickening of the walls by *apposition.

inulin A storage polysaccharide composed of fructose residues linked by $\beta(2-1)$ glycosidic bonds. The inulin chain is usually headed by a glucose residue and is 30–40 residues in length. Synthesis of inulin is from sucrose precursors, hence the glucose at the beginning of the chain. Inulin is an important reserve polysaccharide in the Compositae being found in high proportions in the genera *Helianthus*, *Dahlia*, *Inula*, and *Chicorium*. The tubers of Jerusalem artichoke (*Helianthus tuberosus*) may contain up to 58% inulin.

inversion (chromosome inversion) A structural change in a chromosome in which a length of chromosome has detached and then rejoined the opposite way round. If the centromere is included in the inverted segment the inversion is termed *pericentric*. If the centromere is not in the centre of the inverted segment then the relative lengths of the chromatid arms either side of the centromere will be changed, altering the karyotype. *Paracentric* inversions, in which the centromere is not included, are the more common. An inversion is homozygous if the same segment is inverted in both homologous chromosomes. Apart from the changed linkage arrangements of the genes and any *position effects, homozygous inversions behave much as normal homologous chromosomes. Often though the inversion only occurs in one chromosome giving an *inversion heterozygote*. These can be detected by the characteristic looped appearance during meiosis. Inversion heterozygotes often show reduced fertility because if a crossover occurs within the loop then a dicentric chromosome (a chromosome with two centromeres) and an acentric fragment will result (*see* diagram). Such pieces cannot move normally on the spindle and thus only half the gametes (those containing either of the two chromatids not involved in the crossover) will be viable. As the viable gametes will have the same linear sequence of genes as the parent, inversion heterozygotes have been termed *crossover suppressors*. However they do not suppress crossing over but the resultant recombinants, being inviable, are not apparent. Viable recombinants may be obtained if two crossovers occur in the inversion loop. Inversion heterozygotes may have an adaptive advantage in that any useful combination of alleles in the inverted segment are likely to be held together.

invert sugar *See* sucrose.

in vitro Describing experiments on biological processes that are carried out in laboratory apparatus. The literal meaning is 'in glass'. *Compare* in vivo.

in vivo Describing experiments investigating biological processes that are carried out in living organisms. The literal meaning is 'in life'. *Compare* in vitro.

involucre **1.** A whorl of bracts around or beneath a condensed inflorescence, such as a capitulum or umbel. It resembles and performs the function of the calyx of a single simple flower.
2. The sheath surrounding a group of archegonia or antheridia in certain liverworts, such as *Sphaerocarpus* and its allies.
3. A tubular upgrowth of the thallus surrounding the base of the sporophyte in members of the Anthocerotae (hornworts).

ion-beam etching A method of specimen preparation for scanning electron microscopy in which the specimen is placed under high vacuum and the surface is then eroded by a beam of ions to expose subsurface features. The specimen is then coated and examined. The technique is very successful with some materials but produces artefacts by uneven erosion of the surface of others.

ion-exchange chromatography A chromatographic method in which the components of mixtures are separated by differences in their acid–base behaviour. The chromatogram column is filled with a charged resin that is easily able to exchange ions for any charged molecules that are passed through the column. Amino acid separation is often achieved by this method and the resin commonly used is sulphonated polystyrene in which the sulphonic acid groups are charged with sodium ions. If an acidic mixture of amino acids is added to the column then, because the amino acids are mostly cations at acid pH, they will tend to displace the sodium ions. The more basic amino acids, e.g. histidine, will be held more tightly to the resin than the more acid amino acids, e.g. glutamic acid. By gradually increasing the pH and sodium concentration of the mobile phase the amino acids are displaced from the resin. They emerge from the column in order of decreasing acidity.

IPA (isopentenyl adenosine) A natural *cytokinin, similar in structure to *zeatin. In addition to being a very active cytokinin, it has been found as a minor base in certain transfer RNA molecules. In the tRNA specific for the amino acid serine, IPA is found following the anticodon and thus may serve to mark the end of the anticodon.

iron Symbol: Fe. A metal element, atomic number 26, atomic weight 55.85, required by plants as a macronutrient. It is important as a constituent of the *cytochromes, ferredoxin, and of certain enzymes, e.g. succinate dehydrogenase. Deficiency of iron leads to chlorosis since iron plays a part in chlorophyll formation. Iron is the second most abundant metal in the soil being present in such minerals as haematite, magnetite, and limonite. However in alkaline conditions iron may be precipitated as an insoluble ferric salt making it unavailable to plants. This can be overcome by adding sequestrols containing chelated iron. Ferrous compounds in the soil are also used by the filamentous iron bacteria, which derive energy by oxidizing them to ferric compounds.

iron bacteria *See* Chlamydobacteriales.

irritability (sensitivity) Response by a living organism to external stimuli. In plants, the responses often take the form of various types of movements caused by the bending of the growing region. In addition, certain aquatic microorganisms and some motile gametes exhibit irritability by changing their physical position. *See* tropism, nastic movements, taxis.

isidium An outgrowth from a lichen thallus containing both algal and fungal cells that may break off and propagate the lichen vegetatively. Isidia show a variety of forms, e.g. the coral-like outgrowths of *Umbilicaria* or the budlike projections of *Collema*.

isochromosome A metacentric chromosome in which the two arms are genetically identical. *See also* trisomic.

isoelectric point The pH value of a medium that results in a molecule having no net electric charge. At the isoelectric point the molecules will coalesce and precipitate, because there is no repulsion between them. A mixture of amino acids or proteins may be separated into its components by varying the pH, as different molecules have different isoelectric points.

isoenzyme (isozyme) Any one of the multiple forms of a given enzyme, each with different kinetic characteristics. The different isoenzymes are usually formed by different combinations of the same two or more subunits.

Isoetales An order of the *Lycopsida containing two genera, *Isoetes* (quillworts), with about 70 species, and the monotypic *Stylites*, both of which contain rushlike predominantly aquatic perennial plants. The stem is condensed into a fleshy rootstock from the top of which arises a bunch of long narrow ligulate leaves. Seasonal cambial activity occurs in a zone around the primary vascular tissue but is highly unusual in that it differentiates cortical rather than stelar tissue. Thus the oldest tissue occurs at the outer edge of the stem and is continually sloughed off as new material is generated from within.

The megasporangia are borne in the axils of leaves formed early in the season while the later formed leaves bear microsporangia. The antherozoids differ from those of the Selaginellales and Lycopodiales in being multiflagellate rather than biflagellate.

isogamy The production or fusion of morphologically identical *gametes. It is found only in the more primitive algae and fungi and constitutes the simplest form of sexual reproduction. Although the gametes appear identical it has been found that two gametes produced by the same parent usually will not fuse (*see* heterothallic). Thus physiological differences exist and these are recognized by ascribing such gametes with plus and minus signs depending on their compatibility relationships (*see* mating strain). *Compare* anisogamy, oogamy.

isokont Describing an organism having flagella similar in form and length, as have the motile species of algae in the Chlorophyta, hence the former name of the taxon, Isokontae. *Compare* heterokont.

Isokontae *See* Chlorophyta.

isolating mechanism *See* reproductive isolation.

isoleucine A nonpolar branched chain amino acid with the formula $C_2H_5CH(CH_3)CH(NH_2)COOH$ (*see illustration at* amino acid). Isoleucine is synthesized from threonine. The pathway is similar to that for valine and leucine and many of the enzymes are common to all three. Degradation of isoleucine is also closely related to that of leucine and valine, leading to succinyl CoA. The final part of the degradation, from propionyl CoA, is common to valine, isoleucine, and methionine.

isomerase Any *enzyme that catalyses the conversion of a molecule from one isomeric form to another. An example is glucose phosphate isomerase, which catalyses the conversion of glucose 6-phosphate to fructose 6-phosphate in glycolysis. *See also* epimerase.

isomorphic 1. (homologous) Describing a life cycle in which the alternating generations are morphologically identical as, for example, in the brown alga *Dictyota*. Individuals may be identified as sporophyte or gametophyte by observation of the reproductive process. Isomorphic life cycles are seldom seen in archegoniate plants. An exception is *Psilotum*, in which the subterranean gametophyte resembles, in anatomy, branching pattern, and in the production of multicellular gemmae, the rhizome of the sporophyte. **2.** Describing an organ or organism that exists in one form only. *Compare* dimorphism, polymorphism.

isonome method A technique used to study the distribution of plants in a particular area where there are obvious variations in species distribution. Continuous quadrats are laid out to cover the chosen area and the abundance of each species in each quadrat is recorded. The abundance values of each species are plotted on separate pieces of squared paper. Isonomes (which resemble contour lines) are then constructed by joining up roughly equal abundance values. Isonomes of certain environmental factors, such as the topography of the area, can also be constructed. By superimposing the isonomes of one species onto those of another it can be seen if there are any correlations between the distributions of the two species. Similarly, by superimposing the isonomes of a species onto the isonomes of an environmental variable it may be seen whether that particular environmental factor is correlated with the species distribution.

isopentenyl adenosine *See* IPA.

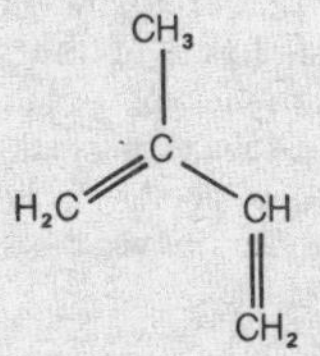

Isoprene subunit.

isoprene subunit The fundamental unit from which many different hydrocarbons, particularly the terpenes, are constructed. It has the formula $CH_2C(CH_3)CHCH_2$ (*see* diagram).

isoquinoline alkaloids A group of alkaloids, the structures of which are based on the isoquinoline nucleus. They are derived from the amino acids tyrosine and phenylalanine. Examples are morphine, from the opium poppy (*Papaver somniferum*), and curare, from the bark of *Strychnos* species.

isotonic Having the same *osmotic pressure as the solution under comparison. If two isotonic solutions are separated by a permeable membrane there will be no net migration of ions from one to the other. *See also* hypertonic, hypotonic.

isotopic tracer A stable or radioactive isotope that can be used to label a metabolite and consequently follow its fate in an intact organism. Many elements found in living organisms have rare isotopes that are useful as tracers. For example, the most abundant form of carbon, carbon-12 (^{12}C), has six protons and six neutrons in its atomic nucleus. About 1.1% of carbon exists as the stable isotope ^{13}C, which has one extra neutron. ^{14}C, which has two extra neutrons, exists in minute amounts, and is radioactive, emitting beta rays. Both of the rarer isotopes are incorporated and function the same way as ^{12}C in an organism. The light-independent reactions of photosynthesis (*see* Calvin cycle) were elucidated using ^{14}C-labelled carbon dioxide and finding the order in which substances incorporated the isotope. Tritium, a radioactive isotope of hydrogen having three neutrons instead of the usual one, has been used to label precursors of DNA, e.g. tritiated thymidine, to examine DNA synthesis.

isotype A specimen collected at the same time and from the same plant or localized population of plants as the *holotype. These duplicate specimens are often separated and deposited in several institutions. *See also* type.

isozyme *See* isoenzyme.

isthmus The narrow central portion connecting the two halves of a placoderm desmid.

J

Jacob–Monod model (operon model) A theory to explain how gene activity is regulated, proposed by F. Jacob and J. Monod in 1960. The theory is well substantiated for prokaryotes but there is little evidence that it is directly applicable to eukaryotic cells. The operon model shows how messenger RNA synthesis, and thus the quantity of enzymes present, may be regulated. By controlling the latter, the system as a whole provides a method of coordinating and regulating cell metabolism at the level of transcription. A parallel model, taking into account the very different organization of chromosomal DNA, has since been proposed for eukaryotes (*see* repeated sequence). *See also* operon.

jelly fungi *See* Tremellales.

Jurassic The middle period of the Mesozoic era between about 195 and 136 million years ago. The climate was uniformly warm and humid and the flora was varied and abundant and dominated by the gymnosperms. *See* geological time scale.

juvenility The condition expressed in immature organisms, which are generally smaller in size than the adults and lack reproductive capability. The seedling represents the juvenile stage of many plants. The term is particularly used when the young plant form is very different from that of the adult. For example in gorse (*Ulex*), the juvenile leaves are flat and trifoliate but the adult plant has spines. In many conifers the seedlings bear needle-like juvenile leaves, and specialized leaves, such as the scale-like leaves of cypress (*Cupressus*) and the dwarf shoots of pine (*Pinus*), are formed later. The change is not always from a simple to more complex form. For example, in ivy (*Hedera helix*) the juvenile leaves are lobed while the mature leaves are entire (*see* diagram). The submerged leaves of many aquatic plants are similar to the juvenile leaves of terrestrial forms of the same species. In some varieties of ornamental plants the juvenile foliage persists in the adult plant (*see also* neoteny). Juvenile plants may also differ in various physiological

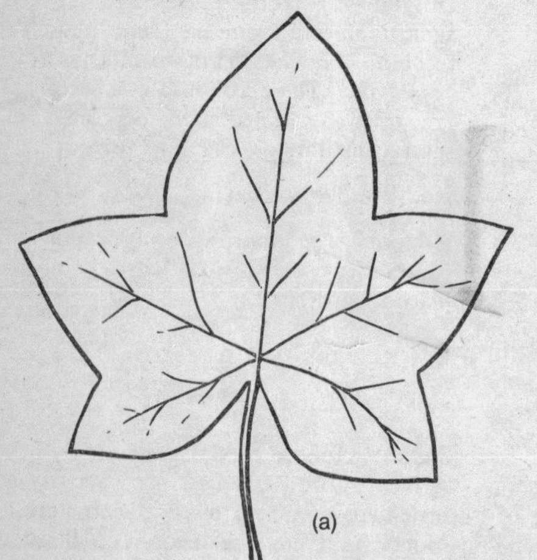

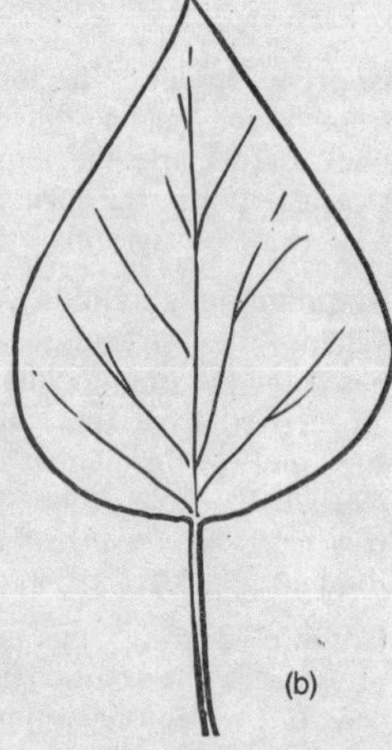

Juvenile (a) and mature (b) leaves of ivy.

ways from mature plants. For example, the immature forms of many deciduous trees, e.g. beech, do not shed their leaves in autumn but retain the dead leaves on the plant throughout the winter.

K

Kampfzone *See* elfin forest.

karyogamy The fusion of two nuclei during *sexual reproduction. Fusion normally occurs between haploid nuclei and follows immediately after *plasmogamy, resulting in the formation of a diploid zygote. In the higher fungi, especially the basidiomycetes, karyogamy may be delayed and occur separately from plasmogamy, resulting in a *dikaryotic mycelium.

karyogram (idiogram) A drawing or photograph of the chromosomes present in an individual in which the chromosomes are arranged in homologous pairs in an agreed conventional sequence. *See also* karyotype.

karyokinesis The process of nuclear division that precedes cytoplasmic division or cytokinesis. *See* mitosis.

karyoplasm *See* nucleus.

karyotype The physical appearance of chromosomes from an individual or species as seen at mitotic metaphase. Cells from the root meristem are commonly used, suitably stained, usually with acetic orcein or acetocarmine. The number, size, and shape of chromosomes in a set is usually highly characteristic of a species and often also constant within a genus. The karyotype is most conveniently studied by forming a *karyogram.

keel (carina) **1.** The two lower fused petals of a 'pea' flower, which form a boatlike structure around the stamens and styles. *See also* standard, wing.
2. Any ridgelike structure.

kelps *See* Laminariales.

α-ketoglutaric acid A five-carbon dicarboxylic keto acid with the formula $HOOCCO(CH_2)_2COOH$. α-ketoglutarate is an intermediate in the *TCA cycle. It is formed by the oxidation of isocitrate, with the concomitant reduction of NAD to NADH. In the next step of the TCA cycle α-ketoglutarate undergoes oxidative decarboxylation to form succinyl CoA in a reaction similar to the formation of acetyl CoA from pyruvate. α-ketoglutarate is also involved in amino acid metabolism (*see* transamination).

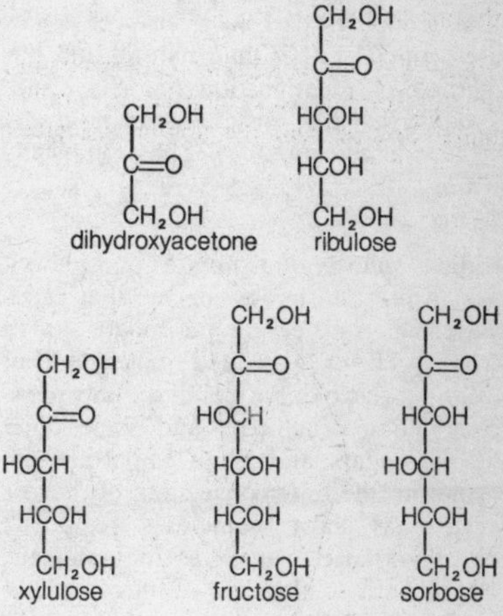

dihydroxyacetone: CH_2OH–C=O–CH_2OH

ribulose: CH_2OH–C=O–HCOH–HCOH–CH_2OH

xylulose: CH_2OH–C=O–HOCH–HCOH–CH_2OH

fructose: CH_2OH–C=O–HOCH–HCOH–HCOH–CH_2OH

sorbose: CH_2OH–C=O–HCOH–HOCH–HCOH–CH_2OH

Some common ketoses.

ketose Any monosaccharide with the carbonyl (CO) group at a position other than on the terminal carbon atom, so forming a ketone group. The simplest ketose is the three-carbon dihydroxyacetone. Other ketoses include the five-carbon ribulose and xylulose and the six-carbon fructose and sorbose (*see* diagram). *Compare* aldose.

key 1. A list of characteristics drawn up to enable speedy identification of a specimen. Prominent contrasting features are given at each stage in the key so that by a process of elimination successively smaller groupings of organisms are split off until the specimen can be identified. Simple keys to flowering plants often rely solely on flower structure. For example, the first step in the key may be to form a number of groups on the basis of petal number. The second stage could be to subdivide these groups on the basis of flower colour and further subdivisions could then be made by petal shape, flower symmetry, etc. The groupings derived from such a key are artificial rather than natural and this method of identification has the disadvantage that if a mistake is made at any stage identification will be completely wrong.
2. *See* samara.

kinase (phosphotransferase, phosphorylase) Any *transferase enzyme that catalyses the transfer of phosphate to or from ATP or a related molecule. For example, hexokinase catalyses the phosphorylation of glucose and some other hexose sugars and is an important enzyme in the interconversion of hexose sugars, as most isomerases require a phosphorylated sugar as a substrate. Hexokinase is also one of three kinases in the glycolytic pathway.

kinetin (6-furfurylaminopurine) A synthetic *cytokinin. It was first isolated from DNA extracts of animal origin but has now also been synthesized commercially.

kinetochore (spindle attachment) A point within a *centromere consisting of paired brushlike filaments, 200–500 nm in length, lying parallel to each other. Kinetochores induce the formation of *microtubules, approximately 50–60 from each, which join to the spindle fibres and so attach the chromosomes to the equator of the spindle during cell division. Formerly the term kinetochore was used synonymously with centromere.

kinetosome (basal body, blepharoplast) A cylindrical structure in the cytoplasm from which a flagellum or cilium arises in motile eukaryotic cells. It is composed of a cylinder of nine microtubules linked by fine crossbridges. It transfers ATP and possibly other materials to the fibres in the shaft of the flagellum or cilium. Striated fibrous extensions, termed *rootlet fibres*, may arise from the basal body and project backwards into the cytoplasm. These probably serve to anchor the flagellum.

kingdom The highest level in the hierarchy of taxonomic ranks. Traditionally all organisms have been placed in either the plant (Plantae) or animal (Animalia) kingdoms but recently several other kingdoms have been described. Thus the fungi are considered by many to be sufficiently distinct from green plants to deserve placement in a separate kingdom, the Mycota. Similarly some authors prefer to place unicellular organisms in their own kingdom, the Protista. Other systems underline the basic differences between prokaryotes and eukaryotes by recognizing the kingdoms (or superkingdoms) Prokaryota and Eukaryota. The plant kingdom is divided into *divisions – the number varying according to the classification.

kinin *See* cytokinin.

klinostat *See* clinostat.

Kranz structure A type of organization of the photosynthetic tissues found in the leaves of *C_4 plants. In such plants the cells of the bundle sheath are large and contain specialized elongated chloroplasts with no grana but capable of forming starch grains. The mesophyll chloroplasts are similar in structure to those of C_3 plants but do not form starch grains. It is within the mesophyll chloroplasts that oxaloacetic acid is formed through the combination of carbon dioxide and phosphoenolpyruvate. This is then transported to the bundle sheath chloroplasts where the carbon dioxide is released and refixed by ribulose bisphosphate carboxylase.

Krebs' cycle *See* TCA cycle.

Krummholz *See* elfin forest.

kymograph A slowly revolving drum, often covered with graph paper or smoked paper, on which a tracer records measurements of such activities as plant growth movements.

L

labelling The tagging of a compound with an *isotopic tracer so that its fate within an organism can be followed.

labellum (lip) **1.** The lower of the three petals of an orchid flower, which differs in morphology and patterning from the two lateral petals and gives the flower its characteristic form. The laterals and the sepals may all be similar but the labellum is always distinct and usually much larger than the other perianth segments. It serves as a platform for pollinating insects, which are attracted by its distinctive colours and markings. *See also* pseudocopulation.
2. The platform formed by the lower petal or group of fused petals in various other lipped flowers, such as those of the Labiatae, Scrophulariaceae, and Leguminosae.

Labiatae (Lamiaceae) A large dicotyledonous family, commonly called the mint family, comprising some 3000 species in about 200 genera. Most species are shrubby or herbaceous. Labiates characteristically have stems that are square in cross section and simple leaves in opposite decussate pairs. They are often covered in aromatic hairs and many species, such as mints (*Mentha*), sage (*Salvia officinalis*), and thymes (*Thymus*), are used as pot herbs. The flowers have five petals, which are usually fused into a tube that terminates in two distinct lips (exceptions being the genera *Mentha* and *Lycopus*). In many species, such as deadnettles (*Lamium*), the flowers appear as whorls (*verticillasters) at each node. In others, such as woundworts (*Stachys*), the flowers are grouped in spikes. The fruit is a *carcerulus.

Laboulbeniomycetes *See* Ascomycotina.

lactic acid bacteria A group of gram-positive anaerobic bacteria that use pyruvic acid as the acceptor for the hydrogen produced in glycolysis, so forming lactic acid. Such organisms (e.g. *Streptococcus lactis*) are used on a large scale in the food industry, especially as starter cultures in cheese making. Most lactic acid bacteria are rod-shaped organisms belonging to the family Lactobacillaceae of the Eubacteriales.

lacuna A space, gap, cavity, or depression, especially: **1.** A cavity, usually air filled, between cells, i.e. an intercellular space.
2. A *leaf gap.
3. A depression in the thallus of a lichen.
Compare lumen.

laevulose *See* fructose.

lag phase The period, following the inoculation of a nutrient medium with microorganisms, in which there is much physiological activity but no increase in numbers. The rate of cell division accelerates when the organisms become adapted to the culture conditions and enters the *logarithmic phase.

LAI *See* leaf area index.

Lamarckism The evolutionary theory put forward by Jean-Baptiste de Lamarck in 1809. He postulated that a characteristic that is acquired during the lifetime of an organism as a result of environmental pressures can be transmitted to the next generation. This is the theory of the inheritance of *acquired characteristics and Lamarck believed that new species could arise in

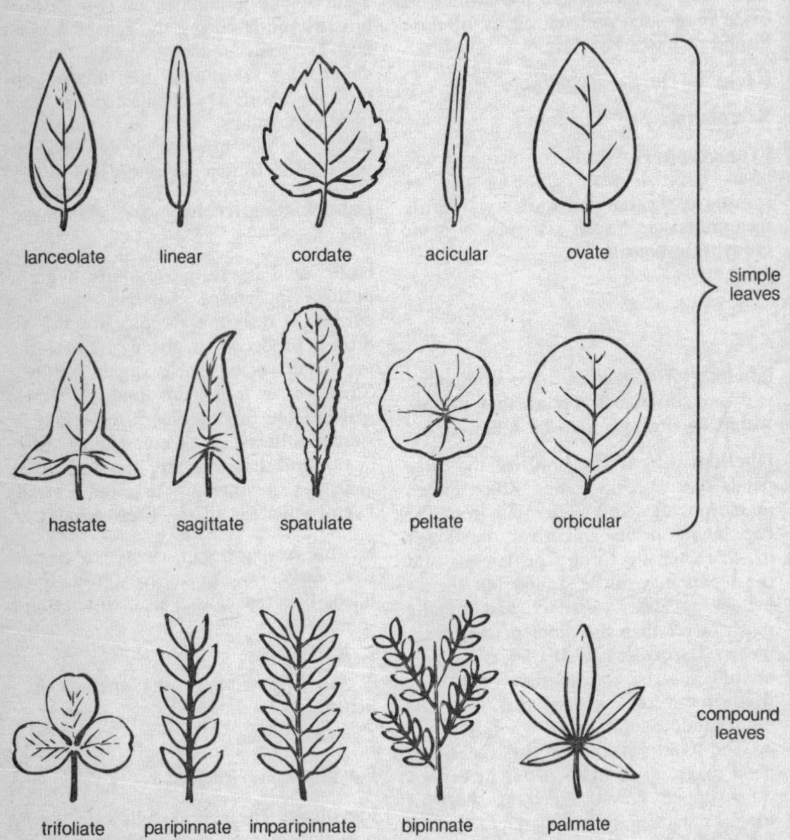

Some common leaf shapes.

this way. Erasmus Darwin also supported this view, but Lamarck was the first to give examples, one of which concerned the seeds of marsh plants. He suggested that if they reached high ground they became adapted to drier conditions by the development of new acquired characteristics, which could be inherited by the succeeding generation resulting in the development of a new species. The theory was later accepted by Darwin (*see* pangenesis) but challenged by Weismann. Today Lamarckism has been almost entirely rejected in favour of Neo-Darwinism. *See also* Weismannism.

lamella A layer within the cytoplasm or within an organelle formed from a flattened membrane-bound vesicle or tubule. The vesicular cavity is therefore narrowed so that the lamella consists of two membranes lying close together. *Chloroplasts have a complex internal system of lamellae.

Lamiaceae *See* Labiatae.

lamina (blade) **1.** The usually flattened bladelike portion of a leaf, as distinct from the petiole and leaf base. The shape of the lamina and the nature of its margin are important taxonomic characters (*see* illustrations). Leaf laminas are usually the main photosynthetic organs and are structured accordingly (*see* palisade mesophyll, spongy mesophyll). They are placed so as to make the best use of incoming light with the minimum amount of either overlapping or wasted space. This is particularly evident in a tree canopy. The pattern so achieved is termed the *leaf mosaic* and is mainly a result of *phyllotaxis.
2. Any thin flat organ, such as a petal

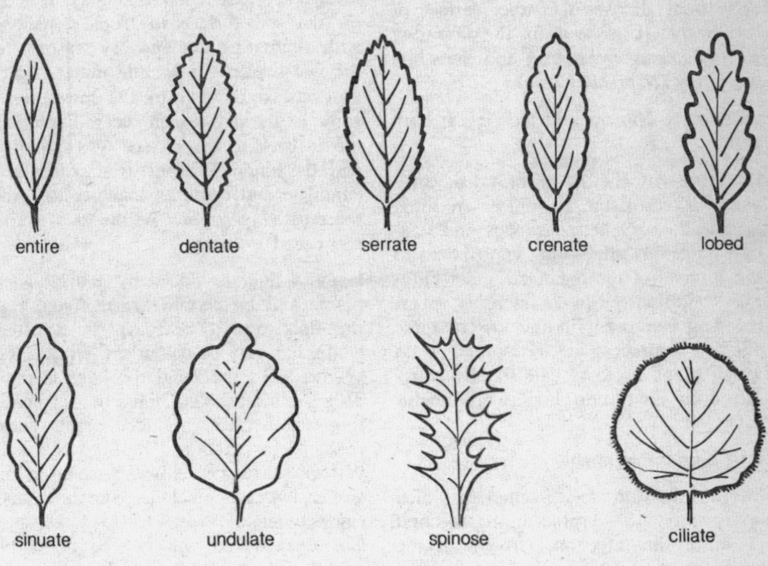

Leaf margins.

or the thallus of many macroscopic algae.

Laminariales (kelps) An order of the Heterogeneratae, a subclass of the *Phaeophyta. It contains the largest known algae, *Macrocystis* and *Nereocystis*. Most species occur below the low-tide mark. The sporophyte thallus is differentiated into a holdfast, stipe, and lamina. The gametophyte generation is much smaller and simpler and shows well developed oogamy. Certain species yield alginic acid, which is commercially important as an emulsifying agent.

laminarin The main assimilation product of algae belonging to the Phaeophyta. It is composed chiefly of glucose units though mannitol may also be present.

laminate placentation A form of *placentation in which the placentae arise from all over the inner surface of the ovary. It is seen in the flowering rush (*Butomus umbellatus*) and the white water lily (*Nymphaea alba*).

lanceolate Narrow and tapering at both ends.

landrace An ancient or primitive cultivar of a crop plant. Landraces are often genetically very heterogeneous and contain numerous alleles that contributed to the survival of the organism under natural conditions. Since intensive plant breeding can result in the loss of these alleles, landraces are a source from which plant breeders can selectively reintroduce them into highly bred cultivars.

late blight *See* blight.

latent infection An infection in which no symptoms are visible during the first phase of the infection. Growth of the pathogen stops soon after penetration but resumes at some later stage. For example, banana anthracnose infection (caused by the fungus *Colletotrichum musae*) originates in the field on unripe fruit but after penetration the fungus becomes latent as a subcuticular hypha for up to five months. The fungus resumes activity as the fruit nears maturity and typical black lesions develop on the ripe fruit. *See also* incubation period.

lateral meristem A *meristem arranged parallel to the sides of the organ in which it occurs and responsible for increase in girth usually by formation of secondary tissues. The *cambia are lateral meristems. Occasionally the term is used of axillary meristems. *Compare* apical meristem. *See also* primary thickening meristem.

lateral root Any *root that originates endogenously from the *pericycle of another root.

laterite A hard crust that may develop on the soil surface in tropical regions with alternating wet and dry seasons. In the wet season the soluble mineral salts are washed down into the lower horizons. In the dry season the soil solution moves back to the surface by capillarity and the aluminium and iron oxides accumulate and combine together to form the crust. *Laterization* results in a sterile soil called a *latosol*.

latex A fluid produced by many higher plants and by certain agaric fungi, e.g. the milk caps (*Lactarius*). It is often white, but may be colourless, reddish, or yellowish. In the fungi the latex is produced in a latex duct consisting of anastomosing hyphae. In green plants latex is stored in laticifers. The fluid contains various substances either in solution or suspension, e.g. alkaloids, starch grains, sugars, mineral salts, etc. In some species there is a high concentration of rubber (caoutchouc) though rubber is limited in occurrence to the dicotyledons. Commercial rubber production utilizes the latex of Brazilian rubber trees (*Hevea brasiliensis*) and, on a smaller scale, Indian rubber trees (*Ficus elastica*). The latex of other species, e.g. *Palaquium gutta*, gives, on coagulation, gutta-percha. The formation of caoutchouc and gutta-percha appears to be mutually exclusive with no plant yet being found to produce both. Chicle, from the latex of *Achras zapota*, and balata, from *Mimusops balata*, are other important latex products.

latex tube *See* laticifer.

laticifer A cell or a linked complex of cells containing a milky liquid called latex and penetrating various tissues in certain plants. A laticifer consisting of a single cell is referred to as *simple*, whereas a joined complex is termed *compound.* Laticifers may be further classified as *articulated* (branched) or *nonarticulated* (unbranched). Sometimes the end walls between the elements of a compound laticifer break down forming a continuous tube, sometimes referred to as a *laticiferous vessel.* A laticifer enclosed by a ray is termed a *latex tube.* Laticifers are present in many plants, such as the rubber tree, *Hevea brasiliensis*, rubber being the latex secreted by the laticifers and tapped by cutting through the bark.

Latin square An experimental design in which the number of treatments is the same as the number of replications, and each treatment occurs once in every column and row. It is analagous to a *randomized block design in two directions. It is often used in fertilizer field trials and has the advantage of eliminating from the total variation environmental differences, such as soil fertility, that exist across and down the square experimental plot. The Latin square generally yields more useful information that would a randomized block design of similar size. However it has the limitation that with a large number of treatments there must conscquently be a large number of replicates and beyond a certain point the labour involved is not worth the information obtained. Also, with small squares, the method is insensitive because the number of degrees of freedom is low. This can be overcome by replicating the squares.

latosol *See* laterite.

layering A method of plant propagation involving the pegging down of runners and stolons to the soil surface. Adventitious roots develop where a node touches the soil and a shoot develops from the lateral meristem. New daughter plants eventually establish. Carnations are commonly propagated in this way. *See also* air layering.

leaching The washing out of soluble substances from the upper layers of the soil by water passing down the soil profile. The substances are either deposited lower down in the B horizon or removed completely. It takes place when the amount of rainfall exceeds the amount of water lost by surface evaporation. It may result in podsolization or the development of an impermeable layer of mineral salts at some point in the soil profile. *Compare* flushing.

leader The tip of the main stem of a plant. The form of the leader (whether erect or drooping) can aid identification of certain conifers. For example, the western red cedar, *Thuja plicata*, has an upright leader, which distinguishes it from the similar Lawson cypress, *Chamaecyparis lawsonia*, which has a drooping leader. The term may also be used of the terminal segment of any main branch in distinction to any lateral branches arising from it. In those conifers in which the leaves are borne in clusters on short side shoots, e.g. cedars (*Cedrus*) and larches (*Larix*), the side shoots are termed spurs while the main shoot is the leader.

leaf The main photosynthetic organ of most green plants, consisting of a lateral outgrowth from a stem and comprising *lamina, *petiole, and leaf base. (*See* lamina for the different kinds of leaf shapes and leaf margins.) There are usually a great number of leaves on any one plant, although these may be lost in the colder or drier months in deciduous plants. A leaf typically consists of conducting tissues and photosynthetic cells (the *mesophyll) often differentiated into *palisade and *spongy mesophyll, surrounded by *epidermis. The epidermis is perforated by *stomata, usually more numerous on the abaxial (lower) side of the leaf. The epidermis is usually covered by a waxy cutinized layer termed the *cuticle. This prevents excessive water loss by transpiration. In many plants, however, leaves may be reduced or even absent, as in many xerophytes. Sepals, petals, and bracts are considered to be modified leaves and many believe the stamens and carpels are also derived from leaves. *See also* microphyll, megaphyll.

leaf area index (LAI) The ratio of the total surface area of a plant's leaves to the ground area available to that plant, i.e. LAI = leaf area/ground area. The LAI is of value when considering the number of plants that can be successfully cultivated on a given area of land. *See also* net assimilation rate.

leaf area ratio A value obtained by dividing the total leaf area of a plant by its dry weight. The ratio is useful in relating total photosynthetic to total respiratory material within the plant, thereby giving information concerning the plant's available energy balance.

leaf buttress A lateral prominence on the shoot apex, destined to differentiate into a leaf. It is the earliest stage in the development of a leaf primordium, and later forms the leaf base, the remainder of the leaf growing upwards and outwards from the buttress.

leaf culture A form of *tissue culture in which excised leaves, leaf material, or leaf primordia are grown on a sterile growth medium. Mature leaves can be kept healthy under culture conditions for considerable periods. Leaf primordia have been used to study growth and differentiation processes. Experiments with the cinnamon fern (*Osmunda cinnamonea*) show that the smallest primordia usually develop into shoots when cultured but, as the size of the excised primordia increases, there is an increasing tendency for them to develop into leaves. This indicates that leaf primordia are not irrevocably committed to becoming leaves until a relatively late stage.

leaf curl A plant disease in which an increase in cells on either side of the midrib and extra growth of the palisade and spongy mesophyll cause curling and puckering of the leaves. In peach and almond this is caused by the fungus *Taphrina deformans* and in tobacco and cotton by tobacco and cotton leaf curl viruses respectively. These viruses are transmitted by whiteflies (*Bemisia tabaci*).

leaf fall The shedding of leaves as a result of the formation of a zone of *abscission at the base of the petiole. Leaves on dead branches do not fall as no such zone is produced. In deciduous plants there is a continual shedding of older leaves throughout the growing season but with the onset of the winter or dry season there is a conspicuous shedding of all the remaining leaves. In evergreens leaves are shed and replaced continually and there is no period when the plant is devoid of foliage. Leaf fall is associated with a drop in the auxin level of the lamina.

leaf gap (lacuna) A parenchymatous area in the *stele of many vascular plants, associated with and positioned immediately above a leaf trace. Leaf gaps are characteristic of angiosperms but are also present in some gymnosperms and ferns.

leaf spot A plant disease in which the principal symptom is limited areas of necrosis on the leaves. There are numerous causes of leaf spots – mineral imbalance, insects, weather conditions, viruses, bacteria, and fungi. The leaf spots may be minor symptoms or a particular phase of a disease that develops other, more characteristic, symptoms. Bacteria and fungi are the usual causal agents of the diseases in which leaf spots are the main symptom. Angular leaf spot of cucumber is caused by the bacterium *Pseudomonas lacrymans*. Fungal leaf spots affect most crop plants and common pathogenic genera are *Septoria*, *Botryodiplodia*, *Colletotrichum*, *Gloeosporium*, *Cercospora*, *Alternaria*, and *Helminthosporium*.

leaf trace A vascular strand leading from the *stele to the leaf. *See also* leaf gap.

Lecanorales An order of the Ascolichenes or of the Discomycetes in which the majority of *lichens (8–10 000 species) are placed. Such lichens bear apothecia on their upper surface from which ascospores are violently discharged. The Lecanorales includes such genera as *Cladonia*, *Lecanora*, *Parmelia*, *Umbilicaria*, and *Usnea*. The fungal partners in such lichens are closely related to the *Helotiales. In certain lichens, e.g. *Sphaerocarpus* species, the asci and paraphyses of the apothecium disintegrate at maturity to form a mass of spores and hymenial tissues. Such lichens may be placed in a separate order, Caliciales. In other lichens the apothecium may be locular and elongate and contain pseudoparaphyses rather than true paraphyses. Such lichens may be placed in the order Hysteriales. *Rocella tinctorum*, a lichen from which the dye litmus is obtained, is an example of this group.

lecanorine Describing a lichen ascocarp in which the margin cannot be distinguished from the rest of the thallus. The margin is said to be thalline and consists of both algal and fungal cells. Lecanorine ascocarps are seen in *Lecanora*. *Compare* lecideine.

lecideine Describing a lichen ascocarp having a thin margin, termed a proper margin, composed solely of fungal cells, as in *Lecidia*. Most lichen ascocarps do have a proper margin but is is often hidden by the thalline margin (*see* lecanorine) when this is present.

lectotype A specimen or other element (description, illustration, etc.) subsequently selected from the original material on which the name of a taxon was based. Very often these original elements will be *syntypes. A lectotype is only necessary when the original author failed to designate a *holotype. *See also* type.

leghaemoglobin A protein found in the centre of root nodules of leguminous plants infected with the nitrogen-fixing bacterium *Rhizobium*. It has been found that the haemoglobin is coded for by a legume gene but that synthesis only occurs in the presence of the bacterium. The haemoglobin is believed to transport oxygen to the bacterium (which respires aerobically) in such a way that the activity of the nitrogen-fixing enzyme, *nitrogenase (which is destroyed on exposure to oxygen), is not affected. Haemoglobin is not found elsewhere in the plant kingdom.

legume (pod) A dry dehiscent fruit containing one or more seeds. It develops from a single carpel, which on ripening splits along the ventral and dorsal sutures to form two valves, each bearing seeds alternately on the ventral margin. Dehiscence is due to differential drying of the carpel wall, which in some species may result in explosive release of the seeds. The valves may also twist during dehydration, dislodging any remaining seeds. This type of fruit is typical of the Leguminosae but may also be found in other families. *See also* lomentum.

Leguminosae (Fabaceae) A large family of dicotyledonous plants, commonly called the pea family, and containing about 17 000 species in about 700 genera. Legumes usually have pinnately compound leaves, and root nodules containing nitrogen-fixing bacteria of the genus *Rhizobium*. The inflorescence is usually a raceme and the individual flowers have five fused sepals and five petals often arranged in a shape fancifully resembling a butterfly (*see also* standard, keel, wing), hence the name of the largest subfamily, Papilionoideae. The three subfamilies are sometimes classified as families; Mimosaceae, Caesalpiniaceae, and Papilionaceae.

The fruit is typically a pod or *legume. Many of the Papilionoideae are important food crops, e.g. *Phaseolus* and *Vicia* (various kinds of bean), *Pisum sativum* (pea), *Lens culinaris* (lentil), and *Arachis hypogea* (peanut). Others such as *Trifolium* (clovers) and *Medicago sativa* (lucerne), are used for forage. Ornamentals include *Wisteria* and *Lupinus*.

lemma (flowering glume) The lower of a pair of *bracts beneath each floret (flower) in the inflorescence of a grass. The lemma is usually membranous to coriaceous, whereas the other bract, the *palea (if present), is usually thinner and more delicate. *Compare* glume.

lenitic Describing a freshwater ecosystem in which there is not a continuous flow of water, such as a lake or pond. *Compare* lotic.

lenticel A small elliptical pore containing loosely packed cells that is the means of gaseous exchange in the *periderm of plant axes. Lenticels are analogous to the stomata of primary tissues and vary in size from being almost microscopic to about 1 cm in length. *Compare* stoma.

Lepidodendrales An extinct order of the *Lycopsida containing arborescent forms, such as *Lepidodendron*, that flourished during the Carboniferous. Species of *Lepidodendron* reached heights of up to 30 m. The trunks, some of which had a girth of over 3 m at their base, were unbranched for the main part but divided dichotomously at the top into numerous branches. The trunk was patterned with diamond-shaped leaf scars and from its base arose four dichotomously branching axes that gave rise to the root system. Strobili, similar, except in size, to those of *Selaginella*, were borne on the ends of the branches and it appears that the method of fertilization was basically similar to that of *Selaginella*.

leptoid An elongated nutrient-conducting cell in the stems of *Polytrichum* (hair mosses) and related bryophytes, analogous to a sieve cell in vascular plants. *Compare* hydroid.

leptosporangium A type of *sporangium that is derived from one initial cell. It is typical of ferns of the order Filicales (the leptosporangiate ferns). The wall of a leptosporangium is usually only one cell thick and does not contain tissue derived from the archesporium. *Compare* eusporangium.

leptotene *See* prophase.

lethal gene A mutant allele that results in the death of an organism. The mutation commonly takes the form of a small chromosomal deletion, so that the genetic code for a key protein is missing or disrupted. Such mutations do not necessarily take effect at fertilization but may manifest themselves at any stage of development. The chlorophyll-less mutant (chl$^-$) of barley, which is typically recessive, only shows its lethal effects after the food reserves in the germinating seed have been used up.

leucine A nonpolar branched chain amino acid with the formula $(CH_3)_2CH CH_2CH(NH_2)COOH$ (*see illustration at* amino acid). It is synthesized from pyruvate and is degraded by a complex pathway leading eventually to acetoacetate. Many of the enzymes involved in the synthesis and degradation of leucine are the same as those involved in the metabolism of isoleucine and valine.

leucoplast A colourless *plastid found in the cells of roots and underground stems and storage organs. Leucoplasts have rudimentary lamellae.

liana (liane) A long-stemmed woody climbing plant that grows from ground level to the canopy of trees. Lianas abound in tropical forest and individual stems can be as long as 70 m. The plants begin their growth in deep shade but by climbing up and over the top of very tall trees they benefit from full sunlight at maturity. They may bind the trees together so if one dies it is held in position until it decays.

lichen A distinct type of organism in which the thallus is composed of both fungal and algal cells in symbiotic association. The fungal partner (the mycobiont) is usually an ascomycete and is dominant to the alga (the phycobiont), which is a green or blue-green alga. Occasionally the fungus is a basidiomycete or a deuteromycete (imperfect fungus). The fungus forms the main part of the thallus, which is usually a stratified structure consisting of an upper and lower cortex of compact fungal tissue with a medulla in between of loosely woven hyphae. The algal cells are in a layer between the medulla and upper cortex and are closely surrounded by hyphae. This stratified type of structure is termed *heteromerous*. In some lichens, e.g. gelatinous lichens, the thallus consists of loosely woven hyphae and algal cells scattered in a jelly-like matrix. This unstratified structure is termed *homoiomerous*. The fungus obtains carbohydrates from the alga, while the alga receives water and nutrients from the fungus. The alga is also protected from desiccation by the surrounding fungal body. Lichen fungi are not found in the free-living state and attempts to culture them independently have not been successful (except in a few cases where the mycobiont is a basidiomycete). Algae very similar to those in lichens are found growing separately though it is uncertain if these are identical to the lichen algae.

Three main types of lichen are recognized by their growth habit. *Crustose* lichens grow closely attached to the substrate and usually lack distinct lobes. *Foliose* lichens are generally attached loosely to the substrate by rhizinae and the thallus has lobed leaflike extensions. *Fruticose* lichens are either erect and bushy or hanging and tassel-like, and are only attached at one point. Some lichens are intermediate in form. For example, *Cladonia* species initially form a basal crust from which arise erect branching structures (podetia).

Reproduction may be by dispersal of fragments of the thallus containing both fungal and algal cells, by special vegetative reproductive bodies (*see* soredium, isidium), or by fungal spores. The nature of the asci and ascospores produced by the fungal component are important in identification.

Lichens are extremely tolerant of almost total desiccation and can thus colonize exposed bare areas where other plants are unable to survive. They are however very slow growing, an average lichen possibly only extending by 1 mm a year. Some are believed to live for up to 4000 years. They have few commercial uses though some yield dyes, while others contain antibiotics, e.g. usnic acid from *Usnea*. Some of the larger species of arctic zones, e.g. reindeer moss (*Cladonia rangiferina*), are important as food for deer. The distribution of certain lichen species is used as an indicator of atmospheric pollution (*see* indicator species). *See also* Lichenes.

Lichenes A division containing the *lichens. It was created before it was realized that lichens are symbiotic associations between fungi and algae. Now lichens are considered to be fungi especially adapted to obtain food from algae living within their tissues. They are consequently classified in the appropriate rank of the Eumycota, according to the nature of the fungus. However, since the affinities of many lichens are still unknown, the division Lichenes is maintained for convenience. It is divided into the three classes Ascolichenes (*see* Lecanorales, Pyrenulales), Basidiolichenes, and Lichenes Imperfecti. The Basidiolichenes, e.g. *Cora* and *Dictyonema*, are limited to the tropics. They superficially resemble bracket fungi and produce basidia in an even layer on the undersurface of the thallus. The fungi involved are considered closely related to members of the Agaricales. The Lichenes Imperfecti, which contain imperfect fungi, include such common genera as *Lepraria* and *Crocynia*. Reproduction in such lichens is wholly by dispersal of fragments of the thallus and asexual conidiophores are not formed.

life cycle The series of events from the production of gametes in one generation to the same stage in the subsequent generation (*see* diagram). In some plants the life cycle only involves the production of one type of individual (*see* haplobiontic). In most however, the life cycle encompasses a haploid and a diploid generation, i.e. there is an *alternation of generations.

life form The overall morphology of an organism, on the basis of which a species may be described as a tree, shrub, herb, succulent, etc. In describing an area of vegetation much information is conveyed by naming the type or types of life form that predominate. The *Raunkiaer system of classification is often used for this purpose.

ligase (synthetase) Any *enzyme that catalyses reactions involving bond formation with concomitant cleavage of ATP. An example is glutamine synthetase, which catalyses the formation of glutamine from glutamic acid.

light (visible radiation) Electromagnetic radiation with wavelengths ranging from roughly 400 nm (extreme violet) to 770 nm (extreme red). Light from the sun provides the energy to fuel the photosynthetic fixation of carbon dioxide by green plants and is thus the basis of life on earth. *See also* action spectrum, red light, far-red light.

light microscope (optical microscope) An optical instrument that contains one or more lenses and is used in the laboratory to enlarge objects that are too small to be examined in detail by the naked eye. Its maximum resolution (the capacity to observe fine detail clearly), is about 0.3 μm (i.e. it can distinguish points only 0.3 μm apart) as compared

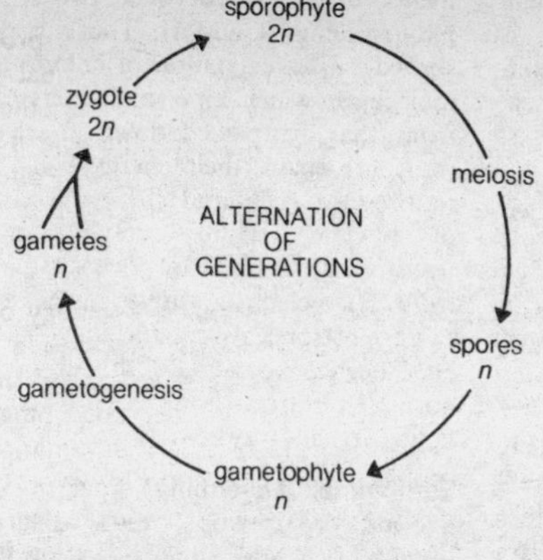

the type of life cycle found in most plants, in which there is a definite development, however reduced, of both sporophyte and gametophyte generations

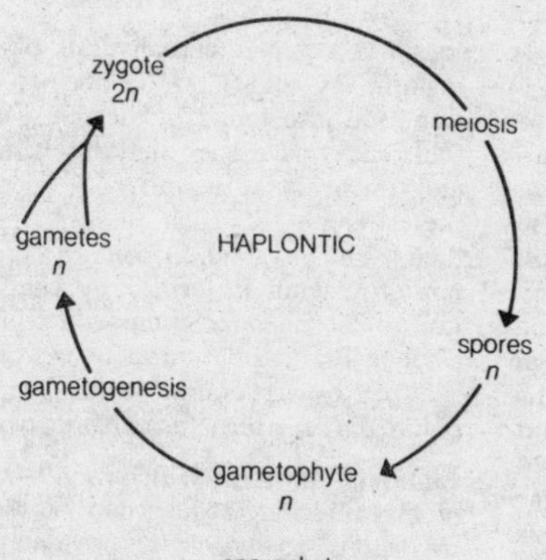

a form of life cycle found in certain algae, e. g. the Ulotrichales, in which there is no sporophyte generation, except as the zygote

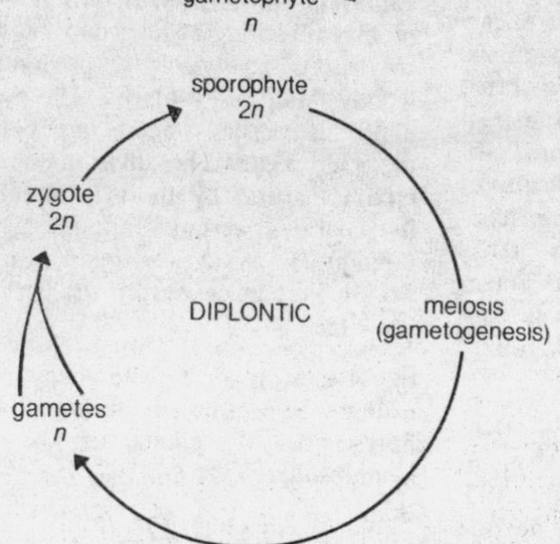

a form of life cycle found in certain algae, e. g. the Fucales, in which there is no gametophyte generation. This cycle is also typical of animals

Three kinds of life cycle found in plants.

to about 80 μm for the average human eye (*see* resolving power). Light is passed through a condenser, which converges the light rays onto the specimen, and then through one or more lenses. The earliest instruments were *simple microscopes* with a single lens (i.e. basically magnifying glasses). Such instruments are little used today as they have to be positioned close to the eye and have a limited field of vision. Also the illumination and mounting of the specimen is very tricky. The *compound microscope* has two lenses, one at either end of the body tube. The objective lens is at the base of the tube close to the specimen and the eyepiece or *ocular* at the other end close to the eye. The image is focused either by moving the body tube or by moving the microscope stage to which the specimen is secured. Most modern compound microscopes have several objective lenses of different magnifications, fitted to a revolving nosepiece. Some microscopes have a built-in light source while others are fitted with a mirror to reflect light from a lamp. The adjustment of an iris diaphragm reduces glare by limiting the part of the specimen that is illuminated. The modern compound microscope is used extensively for observing fine detail of micro-organisms and thin sections through tissues and organs. Sections are often stained to increase contrast. The *stereoscopic binocular microscope* is used for work requiring lower magnifications, such as dissection, or for viewing detail of comparatively large specimens. It has two eyepieces that give a three-dimensional image, and the specimen is usually placed on a contrasting background.

light reactions The light-dependent sequence of photosynthetic reactions that, taking place on the thylakoid membranes of chloroplasts, produce the ATP and NADPH used in the carbon dioxide fixation of the subsequent *dark reactions. Incident light is absorbed by pigments, distributed between two separate photosystems, I and II. The energy absorbed causes excitation of chlorophyll *a* molecules, which emit high-energy electrons that are passed down an electron transport chain, their energy being used to produce ATP and the reduced form of $NADP^+$. Unlike respiration, the transfer of electrons in photosynthesis is from less electronegative molecules to more electronegative molecules. The energy needed to achieve this reversal of normal electron flow is provided by light. *See* photosystems I and II.

lignicolous Describing organisms that live on or in wood, such as bracket fungi.

lignin A complex carbohydrate polymer making up about 25% of the wood of trees and also found in the cell walls of schlerenchyma tissues and vessels, fibres, and tracheids at maturity. It increases the strength of such tissues making them more resistant to compression and tension. Lignin is formed by condensation of the phenolic compound coniferyl alcohol. Its distribution in tissues can be shown by staining with acidified phloroglucin, which turns lignin red.

Lignosae In John Hutchinson's system of classification, a subgroup of the dicotyledons into which the predominantly woody families are placed. The predominantly herbaceous plants are placed in the Herbaceae. The division is considered unnatural by most taxonomists as it separates certain families generally thought of as closely related, e.g. the woody Verbenaceae from the herbaceous Labiatae.

ligulate Strap or tongue shaped as, for example, the outer ray florets of the inflorescences of plants in the family Compositae.

ligule 1. An outgrowth from the top of the leaf sheath in grasses. It may be a scalelike flap of tissue or a ring of hairs and is often of great diagnostic value. *Compare* auricle.

2. The strap-shaped elongation of the corolla tube in certain florets of the Compositae. The five teeth of the ligule represent the tips of the five fused petals. Genera of the tribe Lactuceae, e.g. *Taraxacum*, *Hieracium*, have capitula consisting solely of ligulate florets.

3. An outgrowth from the upper side of the microphyll in species of *Selaginella*. The sporophylls in the strobili also possess ligules.

Liliaceae A large monocotyledonous family, commonly known as the lily family, containing about 3500 species in about 250 genera. Many of the Liliaceae possess swollen underground perennating organs, such as bulbs, corms, and rhizomes. The only important food plant in the family is the onion (*Allium cepa*). Minor food plants include garlic, leek, and asparagus. However horticulturally the Liliaceae are one of the most important families and the flowers display a wide range of form and colour. They may be solitary, as in the tulips (*Tulipa*), in a raceme, as in hyacinths (*Hyacinthus*), or in a cyme, as in the umbel-like cymes of onion. Ornamental genera include *Lilium*, *Aloe*, *Hosta*, *Muscari*, and *Scilla*.

Liliidae A subclass of the monocotyledons containing mainly herbaceous plants often with relatively large flowers. It may contain, depending on the classification scheme referred to, two to five orders. The differences arise because the Zingiberales and Triuridales (sometimes placed in the Liliidae) are alternatively classified in the *Commelinidae and *Alismatidae respectively, and the family Iridaceae (order Liliales) is sometimes raised to the status of an order, Iridales. The order Liliales contains some 14–18 families including: the Iridaceae (which contains the horticulturally important *Crocus*, *Iris*, *Freesia*, and *Gladiolus*); the *Liliaceae; the Amaryllidaceae (which contains the genera *Narcissus* and *Galanthus*); Agavaceae (which includes sisal hemp, *Agave sisalana*); and the Dioscoreaceae (yam family). The order Orchidales contains the Burmanniaceae and the *Orchidaceae.

lime sulphur *See* sulphur dust.

liming The addition of lime to a soil to combat acidity, to improve soil structure, or to remedy calcium deficiency. Quicklime (calcium oxide), slaked lime (calcium hydroxide), or ground chalk or limestone may be used. Liming makes heavy clay soils more workable as it encourages the formation of soil crumbs by a process termed *flocculation*.

limiting factor A factor in the environment that by its presence or absence, or increase or decrease will govern the behaviour of an organism or a metabolic process within an organism. Most metabolic processes depend on more than one factor being present in order to proceed. When all other conditions are favourable, the factor nearest its minimum value is the limiting factor. For example, when a plant photosynthesizes on a warm sunny day in a moist environment, the amount of carbon dioxide available in the air will be the limiting factor. As evening approaches, light will become the limiting factor.

limnology The study of inland aquatic ecosystems.

linear Describing leaves, such as those of grasses, that are elongated and parallel sided for much of their length.

linkage The tendency for different genes to segregate together during meiosis because they occur on the same chromosome. For example, if the genes Aa and Bb occur on the same chromosome then a double heterozygote AaBb formed from Ab and aB gametes would yield a significantly larger number of Ab and aB gametes than AB or ab gametes.

Similarly, if the parental gametes were AB and ab, the heterozygote would generate AB and ab gametes frequently, and Ab and aB gametes rarely. This is in complete contrast to *independent assortment where the genes are on different chromosomes and therefore produce all types of gamete equally frequently. When genes are linked, the two less frequent gametes are produced as a result of *crossing over. As crossing over is relatively rare the genes on one chromosome tend to be inherited together. A group of linked genes is called a *linkage group*, and in general one chromosome corresponds to one linkage group. However, two widely separated genes on a long chromosome with frequent chiasmata may behave as though they are on different chromosomes.

linkage map A diagrammatic representation of the order of genes on a chromosome. Linkage maps can be constructed by crossing appropriate strains carrying different alleles on homologous chromosomes. Results show that the arrangement of genes is linear, although the distance between the genes is expressed not in units of length but in *crossover units* or *Morgan units* (named after T. H. Morgan who discovered linkage). Infrequent crossing over is interpreted to mean that genes are near each other (closely linked), and is expressed quantitatively by a small crossover value (% crossing over). Conversely, genes are regarded as being further apart if the frequency of crossing over is higher, as indicated by a large crossover value.

linoleic acid A diunsaturated eighteen-carbon fatty acid found in various vegetable oils, especially linseed oil. In plants it is formed from oleic acid. However animals cannot synthesize it and, as a precursor of arachidonic acid and the prostaglandins, it is essential in their diets.

linolenic acid An eighteen-carbon triunsaturated fatty acid. It occurs in many plant lipids but is especially prevalent in the chloroplast, where it is the predominant fatty acid. Like *linoleic acid it is essential in animal diets.

lip *See* labellum.

lipase Any *hydrolase enzyme involved in the breakdown of storage fats. Lipases catalyse the breakdown of triglycerides into free fatty acids and glycerol. The fatty acids are then further degraded in the *β-oxidation cycle. Lipase activity is very high in germinating oil-rich seeds, such as that of the castor bean.

lipid A water-soluble hydrocarbon that can be extracted from cells and tissues by nonpolar solvents. Two major classes of lipid can be recognized. *Complex* or *saponifiable* lipids comprise all those lipids that yield soaps on alkaline hydrolysis. These include *acylglycerols, *waxes, *phosphoglycerides, and *sphingolipids. *Simple* lipids are a heterogeneous group of compounds, including the sterols, carotenes, and xanthophylls. Although they are called simple lipids they can be very complex structurally; natural rubber is a simple lipid but it is an extremely large and complex polymer.

The functions of lipids within the plant are varied. Storage triacylglycerols provide an energy store, especially in oil-rich seeds such as the castor bean. The phosphoglycerides are important structural components of membranes, while waxes and cutin, components of the cuticle, help control water loss from the plant.

lipopolysaccharide Any molecule containing both lipid and carbohydrate elements. Lipopolysaccharides are the principal components of gram-negative bacterial cell walls. Such molecules have a repeating trisaccharide backbone, to which are attached side chains of oligosaccharides and the fatty acid 3-hydroxymyristic acid.

lipoprotein An association of lipids and proteins in a system that combines aspects of the properties of both types of molecule. Most membrane systems (*see* plasma membrane) are composed primarily of lipoprotein. The major lipids in plant membranes are phosphoglycerides and glycolipids. (Glycolipids are especially abundant in the chloroplast membranes.) The protein component is not limited to any one type of protein but membrane proteins often have a slightly modified structure to improve their interaction with lipids. For example, there are relatively few disulphide bridges in membrane proteins.
The forces binding lipid and protein together in a lipoprotein system are not conventional chemical bonds. The most important binding forces are hydrophobic interactions but polar and electrostatic forces are also important.

lithocyst A cell that contains a *cystolith.

lithophyte **1.** A plant that grows on rock outcrops or on rocky or stony ground. An example is the fern *Asplenium ruta-muraria* (wall-rue).
2. An organism partly composed of siliceous or calcareous material, e.g. the stoneworts (Charophyta).

lithosere *See* xerosere.

littoral **1.** Describing the seashore between low and high tide marks that is exposed alternately to air and water. It may be muddy, sandy, or rocky. The size of the zone varies with the slope of the coast and the height of the tides. Typical plants are brown algae of the order Fucales.
2. Describing the area of a lake or pond that extends from the edge to the lower limit of rooted aquatic plants. There may be three distinct zones of plants: emergent plants at the water's edge; plants with floating leaves in deeper water; and finally completely submerged plants. Certain mosses, notably *Fontinalis*, can grow at lower light intensities than higher plants, and may form meadows on the lake bed.
3. Describing organisms that live in the littoral zone.
Compare sublittoral.

liverworts *See* Hepaticae.

living fossil A present-day species that has certain characteristics only found elsewhere in extinct groups of organisms. Such species have usually lived in relatively unchanging environments and have evolved very slowly. The maidenhair tree (*Ginkgo biloba*) is an example. It was discovered by Western botanists in Japan in the seventeenth century and subsequently in China but only in cultivation. The fossil members of the Ginkgoales were widely distributed in the Mesozoic era. Similarly the dawn redwoods (*Metasequoia*) were only known from fossil remains until *M. glyptostroboides* was discovered in a remote part of China in the mid-1940s. *See also* bradytelic.

loam A type of *soil in which there is an even mixture of fine clay particles and coarser sand particles. It is the best type of soil for cultivation as the sand helps drainage and aeration but the clay prevents excessive water loss and binds the organic material. *See also* soil texture.

lobed Describing a leaf that is divided into curved or rounded parts connected to each other by an undivided central area.

locule (loculus) A cavity within which specialized organs may develop, most usually the ovules or pollen grains. In an anther there are normally four locules. In a simple ovary there is one locule while in a compound ovary there may be one or many locules depending on how the carpels are fused.

loculicidal Describing fruit dehiscence in which the slits in the pericarp arise along the dorsal suture of each carpel.

Loculoascomycetes A class of the *Ascomycotina in which the asci are two-walled (bitunicate) and are contained in perithecia enclosed in a stroma. It includes some 530 genera and over 2000 species, some of which are important plant parasites, e.g. *Venturia inaequalis*, which causes apple scab.

loculus *See* locule.

locus The position occupied by a gene on a chromosome. Alleles of a gene occupy the same locus on homologous chromosomes.

lodicule Either of the minute scales between the stamens and the glume of a grass flower. They are believed to represent the reduced perianth parts and may, through changes in turgor, be involved in the opening of the flower. Rarely there are three lodicules, while in a few species they are absent.

loess A fine-textured yellowish azonal *soil that is widespread in central Europe, southern Russia, northern China, and Argentina. It consists of clay and silt particles that were deposited at the edge of the ice sheets during the last ice age. It is a fertile, often calcareous, soil and is the parent material of chernozem soils.

logarithmic phase The period during the growth of microorganisms in culture when the cells are dividing rapidly and there is a huge increase in cell number. If the logarithm of cell numbers is plotted against time then a straight line is obtained on the graph. This period usually ends quite abruptly when nutrient levels fall, the pH of the medium changes, or toxic wastes build up. The logarithmic phase may be maintained by transferring cells to a fresh culture medium or by adding more nutrients to the existing medium. *Compare* lag phase. *See also* exponential growth.

lomasome A complex invagination of the plasma membrane into the cytoplasm. Lomasomes have been identified in some fungal hyphae and spore-producing structures, algal cells, and in some cells of higher plants. In the prokaryotic cells of blue-green algae, lomasomes, like the mesosomes of bacteria, are thought to be the site of biochemical processes that in eukaryotic cells are associated with the membrane-bound organelles. *See* chromatophore.

lomentum A dry dehiscent fruit, developed from a single carpel, that contains one or more seeds. It resembles a *legume but on ripening false septa divide the pod into one-seeded units or valves that fracture at maturity. Such fruits are seen in sainfoin (*Onobrychis viciifolia*).

long-day plant (LDP) A plant that appears to require long days (i.e. days with more than a certain minimum length of daylight) before it will flower. In actual fact, it requires a daily cycle with no long dark periods. For example, henbane (*Hyoscyamus niger*) does not flower when given cycles of 12 hours light followed by 12 hours of darkness, but will flower if given cycles of 6 hours light followed by 6 hours darkness. *Compare* day-neutral plant, short-day plant. *See* photoperiodism.

loose smuts *See* smuts.

lophotrichous Describing a bacterium that has a group of flagella at one end of the cell. *Compare* monotrichous, peritrichous.

lotic Describing a freshwater ecosystem in which there is a continuous flow of water, such as a river. *Compare* lenitic.

low-temperature scanning electron microscopy The preparation of specimens for the scanning electron microscope, and their subsequent examination, at very low temperatures. The specimen is frozen with liquid nitrogen (the melting point of nitrogen is −210°C), coated at liquid nitrogen temperatures, and examined in a special specimen chamber similarly cooled. In this way frozen hydrated material can be examined. The results are similar to those obtained by *critical point drying but with fewer artefacts. The technique is particularly useful for delicate plant specimens.

lumen A cavity enclosed by a cell wall, such as the centre of a xylem vessel. *Compare* lacuna.

luminescence *See* bioluminescence.

lyase Any *enzyme that catalyses the nonhydrolytic cleavage of its substrate. An example is pyruvic decarboxylase, which catalyses the formation of acetaldehyde and carbon dioxide from pyruvic acid.

Lycoperdales An order of the *Gasteromycetes containing the puff balls and earth stars. It includes about 275 species in some 48 genera. The basidiocarp remains closed until maturity and its contents become dry and powdery. In the earth stars (e.g. *Geastrum triplex*) the outer wall of the basidiocarp splits into a number of rays, which peel back to surround the spores still enclosed by the thin inner wall. The whole fruiting body thus fancifully resembles a flower or star. The puff balls (e.g. *Lycoperdon pyriforme*) and earth stars discharge their spores in a series of puffs when disturbed.

Lycopodiales An order of the *Lycopsida containing two genera, *Lycopodium* (club mosses) and *Phylloglossum*, both homosporous. There are some 200 species of *Lycopodium* distributed worldwide. They are all herbs and have long dichotomously branching stems bearing numerous small leaves with no ligules (*compare* Selaginellales). The unilocular sporangia are borne singly in the axils of sporophylls, which in many species occur together in distinctive club-shaped strobili. The free-living gametophyte may, depending on species, be either photosynthetic or saprophytic. It is always found in association with an endotrophic mycorrhiza. The antheridia, which produce many biflagellate antherozoids, are located in the centre of the apex of the prothallus and are surrounded by a ring of archegonia.
Phylloglossum contains the single species *P. drummondii*, which is very different from *Lycopodium* in habit, consisting of a basal whorl of leaves and a single strobilus borne on a long stalk.

Lycopsida A subdivision or class of *vascular plants containing only five living genera but having a rich fossil record, being especially abundant in the Carboniferous. The dichotomously branching sporophyte is differentiated into a shoot and root and the vascular tissue contains properly differentiated phloem. The shoot bears small usually spirally arranged leaves that leave no leaf gap in the stele. The sporangia are either borne singly in the axil of the sporophyll or on the upper surface of the sporophyll near the base. Both homosporous and heterosporous forms exist, the spores giving rise to a subterranean or terrestrial gametophyte, which is very reduced and short-lived. Embryogeny is endoscopic.
The Lycopsida is divided into three extant orders: *Lycopodiales, *Selaginellales, and *Isoetales. There are two extinct orders: *Lepidodendrales; and Pleuromeiales, known from the Triassic and intermediate in form between the Lepidodendrales and the Isoetales.

lysigeny The formation of a space by the destruction of cells. This is often achieved by enzymatic dissolution. *Compare* schizogeny.

lysine A basic amino acid with the formula $NH_2(CH_2)_4CH(NH_2)COOH$ (*see illustration at* amino acid). Lysine is synthesized from aspartic and pyruvic acids via an aldol condensation reaction; control of synthesis is through feedback inhibition of this condensation reaction by lysine. Degradation of lysine occurs via acetyl CoA and the TCA cycle.

Lysine is a precursor in the synthesis of the pyridine and piperidine alkaloids, e.g. the nicotine derivative anabasine.

lysis *See* autolysis, lysogeny.

lysogeny The lysis of a bacterial cell following infection with a bacteriophage. All RNA phages and some DNA phages induce lysogeny. The *lytic cycle* involves: adsorption of the phage to a specific site on the bacterial cell wall; penetration of the cell by the phage; transcription and translation of the phage genetic material by the bacterium so that many new phages are formed; and lysis of the cell to release the newly synthesized particles. *Compare* temperate phage.

lysosome An *organelle, bounded by a single membrane, that contains hydrolytic enzymes (e.g. acid phosphatase, ribonuclease, β-galactosidase, protease) capable of degrading cellular components. Phago-lysosomes are formed in association with vacuoles containing ingested material (*see* endocytosis). Autolysis occurs when components of the cell, e.g. mitochondria or fragments of endoplasmic reticulum, are degraded by the activity of lysosomal enzymes. This process is important in the recycling of valuable nutrients from ageing plant tissue. Vesicles from the *golgi apparatus are also involved in the intracellular translocation of lysosomal enzymes.

M

macchia *See* maquis.

maceration In microscopical preparation, the chemical dissolution, often by the application of strong acids, of the matrix binding parts of a specimen. The isolated pieces of the specimen may then be examined or subjected to further preparation.

macrandry *See* nannandry.

macrofibrils Fibrils within the cell wall that are large enough in some cases to be visible under the light microscope. They are formed from microfibrils lying parallel to each other.

macronutrient (essential element) Any chemical element required by a plant in relatively large quantities for successful growth. Macronutrients include *carbon, hydrogen, and oxygen, which are obtained from carbon dioxide and water. The remaining macronutrients, *nitrogen, *phosphorus, *potassium, *sulphur, *magnesium, *calcium, and *iron, are obtained from the soil. In a culture medium these are provided by calcium and potassium nitrates, potassium and iron phosphates, and magnesium sulphates. *Compare* micronutrient.

macrophyll *See* megaphyll.

macrosclereid A relatively short sclerenchyma cell (*sclereid), somewhat columnar in shape. Macrosclereids form the outer layer of the seed coat in many plants. *See also* Malpighian cell.

macrosporangium *See* megasporangium.

macrospore *See* megaspore.

macrosporophyll *See* megasporophyll.

magnesium Symbol: Mg. A metal element, atomic number 12, atomic weight 24.3, important as a macronutrient for plant growth. It forms part of the chlorophyll molecule and deficiency thus leads to chlorosis. It is also a cofactor for phosphotransferase and phosphohydrolase enzymes.

Other symptoms of magnesium deficiency are necrosis, stunted growth, and in some plants a puckering and whitening of the leaves at the edges. Magnesium is added to soil as magnesium sulphate or magnesium oxide.

Magnoliidae (Ranales) A subclass of the dicotyledons containing plants in which the flower parts are numerous and often inserted spirally, and the gynoecium is apocarpous. The number of orders recognized varies considerably between classifications. The Magnoliales, including the Magnoliaceae, Annonaceae, and Myristicaceae; the Laurales, including the Monimiaceae and Lauraceae; the Aristolochiales, including the Aristolochiaceae and (sometimes) the Nepenthaceae; and the Nymphaeales, including the Nymphaeaceae (water lilies), are usually all included in the Magnoliidae. The following orders may also be placed in the Magnoliidae or may be allocated to a second subclass, the Ranunculidae: the Illiciales; the Ranunculales, including the Berberidaceae, *Ranunculaceae (buttercup family), and Menispermaceae (which includes curare); the Papaverales, including the Papaveraceae (poppy family) and Fumariaceae; and the Sarraceniales, including the Sarraceniaceae. Other orders sometimes recognized include the Nelumbonales, which contains one family Nelumbonaceae. However lotus (*Nelumbo*) is often placed in the Nymphaeaceae or Berberidaceae. The Rafflesiales may also be placed in the Magnoliidae but are sometimes allocated to the Rosidae.

In those classifications that recognize the subclass Ranuncilidae, this is separated from the Magnoliidae on the basis that its members are mostly herbaceous rather than woody plants and generally more advanced than the Magnoliidae. They lack the oil cells common to many members of the Magnoliidae and have tricolpate pollen as compared to the monocolpate pollen of the Magnoliidae. Their stomata usually lack subsidiary cells whereas those of the Magnoliidae often have two subsidiary cells.

Magnoliophyta In certain classifications, a division containing the flowering plants. It is divided into the two classes Magnoliopsida (dicotyledons) and Liliopsida (monocotyledons).

major gene A gene whose effects are readily identifiable in the phenotype. The existence of different alleles of such genes results in instances of qualitative variation in the phenotypes. All Mendelian genes are examples of major genes. *Compare* polygenes.

malate shuttle A method that has been proposed to explain the transfer of reducing power across the chloroplast membranes. NADPH, produced by the light reactions of photosynthesis, is not able to pass through the chloroplast membranes. However there is evidence that reducing power produced by photosynthesis is nevertheless used in the cytosol. It is suggested that oxaloacetate in the chloroplast stroma is reduced by NADPH to malate, which is readily transported across the chloroplast membranes. In the cytosol malate is oxidized to oxaloacetate with the concomitant production of NADH from NAD. The oxaloacetate then passes back into the chloroplast stroma to begin the cycle again.

male Describing either the reproductive parts or a whole organism that bears the microspore-producing apparatus and does not nurture the developing embryo. *Compare* female. *See also* anther, antheridium, antherozoid.

maleic hydrazide A growth inhibitor often used as a herbicide or to inhibit sprouting. In some species it promotes flowering by inhibiting vegetative growth.

male sterility A condition in which pollen production is prevented by mutation of one or more genes governing its formation. Male sterility has been employed by plant breeders as a method of ensuring cross pollination and hence F_1 hybrid production. Male-sterile cultivars can be maintained by crossing ss (female parent) × SS (male parent) giving 50% ss and 50% Ss offspring (where s = male sterile).

malic acid A four-carbon dicarboxylic acid with the formula HOOCCH(OH) CH_2COOH. Malate is a TCA cycle intermediate. It is formed by hydration of fumarate and is oxidized to oxaloacetate with concomitant formation of NADH. Malate can also be formed from pyruvate by malic enzyme. Malate can be transported across certain membranes (*see* malate shuttle). This is important in C_4 plants, where malate is an intermediate in the Hatch–Slack pathway.

malonyl ACP The three-carbon dicarboxylic acid malonic acid bound to *acyl carrier protein (ACP). Malonyl ACP, with *malonyl CoA, is central to fatty acid synthesis. It is formed from malonyl CoA and ACP and then reacts with enzyme-bound acetate to form acetoacetyl ACP, with concomitant release of carbon dioxide. Acetoacetyl ACP is then reduced to butyryl ACP before the addition of another molecule of malonyl ACP to form a six-carbon chain. Malonyl ACP thus provides the two-carbon building blocks with which long-chain fatty acids are progressively synthesized.

malonyl CoA The coenzyme A derivative of the dicarboxylic acid malonic acid. It is synthesized from acetyl CoA and bicarbonate in a reaction catalysed by the enzyme acetyl CoA carboxylase. Malonyl CoA provides the carbons for the synthesis of fatty acids – the malonyl moiety reacts with acyl carrier protein (ACP) to form *malonyl ACP, which in turn combines with the growing fatty acid molecules, with elimination of carbon dioxide, to add two more carbons to the chain.

Malpighian cell An alternative term for a *macrosclereid when present in the testa of a leguminous seed.

maltose (malt sugar) A disaccharide consisting of two glucose units linked by an $\alpha(1-4)$ glycosidic bond. Maltose is widely distributed in plants but does not seem to have a specific function; rather it is an intermediate product in the breakdown of starch to glucose. Concentrations of maltose are particularly high during seed germination when starch reserves are being rapidly broken down. Malt, essential to the brewing industry, is produced by allowing barley seeds to germinate then drying them slowly in a kiln. *See also* amylase.

manganese Symbol: Mn. A metallic element, atomic number 25, atomic weight 54.94, needed in trace amounts by plants for successful growth. Manganese ions are required as cofactors by certain enzymes, e.g. kinases and IAA oxidase. Certain deficiency diseases, e.g. blight of sugar cane and grey speck of oats, are attributed to manganese deficiency. More general symptoms include dwarfing and mottling of the upper leaves. Manganese is often added to the soil as manganous sulphate to treat such problems.

mannan A polysaccharide in which the major monosaccharide subunit is mannose. Galactose and glucose are also often present, forming galactomannans, glucomannans, and galactoglucomannans. Mannans frequently occur in hemicellulose. They are also found as reserve polysaccharides in some higher plants, e.g. the ivory palm (*Phytelephas macrocarpa*), in which the extremely hard endosperm (known as vegetable ivory) is composed of mannans.

mannitol A common sugar alcohol formed by reduction of the carbonyl group of mannose or fructose to an alcohol. Mannitol is the principal soluble sugar in fungi and lichens and it accumulates in some algae much as sucrose accumulates in vascular plants. Mannitol is a major photosynthetic product in the brown algae, lichens, and some higher plant species.

mannose An aldohexose sugar (*see illustration at* aldose). In some plants, notably members of the Leguminosae, mannose rather than glucose is the monosaccharide building block for reserve polysaccharides: the polysaccharides formed are known as *mannans. Mannose is also found as a component of some hemicelluloses. The reduction product of mannose, *mannitol, is an important sugar in lichens and some algae.

manubrium A tubular structure, a number of which arise on the inner walls of the antheridium of algae in the Charophyta. A mass of filaments form from specialized cells at the tip of the manubrium. The cells of the filaments subsequently form antherozoids.

manure Animal excreta, usually mixed with other material, especially straw, used to improve soil fertility. The term may be used in a wider sense to mean any sort of *fertilizer.

maquis A stunted form of woodland (scrub woodland) found in semiarid regions that have been deforested for agricultural purposes or by fire. The poor soil supports a mass of tangled bushes and shrubs growing to about three metres, with scattered twisted dwarfed trees. The species include rockroses (*Cistus*), broom (*Cytisus scoparius*), gorse (*Ulex*), etc., together with herbs, such as thymes (*Thymus*), and heathers (*Erica*). Maquis is widespread in countries around the Mediterranean. In Italy it is called *macchia*, and in Spain, *mattoral.* Similar vegetation in California is called *chaparral. See also* garigue.

Marattiales An order of the Filicinae consisting entirely of large fleshy tropical ferns, commonly known as giant ferns. There are about 100 species in 6 genera. In most species the sporophyte consists of a broad short stem giving rise to large compound fronds. The sporangia are formed on the abaxial surface of fertile fronds that otherwise resemble the sterile fronds. The spores germinate to form a thallose gametophyte, somewhat similar to a liverwort, which may be quite long lived. The Marattiales have a rich fossil record going back to the Carboniferous.

Marchantiopsida *See* Hepaticae.

marginal effect A phenomenon exhibited by the plants at the edge of a stand of vegetation whereby they grow more vigorously than those in the centre. It has been well documented in advancing areas of bracken (*Pteridium*), where the fronds at the edge are significantly taller than those behind them. This is believed to be due to increasing rhizome age at the centre rather than to depletion of soil nutrients. Marginal effects may play a part in the *hummock and hollow cycle of vegetation.

marginal meristem A region of meristematic tissue located along the edges of a leaf primordium. Repeated divisions of initials in this area give rise to the mesophyll and epidermal tissues of the blade.

marginal placentation (ventral placentation) A form of *placentation in which the placentae develop along the *ventral suture of a simple ovary. It is seen in the pods of the Leguminosae.

margo 1. *See* pit.

2. In a pollen grain, the zone of thickened sexine around a colpus.

marsh A region of vegetation where the water table is at or just beneath the soil surface. The soil is neither highly acid nor alkaline (*compare* bog, fen). Common marsh plants are water lilies, reeds, sedges, and various grasses. A marsh is a stage in the development of a *climax vegetation from a *hydrosere.

Marsileales *See* Filicales.

mass flow hypothesis The theory that translocation of sugars in the phloem is brought about by a continuous flow of water and dissolved sugars between sources and sinks. (A source is the site of production of sugars, usually leaves, and the sink is the site of their utilization, for example the root system.) At the source *osmotic pressure is high due to the continuous formation of sugars and at the sink osmotic pressure is low as the sugars are used up. Thus water from the xylem enters the phloem at the source and leaves it at the sink, returning to the source via the xylem. This tends to drive the contents of the phloem towards the sink. The hypothesis has been challenged, notably by those who believe there is a metabolic component, i.e. that active diffusion plays a part in phloem translocation.

massulae Mucilaginous extensions of the tapetum that surround the microspores and megaspores of water ferns of the genus *Azolla*. Four massulae surround the megaspore and are believed to aid buoyancy. A variable number of massulae are developed in the microsporangium, each of which contains a number of microspores. *See also* glochidium.

Mastigomycotina A subdivision of the *Eumycota consisting predominantly of aquatic fungi or fungi that flourish under particularly damp conditions. It contains about 190 genera and some 1280 species. Its members characteristically produce motile cells and are thus called zoosporic fungi. The number and type of flagella possessed by the zoospores is used to divide the Mastigomycotina into three classes: the *Chytridiomycetes, whose zoospores have one posterior whiplash flagellum; the Hyphochytridiomycetes, whose zoospores have one anterior flimmer flagellum; and the *Oomycetes, whose zoospores have an anterior flimmer flagellum and a posterior whiplash flagellum. The diversity of zoospore structure implies the Mastigomycotina are polyphyletic and hence not a natural group.

mastigonema Any of numerous fine hairlike rodlets attached to a *flimmer flagellum. Mastigonemata are arranged in longitudinal rows and are orientated at an angle, giving the flagellum a feathery appearance at high magnification.

mating strain A group of organisms within a species that are characterized by not being able to interbreed with each other. However they are able to breed with members of other physiologically different but morphologically identical groups. Different mating strains are usually denoted by the symbols + and – since male and female forms cannot be distinguished as such. The differences are genetically determined. *Compare* physiological race. *See* heterothallic.

matric potential (Ψ_m) That component of the *water potential of plants and soils that is due to capillary and imbibitional forces. Thus the water potential of cell walls and intercellular spaces is largely due to matric potential. Values of matric potentials are always negative and range from 0 bar in fully turgid tissues to –10 bar in slightly wilted plants. *See also* osmotic potential, pressure potential.

mattoral *See* maquis.

m chromosome A very small chromosome, a number of which may be found in the nuclei of mosses. Their function is obscure.

MCPA (2-methyl-4-chlorophenoxyacetic acid) A synthetic *auxin of the *phenoxyacetic acid group that is used as a selective weedkiller.

meadow A region of moist grassland maintained at the subclimax stage by mowing. Similar areas maintained at the subclimax stage by grazing are often termed *pastures*.

mean (arithmetic mean) The average of a series of quantities, obtained by adding up all the observed values and dividing by the total number of observations. If a normal curve were plotted from an infinitely large number of observations, then the mean (and the *median and *mode) would be the distance from the origin of the axes to the centre point of the curve, represented by the symbol μ. However when sample size is limited, as always occurs in practice, then the estimated mean, rather than the median or mode, gives the best estimate of the true mean.

mean deviation The average magnitude of deviations from the centre of a normal curve. *Compare* standard deviation.

mean square The estimated square of the *standard deviation.

mechanical tissue (strengthening tissue, supporting tissue) Any tissue consisting of cells with thickened cell walls, such as *collenchyma and *sclerenchyma.

median In a series of observed values, the one quantity that has an equal number of observations on either side of it. *Compare* mean, mode.

medium The surrounding substance in which an organism exists, such as air or water, or (in an experimental situation) a mounting medium, staining medium, culture medium, etc.

medulla *See* pith.

medullary ray (primary ray) Any of the radial extensions of the *pith (medulla), consisting of parenchyma and penetrating between the vascular bundles in the primary tissues of the stem. *Compare* ray.

medullated protostele A *protostele in which the central core of xylem consists mainly of tracheids interspersed with numerous parenchyma cells. It thus resembles the medulla (*pith) of a *siphonostele (hence the name). The medullated protostele shows how the siphonostele may have evolved from a protostele, with the pith originating in the xylem itself. Medullated protosteles are exhibited in some primitive ferns, e.g. *Gleichenia*.

megaphyll (macrophyll) A leaf typical of seed plants and ferns, usually relatively large and usually with *leaf gaps associated with the *leaf traces. Megaphylls are thought to have evolved from early leafless plants, such as *Rhynia*, by the development of unequal dichotomies of the axis and 'overtopping' or dominance of the longer indeterminate shoot over the shorter determinate shoot. This is thought to have been succeeded by flattening (planation) of the branches of the determinate shoot and subsequent webbing between them by ground tissue to form typical megaphylls, although the precise order of these events remains somewhat unclear. *Compare* microphyll. *See* telome theory.

megasporangiophore A stalklike structure that bears megasporangia. The ovule-bearing structures (*megasporophylls) making up the female cones of cycads are termed megasporangiophores.

megasporangium (macrosporangium) A structure in which megaspores are formed. In the seed plants it corresponds to the *ovule. *Compare* microsporangium.

megaspore (macrospore) The larger of the two types of haploid spores formed after meiosis in *heterosporous species and usually designated female. It is immobile and contains food reserves for the gametophyte. In angiosperms, the cell that gives rise to the nuclei of the *embryo sac is termed a megaspore, though it is usually smaller than a pollen grain (*microspore).

megaspore mother cell A diploid cell that gives rise by meiosis to four megaspores. Often only one develops and the rest abort. *Compare* microspore mother cell.

megasporophyll (macrosporophyll) A leaflike structure that bears the *megasporangia. In angiosperms and gymnosperms it is represented by the carpel and ovuliferous scale respectively. *Compare* microsporophyll. *See also* sporophyll.

meiocyte A cell that divides by meiosis to produce *haploid spores. The microspore mother cells (pollen mother cells) in anthers and the megaspore mother cells in nucellar tissue are the meiocytes of flowering plants.

meiosis (reduction division) The process by which a diploid cell divides to form four haploid cells. The process consists of two consecutive divisions, each with a sequence of stages similar to those of *mitosis. During the first division, which is the actual reduction division, the pairing (*synapsis) and subsequent separation of *homologous chromosomes into separate nuclei results in the reciprocal exchange of portions of maternal and paternal *chromatids (*see* crossing over). It is in this pairing and separation of homologues that meiosis essentially differs from mitosis (*see* diagram). The two haploid nuclei resulting from the first division then divide for a second time, during which the chromatids are separated as in mitosis. Four haploid cells are therefore formed. Meiotic division is the process by which the haploid gamete-producing (gametophyte) phase in

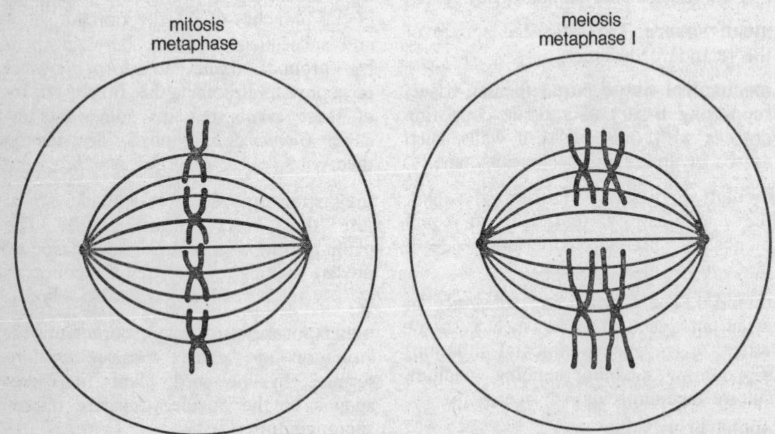

The basic differences in chromosome behaviour in mitosis and meiosis.

the life cycle of plants is established. *See* prophase, metaphase, anaphase, telophase, cytokinesis.

meiotic drive A process occurring during meiosis that results in the two kinds of gametes produced by a heterozygote not being equally common. It is more often seen in the formation of megaspores than microspores. An example is seen in maize plants that are heterozygous for a chromosome mutation, in which one of the homologues of chromosome 10 has an abnormal terminal knob. The knob acts as a centromere and causes the abnormal chromosomes to move to the poles of the spindle earlier in both the first and second divisions of meiosis. The product of meiosis is a linear tetrad in which the abnormal chromosomes are more likely to be in the outer two cells. Since all four cells of a tetrad normally develop into pollen grains, the abnormal chromosome is recovered in equal numbers among the microspores. However only the basal cell of the tetrad develops into the megaspore. Thus the abnormal chromosome is recovered at higher frequencies (about 70%) among the megaspores.

Melanconiales An order of the *Coelomycetes containing many fungi parasitic on plant stems and leaves, e.g. *Gloeosporium* and *Colletotrichum*, both causal agents of anthracnose. It numbers about 1000 species in some 120 genera. The conidiophores erupt in a mass through the host epidermis forming an *acervulus.

membrane *See* cell membrane.

Mendelism The theory that inherited characteristics are governed by discrete factors or genes (*particulate inheritance), which are transmitted to the offspring in a regular and predictable way as described by *Mendel's laws.

Mendel's laws Two laws of inheritance attributed to Gregor Mendel in recognition of his experimental investigations with peas. Mendel's first law, the Law of Segregation, states that while an organism may contain a pair of contrasting alleles, e.g. Tt, these will segregate (separate) during the formation of gametes, so that only one will be present in a single gamete, i.e. T or t (but not both or neither). Mendel's second law, the Law of Independent Assortment, states that the segregation of alleles for one character is completely random with respect to the segregation of alleles for other characters. Thus a TtGg individual will produce equal numbers of all four possible kinds of gametes: TG, Tg, tG, tg. Mendel's first law is still applicable for all chromosomal genes. Mendel's second law is only true if the genes involved are on nonhomologous chromosomes. *See also* linkage.

mericarp Any of the one-seeded portions that result when a compound fruit divides at maturity. Mericarps may be dehiscent or indehiscent. *See also* cremocarp, regma.

meristele One of a number of amphicribal vascular bundles constituting a *dictyostele.

meristem A region containing actively or potentially actively dividing cells, including *initials and their immediate undifferentiated derivatives. Some consider a meristem to consist only of initials but the difficulty in recognizing a boundary between initials and their immediate derivatives has led many to consider the term in the broader sense including undifferentiated nondividing cells. Some meristems, such as axillary buds, are inactive for much of the time, their cells being potentially capable of active division (*see* apical dominance). Meristems may be *apical, axillary, *lateral, *marginal, or *intercalary.

meristem culture (shoot-tip culture) The culture of excised meristems on suitable nutrient media under aseptic conditions. The stem apex is usually used though axillary meristems may also be taken. Often gibberellic acid must be added to the medium to promote normal growth into a plantlet. Once a suitable medium has been found for the regeneration of plantlets, the technique can be used as a means of rapid propagation since the plantlets can be divided at intervals into segments that can be grown on individually. Meristem culture may be used to propagate infertile plants, nursery stock, or F_1 hybrids that do not breed true. As few virus diseases affect the plant apex, the technique can be used to produce virus-free stock.

meristoderm The outer region of the stipe of certain members of the Phaeophyta (brown algae) in which *meristem-like activity occurs. Continued division in the region below the lamina serves to replace tissues that are worn away by tide action.

mesarch Describing xylem maturation in which the older cells (protoxylem) are in the centre of the xylem strand because maturation has progressed both centrifugally and centripetally. *Compare* endarch, exarch.

mesocarp The middle layer of the *pericarp of an angiosperm fruit, positioned between the exocarp and endocarp. In many fruits, such as the peach (*Prunus persica*), the mesocarp is the fleshy part of the fruit. However, in some fruits no mesocarp is present, the pericarp consisting only of exocarp and endocarp.

mesogenous Describing an angiosperm stomatal complex in which the subsidiary cells are derived from the same initial as the guard cells. *Compare* perigenous. *See also* syndetocheilic.

mesophilic Describing microorganisms that require moderate temperatures (30–40°C) for optimal growth. *Compare* thermophilic, psychrophilic.

mesophyll The ground tissue of a leaf, mostly differentiated as photosynthetic *chlorenchyma. In many plants, for example the mesophytes of temperate climates, the mesophyll is differentiated into *palisade and *spongy mesophyll.

mesophyte A plant without adaptations to environmental extremes. *Compare* hydrophyte, xerophyte.

mesosome A structure found in bacterial cells formed by intrusions of the plasma membrane into the protoplasm. The respiratory enzymes are associated with the mesosome and it has the same function as the mitochondrion of eukaryotic cells. It is also attached to the DNA and is thought to control the separation of replicated DNA molecules during cell fission.

Mesozoic An era of geological time between about 225 and 65 million years ago. It is divided into the *Triassic, *Jurassic, and *Cretaceous periods. *See* geological time scale.

messenger RNA (mRNA) A linear single-stranded polynucleotide of uridine, adenine, guanidine, and cytidine monophosphates. mRNA is synthesized on DNA by mRNA polymerase enzyme, and the sequence of bases in it is a complementary copy (transcript) of the bases in one strand of DNA, the *sense strand*. The molecule initially manufactured on DNA now undergoes modification prior to its involvement in *protein synthesis. This modification, called *post-transcriptional processing*, occurs in the nucleolus and involves the addition of adenine polynucleotide (poly A) to one end and a guanine derivative to the other end. In eukaryotic cells, post-transcriptional processing may also include excision of *introns. Sometimes the term *heterogeneous RNA* (hnRNA) is given specifically to newly formed intron-containing mRNA, and the term mature mRNA is reserved for the molecule produced as a result of post-transcriptional processing. Mature mRNA moves to the ribosomes in the cytoplasm, where the triplets of bases in it act as a template and correspond to amino acids in proteins.

mestom sheath (mestome sheath) A thin-walled type of *bundle sheath present in some grasses.

metabolism The sum total of the enzymatic reactions occurring in a cell, organ, or organism. Metabolism serves four major functions: it obtains chemical energy from fuel molecules or from light; it converts exogenous nutrients into the precursors of macromolecular cell components; it assembles these precursors into macromolecules; and it synthesizes molecules to carry out specialized functions in a particular cell.

Metabolism can be broadly divided into catabolic and anabolic processes. *Catabolism*, the breakdown of complex biomolecules to release energy, occurs in three stages. Macromolecules are broken down in the first stage into their constituent subunits. These subunits are then converted into a few simple molecules ready for stage three, the complete oxidation of the simple molecules to carbon dioxide and water. Stage three, of which the TCA cycle is an important component, is also the first stage of anabolism (*see* biosynthetic pathway); hence stage three is both anabolic and catabolic and is termed *amphibolic*.

Particular metabolic processes are located in specific areas of the cell. Thus protein synthesis takes place on the ribosomes and the enzymes of the TCA cycle are found in the mitochondria. There is a continuous turnover of the end products of cell metabolism. This metabolic turnover is a characteristic of all living systems.

metacentric Describing a chromosome in which the chromatid arms are of a similar length either side of the centromere so that the centromere appears to be near the centre of the chromosome. *Compare* acrocentric, telocentric.

metachromatic stain A stain that dyes certain tissues a different colour to the dye solution. Examples are methyl violet and thionine. Thionin violet produces colours in sections ranging from blue to reddish violet and is particularly useful for distinguishing chromatin and mucin. *See also* staining.

metaphase The stage in nuclear division following *prophase that commences with the disintegration of the nuclear membrane and the formation of the spindle. In *mitosis individual chromosomes, each divided into chromatids, gather around the equator of the spindle. At metaphase of the first division of *meiosis the bivalents formed during prophase gather at the equator. The two centromeres of each bivalent are finally situated one on either side of the equator. The placing of any one homologue on one or the other side of the equator is completely random, no factor operating to discriminate between 'maternal' and 'paternal' chromosomes. At metaphase of the second division of meiosis, spindles form in the haploid daughter cells resulting from the first division and the individual chromosomes become attached to the spindle equator, as in metaphase of mitosis.

The end of metaphase and the beginning of *anaphase, in mitosis, is marked by the division of the centromeres so that the chromatids are no longer held together. In meiosis, division of the centromeres does not occur until the end of metaphase of the second division.

metaphloem Late *primary phloem. The metaphloem completes its elongation after the surrounding tissues have ceased elongating and therefore, unlike the *protophloem, is not obliterated. However, in plants exhibiting secondary growth, the growth of the secondary phloem often results in obliteration of the metaphloem. Unlike those in the protophloem, these obliterated cells do not usually differentiate into fibres, although they may undergo sclerification. The *sieve elements of the metaphloem are usually wider and longer than those of the protophloem, and their *sieve plates are usually more distinct. *Compare* metaxylem.

metaxylem Late *primary xylem. The metaxylem completes its elongation after the organ has ceased longitudinal growth and therefore its tracheary elements are not destroyed. Like the *protoxylem, the metaxylem is composed mostly of tracheary elements and parenchyma cells, but sclerenchyma fibres may also be present. The vessels or tracheids of the metaxylem, however, are generally wider and more numerous than those of the protoxylem and have pitted secondary cell walls. In plants not exhibiting secondary growth at maturity, the metaxylem is the only water-conducting vascular tissue. *Compare* metaphloem.

methionine A nonpolar sulphur-containing amino acid with the formula $CH_3S(CH_2)_2CH(NH_2)COOH$ (*see illustration at* amino acid). Methionine is synthesized from aspartate via the nonprotein amino acid homoserine. Control of methionine synthesis is by feedback inhibition. Methionine is broken down to succinyl CoA, which can be further oxidized in the TCA cycle.

The methyl group of methionine is important in the formation of a large number of methylated biomolecules. The actual methyl transferring molecule is not methionine itself but its ATP-activated derivative, *S*-adenosyl methionine. Methionine has also been implicated in the formation of the growth substance ethylene.

methylene blue A blue dye that is particularly important as a bacterial stain as it has an affinity for bacterial protoplasm.

mevalonic acid A precursor in the formation of terpenoids and steroids. It has the formula $CH_2OHCH_2C(OH)(CH_3)CH_2COOH$. Biosynthesis of gibberellins, cytokinins, and abscisic acid is from mevalonic acid. Certain growth retardants, e.g. Amo-1618 and CCC (chlorcholine chloride), are believed to act by blocking the step from geranylgeranylpyrophosphate to kaurene in the mevalonic acid synthesis pathway, so preventing gibberellin synthesis.

micelle 1. An area within the microfibrils of a cell wall, about 50 nm in length, where the cellulose chains lie very closely parallel to each other. The resulting orderly organization of the cellobiose units (*see* cellulose) results in a very regular arrangement of atoms in these areas so that the cellulose is essentially crystalline.

2. A clay-humus particle that is formed when finely divided mull humus complexes with the clay particles in the soil. It is negatively charged and thus various positive ions are absorbed on its surface. These may be exchanged for other positive ions and the total amount of ions held in this way is termed the *exchange capacity* of the soil.

Microascales An order of fungi belonging to the *Plectomycetes whose members, numbering about six genera, characteristically have a beaked cleistothecium. It contains the species *Ceratocystis ulmi*, which causes Dutch elm disease.

microbody A small *organelle in the cytoplasm, about 0.5 μm in diameter and surrounded by a single membrane. Microbodies contain oxidative enzymes. *See* glyoxysome, peroxisome.

microfibril A ribbon-like structure about 10–45 nm in width and visible only under the electron microscope.

Microfibrils form the basic structural units of the cell wall. Each microfibril is formed from many long-chain cellulose molecules lying approximately parallel to each other. Hydrogen bonds between hydrogen and oxygen atoms in adjacent chains bind them together. The intermolecular spaces and the interfibrillar spaces are partly occupied by structural substances, e.g. pectic compounds and hemicellulose. As a result of imbibitional and capillary effects, these spaces are also filled with water so that the cell wall is normally heavily hydrated. In primary walls (*see* cell wall) water may account for up to 70% of the volume. In secondary walls, more efficient packing of the microfibrils leaves less room for water between them.

microflora Small plants, such as certain algae and fungi, found in a given area, e.g. on a leaf surface.

micrograph A photograph of a specimen taken through a microscope. *Photomicrographs* and *electron micrographs* are produced using light microscopes and electron microscopes respectively.

micrometer A device used in microscopy to measure the size of objects accurately. An eyepiece micrometer (*graticule*) is a round glass disc with an engraved scale, which is placed inside the eyepiece of a light microscope. Before using it to measure a specimen it is calibrated with a stage micrometer. This is normally a glass slide with a scale etched in, for example, one hundredths of a millimetre. The microscope is focused on the stage micrometer so the scale in the graticule can be seen just above the scale of the stage micrometer. The size of the divisions of the eyepiece micrometer can then be calculated against the known scale of the stage micrometer.

micrometre Symbol: μm. A unit of length (formerly called a *micron*, symbol μ) equal to one thousandth of a millimetre, i.e. 10^{-6} metre. It is used in the measurement of cells, cell inclusions, bacteria, etc.

micron *See* micrometre.

micronutrient (trace element) Any chemical element required by a plant in very small quantities for successful growth. Micronutrients thus do not fulfil major nutritional requirements but act as enzyme *cofactors or as essential components of pigments, enzymes, etc. They include many heavy metals, such as *copper, *zinc, *molybdenum, *manganese, and cobalt. *Boron is also required in trace amounts and it is possible that sodium, chlorine, vanadium, and *silicon may be necessary to some plants. Specific *deficiency diseases may arise if a particular micronutrient is unavailable. *Compare* macronutrient.

microphyll A leaf typical of many lower plants, such as lycopods, that is usually relatively small and is not associated with *leaf gaps in the stele. Microphylls are thought to have evolved from early leafless plants by the development of *enations* or emergences from the axis, each with a single *leaf trace, although the leaf trace probably evolved after the enations. *Compare* megaphyll.

micropyle The small channel that remains between the tips of the integuments at the apex of the ovule. The pollen tube usually enters the nucellus via the micropyle prior to fertilization. The micropyle may persist as a small pore in the seed through which water is absorbed prior to and during germination.

microscope An instrument for producing a magnified image. *See* light microscope, electron microscope.

microsere *See* sere.

microsome A small vesicle that forms from *endoplasmic reticulum. Microsomes separate out during centrifugation of cell homogenates. Fragments of membrane are seldom found after centrifugation, and, with their continuous membranes, microsomes thus demonstrate the ability of membranes to repair themselves.

microsporangiophore A stalklike structure bearing microsporangia. *See also* microsporophyll.

microsporangium A structure in which *microspores are formed. In the seed plants it corresponds to the *pollen sac. *Compare* megasporangium.

microspore The smaller of the two types of spore formed after meiosis in *heterosporous species and usually designated male. In seed plants it is the *pollen grain.

microspore mother cell A cell that gives rise by meiosis to four haploid cells that develop into microspores. Unlike the tetrad formed by meiosis of the *megaspore mother cell, all four cells usually develop into microspores. Pollen mother cells are examples of microspore mother cells.

microsporophyll A modified leaf that bears the *microsporangia. In angiosperms and gymnosperms it is represented by the stamens and male scales respectively. *Compare* megasporophyll.

microtome A device used for cutting thin sections (3–5 μm thick) of plant or animal material for examination under the microscope. The specimen is often supported, either by embedding it in a suitable medium, such as paraffin wax, or by freezing it. There are various types of microtome; all have a means of holding the specimen, a knife, and a mechanism for moving the specimen slowly towards the knife. *Ultramicrotomes* are used for cutting very thin sections (20–100 nm) needed for electron microscopy. Glass or diamond knives are used and the specimen advanced by minute increments by thermal expansion. The sections are floated in a water- or water-and-acetone-filled trough surrounding the knife and collected by transferring them onto a fine copper grid. When cutting frozen sections (*cryomicrotomy*), both the cutting knife and the specimen are enclosed in a cold chamber.

microtubule An unbranched tubule identified by electron microscopy in a wide variety of cells. Microtubule numbers vary considerably in different kinds of cell and through the life of a cell. They may either be arranged in an orderly manner or randomly distributed through the cytoplasm. Their outer diameter is about 25 nm and they have an apparently hollow core, 4 nm in diameter. The walls are composed of subunits of a protein, tubulin, which are arranged linearly to form protofilaments. Thirteen protofilaments arranged in a cylinder form the wall of each microtubule.

Microtubules form the axoneme fibres of *flagella and cilia and the *spindle fibres in dividing cells. They are also thought to organize the microfibrils of cell walls and may be involved in the transport of materials within and between cells, e.g. through sieve tubes.

microtubule organizing centre (MTOC) A general term for *kinetosomes, centrioles, and other structures involved in the formation and organization of microtubules or their derivatives.

middle lamella The first-formed layer of the primary wall (*see* cell wall) formed from the *phragmoplast. Compounds of pectic acid, e.g. calcium and magnesium pectates, are deposited into the cavities of the phragmoplast and these eventually coalesce to form a continuous opaque layer. Cellulose molecules, organized to form *microfibrils, are deposited to build up the primary wall. In mature tissues the middle lamella cements the walls of contiguous cells.

midrib **1.** The vein running down the middle of a leaf from the petiole or leaf base to the leaf tip, often dividing the leaf into similar halves or mirror images. **2.** Any thickened region of cells running down the centre of a thallus or lamina, such as the midrib of wracks (*Fucus*). *See also* nerve.

mildew A plant disease in which growth of the fungus is seen on the surface of the host. *Downy (or false) mildews penetrate deeply into their hosts, while *powdery (or true) mildews live on the surface of their hosts in a similar way to *moulds.

millimicron *See* nanometre.

Millon's test A technique used to demonstrate the presence of the amino acid tyrosine. Millon's A, a mixture of sulphuric acid and mercury(II) sulphate is added to the test solution. A yellow colour indicates the presence of protein. A drop of Millon's B, a sodium nitrite solution, is then added. A red colour indicates a positive reaction.

Miocene *See* Tertiary.

mirror yeast *See* Blastomycetes.

Mississippian *See* Carboniferous.

mitochondrion An organelle, found in all eukaryotic cells, that provides an efficient apparatus for the production of ATP. The number present varies considerably, being greatest in metabolically active cells. Mitochondria vary in form being generally spherical or threadlike and there is consistency of form in any one cell type. They are approximately 1–3 μm across and are generally freely distributed in the cytoplasmic matrix, though tending to concentrate in regions of the cell where the demand for ATP is high. The mitochondrion is surrounded by two membranes. The outer one is smooth but the surface area of the inner one is very greatly increased by infoldings that extend to varying distances into the central compartment, forming shelflike structures or *cristae*. The central compartment contains a colloidal matrix with a fine fibrillar structure. Ribosomes of the smaller 70S type and circular DNA molecules, both characteristic of prokaryotic cells (*see* serial endosymbiotic theory), are present, as are DNA polymerase enzymes and enzymes of the TCA cycle. In cell cultures, mitochondria display twisting and wriggling movements and the cristae change their shape when respiratory activity is stimulated.

The inner surface of the crista has many knoblike structures uniformly distributed in the membrane. These are composed of a protein, F_1, capable of transferring phosphate to ADP (i.e. an ATP synthase enzyme). Electron transfer compounds, required for the eventual formation of water from hydrogen ions and oxygen, are situated within the membrane of the crista. The energy made available by electron transfer is incorporated into ATP molecules formed by the activity of the F_1 protein particles. The outer membrane is freely permeable to water and soluble ions but the inner membrane has limited permeability and contains carrier compounds that transfer specific metabolites. One such specific carrier allows a molecule of ADP to enter only if a molecule of ATP passes in the reverse direction. Another concentrates calcium ions in the matrix.

Mitochondria can only arise by fission of preexisting ones and therefore must be transmitted to daughter cells during *mitosis. The DNA and ribosomes they contain enable them to synthesize many of their constituent molecules.

mitosis The process by which a cell divides to form two daughter cells, each having a nucleus containing the same number of chromosomes, with the same genetic composition, as that of the mother cell. The changes that occur in the structure of the chromosomes and in the cytoplasm are clearly visible with a light microscope and form a sequence of stages; *prophase, *metaphase, *anaphase, *telophase, and *cytokinesis. As a result of mitotic divisions, all the cells of the sporophyte of higher plants have diploid nuclei that are genetically identical to that of the fertilized egg cell. Similarly all the cells of a gametophytic plant body have haploid nuclei genetically identical to the gamete from which the plant arose.

mitotic crossing over A process that results in genetic recombination occurring in somatic cells. It was first discovered as a rare event in *Drosophila* (fruit flies) but has since been found to occur in other organisms, especially fungi. In the latter, it sometimes occurs at a frequency that suggests it is a major source of genetic variation. The mechanisms involved are uncertain, since chromosomal pairing (a supposed requirement of crossing over in meiosis) is normally believed not to occur during mitosis.

mixed lactic fermentation (heterolactic fermentation) A form of *anaerobic respiration found in certain microorganisms, e.g. *Lactobacillus brevis*, in which the end products are a molecule each of lactic acid, ethanol, and carbon dioxide.

mode The class of observations that occurs most frequently, for example the number 3 in the series 9, 3, 3, 1. *Compare* mean, median.

moder *See* humus.

molecular-exclusion chromatography *See* gel filtration.

Molisch's test *See* alpha-naphthol test.

molybdenum Symbol: Mo. A metal element, atomic number 42, atomic weight 95.94, needed (under certain conditions) in trace amounts for plant growth. It is involved in nitrate reduction being a part of the flavoprotein enzyme nitrate reductase. Plants absorbing nitrogen as ammonium ions do not need molybdenum. The nitrogenase system involved in nitrogen fixation also requires molybdenum since one of the two proteins forming the nitrogenase complex is a molybdenum–iron–sulphur protein.

monadelphous Describing stamen filaments that are all fused for the greater part of their length, so forming a tube around the style.

Moniliales A group of imperfect fungi comprising the *Blastomycetes and *Hyphomycetes.

monocarpellary *See* fruit.

monocarpic *See* hapaxanthic.

monochasium (monochasial cyme) A *cymose inflorescence in which only one axillary bud develops into a lateral branch at each node. Laterals may arise on alternate sides of the stem (a *scorpioid* monochasium) or may arise on one side only (a *helicoid* monochasium), which gives an asymmetrical appearance to the inflorescence. *Compare* dichasium. *See also* bostryx, cincinnus, rhipidium.

monoclinous *See* hermaphrodite.

monocolpate Describing a pollen grain having one colpus, as is commonly found in most petaloid monocotyledon species.

Monocotyledonae A subclass of the *Angiospermae containing all the flowering plants having embryos with one cotyledon. Its members only very rarely possess a cambium and hence lack secondary thickening although some families (e.g. Palmae) have arborescent forms. Other general features by which the Monocotyledonae can be distinguished from the *Dicotyledonae include: having narrow parallel-veined leaves; having flower parts inserted in threes, or multiples thereof; having a fibrous root system composed of adventitious roots; and having numerous scattered vascular bundles (*see* atactostele).

In some classifications monocotyledons are placed in the class Liliopsida. The 60 or so families of the monocotyledons have been divided into a various number of superorders (or subclasses) by different authorities. Two widely used modern classifications (those of A. Cronquist and of A. Takhtajan) both recognize four subclasses in the Liliopsida, the *Alismatidae, *Arecidae, *Commelinidae, and *Liliidae.

monoecious Having the female and male reproductive organs separated in different floral structures on the same plant. Monoecy decreases the chances of self fertilization and is often associated with wind pollination, as in hazels (*Corylus*) and maize (*Zea mays*). The genetic control of sexual expression in monoecious plants is very complex and not clearly understood but it appears to involve combinations of genes that either promote or suppress the formation of male or female flowers. *Compare* hermaphrodite, dioecious.

monohybrid An individual that is heterozygous in respect of a single gene. It is obtained by crossing parents that are homozygous for different alleles, e.g. a homozygous tall parent (TT) crossed with a homozygous short parent (tt) will produce monohybrid offspring (Tt). When selfed, a 3:1 ratio of dominant:recessive phenotypes, or some modification thereof, will be produced. This is called the *monohybrid ratio*. *Compare* dihybrid.

monolete Describing a spore that has a simple linear scar marking the point at which it was joined in the tetrad. The microspores of members of the Isoetales are monolete, which sets them apart from other lycopsids. However their megaspores are trilete (*see* triradiate scar).

mononucleotide *See* nucleotide.

monophyletic Describing taxa arising from the diversification of a single ancestor, i.e. natural groups in a classification. In cladistics only taxa that contain *all* the descendants of a common ancestor are considered monophyletic. An example of such a taxon would be the monocotyledons. Taxa that contain some but not all the descendants of an ancestor are termed *paraphyletic. *Compare* polyphyletic.

monoplanetic Describing fungi that produce only one type of zoospore (planospore). *Compare* diplanetic.

monoploid *See* haploid.

monopodial branching A type of growth exhibited by many plants in which secondary shoots or branches arise behind the growing point but remain subsidiary to the main stem, which continues to grow indefinitely. The largest secondary shoots are furthest from the apex of the main stem and the size of the shoots decreases regularly towards the top of the plant. This results in the pyramidal form of growth typical of many conifers, e.g. the spruce (*Picea*). The secondary shoots also tend to show the same pattern of branching along their length. *Compare* sympodial branching.

monosaccharide A carbohydrate with the empirical formula $(CH_2O)_n$; in organic compounds n is between three and seven. The carbons in a monosaccharide are usually arranged in an unbranched chain with all carbons but one being hydroxylated; the remaining carbon is either ketonic (*ketose sugars) or aldehydic (*aldose sugars). The commonest monosaccharides are the *hexoses*, with six carbons, and the *pentoses*, with five carbons.

All monosaccharides except dihydroxyacetone are *chiral* molecules, i.e. they exhibit stereoisomerism. Most naturally occurring monosaccharides are in the D-form, although some L-isomers do occur. In aqueous solution most monosaccharides form a ring structure in which the aldehydic or ketonic carbon links with one of the hydroxylated carbons. Two forms of ring, the *pyranose and the *furanose, are commonly seen. The ring structure of a monosaccharide has two isomeric forms, designated α and β.

monosomic An organism deficient in one chromosome from an otherwise diploid set, i.e. $2n-1$. Monosomy typically arises by fertilization occurring between a normal gamete (n), and one deficient in the said chromosome ($n-1$). Monosomics are more common in polyploids where the deleterious consequences of missing a chromosome can be masked by chromosomes in other genomes. *See* aneuploidy.

monospore A spore formed by certain red algae of the Bangioideae as a means of asexual reproduction.

monotrichous Describing a bacterium that has a single polar flagellum. *Compare* lophotrichous, peritrichous.

monotypic Describing any taxon that includes only one subordinate taxon. Thus a family with just one genus or a genus with a single species, are examples of monotypic taxa. An extreme case of monotypy occurs in the classification of the maidenhair tree, *Ginkgo biloba*. This species is the sole representative of the genus *Ginkgo*, the family Ginkgoaceae, and the order Ginkgoales – each being monotypic taxa.

monsoon forest *See* forest.

moor A region of land that is found in wet exposed conditions where the soil water can seep laterally very slowly but is not stagnant as in *bogs. It has an acid peaty soil and common plant species found there include mat grass (*Nardus stricta*) and purple moor-grass (*Molinia caerulea*). The top soil loses water in summer but the subsoil is permanently waterlogged. The peat layer is rarely more than 30 cm deep and most of the plants have their roots in the mineral soil horizon.

mor *See* humus.

mordant *See* staining.

morphactin Any of a group of synthetic compounds, all derivatives of fluorene-carboxylic acid, that affect many aspects of plant growth and differentiation. Morphactins have little effect on mature tissues but in seedlings and growing shoots apical dominance is removed and internode elongation inhibited, giving a dwarfed bushy appearance. Expansion of the leaf lamina is also prevented and flowers, if produced, may be deformed. Seeds from morphactin-treated plants may contain abnormal embryos. Morphactins applied to seeds slow down germination and inhibit lateral root formation from the radicle. In legumes, root nodule formation is prevented. Other diverse effects of morphactins include the inhibition of phototropism and geotropism, the induction of parthenocarpy, and an alteration in sex expression in dioecious plants.

Morphactin is thought to induce dwarfism by affecting the orientation of the spindle in dividing cells. It also disrupts the formation of a middle lamella between daughter cells. Although morphactins superficially resemble gibberellins in structure their effects are not thought to be brought about by interaction wth gibberellin or any of the other natural growth substances.

Morphactins may be used commercially to slow growth of, for example, mixed plant populations at roadsides.

morphogenesis The developmental changes that give rise to the adult form from the zygote.

morphology The study of form, particularly external structure. *Compare* anatomy.

mosaic 1. An irregular pattern of small light and dark green areas on the foliage of a plant. Infection with a virus is the usual cause. Examples include tobacco mosaic and cucumber mosaic.
2. A leaf mosaic. *See* lamina.

mosses *See* Musci.

mother cell A cell that divides to form other differentiated cells and hence loses its identity, for example, the cell that gives rise to a sieve element and companion cell. *See also* pollen mother cell.

mould A fungus that produces a distinct mycelium or spore mass, which often resembles a velvet-like pad, on the surface of its host. Moulds frequently occur on dead or decaying vegetable matter, such as food or stored fruits. Common examples are white bread mould (*Mucor*) and blue and green moulds of citrus fruit (*Penicillium italicum* and *P. digitatum*). *See also* Mucorales, Hyphomycetales, Eurotiales.

mounting The final stage in the preparation of permanent slides for microscopical examination. The mount, which may be a section, wholemount, squash, or smear, is immersed in a mounting medium, usually a liquid that later solidifies. The medium permeates entirely through the specimen, leaves no air spaces, and does not support the growth of bacteria and fungi. Examples are glycerol, Canada balsam, and osmium tetroxide.

MTOC *See* microtubule organizing centre.

mucilage Any substance that swells in water to form a slimy solution. Mucilages are usually concerned with water retention; for example, the pentosan mucilages produced in the interior of succulent xerophytes serve to increase the water-holding capacity of the cells and hence reduce the transpiration rate. Many seeds have a mucilaginous coating that aids water uptake during germination.
Structurally, mucilages are very complex: linseed mucilage is a mixture of a complex polyuronide, proteinaceous matter, and cellulose. Mild hydrolysis of the polyuronide yields xylose and galactose residues and a more resistant fraction consisting of galacturonic acid and rhamnose residues.

Mucorales An order of the *Zygomycetes containing about 360 species (55 genera) of mostly saprobic fungi, many of which cause spoilage of stored food. Its members generally form a dense mycelium from which arise sporangiophores or conidiophores. The tips of these may become pigmented, hence the common name 'pin mould' for *Mucor mucedo*. In the coprophilous fungus *Pilobolus* a specialized mechanism has developed to disperse the sporangia. Zygospores are formed by sexual reproduction and many species show heterothallism.

mucronate Having a small fine point (a *mucro*) arising abruptly, usually at the tip.

mulch Organic material, such as peat, leaf mould, or shredded bark, that is spread on the ground to suppress annual weeds. It is also applied around the base of trees and shrubs to help absorb and retain water and add nutrients.

mull *See* humus.

multiple alleles A series of alleles of a particular gene. In theory probably all genes can exist in more than two alternative forms since mutations can occur at any number of points along their length. In practice, natural selection may eliminate all but one or two of these in wild populations. Examples of naturally occurring multiple allelism are found in the incompatibility systems of plants. For example, in Brussels sprouts (*Brassica oleracea* var. *bullata*) there are some

35–40 different incompatibility alleles, termed *S* alleles.

multiple-factor inheritance The determination of an inherited characteristic by *polygenes, so that it shows approximately continuous variation from one extreme to another. The size of beans, ear-length in maize, and many other characteristics show multiple-factor inheritance, and do not, at first sight, behave like Mendelian characters. *Compare* single-factor inheritance. *See* quantitative variation.

multiple fruit (composite fruit) A fleshy fruit that incorporates a complete inflorescence and is thus derived from the ovaries of many flowers. It may also incorporate other floral parts and the receptacles. The *coenocarpium, *sorosis, and *syconus are multiple fruits.

multivalent An association of three or more homologous chromosomes observed during meiosis in polyploid or polysomic organisms. *See* trivalent, quadrivalent.

muricate Having a surface covered by sharp points or prickles or hard short projections.

Musci (Bryopsida) The largest and most widely distributed class of the *Bryophyta, containing the mosses. It includes about 15 000 species in about 610 genera. The gametophyte is the dominant generation and exhibits two distinct morphological stages. The first, which arises on germination of the spore, is the filamentous protonema, which, except for its oblique cross walls, resembles a heterotrichous green algae. The protonema produces buds, from which the familiar leafy moss plant arises. In the Sphagnales and Andreaeales the protonoema is thalloid. The mature gametophyte, which is never thalloid, consists of a main axis (caulid) bearing delicate leaves (phyllids) usually only one cell thick, although a thickened central midrib is often seen. The leaves are generally inserted spirally on the stem. The stem also bears multicellular rhizoids, which distinguishes mosses from liverworts where the rhizoids are unicellular. The gametophyte may be *acrocarpous or *pleurocarpous. The sporophyte arises from an apical cell and exhibits complex spore dispersal mechanisms. The seta elongates gradually in contrast to liverworts where growth of the seta is rapid. There are no sterile elaters in the spore mass.

The Musci is divided into three orders on the basis of differences in capsule structure and in formation of the protonema. The Bryales (commonly termed the true mosses) is the largest order and contains about 600 genera including the advanced *Polytrichum*, which shows some internal differentiation. The Sphagnales (bog or peat mosses) contains a single genus, *Sphagnum*, characteristic of waterlogged acid areas. The ability of sphagnum mosses to create vast areas of peat bog arises, in addition to their low pH and nutrient tolerance, from a peculiarity in their leaf structure, which contains many dead porous cells that act as water reservoirs. The third order is the Andreaeales, which again contains just one genus, *Andreaea*, the members of which are known as granite mosses. Certain classifications elevate these orders to class status, the Bryopsida, Sphagnopsida, and Andreaeopsida. The Bryopsida is then further subdivided into some 19 orders.

The earliest fossil mosses are seen in rocks of the Permian but the group generally does not have a very rich fossil record.

mushroom The usually umbrella-like fruiting body or sporophore of fungi in the *Agaricales. It is a pseudoparenchymatous structure composed of numerous fused hyphae, and is differentiated into a stalk or *stipe* and a circular cap, the *pileus*. In immature mushrooms these structures are united by a membrane, the *universal veil*, but this breaks as the stalk elongates leaving a cuplike structure, the *volva*, around the base of the stalk. In some mushrooms the edge of the cap is united with the stalk by a second membrane, the *partial veil*. This ruptures to expose the underside of the cap leaving a ring, the *annulus*, towards the top of the stem. The lower side of the cap is composed either of *gills or pores, which are lined by the hymenium in which the basidiospores are produced. Mushroom is often used in a wider sense to mean any macroscopic fungal fruiting body. The term *toadstool* is essentially synonymous with mushroom in both the narrow and broad senses, but is more often used of inedible species.

muskeg A peat bog characteristic of the northern coniferous forest of North America. Plants commonly found are pitcher plants, sundews, and black spruce (*Picea mariana*), which thrives in the moist conditions.

mutagen Any agent that causes an increased frequency of mutation. Mutagens are typically short-wave electromagnetic radiations (e.g. ultraviolet irradiation, x-rays, and cosmic rays), ionizing radiations (e.g. α- and β-particles), and chemicals (e.g. nitrous acid and proflavin) that react with nucleotides. A fourth category of mutagens, *base analogues, do not alter existing bases but are incorporated in place of normal bases during DNA replication. At subsequent replications base analogues may 'mispair', resulting in substitution of the 'wrong' base into DNA, and thus an altered codon.

The action of the various types of chemicals and radiation is more variable. Sometimes, as with nitrous acid, the base pairing specificity of the nucleotides is altered, subsequently resulting in incorporation of the 'wrong' base, as with base analogues. Sometimes, as with ultraviolet irradiation, adjacent nucleotides are caused to complex with each other instead of bonding to bases in the complementary helix. This may weaken the double helix and result in chromosome breaks.

mutation An inherited change in the genes as inferred from an inherited change in the appearance of the offspring. The term can be extended to include genetic changes occurring in somatic cells (*somatic mutations*). The latter can result in abnormalities of growth such as *chimaeras. They are not inherited unless the mutant tissue gives rise to a reproductive shoot.

There are two main categories of mutations: *gene mutations, which are alterations in a single gene; and *chromosome mutations, which may involve large numbers of genes. Mutations are characteristically rare, harmful, and recessive. The frequency of spontaneous gene mutations differs with different gene loci. For example in maize, the mutation giving rise to shrunken endosperm occurs about once in every million replications of the gene locus. However the mutation giving a colourless aleurone layer occurs more frequently, about 110 times in every million replications of the locus. Reversions of a mutant gene to a normal gene occur less frequently. The low frequency of spontaneous gene mutations may be attributed to the great stability of DNA. One reason for the occurrence of spontaneous mutation is that all the bases can also exist in very rare isomeric forms. Thus while adenine normally pairs with thymine, in its rare isomeric form adenine will pair preferentially with cytosine. Over two replications this could result in a spontaneous switch from an A:T pair to a G:C pair in the DNA. The low spontaneous frequency of mutation can be increased by various agents called *mutagens.

The harmful effect of most mutations is due to the fact that they occur randomly. A purely random change to a normal allele (gene) is more likely to result in a defective protein than one that confers a selective advantage on its owner. For similar reasons most mutations are recessive. In a heterozygote the effect of a mutant allele that produces a nonfunctional protein is likely to be masked by the normal allele, which produces a functional protein. The former would therefore be described as recessive and the latter dominant.

Despite their infrequent occurrence and their frequently deleterious effects, mutations are of considerable biological significance because they are the ultimate source of all genetic variation. The occurrence of different alleles, e.g. T (tall peas) and t (short peas), of the same gene (pea height) may thus be attributed to mutations occurring sometime in the history of the species.

mutation theory The theory, first proposed by Hugo de Vries in 1903, that suggests that new forms of organisms (mutations) arise suddenly in a population, and that evolution proceeds by natural selection operating on these. The theory was based on de Vries' observations of the relatively frequent occurrence of markedly different forms of *Oenothera erythrosepala* (evening primrose). (It was later demonstrated that the different forms were triploids or tetraploids and hence gave an exaggerated idea of the rate and effects of mutations.) The theory was first validated experimentally when it was found that x-ray induced heritable mutations occurred in *Drosophila*. It is now recognized that although new allelic forms can only arise by mutation much of the variation on which natural selection acts is generated by crossing over and recombination during meiosis.

muton *See* gene.

mutualism An intimate relationship between two or more living organisms that is beneficial to all the participants. A *lichen is an example of *obligatory mutualism* between an alga and a fungus since neither can survive without the other. The relationship between the bacterium *Rhizobium* and members of the family Leguminosae is an example of *facultative mutualism* (*protocooperation*), as the bacterium and the plant can survive independently of one another. *See also* amensalism, commensalism.

Mycelia Sterilia *See* Agonomycetales.

mycelium A mass of branching more or less loosely interwoven *hyphae that makes up the vegetative body of most true fungi.

Mycetozoa *See* Myxomycetes.

mycobiont The fungal partner in a *lichen, which is usually an ascomycete fungus, especially one of the class Discomycetes. *Compare* phycobiont.

Mycobionta *See* Eumycota.

mycology The study of fungi.

Mycoplasmatales (pleuropneumonia-like organisms, PPLO) An order containing extremely small parasitic microorganisms, some of which are only 100 nm in diameter. Mycoplasmas are gram-negative, mostly nonmotile, and lack a rigid cell wall. Most species require sterols for growth. They are thought to be responsible for certain yellows diseases of plants and have been implicated in witches broom of alfalfa and maize stunt. They cause various serious diseases in animals, e.g. pleuropneumonia, bovine mastitis, and puerperal septicaemia.

mycorrhiza A symbiotic association between a fungus and the roots of a plant. *See* ectotrophic mycorrhiza, endotrophic mycorrhiza, vesicular-arbuscular mycorrhiza.

Mycota *See* fungi.

mycotrophic Describing a symbiotic association between a fungus and the whole of a plant. Such an association occurs when a mycorrhizal fungus extends into the aerial parts of a plant, as in certain heathers and orchids.

myrmecochory The distribution of seeds or other reproductive structures by ants. For example, the seeds of *Helleborus* species are distributed by ants, which are attracted to the plant by an oil-containing swelling on the raphe.

myrmecophily The condition when an organism (a myrmecophile) lives with a colony of ants. It is often seen in the animal kingdom but some flowering plants have specialized inflated organs in which to house ant colonies. For example, various tropical American *Acacia* species have specialized thorns, which provide shelter for the ants, and extrafloral nectaries, which provide food. The ants in turn control insect pests that would otherwise feed on the *Acacia* leaves. Similarly epiphytes belonging to the genera *Myrmecodia* and *Hydnophytum* in the Rubiaceae have large swellings on their roots, which house ants in a network of cavities.

myxamoeba A *swarm cell that has lost its flagella. Myxamoebae may increase by division before acting as gametes.

Myxobacteriales (slime bacteria) An order containing flexible creeping bacteria that lack a rigid cell wall and flagella, and move by gliding across the substrate. Most of its members are terrestrial. Myxobacters differ from other bacteria in producing resting cells, termed microcysts, that are often borne in fruiting bodies. Many are soil saprophytes, some species (e.g. of *Cytophaga* and *Sporocytophaga*) producing cellulase enzymes.

Myxobionta *See* Myxomycota.

Myxomycetes (slime moulds) A class of the *Myxomycota containing fungus-like organisms that consist of a naked mass of protoplasm termed a plasmodium. They number about 500 species divided among some 70 genera and are usually found growing on decaying vegetation. The various forms encountered in the life cycle (*see* swarm cell, myxamoeba, plasmodium) show certain animal-like characteristics, notably the ingestion of solid particles of food, and this has led some authorities to place them in the animal kingdom as Mycetozoa.

In some classifications the Myxomycetes contains the orders Acrasiales and Plasmodiophorales but these are now usually elevated to class status as the *Acrasiomycetes and *Plasmodiophoromycetes.

Myxomycota (Myxobionta) A division of the Fungi containing those species that have amoeboid, plasmodial, or pseudoplasmodial thalli. It contains some 475 species in about 90 genera and is divided into the four classes *Acrasiomycetes, Hydromyxomycetes, *Myxomycetes, and *Plasmodiophoromycetes.

Myxophyceae The older and hence, strictly, the correct name for the blue-green algae. It is not however in such common use as *Cyanophyta.

N

NAA *See* naphthalene acetic acid.

NAD (nicotinamide adenine dinucleotide, coenzyme I) A pyridine-based nucleotide that functions as a coenzyme in many oxidation–reduction reactions, according to the general reaction: NAD^+ + reduced substrate $\rightleftharpoons$ NADH + H^+ + oxidized substrate. An example of such a reaction is the oxidation of cit-

rate to α-ketoglutarate in the TCA cycle.
Reduced NAD can act as a reducing agent or it may be reoxidized in the *respiratory chain with coupled ATP formation. NAD is not usually tightly bound to the enzyme with which it is acting but instead acts as a second substrate, binding only during the reaction and being released into the medium with the reaction products.
NAD is sometimes called *DPN*, an abbreviation for the trivial name *diphosphopyridine nucleotide*.

NADP (nicotinamide adenine dinucleotide phosphate, coenzyme II) A pyridine-based nucleotide that functions as a coenzyme to several different oxidoreductases. NADP usually acts as an electron donor in enzymatic reduction reactions; in this it differs from its counterpart *NAD, which acts principally as an electron acceptor in biological oxidations. NADP is sometimes called *TPN*, an abbreviation for its trivial name *triphosphopyridine nucleotide*. *See also* pentose phosphate pathway.

nannandry The existence in a species of male individuals that are considerably smaller than the females. For example, within the moss genus *Homaliadelphus* there is a species in which the male gametophyte is sufficiently small to grow as an epiphyte upon the leaves of the female gametophyte. The dwarf male filaments of certain members of the Oedogoniales are further examples. Species in which the male plants are of normal size are termed *macrandrous*.

nanometre Symbol: nm. A unit of length (formerly called a *millimicron*, symbol mμ) equal to one thousandth of a micrometre, i.e. 10^{-9} metre.

naphthalene acetic acid (NAA) A synthetic *auxin that has proved of great use in stimulating the rooting of cuttings and promoting flowering.

NAR *See* net assimilation rate.

nastic movements (nasties) Plant responses caused, but not directed, by external stimuli. Thus the opening and closing of many flowers are nastic movements, the trigger most often being certain conditions of light or temperature. The movements may be caused either by differential growth rates or by changes in water potential in specialized cells. *Compare* tropism, taxis. *See also* haptonasty, photonasty.

natural classification A *classification of organisms in which there is a high level of predictivity, as opposed to artificial classifications, which tend to be very unpredictive. Most taxonomists strive towards the production of natural classifications, although their approaches and interpretations of the data may vary greatly. Natural classifications can best be achieved by exploring the fields of cytogenetics, microanatomy, and phytochemistry in addition to morphology, anatomy, and plant geography, thus broadening the base on which classifications are founded.
Many phylogenists believe that only those classifications based on *monophyletic taxa are natural.

natural order A historical category more or less equivalent to the modern *family. Fifty eight natural orders were listed by Linnaeus in 1764, many corresponding closely to present-day families, e.g. Umbelliferae and Compositae. However, the International Rules of Nomenclature state that natural orders cannot be used in place of family.

natural selection The action of the environment, as opposed to the actions of man, on individual organisms such that those possessing genotypes better suited to the environment will survive and reproduce more successfully than those with less favourable genotypes, which will eventually die out. By this process the characteristics of a population will change according to the nature of the environmental pressures acting on them. Over a number of generations the population may diverge into a number of distinct groups each adapted to a particular microenvironment. This process will be hastened if there are barriers to gene flow between the groups (*see* reproductive isolation). The concept of natural selection is the cornerstone in Darwin's theory of evolution (*see* Darwinism).

neck The slender portion of the *archegonium through which the male gamete has to travel to reach the female gamete prior to fertilization.

necrosis The death of a plant cell or group of cells while the rest of the plant is still alive, particularly when the dead tissue becomes dark in colour. Necrosis is a common symptom of fungus infection and the shape of necrotic areas is often characteristic of the particular disease.

nectar The sugary solution secreted by a nectary. It may contain up to 60% sugar. It is produced by certain entomophilous plants as an alternative attractant to pollen. Honey is made from nectar by bees.

nectar guides *See* honey guides.

nectary A sugar-secreting *gland. Nectaries are usually present at the base of a flower, sometimes in a spur, to attract pollinators. Often scents are also secreted by the nectary as additional attractants. Nectaries may also be *extrafloral*, for example the gland spines of certain cacti. Here they may serve to attract seed dispersal agents, such as ants.

negative staining A *staining technique that relies on the property of certain components of the specimen not to take up the dye or *negative stain*. These then show up light against a dark background. For example, in light microscopy, bacteria can be made visible by mixing them with a dark dye such as nigrosin or Indian ink. The method is used widely in electron microscopy to study viruses and certain large molecules. The material is mixed with negative electron-dense stains, such as potassium molybdate or phosphotungstic acid (PTA). Any proteinaceous material present in the specimen does not take up the stain. In the resulting electron micrograph the electron-transparent proteinaceous material will appear dark against a light background.

Neo-Darwinism The expansion and modification of Charles Darwin's theory of evolution by natural selection in the light of genetic studies and also relevant discoveries in molecular biology. The collection of such data into one coherent theory of evolution has been termed the *modern synthesis*. Genetics explains how variations in a species arise by both chromosome and gene mutations, and how these variations are maintained by the recombination and reassortment of different alleles. Studies in population genetics have emphasized the importance of *genetic drift and isolating mechanisms when considering the formation of new species (speciation). *See also* Darwinism, Weismannism, natural selection, mutation theory.

neoteny The arresting of the normal development of all cells except for those in the germ line, resulting in a sexually mature organism with juvenile characteristics. Such a process, producing large morphological changes, could result from comparatively small genetic changes affecting growth rate. It is a method by which one species may arise virtually instantaneously from another. For example, it is thought most likely that *Lemna* (duckweeds) evolved from the floating water plant *Pistia* in this manner because the seedlings of *Pistia* are very similar to the adult form of *Lemna*. At a more general level it is believed herbaceous plants originated from trees by neoteny. The resemblance of an adult organism to the juvenile forms of its ancestors is termed *paedomorphosis*. The arresting of development at a later stage may lead to large changes in the morphology of a particular organ. Thus the tubelike flower of *Delphinium nudicaule*, which resembles the buds of other *Delphinium* species, may have evolved in this manner.

neotype A plant specimen chosen to act as a standard taxonomic reference point following the irretrievable loss of all *type material. The neotype should match the original description as closely as possible and ideally should come from the same locality as the original material.

neritic *See* sublittoral.

nerve (costa) A narrow thickened strip of tissue running down the middle of a moss leaf. It is sometimes termed a *midrib.

net assimilation rate (NAR) A value that relates plant productivity to plant size. It is obtained by dividing the rate of increase in dry weight by leaf size (usually leaf area).

neutral stain *See* staining.

neutral theory of molecular evolution The theory that evolutionary change at the molecular level is caused by random drift of selectively equivalent mutant genes rather than by selection.

nexine The inner layer of the *exine in a pollen grain.

niacin *See* nicotinic acid.

niche *See* ecological niche.

nicotinamide adenine dinucleotide *See* NAD.

nicotinamide adenine dinucleotide phosphate *See* NADP.

nicotinic acid (niacin) A carboxylic acid, formula C_5H_4NCOOH, so named because it forms part of the alkaloid nicotine. It is a member of the B group of *vitamins and is formed in plants and most animals (but not man) from various precursors, notably tryptophan. In the form of the amide nicotinamide, $C_5H_4NCONH_2$, it is a constituent of the coenzymes NAD and NADP. Certain plant products, notably maize, are very poor in nicotinic acid. If these form the basis of a human diet, pellagra, caused by nicotinic acid deficiency, results.

Nidulariales An order of the *Gasteromycetes containing some 60 species (9 genera) known as the bird's-nest fungi. The wall of the basidiocarp is nestlike and contains a number of separate spore masses (the 'eggs'), which are exposed when the covering membrane dies back. An example is *Crucibulum laeve*.

night-break effect A phenomenon exhibited by many plants whereby a period of light, administered artificially during the night, even if only of short duration or low intensity, will interfere with the normal flowering rhythm of the plant. The precise effects vary with the time that has elapsed since the light phase but generally, under short-day conditions, a light treatment inhibits flowering in *short-day plants and promotes it in *long-day plants. *See* photoperiodism.

night temperature The temperature prevailing during the period of darkness, which can greatly affect the growth of many plants. For example, tomatoes have been shown to grow better if the night temperature is significantly lower than the day temperature. It is thought that low night temperatures inhibit respiration.

ninhydrin A colourless compound that is used in *chromatography to locate amino acids and proteins on the chromotogram. Amino acids with a free amino group turn blue on heating with ninhydrin whereas those with a substituted amino group (proline and hydroxyproline) turn yellow.

nitrification *See* nitrogen cycle.

nitrifying bacteria *See* nitrogen cycle.

nitrogen Symbol: N. A nonmetallic element, atomic number 7, atomic weight 14, important as a macronutrient for plant growth. It is essential for the formation of amino acids and the purine and pyrimidine bases, and consequently for protein and nucleic acid synthesis. It is also found in many other compounds, e.g. porphyrins and many coenzymes. Nitrogen is absorbed by roots as the nitrate (NO_3^-) ion or more rarely as ammonium (NH_4^+) or nitrite (NO_2^-) ions. Deficiency leads to spindly growth and yellowing of the leaves. Soil nitrogen is replenished by various natural processes (*see* nitrogen cycle, nitrogen fixation). It is also boosted by fertilizer applications of such compounds as urea, ammonium nitrate, ammonium sulphate, and nitrochalk. Many high-yielding crop varieties depend on high levels of nitrogen for optimum yields. However excessive nitrogen produces soft tissues with a high water content that are particularly prone to frost damage. Flowering and fruit set may also be reduced at the expense of vegetative growth, while potassium deficiency can be induced by high concentrations of nitrogen.

nitrogenase The enzyme system isolated from certain bacteria and shown to be responsible for nitrogen fixation. In the presence of ATP and a suitable electron donor it reduces molecular nitrogen ($N \equiv N$) to ammonia. It can also reduce other compounds with triple bonds, such as acetylene ($HC \equiv CH$). The enzyme is destroyed by free oxygen and is inhibited by high concentrations of ammonia.

nitrogen cycle The circulation of nitrogen between living organisms and the environment. Atmospheric nitrogen is returned to the soil by nitrogen-fixing microorganisms (*see* nitrogen fixation) and by electrical discharges in storms, which cause nitrogen and oxygen to combine. The oxides so formed dissolve in rain to form nitrous and nitric acids, which, in the soil, combine with mineral salts to form nitrites and nitrates. The nitrites and the ammonia in the soil derived from animal excretion and the decay of organic matter, are converted to nitrates by *nitrification*. Plants usually assimilate nitrogen as nitrates and the activities of nitrifying bacteria are thus essential to plant growth. The oxidation of ammonia to nitrite is carried out by *Nitrosomonas* species and the oxidation of nitrite to nitrate by *Nitrobacter*. The reverse process, *denitrification, is mediated by different bacteria. Nitrates may also be lost by leaching.

Man also affects the nitrogen cycle in various ways. He fixes nitrogen industrially by combining it with hydrogen in the Haber process and adds nitrogen to the soil in fertilizers. He removes nitrogen by overcultivation, by installing efficient sewage systems thus preventing urine and faeces from reaching the soil, and by factory farming, which often results in animal wastes not being returned to the land in manageable form.

nitrogen fixation The fixation of atmospheric nitrogen either by lightning (*see* nitrogen cycle) or by free-living or symbiotic microorganisms. The association between many species of the Leguminosae and the nitrogen-fixing soil bacterium *Rhizobium*, which lives in nodules on the legume roots, has a significant effect on soil fertility. Hence much use has been made of leguminous crops in agriculture. There are different strains of *Rhizobium*, each specific to one or a group of closely related species. The bacteria are attracted to the legume by a growth substance and invade the root hairs. They then divide forming filaments that eventually reach

and infect the plant cortex. The cortex is stimulated to enlarge and form a nodule the centre of which is red due to the presence of haemoglobin (*see* leghaemoglobin). The bacterium fixes nitrogen by means of the enzyme *nitrogenase.
About 250 species of plants other than legumes also form symbiotic associations with nitrogen-fixing microorganisms. For example, bog myrtle (*Myrica gale*) and alder (*Alnus glutinosa*) have nitrogen-fixing root nodules that appear to contain fungi of the Plasmodiophorales. *Gunnera* and the water fern *Azolla* have blue-green algae in their roots. Free-living nitrogen-fixing microorganisms include the bacteria *Azotobacter*, *Klebsiella*, and *Clostridium* and all the blue-green algae and photosynthetic bacteria.

node A point on the stem from which one or more leaves arise. In the mature stem the nodes are usually well separated by internodes, which elongate through the action of *intercalary meristems. However in some plants, e.g. rosette plants and grasses, the nodes remain close together giving these species their characteristic growth form. The pattern of vascular connections between the stem and leaf at the node may be described as unilacunar, trilacunar, multilacunar, etc. depending on how many leaf gaps are left in the stele.

nomenclatural type *See* type.

nondisjunction The failure of paired (homologous) chromosomes to separate during anaphase 1 of meiosis. As a consequence, one daughter cell, and hence the two gametes formed from this, will receive both homologues. By corollary, the other daughter cell, and the gametes formed from this, will be deficient for the chromosome in question. Cells containing abnormal numbers of chromosomes are called aneuploids (*see* aneuploidy). *See also* monosomic, trisomic.

nonpersistent Describing pesticides that break down and become inactive relatively quickly. Such pesticides should be used if spraying soon before harvest. Examples are the natural insecticides derris and pyrethrum.

nonsense triplet A name given originally to triplets of bases that do not specify any amino acid and were thus once considered as having no function. There are three such triplets, UAG, UGA, and UGG, all of which have subsequently been found to act as 'stop-signals', i.e. they define the end of a polypeptide chain. Polypeptide initiation also requires a 'start signal', the signals being AUG or GUG, which also code for valine or methionine (or a methionine derivative) respectively.

normal curve (normal curve of errors) The bell-shaped curve obtained when a series of observations that shows a normal (Gaussian) distribution is plotted on a graph. The mean, median, and mode all occupy the same high middle point of the curve, which is perfectly symmetrical around this point.

normal deviate Symbol *c*. The ratio of the deviation, *d*, to the standard deviation, σ. It is used to assess experiments where the results do not fall into a limited number of classes but show a continuous range, for example the recording of crop yield in response to various fertilizer levels. To find whether any deviation is significant its probability is looked up in a table of normal deviates. *See also t* distribution.

northern coniferous forest *See* forest.

nu body *See* nucleosome.

nucellus A rounded or oval mass of parenchymatous tissue in an ovule, containing the embryo sac. Its size and shape may be used as a diagnostic character. It is almost totally surrounded by the integuments except for a small channel, the *micropyle, through which the pollen tube may grow prior to fertilization. At fertilization the nucellus may be reabsorbed as the embryo develops or it may persist to form a nutritive *perisperm in some seeds.

nuclear membrane (nuclear envelope) A double membrane surrounding the nucleus. Each layer has a typical membrane structure (*see* plasma membrane) and is 4 nm to 6 nm in width. The outer membrane is studded with ribosomes on the cytoplasm side while the surface of the inner membrane next to the nucleus is smooth. Electron micrographs of many cells reveal connections between the outer membrane and the endoplasmic reticulum. Between the membranes there is a clear space, the *perinuclear cisterna*. At various points the two membranes fuse and pores are formed. The number of pores varies with the degree of nuclear activity. Extensions of the outer granular or fibrous region of the nucleus form cylindrical complexes within the pores. Their precise function is obscure but they appear to be involved in the transfer of information from nucleus to cytoplasm.

nucleic acid A complex organic acid comprising polymers of *nucleotides formed by condensation reactions that establish phosphodiester bonds between the component nucleotides. There are two types, *DNA and *RNA. The acidic properties of these compounds are due to phosphoric acid, a component of nucleotides. They are called nucleic acids because they were first associated only with the nucleus. However they have subsequently been found in chloroplasts and mitochondria.

nucleohistone A complex formed between the polynucleotides of DNA and basic proteins called histones, which only occur in the nuclei of eukaryotic organisms. Nucleohistone complexes are visible as *nucleosomes.

nucleolar organizer The region of chromosomal DNA that codes for ribosomal RNA. Such regions can be identified as secondary constrictions (the centromere being the primary constriction) often located towards the end of the chromosome. Like centromeres, nucleolar organizers are uncoiled regions of the chromosome and stain poorly. *Nucleoli in the interphase nucleus have been shown to be associated with these regions.

nucleolus A structure within the nucleus that stains densely with basic dyes and consists of proteins associated with RNA. The number and distribution of nucleoli varies but is usually characteristic for any one cell type. Electron micrographs show a central area of short fibres surrounded by a matrix of protein material with granules embedded in the peripheral region. Nucleoli are closely associated with the regions of chromosomal DNA that code for ribosomal RNA (*see* nucleolar organizer). The transcription of the code is dependent on a specific RNA polymerase found only in the nucleolus. A long precursor molecule is formed initially and is processed to produce two shorter molecules. The longer of these associates with proteins in the nucleolus to form the larger ribosomal subunits. The smaller molecules similarly associate with proteins to form the smaller subunits, possibly in the nucleolus or surrounding nucleoplasm. The formation of complete *ribosomes from the subunits only occurs when the latter reach the cytoplasm.

nucleoplasm *See* nucleus.

nucleoside A general term for the category of substances formed when a purine or pyrimidine base combines with carbon–1 of a pentose sugar. Such base–sugar complexes are much less common than base–sugar–phosphate complexes, which are known as *nucleotides.

nucleosome (nu body) A nodule, some 7–10 nm in diameter, consisting of *histones, around which is wrapped a strand of DNA about 150 base pairs long. One type of histone, H1, is positioned between adjacent nucleosomes and may serve to clamp the DNA in position. By wrapping around histone beads, the total length of DNA is condensed by a factor of some seven times. Supercoiling of the string of nucleosome beads results in a further condensation. Nucleosomes are thus of considerable importance in packaging enormous lengths of DNA into minute chromosomes.

nucleoside diphosphate sugars (NDP-sugars) Compounds formed by the reaction of a phosphorylated hexose sugar with a nucleoside triphosphate (NTP) in the general reaction:

NTP + sugar 1-phosphate ⇌
NDP-sugar + pyrophosphate

NDP sugars are important as high-energy glycosyl donors in the synthesis of most polysaccharides, for example the enzyme starch synthase utilizes uridine diphosphate glucose (UDP-glucose) and adenosine diphosphate glucose (ADP-glucose) as substrates.

UDP-glucose is the most important of the NDP-sugars. It can be synthesized both from glucose 1-phosphate using the reaction shown above, or from sucrose using the enzyme sucrose synthase (*see* sucrose). Besides their importance as glycosyl donors NDP-sugars are important intermediates in the interconversion of monosaccharides and their derivatives. For many enzymes catalysing sugar interconversions, notably the epimerases, NDP-sugars are substrates rather than the sugars themselves or their phosphate esters.

nucleotide A compound consisting of three essential parts – a purine or pyrimidine base linked to carbon–1 of a pentose sugar (either ribose or deoxyribose), which in turn is esterified on carbon–5 to phosphoric acid. *Polynucleotides* are polymers of nucleotides. There are two types: RNA, which contains ribose, and the bases uracil, adenine, guanine, and cytosine; and DNA, which contains deoxyribose, and thymine, adenine, guanine, and cytosine. Both play an informational role in the cell. In contrast, *dinucleotides*, such as nicotinamide adenine dinucleotide (NAD) and flavin adenine dinucleotide (FAD), function as hydrogen carriers, while *mononucleotides*, such as adenosine triphosphate (ATP), serve as energy carriers.

nucleus The part of a eukaryotic cell that contains the genetic material. It is enclosed in a *nuclear membrane. Prokaryotic cells have no nucleus as such. The interphase nucleus of eukaryotic cells contains a network of chromatin fibres, which become organized into *chromosomes prior to cell division. Apart from chromatin the substance of the nucleus (the *nucleoplasm* or *karyoplasm*) contains complexes and enzymes necessary for the replication of DNA and the synthesis of RNA molecules. One or more *nucleoli are present associated with specific regions of the chromosomal DNA, the *nucleolar organizers.

nullisomic *See* aneuploidy.

numerical taxonomy (taxometrics) A classification method based on the numerical analysis of the variation of a large number of characters in a group of organisms. It is assumed that a classification will be more predictive the more characters on which it is based. It is also assumed that, to begin with, each character is of equal weight (*see* weighting) although some characters may later be weighted. Initially, a matrix of data is compiled of *operational taxonomic units (OTUs) against characters so that for every OTU the state of each of perhaps 50 or more characters is recorded. This matrix can be subjected to a variety of mathematical analyses, which provide a measure of the similarity or dissimilarity between all the OTUs. The end product is usually one or more dendrograms.

nurse tissue Metabolically active tissue that is used in tissue culture work to stimulate the growth of single cells that cannot be grown using defined media. The nurse tissue, which is often callus tissue, may be separated by a piece of filter paper through which growth substances and nutrients can diffuse. Pollen grains have been induced to grow into haploid plants by culturing them on filter paper in contact with intact anthers acting as nurse tissue. It may not be necessary to separate the nurse tissue if it is first treated, for example with x-rays, to prevent cell division.
Growth media may be enriched with growth substances by adding nurse tissue enclosed in a semipermeable membrane. Using such *conditioned media* it is possible to achieve satisfactory growth using a much lower inoculum density than would normally be necessary. This is important in cloning work because cells can be plated out at much lower densities so increasing the chances that any resulting group of cells is derived from a single cell.

nut A dry indehiscent fruit that is usually shed as a one-seeded unit. It forms from more than one carpel but only one seed develops, the rest aborting. The pericarp is usually lignified and is often partially or completely surrounded by a *cupule. True nuts include the acorn, hazelnut, and beechnut. The term is often loosely applied to any woody fruit or seed, such as the walnut (which is a drupe) or the Brazil nut (which is a seed). *Nutlets* are small nuts and are typical of the Labiatae.

nutation *See* circumnutation.

nutrient medium *See* culture medium.

nyctinasty A *nastic movement in which plant parts, especially leaves and flowers, assume a characteristic position at night. These *sleep movements* most often result in a folding together of leaflets. In certain *Acacia* species the process seems to be a form of *photonasty, as the assumption of the night position coincides with the onset of darkness. In other species the cycle of folding continues even if the plant is kept in the dark; the process is then clearly *autonomic rather than nastic.

O

ob- A prefix meaning inverted; hence obcordate describes a leaf shaped like an inverted heart.

obdiplostemonous Describing *stamens that are inserted in two whorls with the outer opposite the petals and the inner opposite the sepals. Obdiplostemonous stamens are seen in some members of the Caryophyllaceae. *Compare* diplostemonous.

obligate Having a specific requirement for a particular environmental factor and being unable to survive if it is not available. For example, obligate parasites cannot grow in the absence of their host.

obtuse Having a blunt or rounded leaf apex.

ochrea A tube around the stem formed by sheathing stipules at the leaf base. Ochreas are characteristic of the Polygonaceae giving the familiar swollen joints on the stems of such plants.

Oedogoniophyceae A class of the *Chlorophyta containing one order, the Oedogoniales, comprising three genera of filamentous algae. These algae are characterized by an unusual kind of cell division, which results in parallel rings of closely spaced 'caps' of old cell wall material being formed at the anterior end of dividing cells. Prior to cell division a ring of wall material forms at the anterior end of the cell. Nuclear division then occurs and the two resulting nuclei become separated by a septum, which, however, is not contiguous with the outer wall at this stage. Meanwhile the ring of wall material at the top of the cell enlarges and the old cell wall to the outside of the ring ruptures. The new wall material is then free to extend and forms the outer walls of the new cell. Part of the old wall remains as a 'cap' at the upper end of this cell. The septum between the daughter nuclei moves up to fuse with the base of the new wall material and cell division is completed (*see* diagram).

The zoospores of the Oedogoniophyceae are also peculiar in having numerous flagella forming an apical ring.

offset A short shoot that arises from an axillary bud near the base of the stem and gives rise to a daughter plant at its apex. Examples of offsets are those produced by the houseleek (*Sempervivum tectorum*). The offset is a type of short *runner and, like runners, is a means of vegetative reproduction. The term is also applied to bulbils and cormlets that form at the side of the parent bulb or corm.

oidium (arthrospore) A spore that forms by the organized fragmentation of a hypha as seen, for example, in *Endomyces*.

oil A triacylglycerol that is liquid at room temperature. The major fatty acids in oils are the unsaturated oleic and linolenic acids. Most oils function as energy storage compounds and are especially important in some oil-bearing seeds such as the castor bean. *See also* essential oil.

oil gland A *gland that secretes an essential oil.

oil immersion A technique used in light microscopy in which a special oil-immersion objective lens is employed to increase the resolving power of the mi-

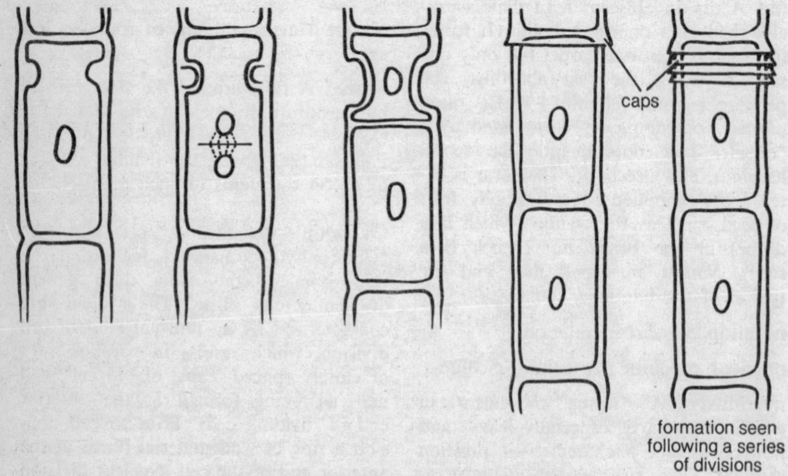

Stages of cell division in *Oedogonium.*

croscope. A specimen mounted on a glass slide is covered by a drop of clear oil placed on the coverslip. The oil has the same refractive index as the glass lens. When the lens is immersed in the oil, the amount of light entering the objective increases (i.e. the effective aperture is increased). Such a technique is particularly useful at high magnifications. Oil is traditionally used but other liquids (e.g. sugar solution) may also be suitable.

oleic acid An eighteen-carbon monounsaturated fatty acid. It is the predominant unsaturated fatty acid in plant cells, except in some specialized organelles such as the chloroplast, where *linolenic acid predominates. Oleic acid is formed from *stearic acid in a desaturation reaction involving molecular oxygen, a reducing agent, and ferredoxin.

Oligocene *See* Tertiary.

oligosaccharide A *carbohydrate composed of between two and ten polymerized monosaccharide units. Most oligosaccharides contain two (disaccharides), three (trisaccharides) or four (tetrasaccharides) sugar units. The most important plant sugar, *sucrose, is a disaccharide. *Cellobiose is another common disaccharide, being the basic repeating unit of cellulose. Other disaccharides include trehalose and gentiobiose. The commonest trisaccharide is *raffinose and the commonest tetrasaccharide *stachyose.
Oligosaccharides may be intermediates in polysaccharide metabolism or may serve as storage compounds.

oligotrophic Poor in minerals and bases. The term is usually used of ponds and lakes with low nutrient levels but it can be applied in other contexts, for example to bog peats, which are usually low in nutrients. In an oligotrophic stretch of freshwater there is limited growth of aquatic plants. There is consequently little decaying vegetable matter and the absence of large numbers of decomposers means there is sufficient oxygen to support fish and other aquatic animals. *Compare* eutrophic.

one gene:one enzyme hypothesis The theory that the principal function of a gene is to determine the structure of an enzyme molecule. This was based on evidence showing that the presence or absence of an enzyme in a biochemical pathway was a genetically determined characteristic, inherited as a single gene. The theory was extended to include proteins other than enzymes. Subsequent analysis showed that where a protein consists of several polypeptide chains, it is more correct to say that one gene codes for one polypeptide chain. The genes coding for the component polypeptides of an enzyme need not be closely linked.

ontogeny All the changes that occur during the life history of an organism. In a plant this would be the development from the zygote through to the production of gametes and, by the fusion of two gametes, a new zygote that will give rise to the subsequent generation. *Compare* phylogeny.

oogamy An extreme expression of *anisogamy, in which fertilization occurs with the fusion of a large nonmotile female gamete or *egg and a small usually motile male gamete. *Compare* isogamy.

oogonium The female reproductive organ of certain algae and fungi. It is unicellular and usually thin walled, and produces one or more nonmotile *oospheres. These may be released prior to fertilization or remain in the oogonium, as in the fungus *Pythium*. *Compare* archegonium.

Oomycetes A class of the *Mastigomycotina containing fungi that typically produce zoospores with an anterior flimmer flagellum and a posterior whiplash flagellum. It contains about 580 species in some 70 genera. The mycelium is coenocytic and, unlike other fungi, has no chitin in its walls. Sexual reproduction involves the formation and fusion of oogonia and antheridia. The oospheres within the oogonia form thick-walled oospores after fertilization. The Oomycetes contains the four orders Lagenidiales, *Saprolegniales, Leptomitales, and *Peronosporales.

oosphere A large nonmotile gamete that is rich in nutrients and is normally designated as female. One or several oospheres may be formed within an oogonium. They may be released unfertilized from the plant, as in the brown alga *Fucus*, or they may be retained within the oogonium and there form a zygote or *oospore following fertilization. The oosphere is equivalent to the egg cells formed in the archegonia or embryo sacs of land plants.

oospore A thick-walled zygote that is formed following the fertilization of an *oosphere. *Compare* zygospore.

operational taxonomic units (OTUs) The entities whose affinities are studied by *numerical taxonomy. Depending on the level and the type of investigation an OTU may be of any taxonomic rank or an individual organism.

operator gene A site on the DNA adjacent to a *structural gene that acts as part of a molecular 'switch' and determines whether or not transcription of the structural gene will occur. The switch may be turned 'off' by a *repressor molecule binding to the operator, so preventing mRNA polymerase reaching the DNA. Conversely, absence of a repressor allows mRNA polymerase to bind to the DNA, i.e. the switch is 'on'. Other methods by which the operator genes function are also known. *See* operon.

operculum **1.** The membranous cap covering the *peristome in the undehisced capsule of many moss sporophytes. It is displaced either by the pressure that builds up within the capsule or by the swelling of the cells of the annulus.
2. A lid covering the aperture in many pollen grains, which is pushed aside by the emerging pollen tube.
3. A cap covering the ostiole in certain ascomycete fungi.

operon A group of adjacent genes coding for a set of enzymes in a particular biochemical pathway. They act as a single unit in that they are either all transcribed together or none are transcribed at all. In order for transcription to occur, mRNA polymerase, which catalyses mRNA synthesis, must first bind with the DNA at a site called the *promoter*. mRNA polymerase activity is prevented (negative control) or promoted (positive control) by the binding of a regulator protein between the promoter and the structural genes (the genes that actually code for the enzymes) at a site called the operator. The regulator protein is itself the product of a gene, the regulator, which may be some distance from the operon. A classic case of negative control is the lac (lactose) operon in the bacterium *Escherichia coli*. Here the regulator protein prevents transcription in the absence of an inducer (a substrate molecule, i.e. lactose, or some derivative). However, if an inducer is present, it complexes with the regulator, the protein formed by the latter no longer binds to the operator, mRNA polymerase activity is permitted, and the enzymes are synthesized. *See also* Jacob–Monod model.

Ophioglossales An order of the Filicinae containing a small group of morphologically distinct ferns. The spores are not borne on the undersurface of the frond but in a separate stalked spike, and unlike all other ferns the fronds do not show *circinate vernation. The order contains three genera: *Ophioglossum* (adder's tongues) containing about 30 species; *Botrychium* (moonworts) containing about 25 species; and *Helminthostachys*.

opposite Describing a form of leaf arrangement in which the leaves arise in pairs at each node. If each pair is at right angles to the pairs above and below it, as is usually the case, the arrangement is termed *decussate*. If the leaf pairs all arise in the same plane a *distichous* arrangement results. *Compare* alternate, whorled. *See illustration at* phyllotaxis.

optical microscope *See* light microscope.

orbicular Almost circular and flattened, as are the leaves of the fringed water lily (*Nymphoides peltata*).

Orchidaceae A very large monocotyledonous family, the orchid family, containing about 18 000 species in about 750 genera. Orchids are widely cultivated for their distinctive flowers, which may be recognized by the characteristic lower lip or *labellum. The anthers and carpels lie opposite the labellum and are fused into a structure called the column. The column is very diverse in form, the various modifications being related to the different pollinating animals. The pollen grains are clumped together into a definite number of pollen masses, termed *pollinia, on the column. Following pollination and fertilization, seeds are produced in vast numbers, but each contains only a rudimentary embryo. Seed maturation and germination is thus a lengthy process, which may take a number of years. In addition, orchids live in symbiotic association with various fungi and the appropriate fungus is needed before a seed can germinate. These factors have hindered the commercial multiplication of orchids though orchid growers have been quick to utilize tissue culture methods of multiplication. Orchid breeding is facilitated by the 'promiscuous' nature of orchids and the ease with which they form viable hybrids. This ability is demonstrated by the number of naturally occurring orchid hybrids, which exceeds those of all other plant families added together. About half the species of orchids (including most of the cultivated species) are epiphytic and thus need special compost mixtures.

order A major category in the taxonomic hierarchy, usually comprising groups of families thought to possess a degree of *phylogenetic unity. Groups of similar orders are placed in classes. The Latin names of orders usually terminate in -ales, e.g. Rosales and Geraniales. However, some orders, which were erected prior to the compilation of the International Code of Botanical Nomenclature, end in -ae, e.g. Glumiflorae and Tubiflorae.

ordination A technique used in ecological work to relate the composition of different stands of vegetation to each other. It is used by those who consider that no two stands of natural vegetation are ever the same and thus they cannot be classified into distinct associations (*compare* classification). Certain properties of each stand, namely data on the species present, are plotted against one or more axes, each axis representing some environmental gradient. The end result should be an arrangement in which the different vegetation stands are ordered in a manner that best reflects their similarities and differences.

Ordovician The second period of geological time in the Palaeozoic era, about 500–440 million years ago. There are no known terrestrial organisms in this period. *See* geological time scale.

organ culture The culture of excised organs in a suitable aseptic medium. Roots, leaves, embryos, meristems, and many other plant structures have been successfully maintained in culture. By observing growth in the presence and absence of various nutrients, valuable information about the physiology of the organ can be obtained. For example, chemical factors affecting sex expression have been studied using cultures of isolated flowers, and root nodule formation has been investigated in cultures of legume roots. *See also* meristem culture, ovule culture, embryo culture.

organelle A membrane-enclosed structure in the cytoplasm organized to carry out a specific process essential to the life of the cell. Examples are the *mitochondria, *chloroplasts, and *golgi apparatus.

organogenesis The developmental changes that occur during the formation of a particular organ.

organomercurial fungicide An organic *fungicide containing mercury. Most of this group are phenylmercury derivatives, e.g. phenylmercuric acetate, and are used as *seed dressings. Mercury-containing fungicides are highly toxic to mammals and treated grain must be clearly labelled or even dyed to distinguish it from edible grain.

organophosphorus insecticide An organic *insecticide, such as malathion and parathion, often used to control aphids, mealybugs, mites, etc.

origin of life The way in which living organisms have developed from inorganic matter. It is generally believed that the earth is about 4600 million years old and that the earliest evidence of life dates from 3000–3500 million years ago. Fossils of this age are interpreted as bacteria-like organisms. It is thought that in the early stages the earth's atmosphere was chemically reducing, composed of such gases as hydrogen, nitrogen, methane, ammonia, and water vapour. Such an atmosphere would be more favourable for the formation of complex organic molecules than the present oxidizing atmosphere (largely nitrogen and oxygen). Laboratory experiments in which mixtures of such gases are subjected to electric sparks or ultraviolet radiation show that simple organic molecules (e.g. amino acids and nitrogenous bases) can be formed. Most theories on the origin of life involve starting with a 'primaeval soup' containing such compounds. Possibly long-chain molecules (e.g. proteins and carbohydrates) were formed by catalytic reactions in which simple molecules adsorbed on regular mineral surfaces (e.g. mica clays). The way living organisms evolved from such molecules is still in question. Possibly it first involved the formation of primitive cells (*protobionts) leading to simple heterotrophic prokaryotic organisms. From these came photosynthetic organisms resembling the blue-green algae, which gradually formed an oxidizing atmosphere enabling aerobic organisms to develop. Aerobic eukaryotic organisms then appeared, possibly arising as symbiotic associations of prokaryotes (*see* serial endosymbiotic theory), and from these came multicellular organisms.

ornithine A nonprotein amino acid, formula $NH_2(CH_2)_3CH(NH_2)COOH$. It is an intermediate in the synthesis of arginine, and many alkaloids, e.g. the tropane alkaloids, are derived from ornithine. It also occurs in the cell walls of some bacteria.

ornithophily Pollination by pollen carried by birds. It is an important form of pollination for many tropical species, the most common pollinators being hummingbirds (in South America) and sunbirds (in Africa). Many of the less specialized species of *Fuchsia* and *Lopezia* are bird pollinated. Bird-pollinated flowers are often red in colour. Pollination by insects (entomophily) is believed to be derived from ornithophily.

orthotropism A *tropism in which the growth response is directly towards or away from the source of the stimulus. Thus the vertical growth of tree trunks may be described as negative orthogeotropism and the growth of the gills of agaric fungi positive orthogeotropism.

orthotropous (atropous) Describing a form of ovule orientation in the ovary in which the ovule develops in an upright position from the placenta (*see illustration at* ovule). This arrangement is rare but may be found in some members of the Polygonaceae. *Compare* anatropous, campylotropous.

Oscillatoriales *See* Hormogonales.

osmium tetroxide A fixative (preservative) for proteins and phosphoglycerides that is used in the preparation of biological material for light microscopy and electron microscopy. It is fat soluble and is reduced to black osmium dioxide by unsaturated fats. Saturated fats in a specimen may be dissolved out in dehydration or embedding, the blackened unsaturated fats remaining. It is therefore used to stabilize and stain cell constituents. Often a preparation is prefixed with glutaraldehyde and post-fixed with osmium tetroxide to increase the extent of preservation.

osmometer An apparatus designed to demonstrate *osmosis and to measure osmotic pressure (water potential). One form of osmometer consists of a porous pot, rendered semipermeable by impregnation of its pores with copper ferrocyanide. The pot is filled with the solution under investigation and a piston placed on the solution. The pot is then immersed in a container of distilled water. Water passes by osmosis into the porous pot and lifts the piston until the force of the water entering the pot equals the hydrostatic pressure created by the weight of the piston.

osmosis The process by which solvent molecules migrate across a *semipermeable membrane from a region of low solute concentration to a region of higher concentration. In biological systems, the solvent is invariably water and the migration tends to equalize solute concentrations on the two sides of a cell membrane. Osmosis is of great importance in the maintenance of cell *turgor and also in vascular transport. *See also* osmotic pressure, water potential.

osmotic potential (solute potential) The component of *water potential that takes into account the concentration of solutes in the cell. It is represented by the symbol Ψ_p and is always negative. Osmotic potential includes a component due to the *matric potential of colloidal substances and organelles in the cytoplasm.

osmotic pressure (Π) The force that has to be applied to a solution to prevent *osmosis occurring. It is equal to the *diffusion pressure deficit plus *turgor pressure. The term is going out of use because it does not take capillary and imbibitional forces into account, which can predominate in certain circumstances. Thus the term *water potential is now preferred.

ostiole A small pore found in the reproductive bodies of certain algae and fungi (e.g. the conceptacles of *Fucus* or the perithecia of certain ascomycete fungi) through which the spores are released.

OTUs *See* operational taxonomic units.

outbreeding The production of offspring by the fusion of distantly related gametes. The consequences of outbreeding are opposite to those of *inbreeding and mechanisms have been developed by the majority of plants to promote it. The main outbreeding mechanisms are *incompatibility factors, heteromorphism (e.g. *heterostyly), *protogyny, *protandry, and *dioecism.

ovary The swollen basal part of the carpel in angiosperms, which contains the ovules. The ovary is hollow and may contain one or many ovules in its locule, each attached by a funiculus. When the carpels are partially or totally united a compound ovary is formed, which may be either unilocular or multilocular depending on whether or not the fused carpel walls break down. Depending on its position in the flower an ovary may be described as *inferior*, when the other floral organs are inserted above it, or *superior* when the other floral organs are inserted below (*see* epigyny, perigyny, hypogyny). The ovary wall is usually thick and serves to protect the developing ovules. After fertilization the ovary wall forms the *pericarp of the fruit.

ovate Describing an organ, such as a leaf, that is egg shaped, with the broadest part nearest the point of attachment.

ovule The female gamete and its protective and nutritive tissue, which develops into the dispersal unit or *seed after fertilization in seed plants. In angiosperms the ovule comprises a central *embryo sac containing the gamete and other haploid nuclei, the surrounding *nucellus, and one or two protective *integuments interrupted by a small opening, the *micropyle. The ovule is attached to the placental tissue (*see* placentation) by means of the *funiculus. In most plants the ovule is completely bent around so that the micropyle faces towards the placenta. Such ovules are described as anatropous (*see* diagram). In some plants, e.g. certain members of the Caryophyllaceae, the ovule is at an angle to the funiculus, which appears to join the ovule midway between the micropyle and chalaza. This is the campylotropous form of ovule orientation. In a few plants, e.g. *Polygonum*, the ovule is erect or orthotropous. There may be one or many ovules in the locule of an ovary. In gymnosperms the ovule tends to be larger and is borne naked on an *ovuliferous scale rather than within an ovary. The formation of a cuticle around the ovule of seed plants is a significant development in the evolution of the terrestrial habit.

ovule culture The culture of excised ovules on suitable media in vitro. Ovules have been cultured with pollen grains of the same species in order to observe the processes of fertilization and embryo development. Culture with pol-

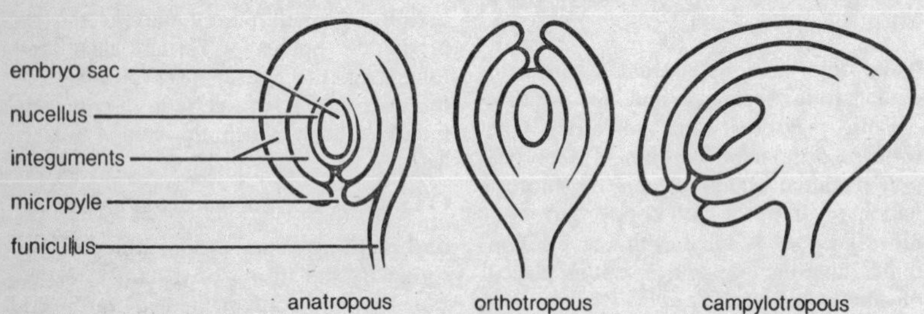

Types of ovule orientation.

len of different species may enable certain crosses to be made that are normally impossible due to incompatibility factors in the stigma. Some ovules in culture give rise to callus tissue that subsequently forms numerous embryo-like structures (embryoids). These can be separated and grown on to mature plants, so providing a means of plant propagation.

ovuliferous scale A scale that bears ovules and then seeds in the female *strobilus or cone of the conifers. It is borne in the axils of the spirally arranged woody bracts that constitute the cone. The ovuliferous scale is a highly specialized *megasporophyll and is homologous with a carpel.

oxalic acid A dibasic acid, formula $(COOH)_2$, found in many plants e.g. rhubarb (*Rheum raponiticum*) and dumbcanes (*Dieffenbachia*). It helps remove excessive amounts of various cations, e.g. calcium, sodium, and potassium, by forming the respective oxalates.

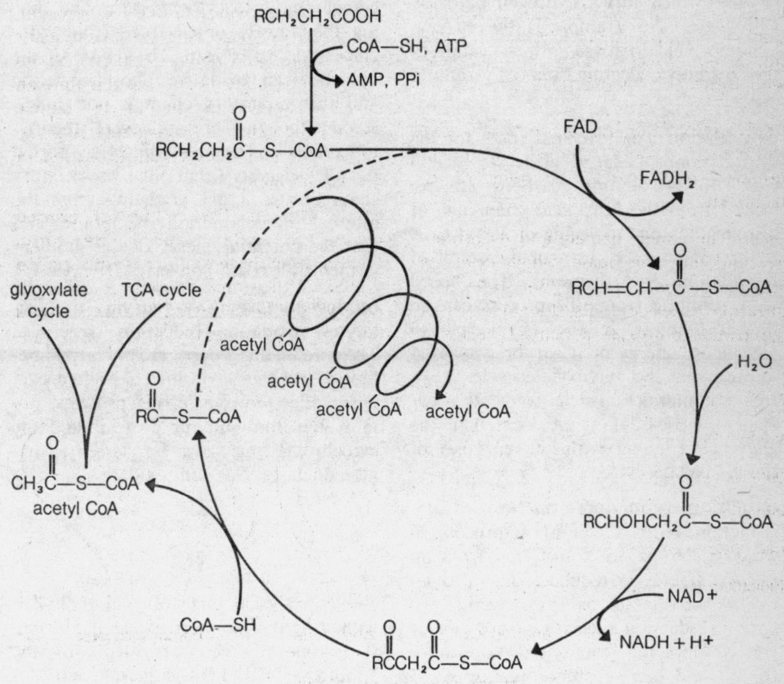

The β-oxidation spiral.

oxaloacetic acid A four-carbon dicarboxylic keto acid with the formula $HOOCCOCH_2COOH$. Oxaloacetate is a TCA cycle intermediate formed by oxidation of malic acid. It reacts with acetyl CoA to form the six-carbon acid citrate. Oxaloacetate is also an intermediate in glucose synthesis, and in C_4 plants of the Hatch–Slack pathway. The carboxylation of pyruvate to form oxaloacetate is a major reaction for replacing TCA cycle intermediates that have been withdrawn to take part in synthetic reactions. The reverse of this reaction is an important route of glucose synthesis.

oxidase Any of the flavin-linked *dehydrogenases in which the reduced coenzyme can be reoxidized by molecular oxygen, which forms hydrogen peroxide with water. An example is the enzyme L-amino acid oxidase, which catalyses the oxidative deamination of L-amino acids.

β-oxidation The principal route for the degradation of fatty acids to acetyl CoA. In β-oxidation, carbons are removed from the fatty acid chain two at a time; thus one molecule of acetyl CoA is formed on every turn of the spiral of β-oxidation (*see* diagram). The acetyl CoA resulting from this process can be further oxidized in the TCA cycle to yield more energy or it can be converted to sugar via the *glyoxylate cycle.

In germinating oil-rich seeds the enzymes of β-oxidation are located in the glyoxysome along with the enzymes of the glyoxylate cycle.

oxidation–reduction reaction (oxidoreduction, redox reaction) A process in which electrons are transferred from an *electron donor* or reducing agent to an *electron acceptor* or oxidizing agent. Oxidation–reduction reactions are major energy supplying reactions in both autotrophic and heterotrophic organisms.

In aerobic organisms the final electron acceptor for oxidation–reduction reactions is oxygen. Anaerobic organisms however use an organic compound as a final electron acceptor, e.g. in yeasts acetaldehyde accepts electrons to form ethanol.

In all organisms there are intermediate oxidation–reduction enzymes and coenzymes between the fuel molecules and the final electron acceptor. The most important of these are the pyridine nucleotides (*see* NAD, NADP).

oxidative phosphorylation The formation of ATP from ADP and inorganic phosphate coupled to the movement of electrons down the *respiratory chain. ATP formation occurs at three sites on the respiratory chain: during the transfer of electrons from NAD to coenzyme Q; during the transfer of electrons from cytochrome *b* to cytochrome *c*; and during the transfer of electrons from cytochrome aa_3 to oxygen. The nature of the link between oxidative phosphorylation and the respiratory chain is not understood although there are several theories to explain it. The most widely supported theory suggests that the respiratory chain creates a pH gradient across the internal mitochondrial membrane and that the potential energy of this gradient is then utilized to convert ADP to ATP.

oxidoreductase Any *enzyme that catalyses oxidation–reduction reactions. The two main groups are the pyridine-linked and the flavin-linked *dehydrogenases. Electron-transferring proteins, notably the iron–sulphur proteins and the cytochromes, may also be classed as oxidoreductases. *See also* oxidase.

P

P680 A special form of *chlorophyll *a* with a light absorption peak of 684 nm. It is one of the constituents of the *photosystem II light-gathering centre of photosynthesis. The absorption of light energy by the P680 molecules causes, among other things, the formation of a strong oxidant that can then oxidize water. This results in the evolution of molecular oxygen and initiates electron flow, which continues, via *photosystem I, through to the production of reduced *NADP. *See also* P700.

P700 A special form of *chlorophyll *a* with an absorption peak for light of 700 nm wavelength, found within the *photosystem I light-gathering centre of photosynthesis. Other ordinary chlorophyll *a* molecules are also present and these absorb light energy and pass it to P700, creating a charged form, $P700^+$. Electron transfer occurs from photosystem II, electrons being passed on ultimately to produce NADPH, the reduced form of *NADP. *See also* P680.

pachytene *See* prophase.

paedomorphosis *See* neoteny.

Palaeocene *See* Tertiary.

palaeoecology The study of extant and fossil organisms with a view to establishing the nature of the life of past ages and the environmental conditions that prevailed. Details of the environment in which given fossil organisms lived can be assumed from a knowledge of the prevailing climatic conditions in which the rocks that contain them were formed. However difficulties arise in distinguishing between the environment in which the organism lived and the environment in which it was buried.

palaeontology The study of *fossils. The branch dealing with the study of plant fossils is termed *palaeobotany*, which includes *palaeoethnobotany*, the study of fruit and seed remains found in archaeological sites. Palaeontology requires a knowledge of the way in which fossilization takes place and of the type of rock in which the fossil is found, a knowledge of the environmental conditions that prevailed when the rock was laid down, and an appreciation of the geological time scale so that the fossil can be dated.

Palaeozoic The era of geological time between about 570 and 225 million years ago that succeeds the Precambrian and precedes the Mesozoic. It is subdivided into six periods, the *Cambrian, *Ordovician, and *Silurian (lower Palaeozoic), and the *Devonian, *Carboniferous, and *Permian (upper Palaeozoic). *See also* geological time scale.

palea (pale, valvule) The upper of the pair of bracts beneath each floret (flower) in the inflorescence of a grass, the other being the *lemma. The palea is usually thin, narrow, and parallel sided. It usually has two ribs, which may project as prominent keels. *Compare* glume.

palindrome *See* repeated sequence.

palisade mesophyll Photosynthetic *chlorenchyma tissue composed of more or less elongate cells arranged in radial columns. Palisade mesophyll occurs with *spongy mesophyll as a ground tissue in the leaves of many mesophytes. Here the palisade mesophyll is usually on the adaxial (upper) side of the leaf, whereas in many xerophytes it forms the bulk of the mesophyll and is present on both sides of the leaf. When the stem has taken over much or all of the photosynthetic work of the plant, the cortex may consist largely of palisade parenchyma.

Palmae (Arecaceae) A monocotyledonous family of mainly tropical trees, the palms. It contains about 2780 species in about 210 genera. The growth form of a palm is characteristic, the plant usually consisting of an unbranched trunk with a crown of large leaves at the apex. When branching does occur, e.g. in *Hyphaene* (doum palms), it is dichotomous. The leaves may be pinnate, as in the feather palms, or the leaflets may all arise from the tip of the midrib, as in fan palms. Palm inflorescences are large and may contain thousands of flowers in a panicle or spike. In some palms, e.g. *Metroxylon sagu* (sago palm), the inflorescence is terminal, and the plant dies after flowering. The fruits are usually one-seeded drupes or berries, which may attain a considerable size. Many palms are of economic importance, examples being the coconut palm (*Cocos nucifera*), the date palm (*Phoenix dactylifera*), and the oil palm (*Elaeis guineensis*).

palmate (digitate) **1.** Describing a compound leaf having four or more leaflets arising from a single point. The leaves of the horse chestnut (*Aesculus hippocastanum*) and of lupins (*Lupinus*) are examples.
2. Describing a form of venation in which several equally prominent veins branch out from the base of the leaf blade. *See also* actinodromous.

palmelloid Describing a type of colonial growth form in which an indefinite number of cells are held together after division in a mucilaginous matrix. Palmelloid forms are characteristic of certain blue-green algae, e.g. *Merismopedia*, and various eukaryotic algae.

palmitic acid A sixteen-carbon saturated fatty acid. It is the commonest saturated fatty acid in plants, forming a high proportion of plant fats. It is also on the synthetic route to other fatty acids such as *stearic and *oleic acids.

palynology The study of living and fossil pollen grains, spores, and similar structures. The diversity of such structures has led to an increasing use of palynological data in systematics, while their resistance to decay has made them an important feature of palaeobotanical and palaeoecological work.

pampas South American *grassland.

pangenesis A now disregarded theory formulated by Charles Darwin that suggested a mechanism by which acquired characteristics may be inherited. He postulated that there were particles (pangenes) carried in the body fluids from all the organs of the body to the reproductive cells. These particles then influenced the gametes that in turn influenced the characteristics of the succeeding generation.

panicle A *racemose inflorescence in which the flowers are formed on stalks (peduncles) arising alternately or spirally from the main axis. Each stalk is a raceme. The peduncles may be long and spreading, as in oats (*Avena*), or short, resulting in a spikelike inflorescence, as in timothy grass (*Phleum pratense*). The panicle is thought to be the most primitive form of racemose inflorescence from which the other types have probably evolved. The term is also sometimes used to describe any type of branching inflorescence.

panmixis Unrestricted random crossing. It is probably fairly rare but the term may be applied to crossing that appears to be random with respect to a particular character or gene. *Compare* assortive mating.

pantocolpate Describing a pollen grain having many *colpi.

pantonematic flagellum *See* flimmer flagellum.

pantothenic acid A compound, formula $CH_2OHC(CH_3)_2CHOHCONHCH_2CH_2COOH$, that is the precursor of coenzyme A. It is synthesized by plants and bacteria but not by vertebrates, which consequently require it as a *vitamin in the diet.

paper chromatography A widely used chromatographic technique for separating the components of mixtures using absorbent paper as the stationary phase. A sheet or strip of paper, with a concentrated spot of the mixture on a pencil-drawn base line, is vertically suspended in a suitable solvent (the mobile phase), that seeps slowly upwards through the paper by capillary action. The compounds in the mixture travel upwards with the solvent and separate out at different levels. When the solvent reaches a level near the top of the paper, this point (the solvent front) is marked. The paper is removed and dried. Colourless compounds may then be developed either by spraying the paper (the chromatogram) with a suitable chemical, such as ninhydrin, or by viewing the paper in ultraviolet light. The compounds can then be identified either by comparing them with chromatograms of known standard solutions run at the same time, or by calculating the *Rf values. *See* partition chromatography, two-dimensional analysis.

papilla A projection from a cell, usually of the *epidermis, often regarded as a kind of *trichome. Papillae are often swollen and covered with wax and in xerophytes may serve to protect from sunlight and excessive water loss.

pappus A modified calyx made up of a ring of fine hairs, scales, or teeth that persist after fertilization and aid the wind dispersal of the fruit, often by forming a parachute-like structure. It is seen in members of the Compositae, e.g. dandelions (*Taraxacum*) and thistles (*Carduus*). The pappuses of an inflorescence may form a 'clock'.

paraflagellar body *See* photoreceptor.

parallel evolution (parallelism) The development of similar adaptations in related organisms as a result of being subject to similar selection pressures. Thus the different species of a genus with a wide distribution may show the same adaptations to similar environments in widely separated regions. The development of divided submerged leaves in various species of crowfoot, e.g. water crowfoot (*Ranunculus aquatilis*) and river crowfoot (*R. fluitans*), is an example. However the phenomenon is not of widespread occurrence and some authorities deny its existence.

parallelodromous Describing a form of leaf venation in which two or more primary veins originate at the base of the lamina and run to the apex in an essentially parallel manner. The leaves of many monocotyledons have this type of venation. In traditional terminology such venation is simply termed *parallel. See illustration at* venation.

parameter A quantity that characterizes and describes a population.

paramylum A polysaccharide that is the assimilation product of both green and colourless algae of the Euglenophyta.

paraphyletic Describing a group of organisms that does not contain all the descendants of a common ancestor. In plant classification, the dicotyledons are a paraphyletic group because their ancestors also gave rise to the more advanced monocotyledons.

paraphysis A sterile unbranched usually multicellular hair borne amongst the reproductive structures of many bryophytes, algae, and fungi. Paraphyses are protective and in some cases are thought to aid in dehiscence.

parasexual recombination A form of recombination seen in the progeny of certain heterokaryotic fungi, which does not involve meiotic segregation. The sequence of events involved is termed the *parasexual cycle* and is as follows: two genotypically distinct nuclei in a heterokaryon fuse to give a diploid heterozygous nucleus; the diploid nucleus multiplies alongside the haploid parent nuclei in the heterokaryon; a homokaryotic diploid mycelium is established from a diploid conidium; recombination occurs during crossing-over in mitosis; a haploid nucleus is formed by progressive shedding of chromosomes from the diploid nucleus.

Such recombination has been observed in various ascomycetes, basidiomycetes, and imperfect fungi. It accounts for the occurrence of new strains of pathogenic imperfect fungi. It has also made feasible the study of the genetics and 'breeding' of imperfect fungi.

parasitism The temporary or permanent relationship between two different species, in which one, the *parasite*, benefits by obtaining food and/or shelter at the expense of the other, the *host*. Some parasites have little effect on the host, some cause serious diseases, and some kill the host. *Ectoparasites* live on the surface of the host, while *endoparasites* live either inside or between the host cells. An *obligate parasite*, e.g. *Phytophthora infestans*, which causes late blight of potato, can only grow on its host. However, a *facultative parasite*, such as the fungus *Pythium*, can exist by feeding in a different way after the death of the host. A species of parasite may live on one particular host or a number of similar hosts, or may alternate between two or more different species. Some flowering plants are partial parasites, being able to live independently but becoming parasitic in certain circumstances. For example, some members of the Scrophulariaceae, such as eyebrights (*Euphrasia*), will photosynthesize and live independently but when their roots come into contact with those of grasses they become attached and absorb food.

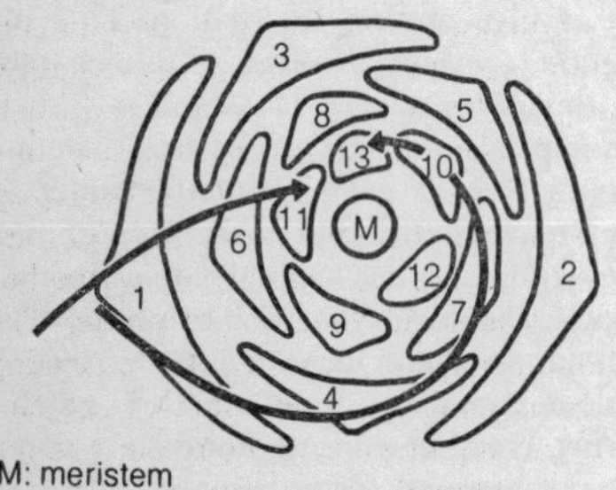

Opposite parastichies at the stem apex.

parastichy An imaginary spiral around a stem apex joining adjacent leaf primordia. Two parastichies can be visualized running in opposite directions (*see* diagram). If a plant is described as having, for example, (3+5) or 3/5 *phyllotaxis, this means that the shallower of the two parastichies joins every third primordium formed at the apex (1, 4, 7, 10, etc.) and the steeper parastichy joins every fifth primordium (1, 6, 11, 16, etc.).

paratonic movement A movement exhibited by a plant or plant part in response to an external stimulus. *Taxes, *tropisms, and *nastic movements and mechanical hygroscopic and turgor movements are all examples of paratonic movements. *Compare* autonomic movement.

paratype A specimen other than the *isotype or *holotype cited in the original publication of the name of a taxon. A paratype can also be one of a collection of *syntypes, following the selection of a *lectotype. Although of little significance nomenclaturally, paratypes are of considerable importance to the taxonomist, who may not be able to see other type material of a particular taxon. *See also* type.

parenchyma Relatively unspecialized tissue usually composed of more or less isodiametric polyhedral cells with thin nonlignified cellulose cell walls and living protoplasts. Parenchyma cells are often present in great numbers, forming a *ground tissue* (ground parenchyma) in which the other tissues are embedded. *See also* collenchyma.

parichnos An area of tissue made up of disintegrated cells seen at the base of the sporophylls in the strobili of *Lycopodium* and in the leaves of various fossil relatives of *Lycopodium*, such as *Lycopodites* and *Lepidodendron*.

parietal placentation A form of *placentation in which the placentae develop along the fused margins of a unilocular compound ovary, as in violets (*Viola*).

paripinnate Describing a pinnate leaf in which all the leaflets are paired. Such leaves are seen in vetches (*Vicia*), the terminal leaflet having been replaced by a tendril. *Compare* imparipinnate.

parthenocarpy The production of a fruit without the setting of seeds. The process is an irrelevance in the life history of a plant and is unusual in that it appears not to benefit the plant in any way. Parthenocarpy is most often seen in plants with numerous ovules in the fruit, e.g. fig and melon. In some species, e.g. cucumber, pollination is not required to stimulate parthenocarpy. Where pollination is necessary parthenocarpy may result either because the pollen tube does not reach the ovule or because the embryos abort following fertilization, as in seedless grapes. Triploidy, which results in sterile plants, can also lead to parthenocarpy, as in the banana.

parthenogenesis The development of an egg cell into an embryo without fertilization. This is usually due to defective meiosis, which results in an egg nucleus with an unreduced number of chromosomes, as is found in dandelions (*Taraxacum*). In some species, e.g. hawkweeds (*Hieracium*), the megaspore is replaced by a cell of the nucellus.

parthenospore (azygospore) A thick-walled resting spore that develops parthenogenetically from an unfertilized gamete. Parthenospores are formed by certain algae, e.g. *Protosiphon*, and fungi, e.g. *Entomophthora muscae*. They may give rise to more gametes or to a haploid individual. *Compare* zygospore.

partial dominance *See* incomplete dominance.

partial veil *See* mushroom.

particulate inheritance The determination of inherited characteristics by stable discrete factors that remain unchanged from one generation to another. Differences observed between parents and offspring are thus due to the recombination rather than modification of such factors. The particulate nature of inheritance was established with Mendel's work. Earlier theories formulated without experimental evidence tended to the idea that the characteristics of parents blended together in the offspring (*see* blending inheritance).

partition chromatography A chromatographic method in which the components of mixtures are separated according to their different partition coefficients. Any substance dissolved, at a given temperature, in two immiscible solvents has, at equilibrium, a certain characteristic proportion dissolved in each solvent. The ratio of the amount of solute in one solvent to that in the other is the *partition coefficient*. In column partition chromatography the column contains some hydrophilic substance, e.g. silica or starch granules. Water is tightly bound to the surface of the granules and acts as the stationary phase. A solvent immiscible with water, e.g. butanol or phenol, and carrying the mixture to be separated is then passed through the column. This is the mobile phase. Those components of the mixture that are more soluble in water will pass through the column more slowly than those that are less soluble in water but more soluble in the mobile phase. The components are separated by collecting and analysing small amounts of the eluate (the liquid passing from the bottom of the column) as it emerges from the column. Mixtures of amino acids are often separated in this way.

*Paper chromatography is a particular kind of partition chromatography in which the hydrophilic cellulose of the paper holds the water and acts as the stationary phase.

passage cell Any of the thin-walled cells found at intervals in the endodermis, generally opposite the protoxylem poles. It is believed that passage cells provide an easier route for water and dissolved solutes to pass through from the cortex to the protoxylem.

pasture *See* meadow.

pathogen An agent able to cause disease. The term is often restricted to agents that are themselves living organisms.

pathogenicity *See* virulence.

patristic Describing similarity between two or more taxa that is due to common ancestry. The concept of patristic relationship is purely *phylogenetic and thus not used in the construction of *phenetic classifications. *See also* cladistic.

peat Partially decomposed plant material. It builds up in areas with poor drainage, namely bogs and fens. Acid bog peat (or peat moss) is composed primarily of the remains of bog plants, such as *Sphagnum* mosses and sedges. It is used as a mulch and to improve soil and is also a constituent of potting composts. Fen peat contains considerably more nutrients than bog peat and if drained provides fertile agricultural land.

pectic substances Pectins and related polysaccharides, in which galacturonic acid, galactose, or a derivative of these is a major component. Besides the pectins found in cell walls and middle lamellae, large quantities of pectic substances are found in some fruits, e.g. apples, pears, and citrus fruits. Pectins are used in the food industry as gelling agents in jams, etc.

pectin A structural polysaccharide found in plant cell walls. It is rich in α (1–4) linked galacturonic acid residues, though other sugars, e.g. rhamnose, are present in small quantities. The carboxyl groups of the galacturonic acid residues in pectin may be methylated. Methyl esterification takes place after polymerization, the methyl donor for the reaction being *S*-adenosyl methionine (*see* methionine).

ped *See* soil structure.

pedalfer A *soil in which the soluble lime has been leached from the surface layers by heavy persistent rain. It is thus acid and the rate of decomposition of soil litter is consequently slow. Iron and aluminium hydroxides are abundant. Pedalfers are one of the two main types of zonal soils (*compare* pedocal) and include brown earths, podsols, and tundra soils.

pedicel The stalk attaching individual flowers to the main axis (peduncle) of the inflorescence. It often develops in the axil of a bract and its internal structure is typically stemlike. The pedicel may act as a temporary storage organ for sugars prior to seed development.

pedigree method A plant breeding technique that is used to create entirely new varieties of plants that combine the best qualities of selected existing varieties. The method is limited to self-pollinating species. Parents are selected and artificially crossed. A few (say six to ten) of the resulting F_1 hybrid seeds are sown widely spaced to encourage prolific seed set. Assuming the parents were homozygous, the F_1 plants should all be identical. The seed from each plant is harvested separately and grown, again widely spaced, in separate 'family' plots. The genotypic variability becomes apparent in the F_2 and the characteristics of each plant are noted. The seed from a small number (perhaps only 1%) of promising plants is harvested to grow on to the F_3. The seeds from each plant are again sown in separate plots and in this way the 'pedigree' of each plant can be recorded. By noting the variability within each of the F_3 plots it may be determined which plants of the previous generation depended excessively on heterozygosity for vigour. At this stage whole families of plants may be discarded. As heterozygosity is approximately halved each generation, increasing emphasis is placed in succeeding generations on selection between rather than within families. If breeding a crop plant, then at about the F_6 generation seeds from selected plants are grown closely spaced as would occur in normal field conditions. Yield may then be determined. From this point selection is entirely between families. By the F_8 generation the remaining selected lines may be assumed to be over 99% homozygous and they are bulked up for variety trials. *Compare* backcrossing.

pedocal A *soil in which lime accumulates in the surface layers, which consequently become alkaline. Pedocals occur in regions with light rainfall and are usually fertile as they may be rich in potash and phosphates as well as lime. In persistently dry conditions calcium salts may be deposited on the soil surface. Pedocals are one of the two main types of zonal soils (*compare* pedalfer), and include the chestnut-brown soils, chernozems, and red desert soils.

pelagic Describing the organisms that live in the surface waters of a sea or ocean, such as plankton.

pellicle (periplast) A membrane surrounding the protoplast in some unicellular algae that lack a cellulose cell wall, for example *Euglena*. In many cases, it is believed to consist of numerous semirigid overlapping rings. The pellicle may be rigid or flexible.

pellucid Transparent.

peloton *See* endotrophic mycorrhiza.

peltate Describing a structure that is circular, with the stalk inserted in the middle, such as the leaf of the garden nasturtium (*Tropaeolum majus*), the hairs on the undersurface of the leaves of sea buckthorn (*Hippophae rhamnoides*), or the sporangiophores of certain cycads.

pendulous placentation *See* apical placentation.

penetrance The degree to which a gene is expressed in the phenotype. Most dominant alleles show complete penetrance. However a gene may be affected by environmental factors to the extent that not all the individuals with a particular genotype show the phenotype characteristic of that genotype. In some maize plants, for example, anthocyanin pigment only develops in those parts of the plant exposed to high light intensities. In shaded parts of the plant the gene is not expressed. A gene whose expression is affected in this way is said to show *incomplete penetrance*.

penni-parallel *See* craspedromous.

Pennsylvanian *See* Carboniferous.

pentamerous (5-merous) Describing flowers in which the parts of each whorl are inserted in fives, or multiples of five. This is the most commonly found arrangement in dicotyledons.

pentosan A polysaccharide in which the major monosaccharide subunits are pentoses. The most common monosaccharide components of pentosans are xylose and L-arabinose. Many *hemicelluloses are pentosans, although they are usually heteropolymers containing other sugars besides the pentoses.

pentose Any five-carbon sugar. Common pentoses are *ribose, *deoxyribose, *xylose, and *arabinose. Pentoses are synthesized from hexoses in the pentose phosphate pathway, from carbon dioxide in the Calvin cycle, or by decarboxylation of nucleoside diphosphate sugars. Polymers of pentoses are collectively termed *pentosans.

pentose phosphate pathway (phosphogluconate pathway, hexose monophosphate shunt) A complex multifunctional metabolic pathway involving oxidation of glucose and formation of three-, four-, five-, six-, and seven-carbon sugars.

In the pentose phosphate pathway glucose 6-phosphate is first oxidized and decarboxylated to a five-carbon sugar with the concomitant formation of NADPH. The five-carbon sugar then undergoes various combinations and transformations, similar to those of the Calvin cycle, leading eventually to the reformation of glucose 6-phosphate.

Although the pathway can be used for the oxidation of glucose, it is more important as a source of five-carbon sugars, as a pathway for the oxidation of five-carbon sugars, and as a source of NADPH for such purposes as fatty acid synthesis.

Pentoxylales An order of extinct gymnosperms from the Jurassic, the affinities of which are uncertain.

PEP *See* phosphoenolpyruvate.

pepo A type of *berry with a hard exterior derived either from the epicarp or noncarpellary tissue of the plant. In members of the Cucurbitaceae the hard exterior is formed from the receptacle of the flower.

peptidase Any *hydrolase enzyme that catalyses the hydrolysis of peptides. The distinction between *proteases and peptidases is not a clear one – peptidases often have protease activity and vice versa.

Several types of peptidase can be distinguished. *Exopeptidases* can only attack terminal peptide bonds while *endopeptidases* attack bonds within the peptide chain and are often specific to certain types of peptide linkage. *Aminopeptidases* attack the amino terminal of peptides and *carboxypeptidases* attack the carboxy terminal. *Dipeptidases* attack only dipeptides.

peptide A molecule consisting of a number of amino acids covalently joined by peptide linkages. Small peptides containing only a few amino acids are known as oligopeptides while larger peptides are called *polypeptides. Most oligopeptides arise as breakdown products of proteins but a few have specific functions. For example, the tripeptide glutathione activates some enzymes and protects lipids from autooxidation.

perennial A plant that lives for many years. There are two types. *Herbaceous perennials* survive each winter as underground storage or perennating organs, such as *bulbs, *corms, *rhizomes, and stem and root *tubers. The foliage leaves and flowers die back in winter. *Woody perennials* are trees and shrubs whose aerial stems have woody tissues and persist above ground. They may be *deciduous or *evergreen. *Compare* annual, biennial, ephemeral.

perfoliate Describing a sessile leaf in which the base of the lamina extends either side of the node and joins together on the far side so the stem is completely encircled. This formation is also seen when pairs of sessile leaves unite at their bases, as in the perfoliate honeysuckle (*Lonicera caprifolium*).

perforation plate The remains of the end walls between two adjacent vessel elements in a *vessel of the xylem, forming an opening between the cells, thus facilitating the free movement of water through the vessel. Perforation plates are present in some ferns, some gymnosperms (Gnetales), and most angiosperms, and are believed to have evolved independently in the three groups. *Compare* sieve plate.

perianth The structure that protects the developing reproductive parts of the flower. In dicotyledons it normally consists of two distinct whorls, the *calyx and the *corolla. In many monocotyledons (such as the tulip) these whorls are not differentiated and the individual perianth units are then termed *tepals.

The perianth units tend to be simple, separate, and inserted spirally in the more primitive families, such as the Magnoliaceae and Papaveraceae. In the more advanced families, such as the Labiatae and Compositae, they are often fused and inserted in whorls. The opening and closing of some flowers is controlled by temperature and light, these stimuli being perceived by the perianth segments.

periblem *See* histogen theory.

pericarp The wall of a fruit, derived from the maturing ovary wall. In fleshy fruits the pericarp usually has three distinct layers, the outer toughened *epicarp, a fleshy *mesocarp, and an inner *endocarp, which may be variously thickened or membranous. In dry fruits the pericarp tends to become papery or leathery. If the pericarp opens at maturity the fruit is described as dehiscent but if it remains closed the fruit is termed indehiscent.

perichaetium Any of the leaves or bracts surrounding the sex organs of bryophytes or the structure formed by such a whorl. Those enveloping the archegonia are sometimes termed *perigynia* while those around the antheridia are termed *perigonia*. *See also* pseudoperianth.

periclinal Parallel to the surface. The *periclinal wall* of a cell is thus one that is parallel to the surface of the plant body. A *periclinal division* is one that results in the formation of periclinal walls between daughter cells. Such a division results in an increase in girth of the organ. In cylindrical organs, such as stems and roots, the term *tangential* may be used in place of periclinal, usually in descriptions of cell walls. *Compare* anticlinal.

periclinal chimaera A *chimaera in which tissues of one genetic type completely surround tissues of another genetic type. The ornamental shrub *Laburnocytisus adami* is a periclinal chimaera that originated from the grafting of *Cytisus purpureus* onto a *Laburnum anagyroides* stock. The epidermis is composed of *Cytisus* cells and the internal tissues of *Laburnum* cells. Externally it is not obvious that this *graft hybrid* is a chimaera but occasionally shoots arise consisting solely of tissues from one or other parent. The seeds, which derive from the internal tissues, are always *Laburnum*. Investigations into the stability of this chimaera led to the *tunica-corpus theory of apical organization.

The variegated leaves of certain plants are examples of periclinal chimaeras that have originated from a plastid mutation preventing chlorophyll synthesis. The mutation may occur in either the second layer of the tunica (L2) or the corpus (L3). Differences in colour appear between the leaf margins and the centre of the leaf because the margins are derived from the L2 while the centre is composed of both L2 and L3. Thus if the mutation occurs in the L2 then the margins are white and the centres green (the underlying green L3 tissues are not masked by the white L2 tissues). This type of variegation is seen in *Pelargonium*.

pericycle (pericyclic region) The outermost layer of the *stele, lying immediately within the endodermis. In most dicotyledons the pericycle, which is composed mostly of parenchyma cells, also includes modified *protophloem often with the addition of sclerenchyma fibres. The pericycle is a more obvious structure in roots, lateral roots arising from this region.

periderm A protective tissue of secondary origin, comprising the phellogen, phellem, and phelloderm and replacing the epidermis as the outer cellular layer of the stem and root in plant axes exhibiting secondary growth. In such tissue the functions of stomata are performed by lenticels. Periderm also often develops at the site of a wound or other recently exposed surface, thus helping to prevent the entry of pathogens. *Compare* bark, rhytidome.

peridinin A xanthophyll pigment characteristic of algae in the Dinophyta.

peridium The two-layered outer wall of certain fungal fruiting bodies. It is most obvious in the ascocarp of gasteromycete fungi, in which it consists of an outer exoperidium and an inner endoperidium. In puffballs (*Lycoperdon*), the exoperidium sloughs off or breaks up into a number of scales, while in the earthstars (*Geastrum*) it peels back over the endoperidium to give the characteristic star-shaped fruiting body. A peridium is also seen in the aecium surrounding the aeciospores of rust fungi and in the ascocarp of certain Erysiphales.

perigenous Describing an angiosperm stomatal complex in which the subsidiary cells are not derived from the same initial as the guard cells. *Compare* mesogenous. *See also* haplocheilic.

perigonium *See* perichaetium.

perigynium *See* perichaetium.

perigyny The arrangement of floral parts, intermediate between *hypogyny and *epigyny in which the sepals, petals, and stamens are inserted on the receptacle at about the same level as the ovary. Perigynous flowers, e.g. cinquefoils (*Potentilla*), are seen when the receptacle becomes flattened or concave, in contrast to hypogynous flowers in which the receptacle is convex.

perinuclear cisterna *See* nuclear membrane.

periplasmodium *See* tapetum.

periplast *See* pellicle.

perisperm A nutritive tissue derived from the nucellus (and therefore diploid) that is found in the seeds of certain plants in which the endosperm does not completely replace the nucellus. Many of the Caryophyllaceae contain such tissue in their seeds.

perispore An extra layer that may surround the spore in some species, particularly certain ferns.

peristome The ring of teeth surrounding the opening of a moss capsule, visible when the operculum is removed. It may consist of a double or single ring. The peristome is closed over the opening of the capsule in wet weather but as the teeth dry out they fold back allowing the spores to escape.

perithecium (pyrenocarp) The type of *ascocarp characteristic of the Pyrenomycetes. It is a flask-shaped structure with a pore or ostiole opening to the exterior. *See also* pseudothecium.

peritrichous Describing bacteria that have flagella all over the surface of the cell. Bacteria with flagella at one or both ends of the cell are termed *polar*. *See also* monotrichous, lophotrichous.

permanent quadrat *See* quadrat.

permanent stain A stain that lasts for a long time without fading and does not damage the specimen. A single stain may be used, such as fast green for macerated tissues, or a tissue section may be double-stained, using, for example safranin and the counterstain light or fast green. Aqueous dyes, such as safranin, should be used before the section is dehydrated, and light green on completion of dehydration. *Clearing and *mounting follow. *Compare* temporary staining. *See also* staining.

permanent wilting point The point at which the amount of water in the soil has dropped to such a level that the plants begin to wilt and will not recover, even if moved to a cool and dark place, unless more water is added to the soil. It occurs when the water potential of the soil is the same as, or lower (more negative) than, the water potential of the plant. The amount of water retained in the soil when the permanent wilting point is reached varies depending on the plant species and the type of soil. *See also* field capacity.

Permian The final period of the Upper Palaeozoic era about 280–225 million years ago. The climate supported luxurious vegetation but became increasingly variable with widespread periods of glaciation. During this period, the ferns became less numerous and plants similar to *Ginkgo* and the cycads first appeared. *See also* Permo-Triassic, geological time scale.

Permo-Triassic A period encompassing the Permian and Triassic periods. It is considered a single period by certain authorities, as the prevailing conditions make it difficult to draw a boundary between the rocks of the Permian and Triassic. Intense earth movements and a period of mountain building and volcanic activity occurred resulting in the extinction of many groups of organisms and the appearance of new forms.

Peronosporales An order of the *Oomycetes containing both saprobic and parasitic fungi, many of which are serious plant pathogens. Its members, which number about 300 species in some 16 genera, are found in damp terrestrial habitats and include: *Pythium* (many species of which cause *damping-off of seedlings); *Phytophthora infestans* (late blight of potato); *Plasmopara*, *Peronospora*, and *Bremia* (various downy mildews); and *Albugo* (white blister rusts).

peroxisome A *microbody that contains amino acid oxidases, urate oxidases, and catalase. The latter catalyses the decomposition of hydrogen peroxide, produced by the activity of the other enzymes, to oxygen and water.

persistent Describing pesticides that take a long time to break down and become inactive. Such pesticides give longer protection than nonpersistent pesticides but are not suitable for use soon before harvest. Some are harmful to the environment as they accumulate in food chains and poison wildlife. Examples are the weedkiller MCPA and the insecticide DDT.

pest Any organism that damages crops or reduces their yield or that irritates or injures livestock. The term is more often used of animals that cause physical damage, for example locusts or rodents. Organisms that cause disease are more commonly termed pathogens or parasites.

pesticide Any chemical used to kill pests, such as an *insecticide, *fungicide, or *herbicide.

petal An individual unit of the *corolla. Petals are thought to be modified leaves with a simplified internal structure, having only one vascular bundle compared with the several normally found in leaves and sepals. Insect-pollinated plants tend to have large, often yellow or white, scented petals often with a nectary at the base and honey guides patterning the surface. The petals of wind-pollinated plants, when present, tend to be small and dull coloured. *Compare* sepal.

petalody *See* double flower.

petiole The stalk that attaches the leaf lamina to the stem. The point of attachment is often strengthened by a widening of the base of the petiole. Some leaves (sessile leaves) lack a petiole and are joined to the stem at the base of the lamina. Sessile leaves are characteristic of most monocotyledons.
The petiole is generally similar in structure to a stem except that the vascular and strengthening tissues are asymmetrically arranged so as to bear the weight of the lamina. The different patterns of veins evident in petioles are useful taxonomically. Various modifications of the petiole are seen. Some are flattened and bladelike (*see* phyllode). Others are inflated, as in *Eichhornia crassipes* (water hyacinth), where they aid buoyancy. In many species the base of the petiole sheaths the stem, as in *Heracleum mantegazzianum* (giant hogweed). The base of the petiole may be modified as a *pulvinus. The petioles of some climbing species, e.g. *Clematis*, are haptotropic. The base of a fern rachis is sometimes termed a petiole.

Pezizales (cup fungi) An order of the *Discomycetes in which the asci are operculate and are contained within a hymenium. It contains about 700 species in some 90 genera. The hymenium lines the surface of the cuplike apothecium and may be brightly coloured, as in *Sarcoscypha coccinea*, the scarlet elf cup. The group also contains the edible morels (*Morchella*).

P_{fr} The active form of the plant pigment *phytochrome that has a peak of light absorption in the far red of the spectrum, i.e. at about 725–730 nm. It is interchangeable with the *P_r form.

PGA *See* phosphoglyceric acid.

pH A measure of the hydrogen ion concentration, $[H^+]$, and hence of the acidity or alkalinity of an aqueous solution. It is the logarithm of the reciprocal of $[H^+]$, i.e. $\log_{10}(1/[H^+])$. The concentration of H^+ in a litre of pure water (i.e. a neutral solution) is 10^{-7} moles, giving a pH value of 7. One integer on the pH scale indicates a tenfold difference in H^+ concentration. The lower the pH value, the higher the concentration of H^+ and vice versa.

Phacidiales An order of the *Discomycetes whose members, which are divided between about 250 species and 64 genera, have inoperculate asci. It includes the plant pathogen, *Rhytisma acerinum*, which causes tar spot of sycamore.

Phaeophyta (brown algae) A division consisting predominantly of marine *algae and including many of the large seaweeds, such as the wracks (*Fucales) and kelps (*Laminariales). The organization of the thallus is more advanced than in other algae and growth is truly parenchymatous in many species. Sexual reproduction is usually oogamous. In addition to chlorophylls *a* and *c*, the brown algae contain carotin and xanthophyll pigments, including fucoxanthin. The main storage products are laminarin, mannitol, and fucosan. The Phaeophyta contains one class, the Phaeophyceae, divided into three subclasses: the Isogeneratae, in which there is an isomorphic alternation of generations (e.g. *Ectocarpus*, *Dictyota*); the Heterogeneratae, with a heteromorphic alternation of generations (e.g. *Laminaria*); and the Cyclosporae, with no alternation of generations (e.g. *Fucus*).

phage *See* bacteriophage.

phagocytosis *See* endocytosis.

Phallales An order of the *Gasteromycetes containing the stinkhorns. There are about 70 species in some 21 genera. When mature the basidiocarp wall of the stinkhorn is broken open by an expanding stalk, which rapidly grows carrying the foul-smelling spores at its tip. A common species is *Phallus impudicus*.

phanerogam In early classifications, a plant whose reproductive organs are easily visible either as flowers or cones. *Compare* cryptogam.

phanerophyte A plant with perennating buds situated on upright stems well above soil level. The buds are therefore potentially exposed to wind, extremes of temperature, and drought. Phanerophytes are thus usually found in temperate moist regions. They include trees, shrubs, herbaceous and succulent plants, and vines. *See also* Raunkiaer system of classification.

phase contrast microscope A microscope that enables living, and hence normally transparent, material to be viewed by converting (invisible) differences of phase in the transmitted light to (visible) differences of contrast. Light passing through a specimen changes its phase by different amounts due to differences in thickness or refractive index. The phase contrast microscope converts these differences to differences in amplitude, which can be seen as shades of grey (because the greater the amplitude of a wave the more bright or intense is the light). The effect is produced by using a special condenser together with a glass plate (the retardation plate) in the objective. The condenser produces a certain pattern of light, often a hollow cone. This is matched in the retardation plate by a similar pattern produced by a groove in the plate. Any light that is not diffracted by the specimen will pass through the groove in the retardation plate and remain unchanged in phase. Any light that has been diffracted by the specimen will pass through the thicker part of the retardation plate and hence be slowed even further. Contrast results from combination of the direct and refracted light. If they are out of phase by half a wavelength the difference produces destructive interference of the light so producing contrast. When this technique is used in conjunction with a time-lapse ciné camera it enables such events as cell division to be filmed and later speeded up, a technique known as *cinemicrography*. *See also* interference microscope.

phellem (cork) The compact protective tissue that replaces the epidermis as the outer cellular layer in plants with secondary growth. It is the outer layer of the *periderm and consists of numerous more or less isodiametric cells arranged in radial rows, with thick waxy suberized cell walls. Commercial cork is obtained from the cork oak *Quercus suber*, in which the same phellogen is active each year (*see* bark). It consists of thin-walled unsuberized cork cells (*phelloids*) and is stripped from the tree approximately every nine years.

phelloderm (secondary cortex) Parenchyma tissue of secondary origin, derived centripetally from the phellogen. Cells of the phelloderm may be distinguished from those of the cortex by their arrangement in radial columns, reflecting their origin from the phellogen. *See also* periderm.

phellogen (cork cambium) The *cambium that arises external to the *vascular cambium in many plant axes undergoing secondary growth. It gives rise to the *phellem and *phelloderm. *See also* periderm.

phelloid *See* phellem.

phenetic Describing relationships between organisms or groups of organisms that are assumed on the basis of overall similarities and differences. A phenetic classification takes into account as many characteristics as possible and, in contrast to a *phylogenetic classification, does not attach more weight (*see* weighting) to characters that are assumed to be the result of evolutionary relationships. (Other characters may be weighted, especially those thought to be 'good' characters.) A phenetic classification does not aim to reflect evolutionary histories but may nevertheless well do so. It is most likely to diverge from a phylogenetic classification in instances of convergent and parallel evolution.

Classifications published before or soon after the appearance of Charles Darwin's *Origin of Species* (1859) were obviously phenetic in nature since taxonomists were not thinking then in evolutionary terms. Most classifications drawn up after about 1880 tend to be phylogenetic. One phenetic system still in common use as a basis of herbarium arrangement is that of George Bentham and Joseph Hooker, published as *Genera Plantarum* (1862–83). A main difference between this and more modern systems is the placement of the gymnosperms between the dicotyledons and monocotyledons.

phenocopy A phenotype of one genotype that resembles the phenotype of a different genotype. The change in phenotype is normally brought about by unusual environmental conditions, such as abnormally high temperatures early in development.

phenogram *See* dendrogram.

phenolics (phenols) Compounds containing a benzene ring directly substituted with one or more hydroxyl groups. The simplest member is phenol (carbolic acid), formula C_6H_5OH. Plant phenolics are widespread and extremely diverse and include the *flavonoids. Simple phenolic compounds containing one or two rings are not as common as the flavonoids, though hydroquinone, coumarin, and the cinnamic acids are widely distributed. The amino acid tyrosine also contains a phenolic ring. Less common are such phenolics as catechol, phloroglucinol, and pyrogallol. Lignin is a phenolic, probably being formed from coniferyl alcohol, a derivative of cinnamic acid. Tannins are also derived from phenolics.

phenolphthalein A hydrogen-ion indicator that is colourless in acids and red in alkalis.

phenon line *See* dendrogram.

phenotype The expressed characteristics of an organism. The phenotype is determined by an interaction between the environment and the *genotype and between dominance and epistatic relationships within the genotype.

phenoxyacetic acids A group of compounds some of which show auxin-like activity. Various chlorine substituted forms cause a disruption in metabolism particularly at the meristems and have been widely used as selective weedkillers. An important factor in the use of phenoxyacetic acids as weedkillers is their immunity from the plant's endogenous IAA oxidizing system, which can normally inactivate superfluous auxins. Examples of such compounds are 2,4-D, 2,4,5-T, and MCPA.

phenylalanine An aromatic nonpolar amino acid with the formula $C_6H_6CH_2CH(NH_2)COOH$ (*see illustration at* amino acid). The biosynthetic pathway (*see* shikimic acid) to phenylalanine from erythrose 4-phosphate and phosphoenolpyruvate is similar for all the aromatic amino acids. Phenylalanine and the other aromatic amino acids are precursors of a large number of aromatic compounds, mostly found only in plants. Deamination of phenylalanine yields cinnamic acid, important in the synthesis of lignin. Phenylalanine is also a precursor of such alkaloids as morphine and curare (i.e. the isoquinoline alkaloids).

phialospore A type of conidium found, for example, in many of the Eurotiales and Hypocreales. Phialospores develop at the tips of specialized finger-like cells termed *phialides*.

phloem (bast) A vascular tissue whose principal function is the translocation of sugars and other nutrients. The phloem is composed mainly of *sieve tubes, sclerenchyma cells, and parenchyma cells, including *companion cells. It occurs in association with, and usually external to, the *xylem. *See also* primary phloem, secondary phloem.

phloroglucin A temporary stain that is used to detect the distribution of lignin (which it turns magenta red) in thin sections of plant tissue. It is usually acidified with hydrochloric acid before use.

phosphatide *See* phosphoglyceride.

phosphoenolpyruvate (PEP) An important phosphorylated intermediate of glucose metabolism. In glycolysis PEP is the immediate precursor of pyruvic acid, while in the synthesis of glucose from acetyl-CoA, PEP is formed from oxaloacetate generated in the *glyoxylate cycle. In C_4 plants PEP is an intermediate in the Hatch–Slack pathway. The aromatic amino acids, and hence a large number of other aromatic nitrogenous compounds, are also synthesized from PEP.

phosphoenolpyruvate carboxylase (PEP carboxylase) The enzyme that is responsible for carbon dioxide fixation in *C_4 plants and in plants exhibiting *crassulacean acid metabolism. It catalyses the carboxylation of phosphoenolpyruvate to oxaloacetic acid.

phosphogluconate pathway *See* pentose phosphate pathway.

phosphoglyceric acid (PGA) The phosphorylated derivative of the three-carbon glyceric acid. It may be phosphorylated at the second or third carbon giving 2-phosphoglyceric acid or 3-phosphoglyceric acid respectively. 3-phosphoglycerate is the first product of the dark reactions of photosynthesis, two molecules being formed by the carboxylation and cleavage of ribulose bisphosphate. It is also an intermediate in glycolysis being formed from 3-phosphoglyceroyl phosphate, two molecules of ATP being formed in the process. It is then converted, via 2-phosphoglycerate, to phosphoenolpyruvate. 3-phosphoglycerate is the precursor of the amino acid serine.

phosphoglyceride (phospholipid, phosphatide, glycerophosphatide) Any of a group of complex lipids similar to acylglycerides except that the third hydroxyl group is phosphorylated, i.e. the backbone is glycerol 3-phosphate rather than glycerol. Common phosphoglycerides include phosphatidylcholine, phosphatidylethanolamine, and phosphatidylinositol (all found primarily in seed tissues), and phosphatidylglycerol (found in leaf tissue).

Phosphoglycerides are the major lipids in most plant membranes, although *glycolipids are important components of chloroplast membranes.

phospholipid *See* phosphoglyceride.

phosphorus Symbol: P. A nonmetallic element, atomic number 15, atomic weight 30.9, important as a macronutrient for plant growth. It is essential for the formation of nucleic acids, lipoproteins, and energy-carrying molecules and is involved in many aspects of metabolism, especially carbohydrate metabolism. Phosphorus occurs in soil minerals as the apatites and as calcium phosphate rock. It is absorbed by roots as the PO_4^{3-} ion, this ion acting as an important buffer in the cell sap. Deficiency results in poor root growth, bluish, bronzed, or purple leaves, and poor germination, ripening, and seed set. Phosphorus can be applied to soils as inorganic basic slag or superphosphate or in various organic materials, guano and bonemeal containing a particularly high percentage. It is often applied in combination with nitrogen and potash in compound fertilizers.

phosphorylase *See* kinase.

phosphotransferase *See* kinase.

photic zone The surface waters of seas, oceans, and lakes penetrated by light and inhabited by plankton. In clear water light at the red end of the spectrum is absorbed and only blue light penetrates to lower levels. However in turbid water blue light is preferentially absorbed. The depth at which the rate of production of carbohydrates by photosynthesis is equalled by the rate of utilization by respiration is termed the compensation level. Most of the plankton is found in the top 10 m.

photoautotroph An *autotroph that obtains the energy for synthesizing organic food substances from inorganic components directly from sunlight. Photoautotrophs include most green plants and some photosynthetic bacteria. *Compare* photoheterotroph.

photoheterotroph A photosynthetic organism that requires a supply of organic compounds as a source of hydrogen. The purple nonsulphur bacteria are obligate photoheterotrophs and also require certain organic growth substances. Some of the purple sulphur bacteria are facultative photoheterotrophs in that under certain conditions they use organic acids rather than reduced sulphur compounds as a hydrogen source. *See also* heterotroph.

photolysis of water The photosynthetic splitting of water into gaseous oxygen and reducing equivalents. Two molecules of water are split to produce one molecule of oxygen, four electrons, which go through the electron transport chain, and four protons. The electrons and protons eventually reduce NADP to NADPH. This in turn is utilized in the reduction of carbon dioxide to carbohydrate.

photomicrograph *See* micrograph.

photonasty A *nastic movement induced by external light stimuli. Some flowers, e.g. those of the garden marigold (*Calendula officinalis*), exhibit these movements and open on exposure to light.

photoperiodism The alternation of light and dark periods that affects the physiological activities of many plants. As the day length changes through the year, various responses are shown by plants, depending on whether they are *long-day or *short-day species. An example is the onset of flowering, triggered in a particular plant by a specific photoperiod. The length of the light period is more important than the intensity of the light. It has also been shown that the effect of a long dark period can be nullified by a brief period of light. *See also* phytochrome.

photophile Describing a phase during which light promotes flowering. In one theory advanced to explain photoperiodism it is suggested that short-day plants are in the photophile phase during the day but at night they are in the *photophobe* (or *skotophile*) phase when light inhibits and darkness promotes flowering. The alternation between photophile and photophobe phases is seen as a type of *circadian rhythm. If the period of daylight extends beyond the photophile phase into the photophobe phase, as would occur with a short-day plant on a long day, then flowering is inhibited.

photophobe *See* photophile.

photophosphorylation (photosynthetic phosphorylation) The production of ATP from ADP and inorganic phosphate utilizing light-induced photosynthetic electron transport as a source of energy. Two types of photophosphorylation occur, noncyclic and cyclic. In noncyclic photophosphorylation electrons are passed from water to $NADP^+$ via *photosystems I and II. During the transfer of electrons from the primary electron acceptor of photosystem II (termed C550) to P700 through plastoquinone, plastocyanin, and various cytochromes, one molecule of ATP is formed. In cyclic photophosphorylation only photosystem I is involved. Energy absorbed by P700 releases electrons, which are absorbed by a primary acceptor termed P430. Electrons then pass, via cytochrome b_6, into the chain connecting the primary electron acceptor of photosystem II to photosystem I. A molecule of ATP is formed as the electron passes down to P700. Cyclic phosphorylation does not produce reduced $NADP^+$.

photoreceptor Any light-sensitive region of a plant. The *paraflagellar body*, visible as a swelling on the flagellum of some Euglenophyta, is an example.

photorespiration Respiration that occurs in plants in the light. It differs from dark respiration in that it does not occur in the mitochondria and is not coupled to oxidative phosphorylation. The rate of CO_2 release by photorespiration in C_3 plants can be three to five times greater than that released by dark respiration. Since the process does not generate ATP it appears to be extremely wasteful. It has been estimated that photosynthetic efficiency could be improved by 50% if photorespiration were inhibited. In C_4 plants photorespiration is hardly detectable, possibly because synthesis of *glycolic acid, the substrate for photorespiration, is much lower in C_4 plants (about 10% of that of C_3 plants). This could be because the concentration of CO_2 in the bundle sheath cells is so high that oxidation (instead of carboxylation) of ribulose bisphosphate is prevented.
Some C_3 plants, e.g. the grass *Panicum miliodes*, have a leaf structure comparable in some respects to the Kranz structure of C_4 plants. Such plants also tend to have reduced rates of photorespiration. These observations have stimulated research into the possibility of breeding C_3 plants with slower photorespiration rates.

photosynthesis The sequence of reactions, performed by green plants and photosynthetic bacteria, in which light energy from the sun is converted into chemical energy and used to produce carbohydrates and ultimately all the materials of the plant. The photosynthetic reaction can be summarized as:

$$6CO_2 + 6H_2O + \text{light energy} \rightarrow C_6H_{12}O_6 + 6O_2$$

There are two distinct phases in photosynthesis, the *light (or light-dependent) reactions and the *dark (or light-independent) reactions. In green plants and blue-green algae the light reactions involve the photolysis of water, producing hydrogen atoms and molecular oxygen. This oxygen, given off during photosynthesis, is the main source of atmospheric oxygen, essential for aerobic organisms. The hydrogen atoms produced are used to reduce $NADP^+$ to NADPH and the energy released also forms ATP from ADP and inorganic phosphate (photophosphorylation). This ATP and NADPH is used up during the dark reactions in which carbon dioxide is fixed into carbohydrates (*see* Calvin cycle). *See also* photosystems I and II.

photosynthetic bacteria A group of phototrophic bacteria that contain a green pigment, *bacteriochlorophyll, similar to the chlorophyll of plants, that enables them to photosynthesize. They obtain hydrogen from hydrogen sulphide or other reduced sulphur compounds (or occasionally from molecular hydrogen or organic acids) and fix carbon dioxide by the Calvin series of reactions. There are three groups of photosynthetic bacteria: namely the green sulphur bacteria (Chlorobiaceae or Chlorobacteriaceae); the purple sulphur bacteria (Chromatiaceae or Thiorhodaceae); and the purple nonsulphur bacteria (Rhodospirillaceae or Athiorhodaceae). Together these form a suborder of the *Pseudomonadales. In the purple bacteria additional red and brown pigments are also present.

photosynthetic phosphorylation *See* photophosphorylation.

photosystems I and II (pigment systems I and II, PSI and PSII) Two photochemical systems containing photosynthetic and accessory pigments and electron carriers that operate in sequence to perform the two light reactions of photosynthesis and so bring about photo-

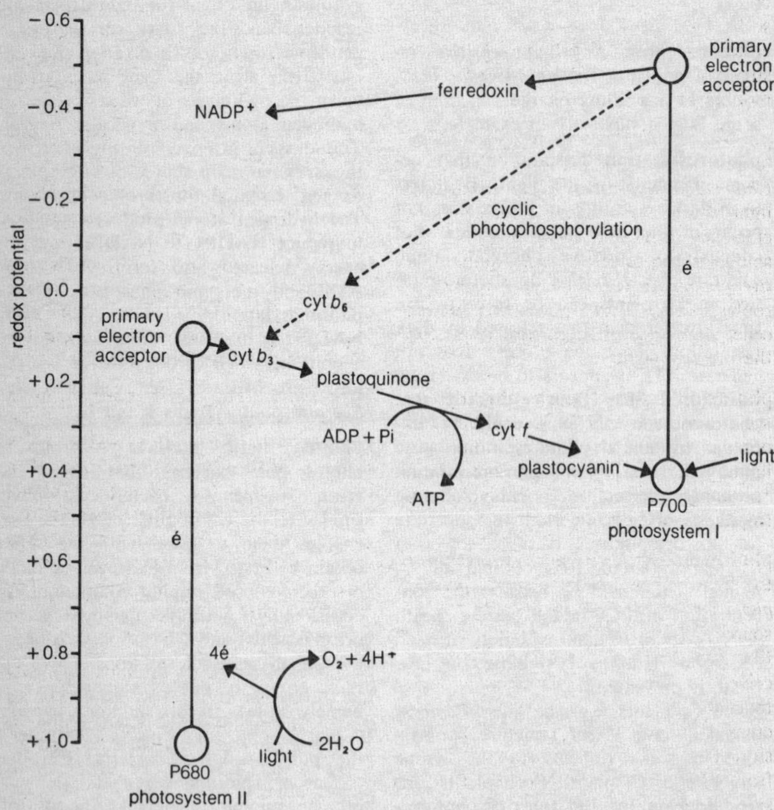

The Z scheme depicting electron transport between photosystems II and I.

phosphorylation and reduction of $NADP^+$ (*see* diagram). PSII contains chlorophyll *a* (*P680) that absorbs light of 684 nm wavelength most efficiently. When activated by light it produces a strong oxidant that oxidizes water to oxygen, hydrogen ions, and electrons. The electrons are transferred via a primary electron acceptor (possibly C550) and plastoquinone and the cytochrome chain to PSI. The energy released when each electron is transferred is used to transform a molecule of ADP to ATP. PSI contains chlorophyll *a* (*P700) that absorbs light of 700 nm wavelength most efficiently. When P700 is activated a strong reductant is produced that reduces $NADP^+$ to NADPH.

phototaxis A *taxis in response to external light stimulation. Cells of *Chlamydomonas* swim freely towards a light source to enhance photosynthetic efficiency but will swim away if the source is too intense (*positive* and *negative phototaxis* respectively). Some chloroplasts also show phototaxis within cells, aligning themselves with respect to the incident light.

phototroph An organism that derives the necessary energy for the synthesis of organic compounds directly from sunlight. The term encompasses both *photoautotrophs and *photoheterotrophs. *Compare* chemotroph.

phototropism A *tropism in response to light. Thus, shoots show *positive phototropism* by growing towards a light source. Work with oat (*Avena*) coleoptiles has shown that the stimulus is received in the tip of the seedling and that the response is effected by increased growth on the relatively shaded side, triggered by *auxins, which pass from the tip more readily on the shaded side. Removal of the tip destroys the phototropic response. Most roots are aphototropic.

phragmoplast A system of fibrils that persists at the outer edge of the equator of the spindle at early telophase. As the *cell plate forms in the equatorial region the fibrils become dispersed. The term is used by some authorities to mean the developing cell plate.

phycobilin (biliprotein) Any of the blue or red pigments found in the blue-green and red algae. Like the *carotenoids they are accessory pigments in photosynthesis, but unlike the chlorophylls and carotenoids they are water soluble. Structurally they are very similar to the porphyrin part of the *chlorophyll molecule, except that they contain no magnesium. The three classes of phycobilins are the *phycoerythrins, *phycocyanins, and allophycocyanins. *See also* phycobilisome.

phycobilisome A small particle, about 30 nm across, many of which are found on the membranes of the lamellar system of blue-green algae. Phycobilisomes are thought to be aggregations of the pigment *phycobilin. They are also found in the plastids of red algae between the photosynthetic lamellae.

phycobiont The algal partner in a *lichen, which is either a unicellular green or blue-green alga and often a member of the genus *Trebouxia.* If a blue-green alga (e.g. *Nostoc*, *Rivularia*, *Gloeocapsa*) is present the lichen is usually gelatinous. The number of algal genera involved in lichens (24 have so far been discovered) is relatively few compared to the large numbers of lichenized fungi. *Compare* mycobiont.

phycocyanin Any of the blue pigments that form one of the two major classes of the *phycobilins. They are found in all blue-green algae and in some red algae and act as accessory pigments in photosynthesis.

phycoerythrin Any of the red pigments that form one of the two major classes of the *phycobilins. They are found in all red algae, and act as accessory pigments in photosynthesis. They absorb the dim blue-green light that reaches the lower levels of the ocean where the algae grow, and pass it onto chlorophyll so that it may be used in photosynthesis. The deeper in the sea an alga lives, the more phycoerythrin it contains in relation to chlorophyll.

phycology The study of algae.

Phycomycetes In older classifications, a class containing all the relatively unspecialized true fungi, i.e. all true fungi except the ascomycetes, basidiomycetes, and imperfect fungi. It is now recognized that the Phycomycetes included a number of only distantly related groups that are better considered as separate classes (*see* Chytridiomycetes, Oomycetes, Zygomycetes) or subdivisions (*see* Mastigomycotina, Zygomycotina). The name is also misleading as it implies a relationship with the algae, which the evidence to date suggests is unlikely.

phyletic *See* phylogenetic.

phyllid A moss or liverwort 'leaf'.

phylloclade *See* cladode.

phyllode A more or less flattened petiole resembling and performing the functions of a leaf. Vertically expanded petioles are seen in a number of *Acacia* species. The leaves of many monocotyledons are believed to be of phyllode origin. *Compare* cladode.

phyllody The transformation of parts of a flower into leaflike structures. This may occur in response to some infections. For example, maize infected with head smut (*Sphacelotheca reiliana*) develops phyllody of both male and female flowers.

phyllotaxis (phyllotaxy) The arrangement of leaves on a stem. In most plants the leaves are inserted singly and spirally on the stem. Sometimes two or more leaves are formed at each node giving an opposite or whorled arrangement (*see* illustration). The phyllotaxis of a given plant may be described by giving the angle of divergence between successive nodes. This varies between species but if many primordia are being formed it is normally not less than 137.5°, the so-called *Fibonacci angle.*

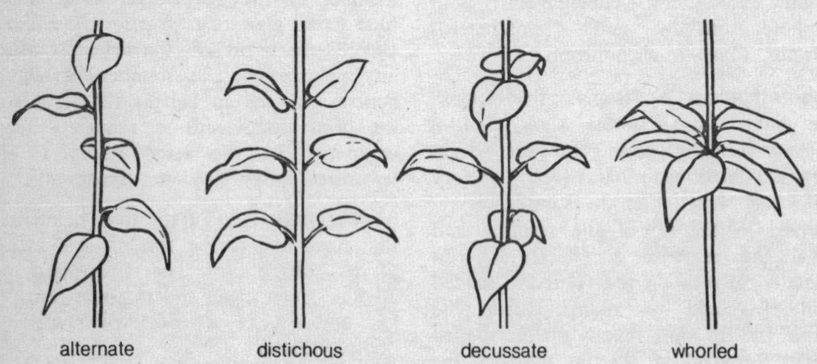

Types of phyllotaxis.

Phyllotaxy may also be described by the *genetic spiral or by *parastichies. Many theories have been advanced to account for the regular nature of phyllotaxis (*see* available space theory, repulsion theory).

phylogenetic (phyletic) Describing the study of the stages in the evolutionary history of groups of organisms. It is based on a study of the fossil record, comparative anatomy, etc., and provides a basis for classification. *Compare* phenetic. *See also* cladistics.

phylogeny The evolutionary history of an organism. A *phylogenetic classification aims to reflect this.

phylum The zoological equivalent of *division, representing the category of highest magnitude within the kingdom. Phylum has recently been used in favour of division by some botanists.

physiographic factor Any of the factors, apart from *climatic, *biotic, and *edaphic factors, that affect the prevailing conditions within a habitat and the distribution of the plants and animals. Such factors include the topography of the area, altitude, drainage conditions, degree of erosion, slope of the land, etc.

physiological drought A drought condition suffered by plants that occurs despite there being sufficient water in the soil. It may occur when the concentration of solutes in the soil water is equal to or higher than that in the root cells so water cannot enter the plant by osmosis. This situation may be found in salt marshes and coastal mud flats. It also occurs during cold spells when there is an increase in the resistance of the root to water movement into the plant. Root resistance increases in response to low temperatures since the permeability of endodermal cells decreases rapidly below 5°C.

physiological race (forma specialis) A population that is physiologically distinct but morphologically indistinguishable from other members of the species. Many fungi have large numbers of physiological races, reflecting considerable genetic diversity, even in nonsexual species. Examples are the various races of certain pathogenic fungi, each of which attack different host species or varieties.

physiology The study of the processes and functions associated with life.

phytic acid The hexaphosphoric ester of *inositol. It is common as a phosphate storage compound in many seeds in which it exists as the calcium–magnesium salt, *phytin*.

phytoalexin A chemical produced by a host plant that inhibits the growth of a pathogenic fungus. For example, the phenolic compound pisatin has been shown to accumulate in pea tissues in the presence of various fungi. Similarly the degree of resistance shown by sweet potato (*Ipomoea batatas*) to attack by the fungus *Ceratostomella* has been correlated with the concentration of the terpenoid ipomeamarone in the plant tissue.

phytobenthos *See* benthos.

phytochrome A protein pigment that mediates in photoperiodic responses and certain other photoreactions, e.g. light-stimulated germination and the removal of the symptoms of etiolation. It exists in two interchangeable forms, P_r, which absorbs in the red part of the spectrum (660 nm), and P_{fr}, which absorbs in the far red (730 nm). Following exposure of a plant to red light P_r changes to P_{fr}, while after exposure to far-red irradiation P_r is reformed from P_{fr}. This reversion of P_{fr} to P_r may also occur in the dark in some plants, a process that is inhibited by low temperatures. P_{fr} may also be lost by decay.

In practice an illuminated plant receives a mixture of red and far-red light and the proportion of P_r to P_{fr}, and hence the response of the plant, will depend on the relative proportions of far-red and red light. Sunlight has a high proportion of red light as compared to far-red and so promotes formation of P_{fr}. Fluorescent light is virtually devoid of far-red light and is thus even more effective in promoting P_{fr} formation. Incandescent lights however have a relatively high proportion of far-red wavelengths, and sunlight that manages to penetrate the shade has had most of the red light absorbed by the chlorophyll in the canopy. Far-red wavelengths consequently predominate and most of the phytochrome of shaded plants will be in the P_r form.

Part of the phytochrome molecule consists of a nonprotein portion, the chromophore. This is similar to the phycobilins of blue-green and red algae. It is thought that the change from P_r to P_{fr} and vice versa is brought about by a shift of two hydrogen atoms in the chromophore. This property of the molecule explains the observation that red and far-red light are mutually antagonistic in their effects, the response of the plant often depending on the last type of radiation received. If the last exposure is to red light then light-requiring seeds will germinate and etiolated plants will begin normal growth. However if the last exposure is to far-red these responses will not occur. This implies that P_r is the inactive form and P_{fr} the active form of phytochrome. The reversal effect of far-red light is only seen if the far-red is given soon after the red light. This is especially true in the case of rapid light-mediated responses, e.g. leaf movements in *Mimosa pudica*.

In experiments on short-day plants given a short period of light during the dark period it was found that red light was effective in inhibiting flowering but far-red irradiation reverses this. The inhibitory effectiveness of red light increases when given towards the end of the dark period being most effective when applied shortly before completion of the critical number of hours. In long-day plants a short period of red irradiation in the dark period promotes flowering.

Phytochrome is found in very low concentrations but is extremely sensitive even to very short flashes of weak red light. Thus exposure of an etiolated plant to moonlight can induce normal growth. It is uncertain how these responses are brought about though it has been noted that gibberellin and cytokinin levels increase following exposure to red light as do the levels of certain enzymes. The transformation of phytochrome may bring about conformational changes in the molecule that expose an active site in the P_{fr} form that is able to bind to membranes and change their function.

phytogeography *See* plant geography.

phytohormone *See* growth substance.

phytoplankton *See* plankton.

phytotron A large plant growth chamber in which many plants can be raised under strictly controlled conditions of light, photoperiod, temperature, humidity, etc. Often many phytotrons are housed in a specially designed building. Entrance to each unit is by a hermetically sealed door.

pigment Any coloured compound. Pigments are coloured because they absorb certain types of light. Thus in solution they will have a characteristic *absorption spectrum. Because light is a form of energy, pigments that absorb light also absorb energy. The first reaction in *photosynthesis is the absorption of energy by accessory and photosynthetic pigments. Pigments also produce the characteristic colours of flowers and fruits.

pigment systems I and II *See* photosystems I and II.

pileus The cap of a mushroom or toadstool. The upper surface is often covered in flaky scales, which are the remnants of the universal veil. The undersurface is composed of gills or pores.

piliferous layer (root-hair zone) The absorbing region of the root *epidermis, covered with *root hairs. It is situated about 4–10 mm from the root tip (beyond the zone of elongation). In young plants this region is responsible for most of the plant's water uptake. However in larger older plants and in conditions of water stress a greater proportion of water moves in across the suberized regions of the root further back from the root tip.

pilose Having soft long hairs. *Compare* pubescent.

pilus (fimbria) A projection from the surface of a bacterium. It is finer than a flagellum and also differs in having a hollow core. Pili are often found in large numbers and are thought to help attach bacteria to other cells rather than serve a locomotive function. Pilus formation may be initiated by a plasmid, in which case it serves as a conjugation tube for the transmission of a copy of the plasmid to a suitable recipient cell. Plasmids with this ability are termed sex factors or F factors.

pin *See* heterostyly.

Pinales *See* Coniferales.

Pinicae *See* Pinophyta.

pinna One of a number of first order leaflets in a compound leaf, such as is typical of many ferns. *Compare* pinnule.

pinnate Describing a compound leaf in which the leaflets (*pinnae*) are arranged in two rows, one on each side of the midrib. Such leaves are common in the Leguminosae. *See also* imparipinnate, paripinnate.

pinnatifid (pinnatisect) Deeply cut into lobes but not so far as the midrib. Examples are the pinnatifid leaves of the fen sowthistle (*Sonchus palustris*).

pinnule One of a number of second order leaflets in a compound leaf, such as is typical of many ferns, where the *pinnae are themselves divided into leaflets.

pinocytosis *See* endocytosis.

pinocytotic vesicle *See* endocytosis.

Pinophyta In certain classifications, a division containing the gymnosperms (*see* Gymnospermae). It is divided into the following subdivisions: Cycadicae (cycadophytes), containing the orders Pteridospermales, Caytoniales, Bennettitales, and Cycadales; Pinicae (coniferophytes), containing the orders Ginkgoales, Cordaitales, Coniferales, and Taxales; and the Gneticae (gnetophytes), containing the orders Ephedrales, Welwitschiales, and the Gnetales. The Cycadicae are differentiated from the Pinicae on the basis that they have palmlike compound leaves, motile antherozoids, radially symmetrical seeds, and loose soft wood. The Pinicae in contrast have simple leaves, nonmotile sperm nuclei (except in *Ginkgo biloba*), bilaterally symmetrical seeds, and denser wood. The Gneticae resemble the Pinicae in having simple leaves, nonmotile sperm, and flattened seeds. However they differ from both the Pinicae and Cycadicae in possessing ovules with two integuments, rather than one, the inner of which is elongated into a tubular micropyle, and in having compound microstrobili.

pioneer community *See* succession, ecesis, sere.

piperidine alkaloids A group of alkaloids, the structures of which are based on the piperidine ring (a six-membered ring with no double bonds and containing one nitrogen atom). Most are derived from the amino acid lysine. An example is nicotine found in many plants, especially *Nicotiana tabacum* (tobacco) and other *Nicotiana* species.

pistil A term used ambiguously to describe either a single carpel (simple pistil) or a group of fused carpels (compound pistil).

pistillate flower A flower possessing female parts (a pistil) but no male parts. *Compare* staminate flower. *See also* monoecious, dioecious.

pit A cavity in the secondary cell wall, allowing exchange of substances between adjacent cells. A pit consists of a *pit cavity* (the aperture in the secondary wall) and a *pit membrane* (the primary wall material adjacent to the cavity). Pits usually occur in pairs (*pit pairs*), the members of the pair being situated in adjacent cell walls. There are two main types of pit, *simple and *bordered. Though a pit in a secondary cell wall is distinct from a cavity in a primary cell wall (*primary pit field), it is often difficult to distinguish between the two and then the term *pitted* is used for either structure. *See also* plasmodesmata.

pith (medulla) A region of parenchymatous tissue found in the centre of many plant stems to the inside of the stele. A pith is also often seen in the roots of herbaceous plants but rarely in the roots of woody plants.

pitted thickening A type of secondary wall patterning in tracheary elements in which the secondary cell wall is more or less continuous, the continuity only broken by *pits and perforation plates (if present). The pits may be arranged in an *opposite* or *alternate* pattern (*see illustration at* tracheary element). If the pits are greatly elongated transversely and parallel to each other, they are said to be *scalariformly pitted.* Like *reticulate and *scalariform thickenings, pitted thickening permits little further extension and is therefore found in tissues that have finished elongating at maturity, such as metaxylem and secondary xylem. *Compare* annular thickening, spiral thickening.

placenta The tissue by which spores, sporangia, or ovules are attached to the maternal tissue. It is usually mostly undifferentiated but contains vascular tissue. *See also* placentation.

placentation The pattern of attachment of an ovule to the ovary wall by the *placenta. If there is only one ovule in the ovary then it is usually attached either at the base (*basal placentation) or the apex (apical placentation). In a simple ovary the ovules may be attached either along the ventral suture (*marginal placentation) or, rarely, all over the inner surface of the ovary wall (*laminate placentation). In a compound ovary the ovules may be attached either on a central axis or on the wall along the junctions of the carpels (*parietal placentation). Placentation on the central axis is termed *axile in a multilocular ovary and *free-central in a unilocular ovary. It is believed that free-central and parietal placentation are derived from axile placentation. The type of placentation is an important diagnostic character in many taxa.

plagioclimax *See* climax.

plagiotropism A *tropism in which the plant part is aligned at some angle to the direction of the stimulus. A plagiotropic response to gravity (*plagiogeotropism*) is exhibited by lateral shoots and roots. *Compare* orthotropism. *See also* diatropism.

plankton The large community of microorganisms that floats freely in the surface waters of oceans, seas, rivers, and lakes. They are moved passively by wind, water currents, or waves, having little or no powers of locomotion themselves. The plant plankton (*phytoplankton*) includes many microscopic algae, particularly the diatoms. They form the base of the food chain in water, being eaten by the animal plankton (zooplankton), which in turn provides food for fish. It has been estimated that 90% of photosynthetic activity is carried out by the phytoplankton. A feature of the plankton is the huge variation in composition at different times of year (*see* bloom).

Plantae The plant kingdom. In most classifications this is taken to include all the green plants (i.e. all organisms containing chlorophyll). In many classifications the fungi and bacteria are also included. Plants are distinguished from members of the animal kingdom (Animalia) by a number of factors. Most are autotrophic, making their food from inorganic starting materials by photosynthesis. Animals by contrast are heterotrophic. However the fungi, most bacteria, and certain parasitic higher plants are also heterotrophic. Plants are usually attached to a substrate and not able to move around freely like animals (this however does not apply to certain flagellate algae and fungi). Plants can generally only respond to external stimuli by growth movements. Response is consequently very slow as compared to animal movements and only occurs if the stimulus is prolonged. Most plant cells are surrounded by cellulose cell walls and starch is a common storage polysaccharide. Animals do not have cellulose cell walls and carbohydrates are commonly stored as glycogen. Perennial plants tend to grow indefinitely while in animals increase in size usually ceases at maturity.
The Plantae has been variously divided into subkingdoms and divisions. Some systems recognize two subkingdoms, the Thallophyta and Embryophyta. *See also* algae, fungi, bacteria, Bryophyta, Tracheophyta.

plant breeding The improvement of plants for agricultural, horticultural, or medical purposes. Probably all cultivated plants are the result of selection practices that originated some 10 000 years ago. Modern breeding techniques include artificial control of pollination, the generation of variability by artificial hybridization and mutation, and selection procedures such as *pedigree breeding and *backcrossing.

plant geography (phytogeography) The study of the geographical distribution of plants and their interrelationships with one another and with the environment. Many aspects overlap with the science of ecology but plant geography places more emphasis on the influence of the environment. *See also* vicariance.

plaque A clear area in a plate of bacteria caused by the lysis of bacterial cells following infection by a bacteriophage.

plasmagel *See* cytoplasm.

plasmagene A gene present in any structure other than a chromosome in the nucleus. The term commonly refers to genes present in organelles such as mitochondria and chloroplasts. *See also* cytoplasmic inheritance.

plasmalemma *See* plasma membrane.

plasma membrane (plasmalemma) The outer layer of the protoplasm below the cell wall. Like most cell membranes it is formed by the orderly orientation of protein and phosphoglyceride molecules. Plasma membranes range from 7.5 nm to 10 nm in thickness and are composed of approximately 60% protein and 40% phosphoglyceride. Membranes of different species contain characteristic types of polar lipids in proportions that are probably genetically determined.
Davson and Danielli, in 1935, proposed that membranes are made up of a central region consisting of phosphoglycerides and an outer denser region composed of proteins. The phosphoglyceride molecules were believed to be arranged in two rows with their hydrophilic polar

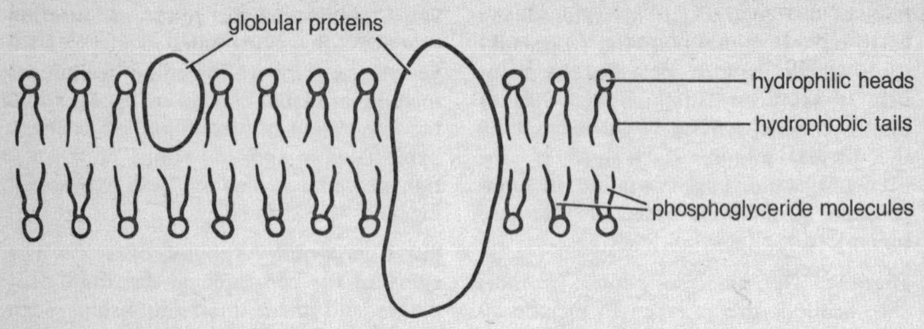

The proposed arrangement of phosphoglyceride and protein molecules in a cell membrane.

heads towards the outer edges and their hydrophobic hydrocarbon tails in the centre. Although it is still accepted that the two rows of phosphoglyceride molecules form the backbone of the membrane it is believed that globular proteins, rather than forming a distinct outer layer, actually penetrate the whole width of the membrane in places (*see* diagram). Both the phosphoglyceride molecules and the proteins are thought to be able to move laterally giving the membrane fluid-like properties. This hypothesis of membrane structure is termed the *fluid–mosaic model*.
Plasma membranes are selectively permeable, controlling the passage of materials into and out of the cell. The proteins of the membrane include enzymes and compounds of the active transport system. Water and nonpolar molecules that dissolve in the phosphoglyceride layer pass readily through the membrane. The membrane is relatively impermeable to charged ions, which enter the cell by means of the active transport system. Permeability varies in different parts of the plant and at different stages of development. Time-lapse photography of living membranes reveals almost constant movement and it is probable that there is continual replenishment of membrane constituents.

plasmasol *See* cytoplasm.

plasmid Any small autonomously replicating piece of DNA found in the cytoplasm of bacteria. Examples of plasmids are the *R factors* (R = resistance), which can carry genes conferring antibiotic resistance. Some plasmids may reversibly insert themselves into the main chromoneme (bacterial chromosome), in which case they are called *episomes*. Some episomes are best regarded as bacterial viruses. Another example of an episome is the sex factor, or *F factor*. Possession of the F^+ factor confers maleness upon a bacterium, as evidenced by the ability to transmit DNA through a sex pilus (conjugation tube) to a female (F^-) bacterium.

plasmodesmata Strands of cytoplasm that pass through *cell walls and connect the *protoplasts of adjoining cells. They may be concentrated in pit pairs or may be distributed throughout the walls. They allow the passage of materials between cells.

Plasmodiophoromycetes A class of the *Myxomycota containing fungi that are obligate parasites in the cells of other plants, in which they often induce hypertrophy. It contains about 35 species in some 10 genera. Two serious parasites of crop plants are *Plasmodiophora brassicae*, which causes clubroot in brassicas, and *Spongospora subterranea*, which causes powdery or corky scab in potatoes. These fungi are often classified as an order (Plasmodiophorales) of the Myxomycetes, or are sometimes classified with the *Mastigomycotina.

plasmodium The motile multinucleate amoeba-like mass of protoplasm that makes up the thallus of fungi in the *Myxomycetes. It develops from the zygote and in favourable conditions forms sporangia. In adverse conditions it may change into a dry resting body, the sclerotium.

plasmogamy The fusion of the protoplasts of two haploid cells during *sexual reproduction. Plasmogamy is usually followed immediately by karyogamy (nuclear fusion), but in some ascomycetes and basidiomycetes the two processes are separated so that binucleate (*dikaryotic) mycelia are formed. *See also* somatic hybridization.

plasmolysis The withdrawal of the cytoplasm from the cell wall because of the outward movement of water from the cell vacuole due to *osmosis. The cytoplasm eventually forms just a small central mass still enclosing a vacuole. Plasmolysis is seen when tissues are placed in solutions of lower water potential than that of the cell. By varying the concentration of external solutions and observing when a solution is just strong enough to induce plasmolysis in 50% of the cells the concentration of solutes in the cell vacuole can be determined. When working with single cells, the concentration that brings about *incipient plasmolysis*, i.e. the first signs of plasmolysis, is taken as being equal to the concentration of the cell vacuole. Plasmolysis normally only occurs in experimental systems where the external solution can pass through the cell to fill the space between the cell wall and the cytoplasm. Thus the cells of a wilted plant do not undergo plasmolysis.

plasticity The capacity of an organism to change its form in response to varying environmental conditions. For example, a plant transferred outside from a warm poorly lit greenhouse may produce smaller paler leaves and shorter internodes. Such changes can only occur in new tissues and the existing mature parts of the plant remain unchanged. In taxonomic work it is important to establish how plastic a particular character can be before placing too much reliance on it. It is also necessary in breeding work to establish what proportion of variation is caused by phenotypic plasticity.

plastid An organelle found in the cytoplasm of the majority of plant cells. Plastids are surrounded by a double membrane and show a wide variety of structure, ranging from minute *proplastids, less than 1.0 μm in diameter, to pigmented *chromoplasts, 10 μm long with a complex internal arrangement of lamellae. These give colour to plant tissues. Plastids are generally interconvertible, one type being derived from another and capable of differentiating into at least one other form. Many plastids, e.g. amyloplasts and *elaioplasts, are storage organelles. *Chloroplasts, the most important plastids, contain chlorophylls, carotenoids, electron transfer compounds, and enzymes, providing a highly organized and efficient photosynthetic organelle.

plastid inheritance *See* plastogene.

plastocyanin One of the electron transfer components of chloroplasts, mediating transfer of electrons from photosystem II to photosystem I. It is a blue copper-containing protein, and it is the reduction and oxidation of the copper atom that allows plastocyanin to accept, then donate, electrons. The position of plastocyanin in the electron transport chain is thought to be very close to photosystem I, probably between cytochrome *f* and photosystem I.

plastogene A *plasmagene present in a chloroplast or other plastid. The transmission of characteristics by such genes is termed *plastid inheritance*. *See* cytoplasmic inheritance.

plastoglobuli Droplets of lipid, 10–500 nm in diameter, found singly or in groups in the stroma of chloroplasts. They do not have an enclosing membrane but nevertheless retain their form when isolated from the stroma. Plastoglobuli from chloroplasts of actively photosynthesizing leaves contain various lipid and lipophilic compounds, e.g. galactolipids, quinones, and polyisoprenols, but no chlorophylls or carotenoids. As leaves become senescent, chloroplasts change to *chromoplasts and changes in the plastoglobuli occur. They accumulate carotenoid pigments and enlarge considerably until they become the dominant features in the chromoplasts.

plastoquinone One of the electron transfer components of chloroplasts, mediating transfer of electrons from photosystem II to photosystem I. It is situated between cytochrome b_3 and cytochrome *f* and the transfer of electrons to cytochrome *f* is accompanied by the formation of a molecule of ATP.

platyspermic Describing the flat bilaterally symmetrical seeds characteristic of the Cordaitales. *Compare* radiospermic.

Plectascales *See* Eurotiales.

plectenchyma A form of 'tissue', commonly found among the higher fungi, composed of a mass of interwoven anastomosing hyphae. It is termed *prosenchyma* when formed from long fused hyphae and *pseudoparenchyma* when it has a cellular appearance due to regular divisions in the hyphae. Pseudoparenchymatous tissue is also seen in the thalli of certain red algae.

Plectomycetes A class of the *Ascomycotina containing those ascomycetes that produce a *cleistothecium. It numbers about 2300 species in some 160 genera. It is not a natural class as it contains both fungi with single-walled asci and others with double-walled asci. Also some are primitive while others are regressive. It contains the orders Ascophaerales, *Eurotiales, *Microascales, *Erysiphales, and Meliolales.

plectostele A *protostele in which the xylem consists (in transverse section) of several plates of tissue surrounded by phloem. This type of stele is exhibited in some species of *Lycopodium* and is thought to have evolved from an actinostele. *See illustration at* stele.

Pleiocene *See* Tertiary.

pleiomorphic *See* pliomorphic.

pleiotropism The control of several apparently unrelated characteristics by a single gene. Thus in tobacco a single gene is responsible for long anthers, calyces, capsules, and petioles. *Compare* polygenes.

Pleistocene *See* Quaternary.

pleomorphic *See* pliomorphic.

plerome *See* histogen theory.

plesiomorphy A primitive character state whose origin can be traced back to a remote ancestor. Plesiomorphies shared by different taxa (symplesiomorphies) are not used in the construction of cladograms or phylogenies. *Compare* apomorphy. *See* cladistics.

pleurocarpous Describing mosses in which the reproductive organs are produced laterally and the main axis is usually creeping. *Compare* acrocarpous.

Pleuromeiales *See* Lycopsida.

pleuropneumonia-like organisms *See* Mycoplasmatales.

plicate Arranged in or having folds as, for example, the mesophyll cells of many gymnosperms and certain angiosperms, which have inward foldings of the cell wall.

Pliocene *See* Tertiary.

pliomorphic (pleomorphic, pleiomorphic) Describing organisms that exhibit two or more different forms in their life cycle. For example, many ascomycetes have two conidial states.

plumule The embryonic shoot, derived from the *epicotyl. In dicotyledons the plumule is situated between the cotyledons. If germination is epigeal the plumule is protected during its passage to the soil surface by the cotyledons. If germination is hypogeal the elongating plumule has a hooked tip.

plurilocular sporangium A sporangium that is divided by septa into many compartments. Plurilocular sporangia are seen, for example, on lateral branches of the diploid generation in the brown alga *Ectocarpus*. *Compare* unilocular sporangium.

pneumatophore (aerophore, breathing root, respiratory root) An erect root that protrudes some distance above soil level. Pneumatophores are formed in large numbers by certain plants, e.g. *Sonneratia* and some mangrove species, growing in areas with waterlogged badly aerated soils. The surface of the pneumatophore is perforated by numerous lenticels that promote gaseous exchange.

Poaceae *See* Gramineae.

pod *See* legume.

podetium A specialized spore-bearing structure that develops on the thalli of certain lichens, e.g. the erect cylindrical podetia of *Cladonia*.

podsol (podzol) A type of zonal acid infertile *soil (a *pedalfer) found in regions of heavy rainfall and long cool or cold winters. The soils of the northern coniferous forest zone of North America and Eurasia are typical examples. The soil is heavily leached by the sudden snow-melt in the spring that washes out lime and iron compounds. Thus beneath the humus layer in the A horizon there is a bleached horizon composed of quartz sand. Iron compounds accumulate in the B horizon, which stains brown and has a clayey texture. These compounds may form an impermeable *hard pan* that, if developed sufficiently, can stop water percolating from above, resulting in the A horizon becoming waterlogged. Such soils may be termed *gley podsols*. The leaf litter is such regions also tends to be highly acid and maintains the acidity of the soil.

Poisson distribution The type of distribution of data characterized by having a variance equal to its mean. It occurs when the probability of an event occurring on any particular occasion is extremely small but the very high number of occasions recorded makes it likely that the event will occur reasonably often overall. An example is the measurement of the density of a particular plant species using quadrat counts. The likelihood of that species occurring in any particular square of the quadrat is small but if sufficient quadrats are thrown the species will be recorded eventually.

polarity The condition that results from the establishment of a definite orientation during the differentiation of a cell, tissue, or organ. Polarity is evident in the early growth of plants, as in the bipolar development of an embryo from the zygote. In later growth it is evidenced in the separate development of roots and shoots, as well as various phenomena at the cellular level.

polarizing microscope A microscope that uses polarized light to illuminate the specimen. Certain crystalline molecules transmit polarized light in a way that depends on the orientation of the molecules in the crystal. The technique has been used, for example, to investigate the arrangement of cellulose molecules in plant cell walls.

polar nuclei The two haploid nuclei found in the centre of the *embryo sac after division of the megaspore. They may fuse to form a diploid definitive nucleus before fusing with the male gamete to form the triploid primary endosperm nucleus. *See also* double fertilization.

pollarding A severe form of pruning in which all the younger branches of a tree are cut back virtually to the trunk. It encourages new bushy growth and is often performed on urban avenues of trees to provide summer shade. The tips of the branches of pollarded trees are characteristically club shaped due to the growth of wound wood. *Compare* stooling.

pollen The *microspores of seed plants, which are produced in vast numbers, usually in a *pollen sac. They are formed as a result of meiosis of somatic *pollen mother cells. In primitive forms they are formed in relatively unspecialized structures with little protection. In intermediate forms they are produced in specialized *stamens, which are numerous and develop symmetrically in the flower. In the most advanced forms, pollen is formed in fewer stamens, which are more strategically placed with regard to the pollination mechanism. Pollen structure may be related to methods of pollination. In insect-pollinated species the pollen is often sticky or barbed, whereas wind-pollinated species usually produce light smooth pollen.

Pollen grains contain concentrated mitochondria, endoplasmic reticulum, and golgi apparatus. The number of nuclei present in the cell at any given time may be diagnostic. On germination of the pollen grain a pollen tube pushes its way through an aperture in the pollen grain wall and the various nuclei migrate into the pollen tube (*see also* generative nuclei, vegetative nucleus).

The pollen grain wall consists of a resistant outer *exine, which may be very highly sculptured, and an inner *intine, which may protrude through pores in the outer layer. These pores, apart from facilitating germination, may also be instrumental in water regulation and in controlling compatibility systems. Although most pollen is short lived, some pollen may be stored at low temperatures for long periods. *See also* palynology.

pollen chamber A cavity at the micropylar end of the nucellus in some gymnosperms in which pollen grains lodge after pollination. The pollen may be immature at this stage but it ripens in the pollen chamber prior to germination and fertilization of the egg. *See also* pollination drop.

pollen mother cell (PMC) A somatic cell that, after meiosis, forms a *tetrad of pollen grains. Many are found closely packed within the pollen sacs of angiosperms and gymnosperms. In angiosperms the pollen mother cell and the resulting pollen grains have been shown to obtain proteins and their precursors from the disintegrating tapetum that surrounds them. *Compare* megaspore mother cell. *See also* spore mother cell.

pollen sac A chamber in which *pollen grains (microspores) are formed in the angiosperms and gymnosperms. It is homologous with the *microsporangia of the pteridophytes. In angiosperms there are usually four pollen sacs in an anther, arranged in two lobes, either side of the connective tissue. In the conifers there may be numerous pollen sacs formed on microphylls in the axils of the male strobilus. *Compare* ovule. *See also* stamen.

pollen tube An outgrowth of the intine of the pollen grain that, on germination, emerges through an aperture in the exine and grows towards the egg, carrying the male gametes with it. It represents the reduced male gametophyte in seed plants. In angiosperms pollen tube growth in compatible stigmatic tissue is usually rapid and may reach 1–3 mm/h. In gymnosperms growth is arrested at the nucellus and may not recommence until the next growing season. In *Cycas* and *Gingko* the gametes are motile. In other gymnosperms and in angiosperms the gametes are nonmotile and the mechanism of their movements down the pollen tube is not fully understood.
In angiosperms the pollen tube may reach the ovule from the stigma either by growing down the stylar canal or by enzymatically digesting its way between or through the individual cells of the style. It then grows through the ovular cavity and reaches the egg apparatus, usually via the micropyle but sometimes through the chalaza (*see* chalazogamy). There, the contents of the pollen tube are discharged. The *vegetative nucleus disintegrates and one of the gametes fuses with the egg cell to form the zygote and the other with the polar nuclei or definitive nucleus to form the endosperm.

pollination The transfer of pollen from the male reproductive organs to the female in seed plants. This involves transfer from the anthers to the stigma in angiosperms and from the microsporangiophores to the micropyle in gymnosperms. The process is usually effected by intermediary agents such as insects (*see* entomophily), wind (*see* anemophily), or water (*see* hydrophily). In flowering plants direct pollination may occur by gravity or contact where the stamens and stigmas are juxtaposed and mature simultaneously (*see* homogamy, cleistogamy).

pollination drop A drop of sugary fluid that is secreted through the micropyle in gymnosperms. Wind blown or insect transported pollen falls into the pollination drop and is drawn through the micropyle to the ovule as the drop is reabsorbed.

pollinium The structure formed when individual pollen grains remain massed together and are transported as a unit during pollination. They may be held together only by sticky secretions or they may be retained within the pollen sac wall, as in many orchids. When several pollen sacs remain together and are transported as a pollination unit they are called a *pollinarium.*

polyadelphous Describing stamen filaments that are fused into many groups.

polyandrous Having separate stamens freely inserted on the receptacle. *Compare* adelphous.

polycarpellary *See* fruit.

polycyclic stele A *stele in which the *meristeles or *vascular bundles are arranged in concentric rings. Polycyclic steles are present in some ferns. *See also* polystely.

polyembryony The formation of more than one embryo in an ovule. These may develop by division of the fertilized zygote (*see* cleavage polyembryony) and thus be a product of sexual reproduction. They may also arise, alongside a zygote, from somatic tissue (*see* adventive embryony). Embryos derived in this manner will have the same genetic constitution as the maternal parent. *See also* apomixis, parthenogenesis.

polygenes Several nonallelic genes all affecting the same character and approximately additive in their effects. Polygenic systems, also called *polymeric systems*, were first noted in red-kernelled maize. The intensity of the colour is governed by three unlinked genes and depends on the number of alleles for redness that are present. *Compare* pleiotropism.

polymeric system *See* polygenes.

polymorphism **1.** The existence of a number of different forms within a species, whether caused by genetic or environmental factors.

2. In population genetics, the existence of many different forms in a population at the same place and time, such that the frequency of the rarest form cannot be explained simply on the basis of recurrent mutation. Though mutation is the ultimate source of all genetic variation, in *stable* or *balanced polymorphism* the frequency of the rare mutant phenotypes is maintained because the recessive allele confers some advantage to the heterozygotes not possessed by either of the homozygotes. In *transient* or *unstable polymorphism*, one or more of the morphs (phenotypes) is eliminated or otherwise lost, so that the population tends towards *monomorphism* (one form).

polynucleotide *See* nucleotide.

polypeptide A *peptide chain containing a large number of amino acid residues (most polypeptides have a molecular weight greater than 5000). *Proteins are made up of one or more polypeptides.

The amino acid sequence of a polypeptide chain determines the three-dimensional conformation of that molecule under physiological conditions. This conformation can be disrupted by heating or extremes of pH (*see* denaturation). *See also* conjugated protein.

polypetalous Having separate petals freely inserted on the receptacle. *Compare* gamopetalous.

polyphyletic Describing taxa derived from two or more ancestral lines. Polyphyletic groups are thought by some to contain taxa whose resemblances are based on shared advanced (derived) characters that have arisen by convergent evolution (*see* cladistics). *Compare* monophyletic, paraphyletic.

polyploidy The condition in which an organism has three or more complete sets of chromosomes (*see* genome) in its nuclei. Polyploids originate when gametes containing more than one chromosome set fuse. Such gametes are formed when chromosomes fail to separate during anaphase 1 of meiosis. Consequently gametes are diploid instead of haploid, and fertilization results in triploid or tetraploid individuals. Polyploid individuals may be produced as a result of multiplication of chromosome sets from one species (*autopolyploidy) or by combining sets of chromosomes from different species (*allopolyploidy). In either case the polyploid offspring may be incapable of reproducing with their parents and so constitute new species. Polyploidy is common in flowering plants (40% of dicotyledons and 60% of monocotyledons are polyploid) and has probably contributed significantly to their evolution. In polyploid organisms harmful recessive alleles are more likely to be masked by normal dominant alleles. Polyploids also have a greater store of genetic variability and thus evolutionary potential is higher in polyploid populations.

In contrast to plants, polyploidy is extremely rare in animals because the sex-determining mechanism often depends on chromosome numbers and ratios. These are upset by polyploidy, and consequently polyploid animals are inviable or sterile. *See also* hybrid, triploid, tetraploid.

Polypodiales *See* Filicales.

Polypodiopsida *See* Filicinae.

Polyporales *See* Aphyllophorales.

polyribosome (polysome) A string of ribosomes attached together by a single molecule of messenger RNA. The strings may be folded or coiled into complex configurations, often spirals. The ribosomes move along the mRNA and as they pass each base triplet a transfer RNA molecule adds the corresponding amino acid to the base of a growing polypeptide chain. When a ribosome moves off the end of the mRNA it releases its completed polypeptide chain and is ready to start at the beginning again. This enables the code to be used highly efficiently.

polysaccharide (glycan) A high-molecular-weight polymer of monosaccharides or monosaccharide derivatives. The major functions of polysaccharides are as energy storage molecules (*reserve polysaccharides*) or as structural elements in cell walls and intercellular spaces (*structural polysaccharides*). *Starch and *cellulose are the most abundant plant polysaccharides.

Glucose is the most commonly occurring monosaccharide residue in polysaccharides; both starch and cellulose are made up exclusively of glucose subunits. Other sugars important in polysaccharides include galactose, mannose, fructose, xylose, and glucuronic and galacturonic acids.

Polysaccharides differ in the nature of their monosaccharide units, in the types of bonding between units, in chain length, and in degree of chain branching. Polysaccharides containing only one type of sugar residue are known as *homopolysaccharides*; these include *glucans* (glucose polymers), **mannans* (mannose polymers), **galactans*, and *fructans*. *Heteropolysaccharides* contain two or more different monosaccharides; examples are the *hemicelluloses and some *pectic substances.

polysepalous Having separate sepals freely inserted on the receptacle. *Compare* gamosepalous.

polysome *See* polyribosome.

polysomy The presence of several additional copies of one particular chromosome in a cell or organism. *See* aneuploidy.

polystely The condition of having a number of independent *steles. For example, the aerial axes of many *Selaginella* species have several protosteles ascending the stem, separated by the cortex.

polytene chromosome A chromosome with multistranded DNA due to repeated replication of the DNA without subsequent separation. The resulting *giant chromosomes* are valuable in cytological and genetical research. They are seen in the salivary glands of the larvae of dipteran insects, e.g. *Drosophila* (fruit flies).

pome A type of fleshy *pseudocarp in which the succulent tissues are developed from a greatly enlarged urn-shaped receptacle, which encloses the real fruit at its core. The pome is typical of the Rosaceae, the apple and pear being examples.

population A local community of potentially interbreeding organisms. In asexual organisms, the term normally refers to a local community of physiologically or morphologically similar individuals of the same species.

population dynamics The study of the changes in the numbers of individuals in a population and the attempted correlation of these with physical or chemical changes in the environment, biotic factors, or the genetical make up of the population. Such studies are carried out over a number of years as there may be seasonal and/or long-term changes. Most studies of plant population dynamics have been carried out on annuals rather than perennials. Factors that have been investigated include seed size, dormancy, thickness of seed coat, method of seed dispersal, leaf size, and stem height.

population genetics The study of the number, variety, and distribution of genes in a population or species, and of the factors that influence these. Population genetics has considerable implications for research interests as diverse as evolution, ecology, and plant breeding. *See also* Hardy-Weinberg law.

porate *See* aperturate.

pore (porus) A circular or slightly elliptic germinal aperture in a pollen grain. *Compare* colpus.

porogamy *See* chalazogamy.

porometer An apparatus designed to measure the resistance to the flow of air through a leaf in different external conditions. An example is Meidner's porometer, which consists of a Perspex clamp attached to a bulb pipette. The bulb is squeezed flat and the leaf is inserted into, and covered by, the clamp. The time taken for the air to pass through the leaf and inflate the bulb is proportional to the resistance of the leaf and gives an estimate of the degree of opening of the stomata.

porphyrin Any compound containing four pyrrole groups joined into a ring by methene (–CH=) groups between their α carbons. Porphyrins form an important group of pigments that includes the chlorophylls, cytochromes, and haemochromes.

position effect 1. A modification in the normal expression of a gene due to a change in its position on the chromosome following an inversion.

2. The determination of the appearance of an organism according to whether two nonallelic mutations of a single gene are in the **cis* arrangement or the **trans* arrangement. The position effect is the basis of the **cis-trans* test.

post-transcriptional processing *See* messenger RNA.

potash *See* potassium.

potassium Symbol: K. A soft alkali metal, atomic number 19, atomic weight 39.09, required as a macronutrient by plants. It is the most abundant cation in plant tissues and is believed to have a role in chlorophyll and protein synthesis and in carbon dioxide fixation. Potassium deficiency is most likely to occur in siliceous and peaty soils, symptoms being poor root growth and a characteristic red or purple coloration of the foliage. Growing points are especially affected and flower and fruit formation is poor. The potassium content of compound fertilizers is called *potash* and is usually measured as the proportion of potassium oxide, K_2O, present.

potassium–argon dating *See* radiometric dating.

potometer An apparatus designed to measure the rate of water uptake by a cut leafy shoot or whole plant and hence, indirectly, the rate of transpiration. One version consists of a glass tube that bends upwards at one end and supports the plant. The other end of the tube is connected to a capillary tube with an attached scale. The whole apparatus is filled with water from a reservoir joined via a tap to the tube and then made airtight. The plant takes up water and the flow of water is measured by observing the progress of an air bubble along the capillary tube. If the diameter of the capillary tube is known the volume of water taken up by the plant can also be measured. The apparatus is usually used to compare the rate of water uptake when the plant is subjected to certain changes in the external conditions, such as moving air, different light intensities, differences in humidity, etc. It may also be used to compare the rate of water uptake by different plants in the same conditions. An *atmometer* is a similar apparatus and is used to measure the rate of evaporation from a nonliving wet surface, such as a porous pot. By comparing water loss from a potometer with that from an atmometer under similar conditions, the rate of water evaporation from a leaf, which is controlled by the leaf to some extent, can be compared to the rate of uncontrolled evaporation.

powdery mildew A plant disease, caused by fungi of the order *Erysiphales, in which the pathogen grows as a white powdery coating (of mycelium and conidiophores) on leaves and stems. Examples include powdery mildew of cereals (*Erysiphe graminis*), gooseberry and blackcurrant mildew (*Sphaerotheca mors-uvae*), and apple mildew (*Podosphaera leucotricha*). *Compare* downy mildew.

PPLO *See* Mycoplasmatales.

P_r The inactive form of the plant pigment *phytochrome that has a peak of light absorption in the red part of the spectrum, i.e. at about 655–665 nm. It is interchangeable with the *P_{fr} form.

prairie North American *grassland.

Precambrian The earliest and longest era of geological time between about 4600 and 570 million years ago. It precedes the Palaeozoic era. Relatively undisturbed Precambrian rocks are found in North America, Australia, and South Africa and what fossil material there is has mostly been collected from these regions. Fossils resembling blue-green algae, fungal spores, and fungal hyphae have been preserved. Calcareous algae have been found in certain late Precambrian rocks in Labrador and Montana. Some cherts of the gunflint formation of Ontario (about 2000 million years old) and older cherts of Australia and Africa have been shown to contain *stromatolites. There is much controversy about the existence of certain of these fossil remains (especially the *chemical fossils) and about the reason for the apparent gap between the few questionable life forms in the Precambrian and the comparative wealth of life forms in the Cambrian period. *See* geological time scale.

prefloration *See* aestivation.

preformation *See* epigenesis.

pressure potential (turgor potential) A component of *water potential, represented by the symbol Ψ_p. Pressure potentials may be negative, in which case they represent tensions. Tensions arise as a result of transpiration and are caused by resistance of the tissues to water flow. In extreme conditions very low (−150 bar) pressure potentials can develop. Pressure potential gradients are responsible for the upward movement of water in the xylem (*see* cohesion theory). Positive pressure potentials are termed hydrostatic pressures.

prickle A short woody pointed outgrowth from the epidermis of a plant. It may be simply protective, as in *Gunnera*, or it may also help the plant become hooked to a support, as do the recurved prickles of roses and brambles. A prickle is a modified multicellular trichome. *Compare* spine, thorn.

primary endosperm nucleus The triploid nucleus that results from fusion of the polar nuclei or definitive nucleus of the embryo sac with one of the male gametes released from the pollen tube. It develops into the endosperm. *See also* double fertilization.

primary growth Size increase due to cell division at the apical meristems and subsequent cell expansion. The tissues so produced are termed the *primary plant body* and comprise all the tissues of a young plant. The gymnosperms, most dicotyledons, and some monocotyledons exhibit increased thickening of stem and root later in life by a process of *secondary growth.

primary phloem *Phloem derived from the procambium in the primary plant body. In nonwoody plants the primary phloem, consisting of the *protophloem and *metaphloem, is the only food-conducting tissue, whereas in mature plants exhibiting secondary growth this function is usually performed by the *secondary phloem.

primary pit field An area of greatly reduced thickness in the primary wall of a plant cell, often penetrated by plasmodesmata. Primary pit fields enable relatively easy transfer of materials between cells, thus having a similar function to *pits. Cells lacking secondary walls can possess only primary pit fields, whereas those possessing secondary walls may also possess pits, which are often, but not always, positioned directly over the primary pit fields.

primary plant body *See* primary growth.

primary ray *See* medullary ray.

primary thickening meristem A *meristem that is found in certain monocotyledons, such as palms, below the leaf primordia at the apex. It serves to increase the width of the stem behind the apex by cutting off rows of cells by periclinal divisions. Prolonged primary growth, due to the activity of this meristem, allows some monocotyledons to attain considerable stature.

primary tissue Tissue formed from a primary meristem, such as the procambium, protoderm, or ground meristem. In woody plants primary tissue is formed before secondary tissue (*see* secondary growth). In herbaceous plants that possess no secondary meristems, all tissues are primary. The primary xylem and primary phloem are examples of primary vascular tissues.

primary xylem (primary wood) *Xylem derived from procambium in the primary plant body. In nonwoody plants the primary xylem, consisting of the *protoxylem and *metaxylem, is the only water-conducting vascular tissue, whereas in mature plants exhibiting secondary growth this function is performed largely by the *secondary xylem. As well as differing from secondary xylem in origin, the primary xylem usually also has longer tracheary elements, which are often arranged in a random fashion, although these differences are not always reliable. Primary xylem also consists of an axial system only and therefore does not contain rays, although rays may be present between the vascular bundles.

primordium Any immature part of a plant destined to differentiate into a certain cell, tissue, or organ. The term is usually used of a part of the *apical meristem that later differentiates further. Thus a *leaf primordium* later differentiates into a leaf. In early stages of development the leaf primordium appears as a microscopic projection (*leaf buttress) from the shoot apex.

prisere *See* sere.

probability Symbol: *m*. The expectation that over a series of observations a certain kind of observation will occur regularly and form a given proportion of the total number of observations. For example, if a plant heterozygous for height, Tt, is selfed, the probability of finding the double recessive, tt, in the progeny is $\frac{1}{4}$. The greater the number of progeny, the more likely it is that the actual number of double recessives will approach 25%. *See also* chi-squared test.

procambium (provascular tissue) The part of an *apical meristem, the derivatives of which give rise to the primary vascular tissues. *Compare* protoderm, ground meristem.

procaryote *See* prokaryote.

procaryotic *See* prokaryotic.

procumbent Describing a plant or parts of a plant that trail loosely along the ground. An example of a procumbent plant is the heath bedstraw (*Galium saxatile*). *Compare* decumbent, prostrate.

producer An organism that is the first stage in a *food chain. Producers include green plants and those bacteria that synthesize organic molecules from inorganic materials by photosynthesis or chemosynthesis. They are eaten by primary *consumers. *Compare* decomposer.

proembryo The young plant individual after fertilization but before tissue differentiation into embryo and suspensor tissue.

profile diagram *See* transect.

profundal Describing the region of a lake or pond below a depth of 10 m, where there is little light, oxygen, or warmth. Heterotrophic organisms, such as bacteria, fungi, molluscs, and insect larvae, live in this region but few green plants. *Compare* photic zone.

progressive endosymbiotic theory *See* serial endosymbiotic theory.

progymnosperm Any of certain plants of the Devonian period that show various characteristics apparently intermediate between nonseed-bearing and seed-bearing vascular plants. They have heterosporous reproduction, and certain anatomical features such as woody fibres in the cortex and secondary phloem, that are also found in the early gymnosperms.

prokaryotic (procaryotic) Describing cells in which the nuclear material is not separated from the rest of the protoplasm by a nuclear membrane. The term also refers to those organisms, namely the bacteria and blue-green algae, that are based on this type of organization. The term is not used of viruses.

As well as lacking a defined nucleus, prokaryotes also lack nucleoli, plastids, mitochondria, vacuoles, golgi apparatus, and endoplasmic reticulum. Ribosomes are present but are smaller (70S) than those of eukaryotes (80S) though similar in size to those of chloroplasts and mitochondria. This observation has led to speculation that eukaryotes may have evolved as symbiotic associations of prokaryotic organisms (*see* serial endosymbiotic theory). The cells themselves are also much smaller (about 1 μm in diameter) than eukaryotic cells (about 20 μm in diameter) and cytoplasmic streaming is not apparent. The genetic material is a circular strand of DNA, which, unlike that of eukaryotes, is not complexed with histone proteins. Cell division is amitotic.

The biochemistry of prokaryotes is essentially similar to that of eukaryotes. However sterols are conspicuous in their absence from prokaryotes and prokaryotic cell walls characteristically contain muramic acid, a sugar acid not found among eukaryotes. Some unusual amino acids, e.g. diaminopimelic acid, are also associated with the cell wall structure and certain familiar amino acids, e.g. alanine and aspartic acid, occur as their D-isomers. Peptides containing D-amino acids are resistant to hydrolysis by peptidase enzymes though lysis of gram-positive bacteria can be brought about by lysozyme enzymes.

The basic differences between prokaryotes and eukaryotes have led many taxonomists to place the prokaryotes in a separate kingdom, Prokaryota.

prolamellar body A three-dimensional regular lattice found in *etioplasts. It is composed of a continuous system of tubules but when exposed to light the symmetrical arrangement is rapidly lost as tubules become pinched off into two-dimensional sections of lattice. These form perforated sheets of membrane that move apart, extend and increase, finally establishing the typical granal and intergranal lamellae of the mature *chloroplast. *Plastoglobuli in the etioplasts are dispersed as their lipid contents are utilized in membrane formation.

prolamine Any of a group of simple plant proteins that are soluble in 70–90% alcohol but insoluble in water and absolute alcohol. Prolamines contain a high proportion of proline and glutamic acid but only small amounts of basic amino acids. Examples are gliadin, hordein, and zein, found as storage proteins in wheat, barley, and maize respectively. *Compare* glutelin.

proline A nonpolar *imino acid, formula C_4H_8NCOOH (*see illustration at* amino acid). Synthesis of proline is from glutamic acid, while breakdown occurs by reversal of the synthetic pathway. Cell walls are rich in 4-hydroxyproline, a derivative of proline.

promoter *See* operon.

promycelium The basidium of the Uredinales and Ustilaginales. It is formed on germination of the resting spores of these fungi and is usually divided by cross walls into a number of cells.

prophage A bacteriophage that has become integrated in the bacterial DNA and is replicated along with it when the bacterium divides. In this state it is quiescent but it may excise itself from the DNA at any time and, following replication, cause lysis of the host bacterial cell.

prophase The initial stage of nuclear division. In both *mitosis and *meiosis, the chromosomes become coiled and recoiled, and, in mitosis, the chromatids can be identified. As they shorten and thicken the distinctive features of the individual chromosomes can be identified with the light microscope. In meiosis, prophase of the first division can be divided into five substages, but there is no clear demarcation between them, the whole process being continuous. The first substage, *leptotene*, is the period during which shortening and thickening occurs, but, although it is known that DNA replication has occurred, the chromosomes do not appear to be divided into chromatids. During the next two substages, *pachytene* and *zygotene*, homologous chromosomes are attracted to each other and *synapsis takes place. This is in contrast to the situation in mitotic prophase where homologues remain entirely separate from each other. The paired chromosomes continue to contract and coil around each other to form a composite structure called a bivalent. *Diplotene*, the fourth substage, begins as the mutual attraction between the chromosomes of the bivalents lapses and is replaced by mutual repulsion, commencing at the centromeres. The chromatids at this stage are clearly visible, and, as the chromosomes separate, it becomes apparent that they are held together at various points, *chiasmata, where chromatids from opposite chromosomes have crossed over and are linked together. By *diakinesis*, the chromosomes are fully contracted. As the centromeres of the homologues continue to move apart, each pulls its attached chromatid pair with it and the regions of cross-over move towards the ends of the chromosomes (*terminalization*). By the end of prophase in both mitosis and meiosis, the nucleoli have dispersed and the nuclear membrane has broken down. In prophase of the second meiotic division, there is only one set of chromosomes i.e. one member only of each homologous pair. *See* metaphase, anaphase, telophase.

proplastid A small *plastid, less than 1.0 μm in diameter, with rudimentary internal structure. The inner of the two surrounding membranes is often extended into finger-like projections. Small membrane-bound vesicles and a few starch grains and plastoglobuli may be present in the matrix. Proplastids are present in the cells of meristematic tissue and are thought to differentiate into mature plastids.

prop root Any of the adventitious roots that arise from the lower nodes of the stem in certain plants and serve to provide additional support. Such roots are seen in maize (*Zea mays*). The woody prop roots formed by certain trees are sometimes termed *stilt roots*. *See also* buttress root.

prosenchyma 1. Any tissue composed of more or less elongated cells with tapering ends. The component cells are called *stereids*. Prosenchyma is an obsolete term, sometimes used in contrast with the more or less isodiametric *parenchyma cells.
2. *See* plectenchyma.

prosthetic group A *coenzyme that is tightly bound to the enzyme with which it acts. An example is the biocytin coenzyme of acetyl CoA carboxylase, an enzyme of fatty acid synthesis. The biocytin has a terminal $CH(NH_2)COOH$ group, which covalently binds to the protein as part of the peptide.

prostrate Describing a plant that grows closely along the ground, such as the white stonecrop (*Sedum album*).

protandry The maturation of the male reproductive organs before those of the female. For example, in many members of the Compositae and Leguminosae the pollen is released from the anthers before the stigma in the same flower is receptive. Protandry is a consequence of the normal centripetal development of the floral parts and is the most frequently encountered form of *dichogamy. *Compare* protogyny.

protease (proteinase) Any *hydrolase enzyme that catalyses the hydrolysis of protein chains. Specific proteases act only on certain peptide (CO–NH) linkages. For example, the bacterial protein thermolysin hydrolyses only those peptide bonds in which the amino group is donated by leucine, isoleucine, or valine. The best known plant protease is *papain*, from the latex of the papaw or papaya tree, *Carica papaya*, which is used commercially as a meat tenderizer. *See also* peptidase.

protein A complex biological macromolecule consisting of one or more *polypeptide chains. Two major classes of protein are recognized: globular proteins, most of which are enzymes; and fibrous proteins, which are usually structural or contractile in function. The structure of a protein can be divided into primary, secondary, tertiary, and quaternary components. Primary structure, the amino acid sequence of the polypeptide chain, determines the three-dimensional shape of the protein. The secondary structure is the coiling or pleating of the polypeptide chain. In globular proteins the coiled polypeptide chain is further folded into a three-dimensional shape, which is maintained by weak hydrophobic and polar interactions between amino acid residues. Quaternary structure is the combination of more than one polypeptide chain to form an *oligomeric* protein molecule. *See also* conjugated protein.

proteinaceous endosperm layer *See* aleurone layer.

proteinase *See* protease.

protein synthesis The multistage process by which information contained in the cell's genetic material is expressed as the amino acid sequence of a protein. All proteins are constructed from about twenty amino acids, the number, type, and sequence of which is unique to a particular protein. The arrangement of amino acids is predetermined by the sequence of bases in the genetic material, DNA (or RNA in some viruses). DNA is mostly confined to the nucleus in eukaryotic cells and contains four types of bases. Sequences of three bases (triplets) correspond to particular amino acids. Hence sequences of triplets (the genetic code) correspond to sequences of amino acids in proteins. The genetic code is first copied (transcription) by the formation of a complementary molecule called *messenger RNA (mRNA). After various modifications in the nucleus, mRNA moves to the *ribosomes in the cytoplasm where protein synthesis occurs. Amino acids in the cytoplasm bind to specific transfer RNA (tRNA) molecules at one end of which are three bases (the anticodon). Which amino acid is incorporated into the protein is determined by whether the three bases on tRNA are complementary to, i.e. can bind with, the triplets on mRNA (the codon). The matching of codons with anticodons and the addition of an amino acid to an existing polypeptide chain is called *translation. *See also* ribosomal RNA.

proteolysis The breakdown of proteins to their constituent amino acids by proteolytic enzymes. Much information about the action of specific *proteases and *peptidases has been gained from studies on animal digestive tracts, but little is known of the mechanisms by which proteins are broken down within the cell. Radioactive tracer studies have shown that turnover of proteins within the cell is extremely rapid, indicating that intracellular proteolysis is a very efficient process.

proteolytic enzyme An enzyme that hydrolyses proteins. *See* protease, peptidase.

prothallial cell The smaller sterile cell formed along with the antheridial cell by the first division of the microspore in certain pteridophytes, e.g. *Selaginella*, and in gymnosperms. The prothallial cell does not usually undergo any further division and represents the only vegetative tissue of the gametophyte generation.

prothallus The free-living gametophyte of certain lower vascular plants, e.g. ferns. It is usually poorly differentiated and in many species superficially resembles a thallose liverwort. The female gametophyte of gymnosperms is sometimes termed a prothallus.

Protista A kingdom containing all the unicellular organisms. It includes bacteria, blue-green algae, unicellular algae and fungi, and protozoans.

protobiont (coacervate) An aggregate of large complex organic compounds bounded by an organic membrane. It is suggested in certain theories on the *origin of life that such aggregates were the precursors of living organisms. It is thought that the membrane selected molecules that could be added to the aggregate and that gradually the chemical component became more efficient and coordinated. Possibly polymeric molecules resembling RNA and DNA developed and became associated with the protobiont, thus enabling it to store and transfer information, a major step in the development of life.

protocooperation *See* mutualism.

protoderm The outermost layer of the *apical meristem, the derivatives of which give rise to the *epidermis and sometimes also to associated subepidermal tissues. *Compare* procambium, ground meristem.

protogyny The maturation of the female reproductive organs before those of the male. For example, in the Rosaceae and Cruciferae the stigma becomes receptive before the anthers in the same flower release their pollen. Protogyny is contrary to the normal centripetal development of the floral parts and is not so frequently encountered as *protandry. *See* dichogamy.

protonema 1. The juvenile form of a moss or liverwort that develops on germination of a spore. In mosses and foliose liverworts it is usually a branched green filament, resembling a filamentous green alga, but in *Sphagnum* it becomes thallose and in thallose liverworts it is not clearly differentiated. The familiar adult form develops from buds on the protonema.
2. The erect green filament that develops on germination of the zygote in algae of the Charophyta.

protophloem Early *primary phloem. The *sieve elements of the protophloem, which often lack companion cells, are functional for a brief period only and are usually later obliterated, being unable to keep pace with the elongation of the surrounding cells. These obliterated cells often differentiate into fibres.

protoplasm The substance of the *protoplast of cells. It is composed of about 90% water in highly active cells although the figure may be as low as 10% in dormant cells. Proteins and amino acids account for approximately 65% of the dry weight, lipid material about 16%, and simple sugars about 12%. There are also small quantities of a wide variety of other organic compounds and mineral salts. Physically protoplasm is a colloidal system of at least three phases; an aqueous solution of organic and inorganic compounds, a disperse phase consisting of oil droplets forming an emulsion, and a framework of protein molecules forming fine fibrils and tubules. Proteins are capable of changing form and so altering viscosity. In addition protein and phosphoglyceride molecules are organized into *cell membranes, which are the basis of the structure of all *organelles.
The substance of the nucleus is termed nucleoplasm, while the remainder of the protoplasm is called *cytoplasm.

protoplast The living part of a cell. In a plant cell this includes the *cytoplasm, *nucleus, *cell membranes, and *organelles in their highly organized condition, but it does not include the cell wall or vacuole.

protoplast fusion The induction of the fusion of naked (wall-less) plant cells under culture conditions. Such cells or protoplasts can be produced by enzymatic digestion of cell walls, osmotic shock, etc. The technique has potential use for hybridizing unrelated or incompatible species.

protostele A *stele consisting of a cylinder of phloem and pericycle surrounding a central core of xylem and lacking a central pith. This type of stele is regarded as being the most primitive both ontogenetically and phylogenetically. The *haplostele, *actinostele, *plectostele, and *medullated protostele are all forms of protostele. *Compare* siphonostele.

protoxylem Early *primary xylem. The protoxylem matures before the organ completes its longitudinal growth and is thus often distorted or destroyed as the surrounding tissues elongate. In higher plants, the protoxylem of the stem occurs at the innermost edge of the vascular bundles whereas in the root it occurs external to the metaxylem as protoxylem poles. The protoxylem is composed mostly of parenchyma cells with a relatively small number of tracheary cells. The tracheary elements of the protoxylem usually possess spiral, annular, or sometimes reticulate thickening and nonpitted secondary cell walls. *Compare* metaxylem, protophloem.

protoxylem lacuna *See* carinal canal.

provascular tissue *See* procambium.

pruinose Having a whitish bloom, e.g. the fruits of the Oregon grape (*Mahonia aquifolium*).

proximal Denoting the region of an organ that is nearest to its point of attachment.

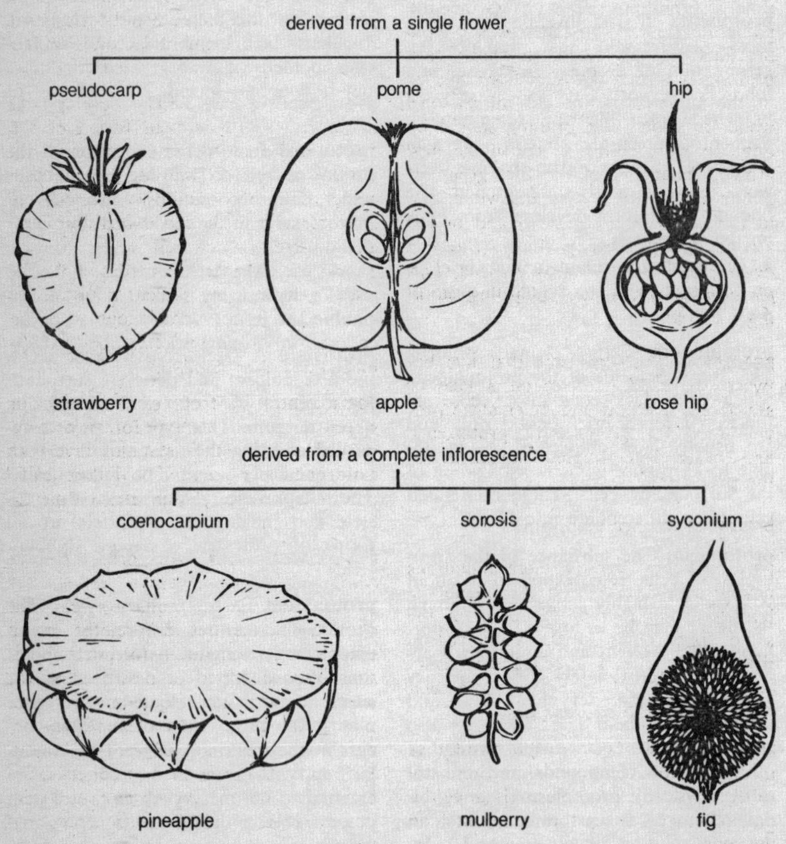

Various forms of pseudocarp.

pruning The cutting back of some or all of the branches of a woody plant. Pruning may be necessary for a number of reasons, e.g. the removal of dead or diseased wood or to train the plant into a special shape. Usually however pruning is performed to promote the vigour of a plant and, in the case of fruit trees, to maintain a balance between vegetative growth and fruit production. The time of pruning depends on whether the plant flowers early in the season on the previous season's wood, in which case pruning is carried out after flowering, or in the summer on the current season's wood, in which case pruning should be done in early spring to encourage new growth. Examples of the first group of plants are *Prunus* species and winter jasmine (*Jasminum nudiflorum*) and of the second group butterfly bush (*Buddleia davidii*) and lemon-scented verbena (*Lippia citriodora*). *See also* pollarding, stooling.

psammon The organisms that live between the sand particles on a lake or sea shore. Diatoms and other algae provide food for the bacteria, protozoans, and other heterotrophic organisms of the psammon.

psammosere *See* sand dune.

pseudoalleles Two or more mutations having similar effects on the phenotype but occurring in different parts of the same gene. Being in the same cistron they will be closely linked and recovery of recombinants from heterozygotes will be rare. *See also cis-trans* test.

pseudocarp 1. (false fruit) A fruit that incorporates tissues other than those derived from the gynoecium. It may be derived from a single flower and include the receptacle or bracts, or may be derived from a complete inflorescence (*see* multiple fruit). *See also* pome, hip.
2. A particular type of pseudocarp consisting of a number of achenes embedded in the outer surface of a fleshy receptacle, e.g. strawberry (*Fragaria*) fruits.

pseudocopulation The attempt by a male insect to mate with a flower, or parts of a flower, which it mistakes, because of the colouring, shape, or scent of the flower, for a female of the same species. During pseudocopulation pollination of the flower is achieved. This pollination mechanism can be observed frequently in the orchid family. *See also* entomophily.

pseudoendosperm 1. The development of an endosperm without fusion of the *polar nuclei or *definitive nucleus with a male gamete. It may arise spontaneously from the definitive nucleus or from somatic tissue and is therefore usually diploid.
2. In gymnosperms, the tissues of the female gametophyte, which nourish the embryo. It is not homologous with the endosperm of angiosperms, being haploid rather than triploid.

pseudogamy A form of *apomixis in which a diploid embryo forms without fertilization though a stimulus from the male gamete is required and thus pollination is necessary. *See also* parthenogenesis.

Pseudomonadales An order mainly containing gram-negative rod-shaped bacteria with, in the motile forms, polar flagella. They do not form spores. Many species are pathogenic to plants, e.g. *Pseudomonas tabaci* and *P. angulata*, which cause wildfire and angular leaf spot of tobacco, and *Xanthosomonas*, which is responsible for various blights and cankers. The order also contains the nitrifying bacteria *Nitrobacter* and *Nitrosomonas* and the *photosynthetic bacteria.

pseudoparenchyma *See* plectenchyma.

pseudoperianth A relatively late-developing membranous sheath surrounding the young sporophyte in certain liverworts, e.g. *Fossombronia*. It is sometimes termed a perigynium (*see* perichaetium).

pseudoplasmodium A small sluglike aggregation of myxamoebae that acts as a unit but within which each *myxamoeba remains distinct. The fruiting structure is a *sorocarp*, which is sometimes borne on a slender cellulose stalk, the *sorophore*. This type of organization is found in the *Acrasiomycetes.

pseudopodium **1.** A leafless stalk that serves to raise the capsule in those mosses (e.g. *Sphagnum*) that lack a seta. **2.** An extension formed by flowing of the cytoplasm that is the means of movement in amoeboid gametes.

pseudothecium The type of *ascocarp characteristic of the Loculoascomycetes. Pseudothecia are superficially similar to *perithecia but contain two-walled asci and also differ in the way they develop. They are often loosely called perithecia.

Psilopsida A subdivision or class of *vascular plants containing two living genera, *Psilotum* and *Tmesipteris*, that together constitute the order Psilotales. The fossil forms, which are known chiefly from rocks of the Devonian period, are placed in the order Psilophytales. These fossils include *Psilophyton* and, from the Rhynie chert, *Rhynia*. The living psilopsids are considered the most primitive of the vascular plants. The sporophyte is dichotomously branching and bears rhizoids rather than true roots. Small leaflike appendages are often present and the vascular tissue consists of tracheids and poorly defined phloem. The spores, which are homosporous, give rise to a subterranean gametophyte. Fertilization of the gametes relies on the presence of a film of water and embryogeny is exoscopic. In some classifications the living and fossil forms are separated into two subdivisions or classes, the Psilotopsida and Psilophytopsida respectively.

psychrophilic Describing microorganisms that require temperatures below 20°C for optimal growth, e.g. the fungus *Cladosporium herbarum*. *Compare* mesophilic, thermophilic.

Pteridophyta In classifications that consider the possession of vascular tissue of equivalent significance to the seed-bearing habit, a division containing all the nonseed-bearing *vascular plants (*compare* Tracheophyta). Its members show a heteromorphic alternation of generations and the gametophyte is often nutritionally independent of the sporophyte. The sporophyte differs from that of seed-bearing plants (*see* Spermatophyta) in lacking vessels in the xylem. Some species exhibit homospory in contrast to spermatophytes, which are always heterosporous. The Pteridophyta contains the subdivisions or classes *Psilopsida, *Lycopsida, *Sphenopsida, and *Pteropsida.

Pteridospermales (Cycadofilicales, seed ferns) An extinct order of gymnosperms known from fossils of the Devonian to Triassic periods. Although they had fernlike fronds, similar to those of the fern order Marattiales, their stems show secondary thickening and more significantly there is evidence of seed formation. The ovules were not borne in strobili but developed along the margins or on the surface of megasporophylls, similar to the foliage leaves. Their seeds were characteristically radially symmetrical, a feature shared by the Cycadales, Bennettitales, and Caytoniales, and thought to indicate that these orders possibly originated from the Pteridospermales.

Pteropsida A subdivision or class of *vascular plants. In some classifications it includes only the ferns (*see* Filicinae), while in others it also includes the *Gymnospermae and *Angiospermae. The Pteropsida differ from the Psilopsida, Sphenopsida, and Lycopsida in having large leaves often with a highly branched vascular system and leaf traces that leave a gap where they depart from the stele.

ptyxis *See* vernation.

pubescent Having fine short hairs. *Compare* pilose.

puff balls *See* Lycoperdales.

pulvinus 1. A prominent swelling at the base of a petiole or pinna. It consists of cells (motor cells) that can rapidly move water into or out of their vacuoles. Such changes in turgor alter the position of the leaf or leaflet and are responsible for certain sleep movements and haptotropic movements. Pulvini are found in many members of the Leguminosae.
2. A thickened region in the stem or leaf sheath of grasses, often containing an *intercalary meristem.

punctuated equilibrium An evolutionary theory proposing that species often arise rapidly, in terms of geological time, rather than by gradual change. It envisages that, following a short period (less than 100 000 years) of rapid speciation, there is a considerable period of stasis lasting several million years.

Punnet square A chequerboard diagram, attributed to the geneticist R. C. Punnet, used to illustrate how the gametes involved in a particular cross result in various genotypes in various frequencies (*see* diagram).

pure culture *See* axenic culture.

pure line (pure strain) A succession of generations recognized by their ability to produce genotypically identical offspring when selfed or crossed between themselves. The members of a pure line are said to *breed true*. By inference, such

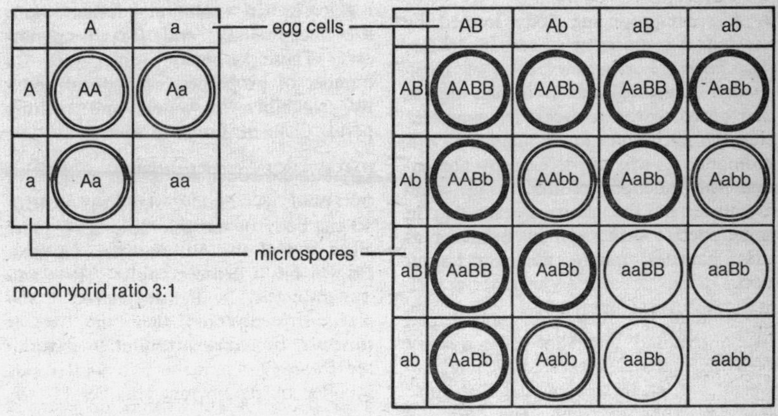

Punnet squares showing the gametes produced by the F_1 and the resulting zygotes of the F_2 in a monohybrid and dihybrid cross.

individuals are deemed to be homozygous.

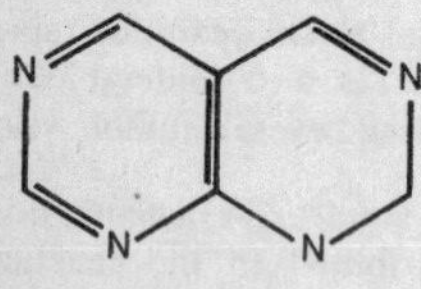

Purine.

purine A nitrogen-containing organic base with a double-ring structure (*see* diagram) synthesized mainly from amino acids. Two purines, *adenine and *guanine, are common constituents of nucleotides.

purple sulphur bacteria *See* photosynthetic bacteria.

pycnidium 1. (pycnium, spermogonium) A flasklike structure produced by certain fungi (e.g. *Puccinia* and *Leptosphaeria*) in which asexual *pycnospores* (*spermatia*) are formed.
2. A spore-producing body formed by the fungal component of certain lichens, e.g. *Parmelia physodes*. It is a flask-shaped structure sunken in the thallus and opens to the surface by a pore. Inside the pycnidium are a number of conidiophores, which produce pycnoconidia. The functions of the pycnoconidia are uncertain though it appears that in some species they may, by behaving as male gametes, initiate ascocarp formation.

pyramid of biomass A diagram shaped like a pyramid that shows the amount of living material, measured by total dry weight, at each feeding stage (trophic level) of a *food chain. The *biomass depends on the amount of carbon that can be fixed by green plants. *See also* pyramid of numbers.

pyramid of numbers A diagram showing the numbers of living organisms present at each feeding stage (trophic level) of a *food chain. The green plants (producers) are usually more numerous and form the base of the pyramid. The successive consumer levels usually decrease in numbers at each stage and form the tiers of the pyramid. There is typically one or a few predators at the top. Normally there are three or four tiers in the pyramid and rarely more than six. If the producers are large (e.g. trees) the pyramid will be partially inverted, as the base is smaller than one or more of the successive tiers. *See also* pyramid of biomass.

pyranose ring A six-membered ring containing five carbon atoms and an oxygen atom. Pyranose rings are formed by any aldose sugars with five or more carbon atoms. In solution the aldehyde group at the first carbon atom reacts with the hydroxyl group on the fifth carbon atom. This reaction renders the first carbon atom asymmetric. The pyranose forms of aldoses can thus exist in two isomeric forms. For example, D-glucose in solution forms both α-D-glucopyranose and β-D-glucopyranose. These isomers differ in a number of properties, e.g. optical activity, solubility in water, and melting point. *Compare* furanose ring.

pyrenocarp *See* perithecium.

pyrenoid A darkly staining proteinaceous body in the chloroplasts of many algae and of the Anthocerotae. In most cases it has a dense granular matrix and is surrounded by tightly packed starch plates. In euglenoid algae the core is traversed by lamellae similar to those in the chloroplast and the pyrenoid is surrounded by paramylum granules.

Pyrenomycetes A class of the *Ascomycotina containing those ascomycetes that produce a *perithecium. With over 6000 species and some 640 genera it is the largest group of ascomycetes and contains the orders Hypocreales,

*Clavicipitales, *Sphaeriales, Coronophorales, Coryneliales, and *Pyrenulales.

Pyrenulales (pyrenolichens) An order of the Ascolichenes or of the Pyrenomycetes containing lichens in which the fungal partner produces perithecia (pyrenocarps). These are buried in the thallus and only visible as dots on the surface. There are over 1500 species in the Pyrenulales, common genera including *Dermatocarpon*, *Pyrenula*, and *Verrucaria*. Lichens with fungi that form bitunicate rather than unitunicate perithecia may be placed in a separate order, the Pseudosphaeriales.

pyridine alkaloids A group of alkaloids, the structures of which are based on the pyridine nucleus (a six-membered ring containing five carbon atoms and a nitrogen atom, and with three double bonds). They include the hemlock alkaloids, e.g. coniine, which is present in *Conium maculatum* (hemlock).

pyridoxine (vitamin B_6) A pyridine derivative, essential in its activated forms, pyridoxal phosphate and pyridoxanine, as a coenzyme in various transamination reactions. *See* vitamin.

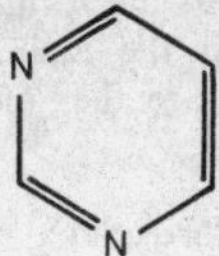

Pyrimidine.

pyrimidine A nitrogen-containing organic base composed of a single heterocyclic ring (*see* diagram), synthesized mainly from amino acids. Two pyrimidines, *cytosine and *thymine, are constituents of DNA, while *uracil replaces thymine in RNA. Thiamin (vitamin B_1) is also partly constructed from a pyrimidine.

Pyrrophyta *See* Dinophyta.

pyruvic acid A three-carbon carboxylic keto acid with the formula CH_3COCO OH. In glucose metabolism pyruvate is an intermediate in glycolysis. Under anaerobic conditions it is reduced to form ethanol or some other fermentation product, while in aerobic conditions it is decarboxylated to form acetyl CoA. In C_4 plants pyruvate is an intermediate of the Hatch–Slack pathway. In amino acid metabolism pyruvate is the degradation product of several amino acids, including alanine and cysteine, while other amino acids are synthesized from pyruvate. Pyruvate can also be carboxylated to form oxaloacetate or malate, both intermediates in the *TCA cycle. These reactions, known as *anaplerotic* reactions, replenish TCA cycle intermediates.

Q

Q enzyme *See* starch.

quadrat A small area, usually 0.5 or 1.0 m^2, marked out with a frame that can be used in an ecological survey to study in detail the distribution and abundance of different species. It can be used for random sampling within an area or for systematic sampling at regular intervals. A *permanent quadrat* can be used to record changes in the distribution and numbers of species over a number of years. A quadrat may be used in conjunction with a metal 'comb' with teeth at regular intervals that allows detailed sampling at equally spaced points. A hundred such points would estimate the percentage of a particular species present. *See also* transect.

quadrivalent The structure formed by the pairing of four homologous chromosomes in a tetraploid during prophase and metaphase of the first division of meiosis. *Compare* bivalent.

qualitative variation (discontinuous variation) The expression of a characteristic in perceptibly different ways by different individuals in a population. Examples include the tall and short forms of garden pea used by Gregor Mendel. Qualitative variation is most obvious in characteristics determined by only one or two genes whose expression is not greatly affected by environmental factors. Such variation may be due either to recurrent mutation or to the existence of a balanced *polymorphism. *Compare* quantitative variation.

quantasome *See* chloroplast.

quantitative variation (continuous variation) The type of variation shown by a characteristic that is exhibited to a greater or lesser extent by all members of a population or species. Often the characteristic has an average value, about which individuals in the population are distributed. An example would be the size of ears from a cereal crop. Continuous variation is common in characteristics that are strongly affected by environmental influences, and in characteristics determined by many genes, each with a small effect. *Compare* qualitative variation.

Quaternary The second period of the *Cenozoic era from two million years ago to the present day. It is divided into the Pleistocene and Holocene (Recent) epochs. At its commencement, the climate was cool and temperate and there were four glacial periods, the evidence for which is borne out by the distribution of fossils. Many plants and animals became extinct. In the Holocene epoch, beginning about 10 000 years ago, the rise of civilization saw the origin of crop plants and their associated weeds. *See* geological time scale.

quiescent centre The region in the *apical meristem of a root where little or no cell division occurs. The cells in the quiescent centre are capable of assuming meristematic activity if the initials in the meristem are damaged, and thus act as a reservoir of potential initials, protected from damage by their relative inactivity. The quiescent centre may also be a site of auxin synthesis.

quillworts *See* Isoetales.

quinone fungicide Any fungicide developed from chlorinated quinones, such as chloranil, which is used as a seed dressing, and dichlone, which is used as a foliar fungicide.

R

race In taxonomic classification, an infraspecific category of uncertain position but occasionally used in floras in place of, or subordinate to, form. The term is also used in lieu of ecotype, implying a category between subspecies and variety covering geographical groupings of plants. Races are often uniform in respect of ecological preference, physiological requirements, and topographical distribution. *See also* physiological race.

raceme A *racemose inflorescence in which the flowers are formed on individual pedicels on the main axis (peduncle). Examples are the inflorescences of lupins (*Lupinus*).

racemose inflorescence (indefinite inflorescence) An inflorescence in which meristematic activity continues at the apex of the main stem and primary laterals and flowers are developed from the axillary meristems. The inflorescence is often pyramidal in shape with older flowers at the base or, in the case of a flat-topped arrangement, on the outside. Different types of racemose inflorescence include the *capitulum, *corymb, *panicle, *raceme, *spadix, *spike, and *umbel (*see* diagram). *Compare* cymose inflorescence.

rachilla (rhachilla) **1.** The secondary axis of a pinnately compound leaf, such as that of a fern.
2. The main axis of a sedge or grass spikelet.
Compare rachis.

rachis (rhachis) **1.** The main axis of a compound leaf possessing pinnae, such as a fern frond.
2. The main axis of an inflorescence.
Compare rachilla.

radial 1. Describing a longitudinal section that passes through the centre of a cylindrical organ (i.e. along a radius or diameter). In a radial section down a woody stem the growth rings appear parallel and flecks of rays may be seen. *Compare* tangential.
2. *See* anticlinal.

radial symmetry The arrangement of parts in an organ or organism such that any cut taken through the centre divides the structure into similar halves. Most stems and roots exhibit such symmetry. Radial symmetry in flowers is usually termed *actinomorphy*. The flowers of relatively primitive angiosperm families, e.g. Ranunculaceae, are usually actinomorphic. In a floral formula, actinomorphy is represented by the symbol ⊕.

radical Describing leaves arising close together at the base of the stem, as in a rosette plant. *Compare* cauline.

radicle The embryonic root, which in the seed is directed towards the micropyle. It is normally the first organ to emerge from the testa on germination. The radicle may persist to form a *taproot. Alternatively it may be replaced either by lateral or adventitious roots.

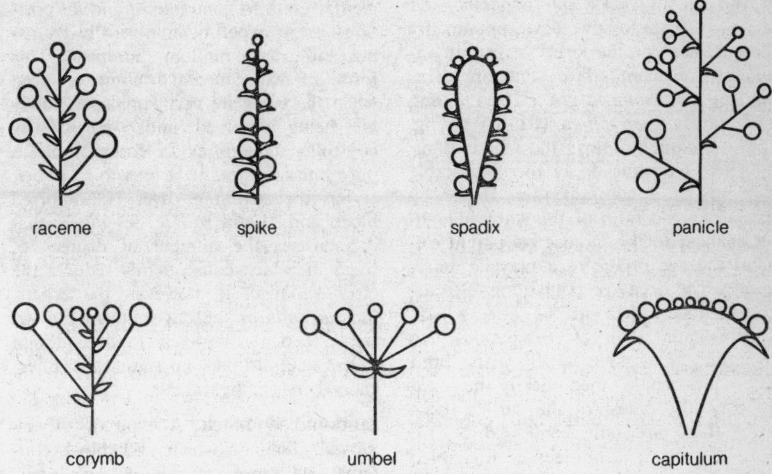

Types of racemose inflorescence

radioactive tracer *See* isotopic tracer.

radiocarbon dating A method used to determine the age of organic materials up to 70 000 years old. It relies on the fact that the ^{14}C isotope of carbon is unstable and decays, emitting beta rays, to ^{14}N, with a half-life of about 5700 years. Plants incorporate $^{14}CO_2$ into their tissues during photosynthesis but when they die the concentration of ^{14}C starts to fall at a rate related to the half-life. By comparing a specimen of unknown age with a sample of zero age, the age of the specimen may be calculated by measuring the amount of ^{14}C using a mass spectrometer.
The method assumes that the ^{14}C:^{12}C ratio in the atmosphere has always remained constant. Certain discrepancies between age determinations based on radiocarbon dating and *dendrochronology show there have been systematic variations in this ratio.

radiometric dating The determination of the age of rocks and minerals, and hence of the fossils they contain, by measurement of the levels of certain radioactive elements. Two common methods are *potassium–argon* (K-Ar) dating and *rubidium–strontium* (Rb-Sr) dating. The technique employs the fact that radioactive elements decay to other stable elements at a constant rate. Hence by measuring the ratio of the stable daughter element to the radioactive parent element the age of the rock may be determined. The decay of potassium-40 to argon-40 has a half-life of 11.8×10^9 years, while that of rubidium-87 to strontium-87 is 48.8×10^9 years. Both techniques could theoretically be used on even the oldest of the earth's rocks since these are only some 4.6×10^9 years old. Such techniques have enabled geologists to construct an absolute geological time scale.

radiospermic Describing the radially symmetrical seeds of the Pteridospermales, Cycadales, Bennettitales, Caytoniales, and Angiospermae. *Compare* platyspermic.

raffinose A common plant trisaccharide found particularly in cotton seed and sugar beet. It is a tasteless nonreducing sugar composed of galactose, glucose, and fructose units.

rain forest *See* forest.

raised bog *See* bog.

ramentum Any of the small brown scales that cover the young fronds of a fern. The ramenta are shed as the frond unfurls but some persist on the rachis.

ramet Any individual belonging to a known *clone. Studies of phenotypic plasticity in relation to environmental factors are more quantifiable if ramets are used rather than genetically mixed individuals.

Ranales *See* Magnoliidae.

randomization The allocation of experimental units to completely random positions in an experiment, normally by using tables of random numbers. This form of experimental design is used when the units are very similar and they are being observed under particularly controlled conditions as, for example, a pure line of plants in a growth chamber. It has the advantage over *randomized block and *Latin square designs in that it increases the number of degrees of freedom, which consequently reduces the error variance. If, however, the experimental units or the environment are variable then a completely random design is less likely to pick up significant differences between treatments.

randomized block A type of experimental design in which each block contains one representative of each treatment, the treatments being allocated to random positions in the block. Each block is thus a complete replication but allocation of treatments to columns is random. The columns are therefore not complete replications since they may contain two or more representatives of one treatment and none of another. Randomized blocks may be used, for example, when dealing with identifiable inherent variation. *Compare* Latin square.

rank The level or position occupied by a category in the taxonomic hierarchy, e.g. class, family, genus, etc. (*see* hierarchical classification). Taxa belonging to ranks between and including the levels subkingdom and subtribe are usually indicated by characteristic suffixes. Hence, taxa of the rank subkingdom end in -bionta, division in -phyta, subdivision in -phytina, class in -opsida, subclass in -idae, order in -ales, suborder in -ineae, family in -aceae, subfamily in -oideae, tribe in -eae, and subtribe in -inae.

Ranunculaceae A large family of dicotyledonous plants commonly known as the buttercup family. It is cosmopolitan in distribution though best represented in temperate and cold latitudes. There are some 50 genera and over 1800 species, most of which are herbaceous. The leaves are often divided or lobed and usually arise either from the base of the stem or in an alternate arrangement. The flowers may be solitary, e.g. *Anemone*, or borne in a raceme or cyme. The flowers are usually actinomorphic with the parts arranged spirally or, less commonly, in whorls. There are numerous stamens and normally five prominent petals. Some genera, e.g. *Consolida* (larkspurs), *Aconitum* (monkshoods, wolfbanes), and *Aquilegia* (columbines), have zygomorphic flowers. The fruit is usually an achene or a many-seeded follicle.
The family contains many ornamentals, e.g. *Clematis*, *Caltha*, and *Nigella*, but there are no important food crops.

Ranunculidae *See* Magnoliidae.

raphe **1.** A longitudinal ridge on the outer integument or seed coat in anatropous ovules where the funiculus becomes fused with the integument.
2. The longitudinal fissure in the cell wall of motile bilaterally symmetrical diatoms. It is believed to play a part in the movement of such organisms by allowing contact between the cytoplasm and external medium.

raphide A needle-shaped crystal, often of calcium oxalate, found in clusters in plant cells. The presence and form of raphide clusters are sometimes useful as diagnostic characters. For example, the

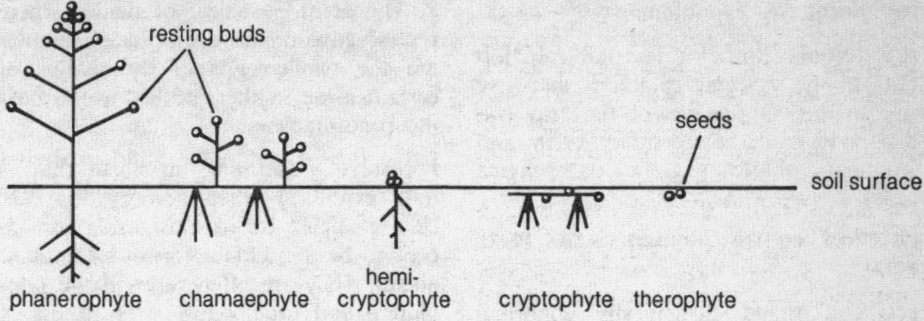

Raunkiaer's life forms.

subfamily Rubioideae of the Rubiaceae is distinguished from the other two subfamilies, Cinchonoideae and Guettardoideae, by the presence of raphides in the leaves.

Raunkiaer system of classification A method of plant classification devised by the Danish botanist, C. Raunkiaer. Plants are divided into groups depending on the position of perennating buds and the degree of protection that they give in cold conditions or drought (*see* diagram). Generally the closer the buds are to the ground the more protected they are. It is particularly useful for dividing given areas of vegetation into groups and comparing the occurrence of these groups in different climatic regions. *See* chamaephyte, cryptophyte, hemicryptophyte, phanerophyte, therophyte.

ray (secondary ray) A radial line of parenchyma cells in the secondary xylem and phloem, derived from the ray initials of the vascular cambium. The ray system of gymnosperms often also contains tracheids, distinguished from the ray parenchyma by their bordered pits and lack of protoplast. Rays may be *uniseriate* (one cell thick in tangential longitudinal section) or *multiseriate* (several cells thick in tangential longitudinal section). *Compare* medullary ray.

ray floret *See* capitulum.

ray initials More or less cubical *initials in the vascular cambium that give rise to the components of the *ray (radial) system of the secondary xylem and secondary phloem (e.g. ray parenchyma cells). *Compare* fusiform initials.

reaction centre pigments *See* P680, P700.

reaction wood Structurally abnormal wood formed in response to various stresses. In gymnosperms it is termed *compression wood* and tends to form on the lower side of branches. Compression wood is dense in structure due to the heavy lignification of tracheid walls. *Tension wood* forms in angiosperms on the upper side of the branches and the fibres tend to be gelatinous rather than lignified.

recalcitrant seed A seed that loses its viability if stored for any length of time, even under conditions that are normally conducive to seed longevity, i.e. low moisture content and low temperatures. Many tropical plants, e.g. coconut, rubber, and tea, have recalcitrant seeds. Such seeds can often only be kept for a year or less.

recapitulation theory A theory formulated by Ernst Haeckel that proposed that ontogeny is a short recapitulation of phylogeny, i.e. an organism in its development goes through a series of stages that resemble the adult forms of its ancestors in evolutionary sequence. The theory has since been discarded.

Recent *See* Quaternary.

receptacle 1. (thalamus, torus) The expanded region at the end of a peduncle to which the floral parts are attached. It is usually convex but may become flattened or concave. Such modifications (*see* hypogyny, perigyny, epigyny) alter the position of the gynoecium in relation to the other floral parts.

2. The point on a leaf or thallus where reproductive organs are borne. Examples are the swollen tips of the thallus of certain algae in the Fucales, which bear the conceptacles.

recessive Describing an allele that is only expressed when homozygous. Recessive alleles are in most cases considered to be the mutant form of dominant alleles. They are often rarer in the population and usually lower the fitness of the individual. Their effect is often due to their failure to produce a normal functional protein. They are masked (i.e. recessive) in a heterozygote because the normal (*dominant) allele will produce a normal protein and hence a normal phenotype.

reciprocal cross A cross in which the source of the male and female gametes is reversed. Thus in the cross yellow peas × green peas, the pollen would originate from the yellow-seeded parent in one cross, then from the green-seeded parent in the reciprocal cross. A reciprocal cross should determine whether there are any maternal (or paternal) factors influencing the inheritance of the characteristic.

reciprocal translocation *See* translocation.

recombinant DNA 1. Genetic material in which crossing over or chromosome reassortment has occurred by any natural means.

2. Genetic material that contains novel gene sequences produced using the techniques of *genetic engineering (recombinant-DNA technology).

recombination The formation of new combinations of genes during meiosis by *crossing over and by reassortment of whole chromosomes into new sets. As a result of recombination, the gametes that an organism produces differ from the gametes from which it arose. Consequently the offspring will vary genetically and hence phenotypically from the parents. Recombination is a major source of variation.

recon *See* gene.

recurrent parent *See* backcrossing.

red/far-red effects Those physiological responses exhibited by many plants as a result of being illuminated by either *red or *far-red light. Usually a period of red light will reverse the effect of a period of far-red light and vice versa. Thus the response elicited in the plant depends on which treatment was given most recently. *See* phytochrome.

red algae *See* Rhodophyta.

red-and-yellow forest soil A type of zonal acid *soil (a *pedalfer) formed in wet subtropical regions. Leaching results in the accumulation of salts in the B horizon. The red colour is due to iron oxides in the A horizon though if the soil is sandy, it tends to be yellow. Such soils contain little humus and are soon exhausted of nutrients when cultivated. They can be easily eroded.

red desert soil A type of coarse zonal *soil rich in salts and lime (a *pedocal) but poor in humus. They are formed in hot deserts and can be cultivated under irrigation.

red light Electromagnetic radiation of approximately 630 nm wavelength. Light in this part of the spectrum is the most effective in initiating various light-dependent reactions in plants. For example, short-day plants are prevented from flowering if given a short burst of red light during the dark period. Similarly red light is best for promoting germination in light-sensitive seeds and for reversing etiolation. *Compare* far-red light. *See* phytochrome.

redox potential The tendency of a reducing agent to lose electrons (or an oxidizing agent to gain electrons) as measured against a known standard, usually hydrogen. By convention the reaction is written in the direction

$$\text{oxidant} + ne^- \rightleftharpoons \text{reductant}$$

where n is the number of electrons transferred. Compounds with a redox potential more positive than that of hydrogen tend to donate electrons to H^+; compounds with a more negative redox potential tend to accept electrons from H_2.

redox reaction *See* oxidation–reduction reaction.

red tropical soil A type of zonal *soil formed in rain forests, and in savanna with alternating dry and wet seasons. It is heavily leached but a balance is maintained between the nutrients produced by decomposition and those taken up by the plants. The red colour is due to the presence of iron, magnesium, and aluminium oxides. Some red tropical soils may form a hard crust of chemicals on the surface.

reducing sugar A sugar capable of reducing an oxidizing agent. To do this it must have a potentially active aldehyde or ketone group. All monosaccharides have such a group and consequently they are all reducing sugars. For a disaccharide to be reducing one of the reducing groups of either of the two component monosaccharides must be left intact. Maltose, which consists of two glucose units linked by an $\alpha(1-4)$ glycosidic linkage, is reducing because the second glucose residue can undergo oxidation, having an aldehyde at carbon five. Sucrose however is nonreducing because the component glucose and fructose units are linked by their aldehyde (carbon one of glucose) and ketone (carbon two of fructose) groups. Various solutions, e.g. Benedict's and Fehling's, are used to detect reducing sugars.

reduction division *See* meiosis.

reflexed Describing a structure that is bent sharply backwards, such as the petals of *Cyclamen*.

regeneration The regrowth of tissues and organs from differentiated tissues of the plant. It is seen following damage or when an organ is removed from the parent plant. The technique of grafting relies on the regeneration of vascular tissue, while vegetative propagation techniques of taking shoot, root, or leaf cuttings depends on the ability of such segments to regenerate complete plants. *See also* tissue culture, wound hormone.

regma A type of schizocarpic *capsule formed from several fused carpels, that breaks explosively into one-seeded units or mericarps at maturity, as in members of the Geraniaceae.

regulator gene (i-gene) A gene whose product can prevent or promote the transcription of other genes. The other genes may or may not be adjacent to the regulator gene. *See also* inducer, operon.

relative humidity The ratio of the amount of water vapour present in a body of air at any given temperature compared to the maximum possible amount that that body can contain at the same temperature. Relative humidity influences *transpiration, which is more rapid when relative humidity is low. It is also an important factor affecting the outbreak and spread of many plant diseases (*see* Beaumont period).

renaturation *See* denaturation.

rendzina (calcisol) A type of shallow intrazonal *soil rich in lime (*calcimorphic*) formed from the underlying limestone or chalk rocks. The A horizon is brown or black and rich in humus. The B horizon, if one is distinguishable, is grey or yellowish with limestone fragments. Brown earths develop in similar climatic conditions from other parent rocks. *Compare* terra rossa.

repeated sequence Any long base-pair sequence found at many different places throughout the chromosomes. This excludes DNA coding for structural genes, which with few exceptions are present as single copies. Repeated sequences account for more than 50% of all the DNA present and fall into three categories.

Repetitive DNA usually consists of widely scattered sequences about 300 base pairs long, which are often adjacent to structural genes. It has been proposed that these sequences are functionally equivalent to the *operator genes of bacteria. Suppose that sequence A is adjacent to three unlinked genes

X, Y, and Z, and that A′ is an *inducer of A. The presence of A′ in the cell would 'switch on' all genes to which A was adjacent, in this case X, Y, and Z. An appropriate mix of repeated sequences, say A, B, C, and D next to structural gene X would make possible an elaborate *transcriptional control system as demanded in a complex organism.

Highly repetitive or *satellite DNA* is often about 20 base pairs long and is usually found near the centromere. *Palindromes*, which are often a few hundred base pairs long, are scattered throughout the chromosomes. Such sequences read the same in both directions, i.e. are palindromic. The functions of satellite DNA and palindromic DNA are uncertain.

repetitive DNA *See* repeated sequence.

replica plating A technique used in electron microscopy to examine the surfaces of specimens that are too electron dense, too delicate, or too robust for sectioning. The specimen surface is coated with a thin film of material, such as carbon, in order to make a cast or replica. This is removed by soaking the specimen in 10% sodium hydroxide solution. It may be subjected to *shadowing procedures before examination. The development of the scanning electron microscope has tended to make replica plating redundant, although it is still used in freeze fracturing.

replication The formation of exact copies (replicas). The term has been applied both to DNA and asexual organisms, both of which produce essentially identical copies/offspring. *See also* semiconservative replication.

repressor A protein product of a *regulator gene that prevents transcription of other genes by binding to the DNA and blocking attachment of messenger RNA polymerase.

reproduction The process by which new individuals of a species are formed and by which the species is perpetuated. *See* asexual reproduction, sexual reproduction, vegetative reproduction.

reproductive isolation The prevention of gene flow between all the members of a population due to the development of reproductive barriers (*isolating mechanisms*), which result in the formation of distinct noninterbreeding groups that may in time develop into separate species. Isolating mechanisms can arise in many ways. Populations may be separated geographically (*see* allopatric). Populations occupying the same area (*see* sympatric) may not interbreed because they have different flowering times or different pollinators, or because they occupy different niches. Such populations may also be isolated by the inviability or sterility of any hybrids that do form.

repulsion *See trans* arrangement.

repulsion theory (field theory) The theory that the pattern of origin of leaf primordia at the shoot apex is regulated physiologically by inhibitory substances synthesized by the apex and the older primordia. A new primordium arises in a position where the concentration of these substances has fallen below a certain threshold level. Although no such inhibitors have yet been isolated various surgical and modelling experiments support the theory. *Compare* available space theory. *See also* phyllotaxis.

resin A mixture of high-molecular-weight compounds, mainly polymerized acids, esters, and terpenoids, exuded by certain plants particularly when wounded. Resins are insoluble in water but soluble in ethanol. On exposure to air the volatile components evaporate leaving a solid or semisolid residue protecting the damaged area. Resins are particularly prevalent in conifers, which contain specialized *resin canals. Pine resin yields the essential oil turpentine ($C_{10}H_{16}$) on distillation and the solid residue rosin, used in lacquers. Other commercially important resins are dammar (from trees of the genera *Shorea* and *Agathis*), kauri (from the New Zealand conifer *Agathis australis*), jalap (from the Mexican convolvulaceous plant *Exogonium purga*), and mastic (from the evergreen anacardiaceous tree *Pistacia lentiscus*). Semisolid mixtures of resins and essential oils are often termed *balsams* and include Canada balsam from *Abies balsamea* and frankincense from species of *Boswellia*.

resin canal (resin duct) A longitudinal resin-containing channel in the *secondary xylem and leaves of many gymnosperms. Resin canals usually form by separation of parenchyma cells, which later form a lining to the canal termed an epithelium.

resistance The ability of a plant to restrict the activities of a pathogen so that its growth is not significantly affected. Complete resistance is termed *immunity. Resistance may be *specific*, i.e. effective against a particular strain of the pathogen, or *nonspecific*, i.e. effective against all strains. Specific resistance is usually controlled by one or a few major genes, e.g. resistance of barley to mildew and of flax to rust are inherited as single genes. Nonspecific resistance is conferred by the combined action of a number of minor genes. Such multigenic or *field resistance* is usually more stable and less likely to be overcome by mutants or recombinants of the pathogen. Environmental factors may influence the expression and degree of resistance. *Compare* susceptibility. *See also* hypersensitivity, phytoalexin.

resolving power The capacity of a microscope to enable clear observation of the fine details of a specimen. This differs from the *theoretical resolution*, i.e. the capacity to distinguish between two individual points despite their close proximity, as the resolving power of a microscope depends on the perfection of the lens, the light source, etc. The theoretical resolution is limited by the wavelength of light and is greater with decreasing wavelength (*see* ultraviolet microscope).

respiration The oxidative breakdown of food substances within the cells of living organisms, resulting in the liberation of energy for subsequent use in growth, etc. The process usually involves the absorption of molecular oxygen, and water and carbon dioxide are typically end products. The reactions involved occur in two stages, *glycolysis and the *TCA cycle. *See also* aerobic respiration, anaerobic respiration.

respiratory chain A series of membrane-linked oxidation–reduction reactions in which electrons are transferred from reduced cofactors, formed in TCA cycle and other *dehydrogenase reactions, to oxygen, which then combines with hydrogen ions to form water.

At present seven different electron-transferring enzymes or cofactors are known intermediates between NAD, the starting point of the chain, and oxygen. The *redox potential is most negative at the start of the chain, and rises as the electrons flow towards oxygen.

For every two electrons travelling down the respiratory chain, three molecules of ATP are generated by the process of *oxidative phosphorylation, which is linked to respiratory electron transfer.

respiratory enzyme *See* cytochrome.

respiratory quotient (RQ) The ratio of the volume of carbon dioxide evolved during respiration to the volume of oxygen absorbed, i.e. RQ = volume CO_2/volume O_2. The complete respiratory breakdown of sugars by *aerobes often gives an RQ value of one, but it rises very considerably under anaerobiosis. When fats or proteins are used in respiration, the RQ value is less than one as less carbon dioxide is produced.

respiratory root *See* pneumatophore.

restitution nucleus *See* aneuspory.

restriction enzyme (restriction endonuclease) An enzyme that recognizes and binds to specific short sequences of double-stranded DNA and cuts the DNA at or near this point. Such enzymes are produced by certain bacteria in response to invasion by bacteriophage – they destroy the virus by cutting up its DNA. Restriction enzymes are a useful tool in genetic engineering work.

reticulate chloroplast A complex plastid that forms a cylindrical network in the peripheral region of the cells of some green algae, e.g. *Oedogonium*.

reticulate thickening A type of secondary wall patterning in *tracheary elements in which the secondary cell wall is laid down to form a network. Unlike *annular thickening and *spiral thickening, reticulate thickening permits little further elongation and therefore occurs in tissues such as metaxylem, which have already completed most or all of their elongation. *Compare* pitted thickening, scalariform thickening.

reticulodromous Describing net or reticulate leaf venation as, for example, seen in *Rhododendron* leaves. *See illustration at* venation.

reverse transcriptase An enzyme present in some microorganisms that reverses the normal transcription sequence and synthesizes DNA from messenger RNA. It is widely used in research for producing *intron-free DNA.

revolute Curved backwards or downwards as, for example, the leaf margins of the white rockrose (*Helianthemum apenninum*). The term is also used to describe this form of vernation (*see illustration at* vernation).

R factor *See* plasmid.

Rf value (relative front) In chromatography, a value that is calculated by dividing the distance moved by the solute spot on the chromatogram by the distance moved by the solvent front. The value is constant for a particular molecule.

rhachilla *See* rachilla.

rhachis *See* rachis.

rhamnose A methylated pentose sugar, rarely occurring free, but common as a constituent of many glycosides, e.g. the flavonol glycoside quercitrin, isolated from the bark of the oak *Quercus tinctoria*. Rhamnose is also found in certain gums (e.g. gum arabic and the gum of flax seeds), mucilages, and in bacterial polysaccharides. It has been identified in the free state in poison ivy (*Rhus toxicodendron*).

rhipidium A scorpioid cyme (*see* monochasium) that is usually flattened in one plane and hence fan shaped. Examples are seen in certain *Iris* species. *Compare* bostryx.

rhizodermis *See* epidermis.

rhizoid A threadlike outgrowth from a thallus, as seen, for example, in the gametophyte generation of mosses, liverworts, and ferns. Usually rhizoids serve to anchor the plant and absorb water and nutrients. In some algae, e.g. *Ulva*, they are incorporated into and strengthen the lamina. Mosses can be distinguished from liverworts by their multicellular, usually brownish rhizoids. In liverworts rhizoids are unicellular and more or less colourless, resembling the root hairs of higher plants. *Compare* root.

rhizome An underground stem that grows horizontally and, through branching, acts as an agent of vegetative propagation. In some plants the rhizomes are cordlike, as in nettle (*Urtica dioica*), while in others, e.g. Solomon's seal (*Polygonatum multiflorum*), they are fleshy and also serve as organs of perennation. *Compare* corm, rootstock, tuber.

rhizomorph A tough cordlike mass of fused hyphae that acts to carry a fungus from one favourable location to another, across unsuitable substrates. Rhizomorphs are seen, for example, in the tree parasite *Armillaria mellea* (the boot-lace or honey fungus) and the dry-rot fungus *Serpula lacrymans*.

rhizophore Any of the leafless branches, seen in many species of *Selaginella*, that arise from the stem at points of forking. Usually two rhizophores form at each fork but only one continues to develop. The rhizophore grows towards the soil, often branching repeatedly before reaching the soil surface. On contact with the ground, roots are produced from the swollen tips of the rhizophore.

rhizopodium A thin outgrowth of cytoplasm formed by certain algae that lack rigid cell walls. Rhizopodial algal cells are seen in the Chrysophyta, Xanthophyta, and Dinophyta.

Rhodophyta (red algae) A division containing nonflagellate usually thalloid *algae, most of which live in tropical seas, many epiphytic or parasitic on other algae. Red algae contain the phycobiliprotein pigments *phycocyanin and *phycoerythrin, which confer the characteristic red colour, although freshwater species are often grey-green.

The Rhodophyta contains one class, the Rhodophyceae, divided into two subclasses, the Bangioideae (or Bangiophycidae) and the Florideae (or Florideophycidae). Sexual reproduction is less complex in the Bangioideae and this group also lacks pit connections between the cells, which are a prominent feature of the Florideae. Certain members of the Florideae secrete a calcareous covering and play a part in the formation of coral reefs.

rhizopodium A thin outgrowth of cytoplasm formed by certain algae that lack rigid cell walls. Rhizopodial algal cells are seen in the Chrysophyta, Xanthophyta, and Dinophyta.

Rhynie chert A siliceous rock of the *Devonian period in Scotland, in which well preserved psilopsid fossils are found. In some, anatomical details can be clearly seen. The plants are similar to *Psilotum*, a living psilopsid, in having aerial branches, tracheids, an epidermis with stomata, and sporangia at the ends of the branches. Examples are *Rhynia* and *Horneophyton*. These extinct psilopsids are placed in the order Psilophytales, named after the Devonian fossil *Psilophyton*, discovered in eastern Canada in 1859. Another important fossil in the Rhynie chert, *Asteroxylon*, resembles a lycopsid.

rhytidome *See* bark.

riboflavin (vitamin B_2) A flavin pigment that has been implicated both in the perception of the phototropic stimulus and in photooxidation of endogenous auxins. It consists of an isoalloxazine ring substituted with the sugar alcohol ribitol. In its active phosphorylated form riboflavin phosphate, otherwise known as *FMN, it acts as a coenzyme for various oxidizing enzymes. Riboflavin is not synthesized by animals. *See also* FAD, vitamin.

ribonucleic acid *See* RNA.

ribose An aldopentose sugar (*see illustration at* aldose). It is a component of nucleotides (e.g. AMP, ATP), dinucleotides (e.g. NAD, FAD), and ribonucleic acid (RNA). Unlike *deoxyribose ($C_5H_{10}O_4$), the empirical formula of ribose ($C_5H_{10}O_5$) is that of a typical five-carbon monosaccharide.

ribosomal RNA (rRNA) A form of *RNA that is confined entirely to the *ribosomes. Synthesis of rRNA takes place on distinctive parts of the chromosomes associated with the nucleolus, called *nucleolar organizers. The base composition of rRNA is very similar in all species. After synthesis it combines with protein to form nucleoprotein. The partly constructed ribosomes then leave the nucleus, via nuclear pores, and enter the cytoplasm where synthesis is completed. The absence of fully constructed ribosomes in the nucleus may account for the absence of protein synthesis there. *Compare* messenger RNA, transfer RNA.

ribosomes Protoplasmic particles that are sites for the assembly of amino acids into the polypeptide chains of protein molecules in the order dictated by the genetic code of messenger RNA. During the process mRNA, formed in the nucleus, becomes attached to the ribosomes. The ribosomes of prokaryotic cells contain 60–65% ribosomal RNA and 35–40% protein while those of eukaryotic cells contain approximately equal quantities of each. In prokaryotic cells, e.g. *Escherichia coli*, they occur freely scattered throughout the protoplasm, but in eukaryotic cells most are associated with the membranes of the endoplasmic reticulum. They have also been identified in mitochondria and chloroplasts. Ribosomes consist of two parts or subunits, unequal in size and each containing RNA and a number of proteins. The ribosomes of prokaryotes are smaller than those in the cytoplasmic matrix of eukaryotes but similar in size to those in eukaryotic chloroplasts and mitochondria. During protein synthesis both the developing polypeptide chain and the ribosome are translocated along the mRNA molecule as the genetic code is translated. One of the proteins identified in the larger ribosomal subunit is peptidyl transferase, an enzyme that catalyses the formation of peptide bonds between amino acids.

ribulose A five-carbon ketose sugar (*see illustration at* ketose). In the phosphorylated form, ribulose 5-phosphate, it is an intermediate in the pentose phosphate pathway and the *Calvin cycle. As *ribulose bisphosphate it is important as the carbon dioxide acceptor in photosynthesis.

ribulose bisphosphate (RuBP) A phosphorylated form of ribulose that acts as the initial carbon dioxide acceptor in C_3 photosynthesis and as the eventual acceptor in C_4 photosynthesis and crassulacean acid metabolism. Addition of carbon dioxide occurs between the second and third carbon atoms and results in the formation of two molecules of phosphoglyceric acid. These feed into the Calvin cycle, which allows the fixed carbon dioxide to be converted into carbohydrate and regenerates the ribulose bisphosphate.

ribulose bisphosphate carboxylase (RUBP carboxylase, carboxydismutase) An enzyme that catalyses the carboxylation of ribulose bisphosphate to two molecules of phosphoglyceric acid. This is the first step in the *Calvin cycle. RUBP carboxylase is the *fraction 1 protein* of chloroplasts. It is very abundant in chloroplasts and may account for up to 15% of the total protein.
This enzyme can also act as an oxygenase, forming a molecule of phosphoglyceric acid and a molecule of phosphoglycollic acid from ribulose bisphosphate and oxygen. This reaction forms the basis of photorespiration.

ribulose diphosphate (RuDP) An outdated term for the initial carbon dioxide acceptor in photosynthesis that has now been replaced by *ribulose bisphosphate. The terms di- and bis- both refer to the presence of two groups in a molecule. In diphosphates the second phosphate group is added directly to the first (e.g.

adenosine diphosphate). In bisphosphates the second phosphate is on a different carbon atom to the first (e.g. ribulose bisphosphate or fructose bisphosphate).

Rickettsiales An order containing obligately parasitic bacteria that are much smaller (0.2–0.5 μm) than other bacteria. They cause a number of serious diseases in animals, e.g. typhus and Q fever.

ringing experiments A means of investigating the routes of water and carbohydrate translocation in plants by removing a ring of the outer tissues of the stem containing the phloem. This results in sufficient water still reaching the upper shoots to keep them turgid but photosynthetic products from the leaves are not translocated to regions below the ring and tend to accumulate above it. Such observations confirm that water is transported within *xylem and photosynthates within *phloem vessels, the former being undisturbed and the latter disrupted by the ringing process.

ring-porous wood Secondary xylem in which there is marked seasonal variation in tracheary-element diameter with wider elements being formed in spring and early summer and narrower elements being formed later in the year. Growth rings are distinct in such wood, e.g. that of pedunculate oak, *Quercus robur*. *Compare* diffuse-porous wood.

riverine forest *See* forest.

RNA (ribonucleic acid) A single-stranded nucleotide polymer, each nucleotide being constructed from phosphoric acid, ribose, and an organic base. The base may be adenine or guanine (the purine bases), or cytosine or uracil (the pyrimidine bases). The polynucleotides are held together by phosphodiester bonds between the phosphate group of one nucleotide and the sugar of an adjacent nucleotide. The RNA molecule may be linear, as in messenger RNA, or fold back on itself to form a three-dimensional clover-leaf-shaped molecule, as in transfer RNA. RNA is the principal agent for the transcription (copying) and translation (conversion) of the genetic code during protein synthesis. Various types of RNA are associated with these processes, namely *messenger RNA, *transfer RNA, and *ribosomal RNA. Viruses are exceptional and may not conform to the above generalizations. In many plant viruses, such as tobacco mosaic virus, the genetic material is RNA not *DNA. In some viruses, such as ϕ6 bacteriophage, the RNA is double stranded, like DNA.

rod cell *See* macrosclereid.

rogue 1. Any plant that varies from the rest of the crop and is consequently not wanted. Examples are wild oats (*Avena fatua*) growing in fields of cultivated oats (*A. sativa*), plants growing from self-pollinated seed in a field of F_1 hybrids, and diseased plants, such as wheat plants affected with covered smut. **2.** To remove and destroy such plants.

root The usually underground part of the plant *axis, specialized for anchorage, absorption, and sometimes food storage. It may usually be distinguished from a stem by the absence of chlorophyll and buds. (Exceptions are the aerial roots of certain epiphytes, e.g. *Taeniophyllum*, that can develop chlorophyll when illuminated. Adventitious buds may form on roots and give rise to suckers.) The root system may include an obvious main root derived from the radicle (*see* taproot) or may be fibrous due to repeated branching of the radicle. Alternatively it may include or consist solely of adventitious roots arising from the base of the stem.

The vascular tissues of the root normally form a solid central stele, which is better able to resist the tensions and pressures exerted on a root than would be a hollow stelar cylinder, typical of most stems. The absorptive properties of the root are enhanced by the formation of *root hairs behind the tip. Beyond these the root branches to form lateral roots. Unlike stem branches, these do not arise superficially but develop from the outer tissues of the stele and grow through the root cortex. Lateral root formation is thus said to be *endogenous*. Roots are usually positively hydrotropic and geotropic. A root cap (*see* calyptra) at the root tip protects the root as it grows down through the soil. Numerous modifications of roots exist (*see* aerial root, climbing root, contractile root, pneumatophore, prop root). Symbiotic associations between plant roots and various fungi, bacteria, and blue-green algae are common (*see* mycorrhiza, root nodule).

root cap *See* calyptra.

root hair A *trichome originating from the *piliferous layer of the root. Root hairs are projections from single epidermal cells in direct contact with the soil and serve to increase the surface area for absorption. They are responsible for the first stage in absorption of water and solutes from the soil. They are also thought to help retain contact between the root tip and the soil when dry conditions tend to cause the soil to contract away from the roots.

root nodule A tumorous growth that develops on the roots of leguminous and certain other plants in response to infection by symbiotic microorganisms. In legumes the symbiont is always a bacterium of the genus *Rhizobium.* In nonleguminous plants with root nodules the symbiont appears to be either a member of the Plasmodiophorales, e.g. in the roots of bog myrtle (*Myrica gale*) and alder (*Alnus glutinosa*), or a blue-green alga, e.g. in the roots of some *Gunnera* species. In legumes, following invasion of the root tissues, the bacteria induce a localized proliferation of the host tissues. Like the induction of crown gall (*see* tumour inducing principle), the ability of *Rhizobium* to induce nodules appears to be controlled by a plasmid. The plasmid also controls the host specificity of different *Rhizobium* strains. *Nitrogen-fixation is carried out by the bacteria, which assume a characteristic shape and size within the host cells (*see* bacteroid).

root pressure The pressure that can build up in the root systems of plants so as to force water upwards through the xylem vessels. The force is a function of the *osmotic potential of the root cell contents and is evidenced by the continuing flow of water from the cut surface of a recently severed stem. Positive root pressures tend to build up at night, when the rate of transpiration is very low, and lead to the process of *guttation. Root pressure is not of great importance in water uptake and varies seasonally, being lowest in magnitude during the summer.

root rot A plant disease in which there is disintegration of the root tissues. Above-ground symptoms are often initially similar to those of nutritional disorders or bad drainage. Annual plants appear sickly and eventually wilt and die as a result of shortage of water and nutrients. Affected trees usually show symptoms of gradual decline and dieback of the crown. Numerous different species of fungus cause root rots. *Armillaria mellea* (the honey fungus) is a ubiquitous root pathogen affecting many forest and garden trees. Other common root-infecting fungi include *Corticium*, *Phytophthora*, and *Fusarium*. *See also* take-all.

rootstock 1. A short erect underground stem, as seen in various angiosperms, e.g. plantains (*Plantago*), and certain pteridophytes, e.g. *Isoetes*, *Osmunda.* It is the equivalent of a vertical rhizome.

2. Any underground part of a plant. *See* stock.

Rosaceae A large cosmopolitan dicotyledonous family containing about 3400 species in some 122 genera. It includes both woody and herbaceous plants. The leaves are usually alternate and, except in *Spiraea* and some other members of the subfamily Spiraeoideae, have two stipules at the base. The flowers are usually actinomorphic and often large and showy. There are numerous stamens arranged in whorls and the carpels are also usually numerous and free. An epicalyx is often present. Many different types of fruit are formed, the pome being characteristic of the subfamily Maloideae, e.g. *Malus* (apples), *Pyrus* (pears), and *Cydonia* (quinces), and a drupe characteristic of the subfamily Prunoideae, e.g. *Prunus* (plums, cherries, apricots).

The family contains most of the important orchard fruits, many bush fruits, e.g. blackberries and raspberries (*Rubus*), and other soft fruits, e.g. strawberries (*Fragaria*). It also includes many important ornamental genera, notably *Rosa*, of which there are estimated to be over 5000 cultivars. Other ornamentals include *Cotoneaster*, *Kerria*, *Potentilla*, and *Spiraea*.

rosette plant Any plant with its leaves radiating outwards from a short stem at soil level. The rosette habit enables such plants to survive grazing and trampling and to be more successful in competing with other species for space. Examples are dandelions, daisies, and plantains.

Rosidae A subclass of the dicotyledons containing both woody and herbaceous species. Its members usually have bisexual flowers that often contain numerous stamens, which develop centripetally. The Rosidae may contain some 14–21 orders depending on the classification scheme referred to. The inclusion of the following 13 orders is generally agreed: Rosales, including the *Rosaceae; Fabales, including the *Leguminosae; Myrtales, including the Rhizophoraceae (e.g. mangroves) and Myrtaceae; Proteales; Podostemales; Santalales, including the Santalaceae (e.g. sandalwoods); Celastrales, including the Celastraceae (e.g. spindle tree) and Aquifoliaceae (e.g. hollies); Rhamnales, including the Rhamnaceae (e.g. buckthorns) and Vitaceae (e.g. grape vines); Sapindales, including the Sapindaceae (e.g. soapberries), Hippocastanaceae (e.g. horse chestnuts), and Aceraceae (maples); Geraniales, including the Geraniaceae, Oxalidaceae (e.g. wood sorrels), and Balsaminaceae; Polygalales; Umbellales (or Cornales), including the Araliaceae (e.g. ivies) and *Umbelliferae; and Haloragales (or Hippuridales).

In addition, the following four orders may be recognized but these are often included in other orders of the Rosidae: Saxifragales, including the Crassulaceae (e.g. stonecrops) and Saxifragaceae, often placed in the Rosales; Connarales (e.g. zebra woods), and Rutales, including the Rutaceae (citrus family), Meliaceae (e.g. mahogonies), and Anacardiaceae (e.g. cashews), both often placed in the Sapindales; and Elaeagnales, containing the Elaeagnaceae (e.g. oleasters), often placed in the Proteales.

Some classifications include the Rafflesiales (often placed in the Magnoliidae) and the Juglandales (often placed in the Hamamelidae) in the Rosidae. The order Oleales, including the Oleaceae (e.g. olives, lilacs), is often seen in the order Gentianales of the subclass Asteridae. The Euphorbiales, including the *Euphorbiaceae (spurge family), is sometimes classified in the Dilleniidae. Finally the two families Droseraceae (e.g. sundews) and Nepenthaceae (pitcher plants) making up the order Nepenthales are sometimes split and allocated to the Rosales and Aristolochiales (subclass Magnoliidae) respectively.

rostellum A flap of sterile tissue that separates the stigmatic surface from the anthers in the column (the structure formed by the fusion of the sex organs) of an orchid flower. It is a modified stigma.

rot A plant disease in which there is disintegration of the tissues. Rots are particularly important as postharvest diseases. *See also* dry rot, wet rot, foot rot, soft rot, root rot.

rough endoplasmic reticulum *See* endoplasmic reticulum.

rRNA *See* ribosomal RNA.

Rubiaceae A large family of dicotyledonous plants containing about 7000 species in some 500 genera. It is mainly tropical in distribution, all the tropical species being woody. Temperate representatives are always herbaceous and include such species as *Galium saxatile* (heath bedstraw) and *Asperula arvensis* (blue woodruff). The temperate species characteristically have square stems and leaves borne in whorls. All members, both temperate and tropical, bear stipules and the flowers are usually actinomorphic and hermaphrodite. The ovary is normally inferior and in most members contains two carpels.

Commercially important products of the Rubiaceae include coffee, from *Coffea arabica* and *C. canephora*, and quinine, from species of *Cinchona*. *Gardenia* species are much planted as ornamentals.

rubidium–strontium dating *See* radiometric dating.

RuBP *See* ribulose bisphosphate.

RuDP *See* ribulose diphosphate.

rugose Wrinkled.

ruminate Describing endosperm that is irregularly grooved or ridged and so appears chewed as, for example, seen in many members of the Myristicaceae.

runner A creeping stem that arises from an axillary bud and runs along the ground, giving rise to plantlets at the nodes, as in the creeping buttercup (*Ranunculus repens*), or apex, as in the wild strawberry (*Fragaria vesca*). Runners are formed by many rosette plants. They often differ greatly from the normal stem of the plant and usually possess greatly lengthened internodes. *See also* stolon.

russetting The development of brown corky patches on certain fruits, especially apple varieties, in response to various chemicals. Russetting of leaves, tubers, and other plant organs may also occur in response to damage or infection.

rusts Plant diseases caused by fungi of the order *Uredinales. Rust diseases are easily recognized by streaks of dark pustules on the leaves or stems. Autoecious rusts complete their life cycle on one host whereas heteroecious rusts have an alternate host for some of their spore stages. Many rust fungi cause economically important diseases. Rust of currant is caused by the heteroecious rust *Cronartium ribicola* whose alternate host is pine. Black stem rust of cereals (*Puccinia graminis*) has barberry (*Berberis*) as an alternate host.

ruthenium red A temporary stain that turns mucilage and certain gums pink. It is used as a test for pectin in the middle lamella of plant cells.

S

saccharide *See* carbohydrate.

saccharose *See* sucrose.

sac fungi *See* Ascomycotina.

safranin A red permanent stain that is used especially to stain nuclei in plant cells. It also stains lignified and cutinized tissues red and chloroplasts pink. It is usually used with a green (e.g. fast green) or blue (e.g. haematoxylin) counterstain.

sagittate Describing a structure shaped like an arrowhead. Sagittate leaves have two barbs extending back behind the point where the petiole is joined to the leaf. They are characteristic of the genus *Sagittaria*. The marsh arrow-grass (*Triglochin palustris*) is so named because of its sagittate dehisced fruits.

***S* alleles** A multiple allelic series governing incompatibility reactions in certain plant species. Plants possessing the same two *S* alleles are incompatible. Plants sharing one *S* allele are semicompatible in that about half the pollen grains will be able to germinate on the style and subsequently achieve pollination. Such plants are usually indistinguishable from fully compatible plants (plants possessing two different pairs of *S* alleles) since pollen is usually produced in sufficient quantity to mask the fact that 50% is inviable.

S alleles have been found in many fruit crops (e.g. *Prunus avium*), in several grasses, in clover, and in brassicas. There may be more than 40 such alleles in a species and complex dominance relationships are found.

salt absorption The uptake of inorganic ions across the *semipermeable membranes of plants. If the absorption is against a diffusion gradient, as is often the case across the tonoplast, then *active transport is involved. In addition, the uptake is often selective. For example, in sea water the concentration of sodium ions is around forty times higher than that of potassium ions. However within the cells of marine algae the potassium concentration is often about six times greater than that of sodium.

salt marsh A type of vegetation found in sheltered river estuaries subject to frequent covering by the tides. Salt marshes are formed from stabilized pioneer communities (*see* halosere).

Salviniales *See* Filicales.

samara A type of *achene with a pericarp extended into a membranous wing, which aids wind dispersal of the seed. The winged fruits or *keys* of the ash (*Fraxinus excelsior*) are an example. The double samara, typical of sycamores and maples (*Acer*), is a kind of schizocarp.

sand Mineral particles, usually consisting mainly of quartz, with a diameter of 0.05–2.0 mm. A sandy soil is defined as one containing at least 85% sand and not more than 10% clay.

sand dune An accumulation of blown sand found on the coast and in inland desert areas. Coastal dunes are formed by inshore winds that carry sand particles and deposit them on various obstacles in their path, such as pieces of seaweed. In regions that are relatively undisturbed by the tides, seeds of xerophytic plants germinate and establish a pioneer community termed a *psammosere*. The plants tolerate abrasion by blown sand, high winds, high temperatures in the day, and some salt from sea spray. Sand twitch (*Agropyron junceiforme*) helps to stabilize the dune with its extensive network of rhizomes. It is also tolerant to limited immersion in sea water. Marram grass (*Ammophila arenaria*) then becomes established, although in some cases it may itself be the pioneer plant. Like sand twitch, marram grass has a network of rhizomes that stablizes the dune but it is intolerant to immersion in sea water. The landward slopes of the dunes are more sheltered and less steep than the seaward slopes and other plants grow including lyme grass (*Elymus arenarius*), sea holly (*Eryngium maritimum*), etc. A dune system is established, the innermost series being stable and termed *fixed* dunes. Behind these, dune pasture occurs with typical dune plants together with species commonly found in pasture. *See also* xerophyte, xerosere.

sap The liquid, consisting of mineral salts and sugar dissolved in water, that is found in xylem and phloem vessels. The term is also used of the fluid in the cell vacuole (*vacuolar sap*).

sapogenin *See* saponin.

saponin Any of a class of bitter-tasting *glycosides in which the aglycone portion (the *sapogenin*) is a steroid alcohol. Saponins are soluble in water, characteristically producing a foam, hence their name. Many are toxic to animals and have been used as fish poisons. Some saponins are important in the commercial production of steroid hormones, notably diosgen, obtained from various yam (*Dioscorea*) species. The function of saponins in the plant is obscure though they may serve to deter predators.

saprobe The preferred term for a *saprophyte when the organism concerned is a fungus.

Saprolegniales (water moulds) An order of the *Oomycetes containing some 32 genera and about 200 species of aquatic, mainly saprobic, fungi (e.g. *Saprolegnia*), often with *diplanetic zoospores. Some are parasitic on fish, e.g. *S. parasitica*, which often attacks salmon and goldfish.

saprophyte A plant that feeds by the external digestion of dead organic material thus bringing about decay. Many fungi and bacteria are saprophytes and play an important part in the recycling of matter, as the inorganic by-products of their digestion can be rebuilt into organic compounds by green plants.

sapwood (alburnum) The outer functional part of the secondary xylem cylinder, as compared to the central nonfunctional *heartwood.

satellite (trabant) A spherical body seen attached to one end of a chromosome arm by a narrow filament at mitotic metaphase. Often a pair of satellites are apparent. Usually satellites are only found on one or two chromosomes but occasionally more are found. The number and position of satellites are one of the features of the *karyotype. The satellite stalk functions as a *nucleolar organizer.

satellite DNA *See* repeated sequence.

savanna (savannah) Tropical *grassland.

scab Any plant disease having conspicuous raised scablike lesions that develop as a result of the formation of cork layers. The host response to the disease is similar to that found in *canker but occurs to a lesser degree. Scab diseases are very varied. Examples include common scab of potatoes caused by the bacterium *Streptomyces scabies* and apple scab caused by the fungus *Venturia inaequalis*.

scalariform Resembling a ladder as, for example, certain secondary xylem elements possessing parallel bands of thickening or some perforation plates that have several pores separated by parallel bars of tissue.

scalariform thickening (scalariform-reticulate thickening) A type of *reticulate secondary wall patterning in *tracheary elements in which the network is broken by elongate unthickened areas arranged in a more or less parallel fashion. Whereas *annular thickening and *spiral thickening permit further elongation, reticulate and scalariform thickenings allow little further extension, and are therefore to be found in tissues, such as metaxylem and secondary xylem, that do not elongate after maturation. *Compare* pitted thickening.

scandent Climbing.

scanning electron microscope *See* electron microscope.

scape The leafless stem of a solitary flower or inflorescence, such as that of the dandelion inflorescence.

scarification The abrasion or chemical treatment of the surface of a *hard seed to make it permeable to water and so hasten germination.

scarious Having a dry membranous appearance, but fairly stiff.

Schiff's reagent A colourless solution that is produced by the reduction of basic fuchsin (a magenta dye) with sulphurous acid. It is used in histochemical tests to detect aldehyde and ketone groups in certain compounds, which oxidize the reagent and restore its magenta colour. *See also* Feulgen's test.

schizocarp A dry fruit that is derived from two or more one-seeded carpels that divide into one-seeded units at maturity. The one-seeded units may be achenes, berries, follicles, mericarps, nutlets, or samaras. This form of fruit is intermediate between the dehiscent and indehiscent types. *See also* cremocarp. *See illustration at* fruit.

schizogeny The formation of a space by the separation of cells. *Compare* lysigeny.

Schizomycetes (fission fungi) Formerly, a class of the fungi containing the bacteria. With the realization that bacteria have no affinities with fungi the name is going out of use.

Schultze's solution (chlor-zinc iodide, CZI) A temporary stain containing a mixture of zinc dissolved in hydrochloric acid added to iodine dissolved in potassium iodide. It is used to detect the presence of cellulose in plant tissue, which it turns blue. Lignified walls stain blue-green, lignin and suberin yellow, and starch blue-black.

scion A shoot or bud taken from one plant and joined by *grafting or *budding onto another plant, the *stock.

sclereid A relatively short *sclerenchyma cell, usually formed when the wall of a parenchyma cell undergoes secondary thickening and, often, lignification. The simple pits of sclereids are often more conspicuous than those of *fibres. Sclereids are often present as idioblasts in other tissues. They may however be present in large numbers, as, for example, in those plants where they form the testa of the seed. *See also* astrosclereid, brachysclereid, macrosclereid.

sclerenchyma Strengthening tissue composed of relatively short cells (*sclereids) and/or relatively long ones (*fibres) with thick, often lignified, cell walls and usually lacking a living protoplast at maturity. Sclerenchyma cells usually possess simple unbordered pits, although *fibre-tracheids may have pits with a slightly raised border. Sclerenchyma may form by thickening (sclerification) of the secondary walls of parenchyma cells, often involving lignification, or it may develop directly from meristematic tissue.

sclerotium **1.** The normally pseudoparenchymatous resting body of certain fungi, which can survive long periods of adverse conditions to produce either a mycelium or fruiting bodies. Sclerotia are important sources of inoculum for various root-rot diseases. The hard dark club-shaped structures seen in the ears of cereals and grasses affected with ergot (*Claviceps purpurea*) are sclerotia. **2.** *See* plasmodium.

SCP *See* single-cell protein.

Scrophulariaceae A large family of dicotyledonous plants, numbering about 3000 species in some 220 genera. The family, commonly called the foxglove or figwort family, is distributed worldwide though the greatest concentration of

species is in northern temperate regions. Most of its members are herbaceous. The leaves occur in a variety of forms but always lack stipules. There is also a range of flower and inflorescence types. The flowers are usually irregular and often two-lipped, as in the snapdragons (*Antirrhinum*, *Misopates*) and toadflaxes (*Linaria*, *Cymbalaria*). However in the mulleins (*Verbascum*) and speedwells (*Veronica*) the flower is almost actinomorphic. In many flowers there is a reduction in the number (assumed originally to have been five) of floral parts. For example, figworts (*Scrophularia*) only have four stamens, and speedwells have four sepals and four petals and only two stamens. The fruit is usually a dehiscent capsule.
Few members of the family are of economic importance though many genera, e.g. *Calceolaria*, *Hebe*, *Mimulus*, *Antirrhinum*, include ornamental varieties. Some species are parasitic on the roots of other angiosperms, especially grasses, and may be serious weeds, e.g. *Striga* (witchweeds). The foxglove (*Digitalis purpurea*) is the source of the drug digitalin (*see* cardiac glycoside).

scutellum An intermediate absorbing organ between the embryo and the nutritive endosperm in the grass fruit (caryopsis). During germination it secretes enzymes involved in the digestion of the endosperm. It is thought to be a modified cotyledon.

seaweed Any of various macroscopic marine *algae found on rocky coasts or free floating in the sea. Species occurring in the intertidal zone are periodically exposed to the air and have developed leathery mucilaginous thalli to prevent desiccation. Most seaweeds belong to the *Phaeophyta or *Rhodophyta. *See also* phytoplankton.

secondary cortex *See* phelloderm.

secondary growth (secondary thickening) The increase in diameter of a plant organ resulting from cell division in a *cambium. Secondary growth results in the formation of *secondary tissues*, for example the secondary xylem, secondary phloem, phellem, and phelloderm. *Primary tissues, on the other hand, are derived from primary meristems. Secondary growth occurs in most dicotyledons and gymnosperms and a few monocotyledons. There are relatively few living representatives of lower vascular plants exhibiting secondary growth but it was common in many extinct species.

secondary phloem *Phloem derived from the vascular cambium in plants exhibiting secondary growth. As in the *secondary xylem, the secondary phloem consists of two systems: the *axial* (vertical) *system*, derived from the fusiform initials of the vascular cambium and consisting mainly of *sieve elements, their associated *companion cells, and some phloem fibres; and the *ray* (horizontal) *system*, derived from the ray initials of the vascular cambium and consisting mostly of ray parenchyma.

secondary ray *See* ray.

secondary thickening *See* secondary growth.

secondary tissue *See* secondary growth.

secondary xylem *Xylem derived from the vascular cambium in plants exhibiting secondary growth. Secondary xylem is regarded as consisting of two systems: the *axial* (vertical) *system*, derived from the fusiform initials of the vascular cambium and consisting mainly of tracheary elements and fibres; and the *ray* (horizontal) *system*, derived from the ray initials of the vascular cambium and consisting mostly of ray parenchyma, forming the *rays. As well as differing from *primary xylem in origin, secondary xylem usually also has shorter tracheary elements, often arranged in regular rows, although these differences are not always reliable. Commercial wood is secondary xylem. There are two main types of commercial wood, *softwood* from gymnosperms and *hardwood* from angiosperms. However, these terms are somewhat misleading as there are hard gymnosperm woods and soft angiosperm woods. Softwoods contain only tracheids as the water-conducting elements, whereas hardwoods usually contain vessels.

section A rank in the taxonomic hierarchy above series but subordinate to genus. Genera containing a large number of species are often subdivided into sections. The section name is printed in italic with a capital first letter and either has the same form as the generic name or, if a plural adjective, agrees in gender with the generic name. For example, the genus *Chenopodium*, which contains about 110 species, is subdivided into the four sections *Agathophytum*, *Chenopodium*, *Pseudoblitum*, and *Morocarpus*. The section name is preceded by the abbreviation sect. to show its rank, thus *Chenopodium* sect. *Pseudoblitum*.

sectorial chimaera A *chimaera that results from a cell mutation in a meristem that subsequently gives rise to a group or sector of mutant cells. Such chimaeras are usually unstable.

sedimentation coefficient Symbol: s. The rate of sedimentation in centimetres per second that occurs when a solution is subjected to a centrifugal force of one dyne (10^{-5} newton). These rates are extremely low and are usually multiplied by 10^{13} to give a reasonable figure. In this modified form the unit is termed the *Svedberg unit*, symbol S. Different molecules and cellular inclusions tend to have characteristic sedimentation rates depending on the weight and shape of the molecule. For example, the ribosomes of prokaryotes have a value of 70S and those of eukaryotes 80S.

seed The structure that develops from the fertilized ovule in seed plants. It usually contains one *embryo together with a food supply, which may be contained in a specially developed *endosperm or in the *cotyledons of the embryo. The whole is surrounded by a protective coat, the *testa. In gymnosperms the seed remains naked and unprotected but in angiosperms it is enclosed within the ovary wall.

A seed may germinate immediately on ripening or may have special *dormancy mechanisms to prevent germination under unfavourable conditions.

Although seed production is only observed today in angiosperms and gymnosperms, there are extinct forms possibly intermediate between these and the nonseed-bearing vascular plants (*see* Pteridospermales). The development of the seed-bearing habit has released seed plants from dependence on the availability of water for their reproductive phase, thus opening up new habitats for colonization. In addition the seed allows for wide dispersal of the plant and, in annual and ephemeral plants, serves as an organ of perennation.

Some plants produce seed without fertilization taking place (*see* apomixis).

seed dressing A chemical applied to seeds to protect them against fungal diseases or insect attack. The chemicals are applied in dusts, slurries, or concentrated solutions. Treated seed is toxic and must be distinguishable from untreated seed to prevent it being used for feed. *See also* organomercurial fungicide, quinone fungicide.

seed ferns *See* Pteridospermales.

seed leaf *See* cotyledon.

seedless fruit A fruit that develops without fertilization occurring so no seed is formed. *See* parthenocarpy.

seedling A young plant. The seedling often differs significantly from the mature plant in morphology and habit. *See* juvenility.

seedling blight A disease of germinating seedlings. The term is sometimes applied to post-emergence *damping-off but generally it is used when well established seedlings are affected. Growth slows and foliage may become yellow or wilt. Roots of affected plants are found to be rotten and there may be lesions at soil level. The causal fungi include *Pythium*, *Phytophthora*, *Rhizoctonia* (*Corticium*), *Fusarium*, and *Helminthosporium*. Correct planting conditions, seed dressings, and soil sterilization are the usual preventative measures.

seed plants *See* Spermatophyta.

segmental allopolyploid *See* allopolyploidy.

segregation The separation of homologous chromosomes at anaphase 1 of meiosis. As a consequence of segregation, alleles that were paired in the somatic cells of the organism become separated, so that only one allele of each kind is present in a single gamete.

segregation ratio The proportion of one type of offspring to another that occurs as a result of separation of alleles at meiosis (*see* dihybrid ratio).

seismonasty A *nastic movement in response to shock. The result, as in the collapse of *Mimosa pudica* leaflets after subjection to shaking or singeing, is often the assumption of the night position as normally found after the completion of *nyctinasty. *See also* haptonasty.

Selaginellales An order of the *Lycopsida containing one extant genus, *Selaginella*, the 700 or so species of which are mainly tropical in distribution. In gross morphology *Selaginella* resembles *Lycopodium* (*see* Lycopodiales) except that it has ligulate leaves and the sporophylls are always grouped in strobili. The leaves are usually inserted spirally but in some species they are arranged in four ranks, the upper two consisting of small leaves adpressed to the stem while the lower two ranks contain expanded leaves. The roots are borne at the end of leafless branches (*see* rhizophore). Internally, selaginellas are distinguished by their trabeculate endodermis, which consists of filamentous cells that traverse a continuous cavity between the cortex and the pericycle, so suspending the stele in the centre of the stem.

The strobili bear both mega- and microsporangia and produce the largest megaspores of any spore-producing plant. The megaspore contains a large food supply, which enables the female gametophyte to develop independently of external food sources. The microspore develops into an extremely reduced male gametophyte consisting of a prothallial cell and an antheridium, which produces many biflagellate antherozoids. Fertilization relies on a megaspore and microspore being in close proximity and a film of water is needed for the antherozoids to swim to the egg cell of the archegonium. Occasionally fertilization occurs before the megaspore is shed from the parent.

selection pressure The intensity with which the environment tends to alter the frequency of a particular allele in a given population. It is not possible to give an absolute value for selection pressure. However, by comparing the survival rates of individuals with different alleles, a measure of the *fitness of one relative to another may be derived. *See* natural selection.

selective Describing herbicides that kill some plants and not others. Usually broad-leaved plants are killed while narrow-leaved plants are unaffected. Selective weedkillers are therefore useful for cereal crops and lawns. Some selective weedkillers, e.g. Simazine, kill germinating plants but not well established shrubs. *See also* systemic.

selective absorption *See* salt absorption.

self fertilization *See* autogamy.

self incompatibility (self sterility) The failure of gametes from the same plant to form a viable embryo. In angiosperms this is usually due to complex interactions between the pollen and stigmatic tissues. Self incompatibility caused by the stigmatic tissues is described as sporophytic while that due to the pollen tube is termed gametophytic. It may be achieved by the prevention of pollen germination, the retardation of growth or the disorientation of the pollen tube, or by failure of nuclear fusion. Although self incompatibility is known to be genetically controlled (*see S* alleles) the exact method by which the genetic information manifests itself is not clearly understood. Enzymes either on the pollen or stigma, antibodies and antigens, growth substances, or variable cell morphology may account for differential treatment of pollen grains on the stigma. Incompatibility is rarely total and usually develops as the stigma matures. Artificial crosses of normally incompatible lines can therefore be made by in-bud pollination, chemical treatment, or by stigma excision.

self pollination The transfer of pollen from the anthers to the stigma of the same flower, or to a flower on the same plant. This is achieved by homogamy and appropriate positioning of the reproductive parts so that pollen can be transferred, usually by insects, gravity, contact, or rain splash. Self pollination does not necessarily preclude *cross pollination except in cleistogamous plants.

self sterility *See* self incompatibility.

Seliwannoff's test A test for ketose sugars, such as fructose, in a solution. The reagent is made by dissolving a few crystals of resorcinol in equal amounts of concentrated hydrochloric acid and water. In the presence of ketoses it forms a red colour on heating.

SEM *See* electron microscope.

semantide An information-carrying molecule. DNA, as the carrier of the genetic code, is termed a *primary semantide* and is regarded by taxonomists and evolutionists as a primary source of information in drawing up classifications and evolutionary trees. RNA, as the transcription of the genetic code, is termed a *secondary semantide*, and proteins as the end product of this information transfer, *tertiary semantides*. Much effort has recently been directed to unlocking the information carried by such molecules. The amino acid sequences of various proteins, e.g. cytochromes and plastocyanins, have been determined and compared for a number of species. It is now also possible, using restriction enzymes, to determine the base sequences of nucleic acids. *See also* DNA hybridization.

semiconservative replication The process by which DNA makes exact copies of itself. During this process, double-stranded DNA uncoils and the separated polynucleotides act as templates for the formation of new complementary polynucleotide chains. Hence any given DNA molecule will consist of one old strand and one new one and the original molecule is semiconserved during the manufacture of an identical copy (replica).

seminal root Any of the adventitious roots that grow from the base of the stem during early seedling growth and take over the functions of the radicle.

semipermeable membrane A membrane that acts as a barrier to certain substances but will allow others to pass through. In many cases a semipermeable membrane will allow the passage of solvents, almost invariably water in biological situations, but not certain solutes. Such a membrane usually lets through small molecules but not larger ones. In such cases the membrane is said to be *selectively permeable*. *See also* osmosis.

senescence The period between maturity and death of a plant or plant part. A gradual deterioration occurs and, in the case of senescent leaves and fruits, the end point is usually *abscission from the plant. Senescence is characterized by an accumulation of waste metabolic products, a decrease in dry weight as reusable substances are withdrawn from the affected part, and a rise in the respiratory rate (*see* climacteric).

sensitivity *See* irritability.

sepal An individual unit of the *calyx. It is usually green and often hairy but in some species, e.g. marsh marigold (*Caltha palustris*), the sepals are brightly coloured and assume the function of *petals. In such plants the petals may be absent, as they are in marsh marigold, or reduced, as in the Christmas rose (*Helleborus niger*), where they form small tubular nectaries. Sepals are supplied by several vascular bundles and are thought to be modified leaves.

separation layer *See* abscission.

septicidal Describing fruit dehiscence in which the slits in the pericarp arise along the junctions of the carpels.

septum Any partition, whether within a cell, as in a septate fibre, or in an organ, such as a fruit. In fruits, such a separating wall is often termed a *dissepiment*.

sequestrol A preparation of trace elements used to treat deficiency diseases, particularly those caused by lack of iron or magnesium. The elements are chelated to promote their uptake by plant roots and prevent them becoming bound to the soil particles. A preparation containing a particular element is named accordingly, e.g. sequestering iron.

sere (seral community, seral stage) Any plant community in a succession leading to a *climax. The nature of a sere is influenced by that of the preceding sere and itself influences the development of the succeeding sere. The initial community (pioneer community) in a succession is known as a *prisere*. *Subseres* are the stages in the development of a secondary succession, while *microseres* are the stages in microhabitats, such as small puddles or animal droppings. *See also* clisere, halosere, hydrosere, xerosere.

serial endosymbiotic theory (progressive endosymbiotic theory) The proposal that plastids, mitochondria, and possibly also cilia and flagella arose from symbiotic prokaryotic organisms living within a eukaryotic host cell. It is thought that plastids probably originated from organisms similar to present-day blue-green algae while mitochondria arose from aerobic bacteria. Such conclusions are based on various similarities between plastids and mitochondria and free-living prokaryotes. For example, such organelles are self replicating and contain DNA, which in addition to having a different base composition from the nuclear DNA, is circular rather than linear. The ribosomes of mitochondria and chloroplasts are smaller (70S) than those in other parts of the cytoplasm (80S) but similar in size to those of prokaryotes. There is evidence moreover, that the origin of plastids could be polyphyletic, i.e. in different groups of plants different types of symbionts have been incorporated. Thus in the Chlorophyta, Rhodophyta, and higher plants, which all have chloroplasts with a double membrane, it is thought that the plastids are derived from a prokaryotic symbiont. However in other groups of algae the chloroplasts may be surrounded by three or four membranes, implying that these are derived from eukaryotic symbionts. Such evidence has been used in revising classifications at the kingdom and division level.

series A rank in the taxonomic hierarchy above species but subordinate to section.

serine A polar amino acid with the formula $HOCH_2CH(NH_2)COOH$ (*see illustration at* amino acid). Serine is synthesized from phosphoglyceric acid. It is also formed during photorespiration and probably during the catabolism of cytokinins. Breakdown of serine occurs by oxidative deamination to pyruvic acid. It is a component of several phosphoglycerides.

serology The study of blood serum. Following invasion of an animal body by foreign materials (*antigens*) the serum contains *antibodies* produced to combat the antigens. Serum containing antibodies produced against a particular kind of antigen is termed *antiserum*. Each antigen stimulates the production of a specific antibody active against that antigen alone or particles very like it. The specificity of the immune response has been exploited in the identification of viruses and in chemotaxonomy, where it is used to compare protein extracts from different plant species. This field of plant taxonomy is termed *serotaxonomy*. A laboratory animal, usually a rabbit, is injected with a plant protein extract. A few weeks later when the animal has produced sufficient antisera, a blood sample is taken. The antisera is then mixed in vitro with antigens from the same species used to raise the antisera (the homologous species extract) and subsequently against antigens from a range of other species. The degree of similarity between antiserum and antigen is measured by the turbidity produced in a liquid by the antibody–antigen precipitation reaction. The turbidity should be greatest when the antisera is mixed with the homologous species extract. Antigens from different species will produce varying amounts of precipitation, the more similar species producing greater turbidity than the less similar species.

serotaxonomy *See* serology.

serrate Describing a leaf margin that is toothed, with forward-pointing notches. Leaf margins finely toothed in this manner are termed *serrulate*.

sessile Unstalked, as a leaf with no petiole or a stigma with no style.

seta 1. The stalk of a bryophyte sporophyte that supports the capsule. At maturity, the seta usually elongates rapidly, bearing the capsule to a height suitable for spore dispersal. A seta often contains vascular tissue, enabling the sporophyte capsule to obtain nutrients from the parent gametophyte. *See also* apophysis.
2. A nervelike extension to a leaf, as seen in certain species of *Selaginella*.

Sewall Wright effect *See* genetic drift.

sex chromosome A chromosome that carries sex-determining genes. All other chromosomes are called *autosomes. Sex chromosomes are widespread in the animal kingdom, associated with the fact that most animals have separate sexes. However most plants are hermaphrodite and sex chromosomes are correspondingly rare in the plant kingdom, although they are found in dioecious plants. In some such plant species sex is determined by one pair of alleles. However in certain angiosperm genera (e.g. *Silene*, *Cannabis*) cytologically identifiable sex chromosomes are seen.

sexine The outer layer of the *exine in a pollen grain.

sex-limited inheritance The restriction of a trait to one sex as a result of incomplete *penetrance rather than allelic differences.

sex linkage (sex-linked inheritance) The tendency for a certain character to appear more often in one sex than the other. It occurs because the gene controlling that character occurs on the same chromosome as the gene or genes involved in sex determination. Sex-linkage is seen in some dioecious plants, e.g. *Silene album* (white campion) in which the male is XY and the female XX, the Y chromosome being considerably larger than the X.

sexual reproduction The formation of new individuals of a species by the fusion of two normally haploid gametes to form a diploid *zygote. The gametes may be derived from the same parent (*see* autogamy) or from two different parents (*see* allogamy). In certain unicellular organisms the whole individual may participate in the process, as seen when two haploid yeast cells fuse to form a diploid cell. In multicellular organisms the gametes are formed and often fuse in specialized organs (*see* gametangium). *Compare* asexual reproduction.

shade plant A plant that is able to flourish in conditions of low light intensity. Some shade species are sensitive to intense light and cannot live in the open. Shade plants usually have thinner epidermal and palisade layers and fewer stomata than normal. They also tend to have a short compensation period, i.e. the food reserves used in respiration at night are quickly replaced by photosynthesis in the day. Shade plants grow in woodlands where the trees, when in leaf, form a canopy, cutting out much of the light. Green light, the part of the spectrum least absorbed by plants, is predominant beneath a closed canopy. Examples of shade plants are mosses, enchanter's nightshade (*Circaea lutetiana*), certain violets (*Viola*), and dog's mercury (*Mercurialis perennis*). Some trees are also tolerant of shade. For example, yew trees can tolerate the deep shade in beech woods but ash trees cannot. However ash trees can survive in oak woods where the shade is not as dense. *Compare* sun plant.

shadowing (heavy-metal shadowing) A technique used in electron microscopy to determine the scale of surface features on a specimen. The specimen is partially coated with a thin layer of a heavy metal, e.g. platinum, palladium, or gold, which is evaporated in a vacuum chamber. The source of the metal ions is situated at a known angle to the specimen. Only the 'windward' side will be coated since metal atoms travel in straight lines. On placing the specimen in a beam of electrons, the uncoated 'leeward' side will allow freer passage of electrons and hence appear darker (a 'shadow') on the photographic plate. By measuring the length of the shadow cast it is possible, knowing the angle of the metal source, to calculate the height of the object casting the shadow. The technique has been used to gain more accurate measurements of very small specimens such as viruses.

shadow yeasts *See* Blastomycetes.

shikimic acid An aromatic acid, abundant in certain higher plants, that is an intermediate in the synthesis of the aromatic amino acids phenylalanine, tyrosine, and tryptophan. It is also an intermediate in the synthesis of lignin and of the electron carriers ubiquinone and plastoquinone. The *shikimic acid pathway* is the normal route for the synthesis of phenolics and is one of the few pathways in which aromatic compounds are formed from aliphatic precursors.

shoot *See* stem.

shoot-tip culture *See* meristem culture.

short-day plant (SDP) A plant that appears to require short days (i.e. days with less than a certain maximum length of daylight) before it will flower. In actual fact it requires a daily cycle with a long period of darkness. Examples of short-day plants are the autumn-flowering (e.g. *Chrysanthemum*) and spring-flowering (e.g. *Fragaria*) species of temperate latitudes. *Compare* long-day plant, day-neutral plant.

sibs Plants derived either by selfing, or by crossing between genetically similar parents. In the production of F_1 hybrid varieties occasional crossing may occur within rather than between the parent lines giving a proportion of nonhybrid seed. The plants that grow from such seed are called sibs.

The term *sibling species* is sometimes used to describe closely related populations having a common ancestry.

sieve area A specialized part of the primary wall of a *sieve element, perforated by a number of pores, derived from a *primary pit field. These pores are lined with *callose and penetrated by cytoplasmic strands. The sieve areas of gymnosperm *sieve cells are relatively unspecialized, whereas those of angiosperm *sieve tubes are usually differentiated as *sieve plates.

sieve cell A relatively primitive vascular cell in the *phloem of the lower vascular plants and gymnosperms, whose main function is the translocation of sugars and other nutrients. Sieve cells have relatively unspecialized sieve areas compared to those of *sieve tube elements and are longer and narrower in shape.

sieve element A vascular cell in the *phloem whose main function is the transport of sugars and other nutrients from the site of production to the site of utilization. This is a general term covering both *sieve cells and *sieve tube elements.

sieve plate A highly specialized *sieve area, or collection of sieve areas, perforated by relatively large pores and usually located on the end walls of *sieve tube elements. A sieve plate composed of a single sieve area is called a *simple sieve plate*. A number of sieve areas may be arranged collectively to form a *compound sieve plate*, as in *Nicotiana*. The pores of a sieve plate, so named for its resemblance to a sieve, are lined with *callose and penetrated by cytoplasmic strands. *Compare* perforation plate.

sieve tube A continuous longitudinal tube composed of numerous *sieve tube elements.

sieve tube element (sieve tube member) A relatively advanced vascular cell in the phloem of angiosperms, characterized by the presence of a *sieve plate, a nonlignified secondary cell wall, and a living enucleate protoplast. Sieve tube elements are joined together at their ends to form *sieve tubes, their end walls having become modified to form the sieve plates. Sieve tube elements usually have specialized parenchyma cells associated with them. These are known as *companion cells, the protoplast of the companion cell being connected to that of the sieve tube member by plasmodesmata. *Compare* sieve element, sieve cell.

silicle *See* silicula.

silicon Symbol: Si. A grey metalloid element, atomic number 14, atomic weight 28. It is not considered essential to plant growth but nevertheless accumulates as silica (SiO_2) in high proportions in the cell walls of some plants. The cell walls of diatoms (Bacillariophyta) are largely composed of silica and certain members of the Xanthophyta and Chrysophyta also accumulate silica. Among higher plants, the stems of *Equisetum* are reinforced with silica and the leaves of many grasses also contain substantial amounts. The presence of silica bodies can be demonstrated by carbolic acid solution, which turns such bodies pink. Other crystals remain colourless.

silicula (silicle) A dry dehiscent fruit derived from two carpels fused together to form a flattened pod with two loculi separated by a false septum. The seeds are exposed on the septum as the two valves separate from it, from the base up. Siliculae are similar to *siliquae, except that they are as broad or broader than they are long. The fruits of shepherd's purse (*Capsella bursa-pastoris*) are an example.

siliqua (silique) A dry dehiscent fruit similar to a *silicula except that it is long and narrow. It is typical of members of the genus *Brassica.*

silt Mineral particles having a diameter of 0.002–0.05 mm. A silt soil is defined as containing more than 80% silt and less than 12% clay and has a characteristic smooth soapy feel. *See* soil texture.

Silurian The third period of geological time in the Palaeozoic era from about 440 to 395 million years ago, during which there is the first evidence of the invasion of the land by plants and animals. Thick-walled spores with triradiate markings have been found. The fossil *Cooksonia* from the Upper Silurian with its slender branched stem, lack of lateral appendages, terminal sporangia, and cutinized spores, resembles *Rhynia*, a psilopsid from the Rhynie chert. The fossil *Nemaphyton* resembles the stipe of a seaweed. Marine life was similar to that of the Ordovician, and the climate is thought to have been uniformly warm, becoming drier towards the end of the period. *See* geological time scale.

simple pit A *pit lacking a border. Simple pits are found, for example, in certain parenchyma cells, extraxylary fibres, and sclereids. *Compare* bordered pit.

single-cell protein (SCP) Protein produced by unicellular organisms for animal and human consumption. Microorganisms are considerably more efficient in producing protein than crops and farm animals. Thus yeast and bacteria can double their mass in a minimum of 20 minutes while cereals would take 7–14 days and young cattle 4–8 weeks to achieve this. Considerable research has consequently been directed towards developing suitable industrial plants for the large-scale cultivation of certain bacteria, fungi, and algae. Many products are already available, e.g. Pruteen from ICI.
The substrates used to grow the microorganisms can be low-cost agricultural wastes, e.g. straw, molasses, and animal sewage. The technology thus has potential applications in pollution control as well as food production.

single-factor inheritance The determination of a character by one major gene, although the gene may exist in various allelic forms. Mendel's genes are examples of single-factor inheritance. *Compare* multiple-factor inheritance.

sink A site within a plant or cell where a demand exists for particular substrates or catalysts. Thus the mitochondria are sinks for oxygen and respiratory substrates while the chloroplasts require carbon dioxide. At a higher level of organization fruits, roots, and shoot apices are sinks for photosynthates. *See* mass flow hypothesis.

sinuate Describing a leaf margin divided by irregularly spaced narrow notches into wide lobes.

siphonaceous Describing the tubular growth habit seen in certain algae, e.g. *Vaucheria.*

Siphonales *See* Caulerpales.

siphonostele A *stele containing a central core of pith internal to the xylem. The pith may be extrastelar in origin, i.e. originating from invagination of cortical parenchyma, or it may originate in the xylem itself. In a simple siphonostele the leaf traces do not leave any gaps in the vascular cylinder. A siphonostele may be *ectophloic*, i.e. possessing phloem external to the xylem, or *amphiphloic*, i.e. possessing phloem both external and internal to the xylem as in a *solenostele. Siphonosteles are found in the stems of many ferns. *Compare* protostele. *See also* medullated protostele.

skotophile *See* photophile.

sleep movements *See* nyctinasty.

sliding growth A type of cell growth in which an area of the cell wall expands and slides over that part of the wall of the adjacent cell with which it was originally in contact. This results in the severance of plasmodesmata between adjacent cells but does not lead to cell disruption. *Compare* intrusive growth, symplastic growth.

slime bacteria *See* Myxobacteriales.

slime moulds *See* Myxomycetes.

smooth endoplasmic reticulum *See* endoplasmic reticulum.

smuts Plant diseases caused by fungi of the order *Ustilaginales, so named because of the black spore masses usually produced on the host. In the *covered smuts*, such as covered smut of barley (*Ustilago hordei*), the mature spore mass remains within the sorus for a while, often until the sorus is dispersed from the host. However in the *loose smuts*, such as loose smut of wheat (*U. nuda*), the spores form an uncovered mass of black powder. Seed treatment with fungicides is the usual control method. *See also* bunt.

society A minor community within a *consociation, with a particular dominant species. It arises because of some variation in conditions within a habitat. For example, in a consociation such as an oak wood there may be a society of bluebells or primroses. *See also* association.

sodium bicarbonate indicator A mixture of sodium bicarbonate solution and the dyes cresol red and thymol blue. It is used as an indicator to detect small changes in pH. It changes from red to orange and yellow with a slight increase in acidity.

soft rot A disease, caused by fungi or bacteria, in which the tissues of the affected parts become soft. Soft rots are most notable as postharvest diseases. Bacteria of the genus *Erwinia* and fungi of the genus *Rhizopus* cause soft rots of stored vegetables and fruits – particularly in humid badly ventilated conditions. The tissues become soft as the enzymes from the pathogen break down the cell walls.

softwood *See* secondary xylem.

soil The superficial layer that covers large areas of the earth's crust. It consists of mineral particles, decaying and decayed organic material, living organisms, air, and water. It is the medium in which plants grow, supporting them and supplying them with nutrients, and is also a habitat for numerous animals and microorganisms.

The mineral content of the soil is derived from the mechanical or chemical weathering of exposed rocks. A soil may be formed from the underlying bedrock (*sedentary soil*) or it may result from the deposition of mineral particles transported by such agents as water or wind, resulting in a mixture of particles of different origins (*sedimentary soil*). The proportions of the various sizes of mineral particles (silt, clay, sand, gravel, etc.) have a profound effect on the *soil texture and structure. This is further influenced by the amount of organic matter present, most of which is concentrated in the top 230–300 mm of soil.

Living organisms affect the soil in a number of ways. The larger animals, notably earthworms, mix and aerate the soil and billions of microorganisms feed on and decompose the dead organic matter, thus making nutrients available to the plants. The water in the soil contains dissolved mineral substances and gases that can be taken up by plant roots. The level at which the soil becomes saturated with water (the *water table*) rises and falls depending on the amount of precipitation. Its level affects soil aeration. The air present in the spaces between the soil particles is generally richer in carbon dioxide and poorer in oxygen than atmospheric air.

There are a number of different methods used to classify soil. Generally soil can be divided into three main groups: *zonal soils*, where the soil type reflects the prevailing climatic conditions; *intrazonal soils*, where some other factor, such as the nature of the parent rock, has more influence on soil development; and *azonal soils*, which are immature newly formed soils, e.g. alluvial soils. Zonal soils may be further subdivided into *pedalfers and *pedocals. Intrazonal soils can be subdivided into *hydromorphic soils* (with excessive moisture), *halomorphic soils* (with a high salt content), and *calcimorphic soils* (rich in lime from the parent rock).

Soil colour may be used as an indication of soil composition, amount of aeration, drainage, etc. For example, badly aerated soils are often bluish grey whereas well aerated soils are reddish because of the high concentration of iron oxides. *See also* soil structure, soil profile, humus.

soil-borne diseases Any plant disease that originates from inoculum in the soil. *Damping-off and *club root are typical soil-borne diseases. Control can be difficult and methods include crop rotation, fumigation, chemical treatment, soil sterilization, the use of seed dressings, and improved cultural methods (e.g. better drainage). *See also* fumigant.

soil factors *See* edaphic factors.

soil horizon *See* soil profile.

soil profile The arrangement of the layers that can be seen in a section of *soil extending from the surface to the bedrock. The layers are *soil horizons* and there are differences in colour, texture, and composition between them. The upper layer, horizon A or the *zone of eluviation*, contains the most humus, is most exposed to weathering, and is the zone in which most decomposition by microorganisms takes place. The underlying horizon B or *zone of illuviation* accumulates salts and colloids washed down from horizon A. Horizon C (the *subsoil*) consists of the mantle rock produced mainly by the mechanical weathering of the parent rock. In horticulture, the term subsoil is taken to mean the soil beneath the humus-rich horizon A (topsoil). Some soils do not have such horizons and may be more or less uniform throughout. They are generally found in mature soil in moist cool temperate climates. In arid and semiarid climates where the rate of evaporation exceeds the amount of precipitation there is a tendency for the upward movement of water by capillarity, and mineral salts, particularly calcium salts, are deposited in the upper layers resulting in calcification.

soil structure The arrangement of the mineral particles in soil. Particles may exist individually, e.g. sand particles, or they may be grouped together in larger units (*peds*). The products of decomposition of organic matter are often sticky and glutinous thus binding the particles together. The aggregates can be grouped in various ways and may be described as crumby, blocky, platy, etc. The structure provided by these aggregates makes the soil more fertile as the channels between the peds aid aeration and drainage. Soil structure can be broken down by overcultivation and removal of too much organic matter. Often a clay soil forms large sticky *clods* making it difficult to work. Clay soils can be improved by adding lime, which causes the individual clay particles to stick together (*flocculate*) in smaller aggregates.

soil texture The relative sizes of the different mineral particles in a *soil that affects its aeration, capillarity, porosity, water absorption, and ease of cultivation. Gravel and stones are defined as being larger than 2 mm in diameter. On the same scale sand particles are 0.05–2.0 mm, silt 0.002–0.05 mm, and anything smaller than 0.002 mm is termed clay. Thus a sandy soil with larger particles has larger spaces and is well aerated and drained but dry. A clay soil is badly drained and becomes waterlogged. However clay particles tend to group together and also chemically attract other substances such as *humus. The *clay–humus complex* is chemically active and thus capable of holding plant nutrients in the soil, maintaining fertility. Cultivation affects soil texture, e.g. ploughing, harrowing, and rolling will break down large aggregates into smaller particles. Over cultivation, however, can lead to soil erosion.

Solanaceae A large family of dicotyledonous plants commonly called the potato or nightshade family. It contains about 90 genera and some 3000 species have been described. However the actual number of species may be closer to 2000 since the great infraspecific variation seen in many species has led to taxonomic confusion and some species being given a variety of different names. The Solanaceae are cosmopolitan in distribution though there are concentrations of genera in Central and South America and Australia. The majority of solanaceous plants are herbaceous. The leaves are exstipulate and the flowers are usually regular and borne in a cyme. Most commonly there are five sepals and petals, more or less fused, and five anthers. In the largest genus, *Solanum* (about 1500 species), the flowers characteristically possess a column of touching, but not fused, stamens, which are particularly prominent because of the downward-turned petals. The fruit is usually a berry, though sometimes a capsule, as in henbane (*Hyoscyamus niger*) and thorn apple (*Datura stramonium*).
Many solanaceous plants are poisonous due to their possession of alkaloids, e.g. atropine found in deadly nightshade (*Atropa belladonna*) and nicotine found in tobacco (*Nicotiana tabacum*). Others are important as food plants, especially the potato (*Solanum tuberosum*) and tomato (*Lycopersicon esculentum*). The aubergines and the various capsicums also belong to this family. The genus *Nicotiana* includes the important cash crop tobacco and various ornamental species. Other ornamentals include species of *Petunia*, *Schizanthus* (butterfly flowers), and *Salpiglosis* (velvet flowers).

solenostele (amphiphloic siphonostele) A *siphonostele in which there is also a cylinder of phloem internal to the xylem. When an endodermis is present, there is also an endodermal layer internal to the internal phloem. Leaf gaps are present, but are sufficiently spaced longitudinally so that only one is visible in any one transverse section. This type of stele is found in certain ferns, e.g. *Marsilea* and *Adiantum*, and is often quoted as evidence for the hypothesis that siphonosteles originated through invasion of the xylem by other extrastelar tissues. *Compare* dictyostele. *See illustration at* stele.

solonchak (white alkali soil) A type of intrazonal infertile *soil with a high salt content (*halomorphic*). It is found in arid and semiarid inland regions and in areas where sea spray has blown inland resulting in the deposition of salt. The evaporation of the surface water results in salts from the underlying layers being drawn upwards and deposited on the surface. Solonchaks are usually associated with lime-rich soils (*pedocals) and are commonly found in dry continental interiors. If leaching occurs either through heavier rainfall or by irrigation, the salts are washed down into the B horizon thus giving rise to a *solonetz* soil. If there is continued leaching and improved drainage a *soloth* or *solod* is formed, which is weakly acid.

solute potential *See* osmotic potential.

somatic cell Any cell of the body, i.e. any cell other than the spores, gametes, or their precursors.

somatic hybridization The production of cells, tissues, or organisms by fusion of nongametic nuclei. The phenomenon may be induced under laboratory conditions in cells that never normally fuse together and used as a plant breeding or genetic tool. It may also occur naturally, especially in fungi. *See also* parasexual recombination, protoplast fusion.

somatic mutation *See* mutation.

soralia *See* soredium.

sorbitol A common sugar alcohol found especially in certain algae and in the fruits of many higher plants, e.g. mountain ash (*Sorbus aucuparia*). It is formed by the reduction of D-glucose or L-sorbose. In some plants, e.g. crab apple (*Malus sylvestris*), sorbitol, rather than sucrose, is the main form in which sugar is translocated in the phloem.

soredium A small segment of a lichen thallus consisting of a number of algal cells loosely surrounded by a few fungal filaments. Soredia serve as agents of vegetative propagation and are seen as a powdery dust on the lichen surface. They are often formed in specialized structures termed *soralia*.

sorocarp *See* pseudoplasmodium.

sorophore *See* pseudoplasmodium.

sorosis A *multiple fruit derived from the ovaries of several flowers, as in mulberries (*Morus*). *Compare* coenocarpium.

sorus A reproductive structure made up of a collection of sporangia. Sori are usually prominent on the undersurface of fertile fern fronds, where they are often borne in a regular pattern, each sorus being covered by an *indusium. The reproductive areas, consisting of a mass of unilocular sporangia, found on the blades of the brown alga *Laminaria* are also termed sori as are the various spore-forming bodies of the rust fungi (e.g. the aecium, teleutosorus, and uredosorus).

spadix A *racemose inflorescence in which the flowers are sessile and borne on an enlarged fleshy axis. The spadix is typical of the Araceae (arum family) in which the fleshy axis secretes sticky insect-attracting compounds and is usually surrounded by a large bract or spathe. The female maize inflorescence is also a spadix.

spathe A large bract enclosing a spadix, which may be highly coloured and petaloid, as in *Anthurium*. In such cases it functions in attracting insects to pollinate the plant.

spatulate (spathulate) Describing structures that have a broad apex and a long narrow base, such as the leaves of the daisy (*Bellis perennis*).

special creation *See* creationism.

speciation The formation of new species. This typically involves the establishment of barriers (isolating mechanisms) that prevent two populations from interbreeding (*see* allopatric, sympatric). Natural selection may then occur, taking the two populations along different evolutionary paths so that they become progressively different from each other. Speciation is often regarded as a relatively slow process, especially in the animal kingdom (but *see* punctuated equilibrium). In the plant kingdom, however, a major contribution to speciation is polyploidy. This may establish a new population (species) in a single generation that is incapable of reproducing with either parent. Polyploidy has thus been cited as an example of 'instant evolution'. *See also* adaptive radiation, neoteny.

species The fundamental unit of study in taxonomy, comprising all the populations of one breeding group that normally are permanently separated from other such groups by marked discontinuities. If crossing between species does occur then the resulting hybrids are normally sterile so maintaining the reproductive barrier between species. This broadly genetic definition of a species cannot be applied to species that reproduce by self-fertilization or by asexual means or to extinct species. In such cases and when breeding patterns have not been studied, a species is delimited by observation of the similarities between its members and dissimilarities between it and other species.
In botanical nomenclature the *specific epithet* forms the second part of the binomial (the first being the generic name) and it is always written lower case. (*See* Appendix for the meanings of some common specific epithets.) The ending of the specific epithet always agrees with the gender of the generic name. Unlike the situation in zoology, it is not legitimate for the specific epithet to repeat the generic name. There are several infraspecific categories in the taxonomic hierarchy, but *subspecies is probably the most widely used. Groups of similar species are placed in genera (*see* genus).

spectrophotometer An instrument for investigating quantitatively the way in which electromagnetic radiation of different wavelengths is absorbed by a specimen. Typically, it consists of a source of the radiation (infrared, visible, or ultraviolet) from which a particular wavelength can be selected by means of a monochromater. A beam of this radiation is passed through the specimen (or reflected from it) and the intensity measured by a detector (e.g. a photocell). As the wavelength of incident radiation is changed, an *absorption spectrum is produced showing how the intensity of transmitted (or reflected) radiation varies with wavelength. Typically, substances absorb the radiation over particular bands of wavelength, corresponding to energy-level changes in their molecules. Spectrophotometry is used extensively as a means of analysis and in the study of how radiation interacts with matter (e.g. the absorption of light in photosynthesis). The speed of a chemical reaction can also be followed in cases where the products of the reaction absorb radiation of a particular wavelength that the reactants do not.

spermatium **1.** A small nonmotile male cell shed from the mycelia of certain ascomycete fungi and capable of fertilizing an ascogonium.
2. *See* pycnidium.

spermatocyte *See* antherocyte.

Spermatophyta (seed plants) In classifications that consider the seed habit of equal or greater significance than the possession of vascular tissue, a division containing all the seed-bearing *vascular plants. It contains the two classes *Gymnospermae and *Angiospermae. *Compare* Tracheophyta.

spermatozoid *See* antherozoid.

sperm cell One of the two male gametes formed in the pollen tube of gymnosperms after mitotic division of the *body cell. In some species, e.g. *Cycas* and *Ginkgo*, they are flagellate and swim from the pollen tube to the archegonia. Such antherozoids, in contrast to the nonmotile *generative nuclei of angiosperms, are associated with a large amount of cytoplasm and in *Cycas* reach a diameter of 300 μu. In other gymnosperms, e.g. the Gnetales and Coniferales, the sperm cells are not motile. The term sperm cell is sometimes applied to generative nuclei.

spermogonium *See* pycnidium.

Sphaeriales An order of the *Pyrenomycetes mainly containing saprobic fungi though some are parasitic on plants. It contains over 5000 species in some 500 genera. They typically have hard dark perithecia, as seen, for example, in *Daldinia concentrica.*

Sphaeropsidales An order of the *Coelomycetes containing many fungi that are plant parasites, e.g. *Septoria apiicola*, which causes leaf spot of celery. It includes about 6000 species in some 750 genera. The conidiophores are formed within pycnidia.

sphaeroraphide *See* druse.

Sphagnales *See* Musci.

S phase The period of the *cell cycle when DNA replication and histone synthesis occurs as chromosomes divide into chromatids.

Sphenopsida A subdivision or class of *vascular plants containing a single living genus, *Equisetum* (horsetails) with about 23 species, but having a rich fossil record. The most conspicuous feature of sphenopsids is the aerial jointed stem, which bears whorls of leaves inserted at each joint. The stems arise at intervals from a perennial creeping rhizome and may either be all photosynthetic or differentiated into two types, green sterile stems and colourless fertile stems. The stems are grooved and consist of a central cavity surrounded by a ring of smaller cavities in the cortex (*see* vallecular canals). These alternate with still smaller cavities associated with the protoxylem (*see* carinal canals). The sporangia are borne on sporangiophores in terminal strobili.

The spores are all approximately the same size but give rise to separate male and female gametophytes. The antherozoids are multiflagellate. The Sphenopsida is divided into four orders, the Equisetales and the extinct Calamitales, Sphenophyllales, and Pseudoborniales. The arborescent Calamitales, which flourished in the Carboniferous, most closely resemble the Equisetales.

spherosome An *organelle 0.5–1.0 μm in diameter with a single membrane and a fairly granular matrix containing triglycerides. Spherosomes are abundant in cells in which lipids are stored and they contain the hydrolytic enzyme, lipase. They probably have a role in the mobilization of stored lipids when they are required for cell metabolism.

sphingolipid A type of complex *lipid containing a long-chain amino alcohol, such as sphingosine, joined by an amide linkage to a long-chain fatty acid. Sphingolipids are important as membrane components in both animal and plant cells. In yeasts and in higher plants the amino alcohol is commonly phytosphingosine.

spike A racemose inflorescence in which the flowers are sessile and borne on an elongated axis, as in wheat. The *catkin and *spadix are types of spike.

spikelet The basic unit of a grass inflorescence. It consists of a short axis or *rachilla, two bracts or *glumes, and one or more florets and their bracts (the *lemma and *palea).

Spikelet characteristics such as size, number of florets, or how it fractures at maturity may be used as diagnostic traits for the genus. The whole spikelet is the unit of dispersal for such grasses as *Holcus* and *Setaria.*

spindle A fibrillar structure formed in the cytoplasm from microtubules at the commencement of *metaphase. When present, *centrioles move so they are diametrically opposite each other on the disintegrating nuclear membrane. In the majority of plant cells, where there are no centrioles, poles still become established at opposite points. The fibrils radiate from each pole towards the centre or equator. The attached microtubules direct the chromatids (or chromosomes in the first meiotic division) towards opposite poles. Evidence tends to support the view that this is achieved by the microtubules becoming progressively shortened as a result of depolymerization in the regions nearest to the poles.

spindle attachment *See* kinetochore.

spine A modified leaf or part of a leaf forming a sharp pointed structure, often with a vascular trace leading to its base. If some or all of the leaves are modified into spines then leaf area, and hence transpiration, is reduced. Spines are thus common in xerophytes, such as cacti, their points sometimes acting as nuclei on which water droplets can condense, run down to the soil, and provide moisture for the plant. In such plants the functions of the leaf may be taken over by the stem (*see* cladode). Spines also protect against herbivores and excessive sunlight. *Compare* thorn, prickle.

spiral thickening (helical thickening) A type of secondary wall patterning in *tracheary elements in which the secondary cell wall is laid down in a helical pattern. Spiral thickening, like *annular thickening, permits further extension of the cell. It is therefore found in tissues such as protoxylem, which develop in young organs that are still elongating. *Compare* pitted thickening, reticulate thickening, scalariform thickening.

spirillum Any corkscrew-shaped bacterium. *Compare* bacillus, coccus, vibrio.

Spirochaetales An order containing bacteria with slender flexible spiral-shaped cells. They lack flagella but are nevertheless motile. Many are animal parasites causing such diseases as syphilis and yaws.

spongy mesophyll The part of the leaf *mesophyll that is composed of variously shaped, often conspicuously lobed, cells with large intercellular spaces. It is sometimes regarded as a type of aerenchyma. A humid atmosphere is maintained in the intercellular spaces, thus facilitating gaseous exchange between the atmosphere and the mesophyll for photosynthesis and respiration.

spontaneous generation The theory that living organisms form directly and spontaneously from nonliving material. It was conclusively disproved by Louis Pasteur in 1862. A distinction should be drawn between this idea and the concept of abiogenesis and the origin of life.

sporangiophore A structure bearing one or more *sporangia. It may be a simple stalk, such as that formed on germination of the zygospore of such fungi as *Rhizopus* and *Mucor*. Alternatively it may be multicellular or branched as in certain pteridophytes. *See also* sporophyll.

sporangium A structure in which spores are formed. The sporangium may be simple and unicellular, as in the algae and fungi, or multicellular, as in the bryophytes and vascular plants. In higher plants the sporangium may be protected by a thickened outer wall. The entire protoplasmic contents may be converted into motile or nonmotile spores, which are usually liberated by rupture of the sporangium wall. *See also* megasporangium, microsporangium.

spore A simple asexual unicellular reproductive unit. Spores are produced by the sporophyte generation following meiosis and are thus usually haploid. On germination they develop into the haploid gametophyte. Sometimes the suffix 'spore' is seen in terms, e.g. *carpospore, *zygospore, that describe a diploid cell. *See also* zoospore, megaspore, microspore.

spore mother cell (sporocyte) A cell that undergoes meiosis to form haploid spores. The term is usually used of homosporous species in which megaspore and microspore mother cells are not distinguished.

Sporobolomycetales *See* Blastomycetes.

sporocarp A hard usually globose multicellular structure that contains the spores in water ferns, e.g. *Pilularia*. It is formed by the fusion of the margins of a fertile frond. The spores are not released until the structure decays.

sporocyte *See* spore mother cell.

sporogenesis The formation of *spores. *See also* spore mother cell.

sporophore The fruiting body of ascomycete and basidiomycete fungi.

sporophyll A modified leaf that bears the *sporangia. In the lower plants, e.g. *Lycopodium*, it may be a leaf that has retained its normal structure and function and shows no great specialization. In the seed plants the leaf has become highly modified and lost its photosynthetic function, as in the carpels and stamens of the angiosperms. *See also* megasporophyll, microsporophyll.

sporophyte An individual of the diploid generation in the life cycle of a plant. The sporophyte arises from the fusion of two haploid gametes from the *gametophyte or haploid generation. In bryophytes, the sporophyte constitutes the minor part of the life cycle, and is either partially or completely dependent on the gametophyte for anchorage and nutrition. In the vascular plants the sporophyte is the dominant generation and is independent of the gametophyte. The sporophyte is more specialized than the gametophyte and by the development of such structures as a cuticle, conducting tissue, and stomata is adapted to a wider range of environments. *See also* alternation of generations.

sporopollenin The highly resistant material making up the exine of spores and pollen grains. It is a polymerized carotenoid, capable of withstanding temperatures approaching 300°C and strong acids.

sport An atypical form of an individual or part of an individual, due to mutation or segregation. The term is normally restricted to new forms and is most commonly used in horticulture.

spur 1. A short thin side shoot from a branch, especially one that bears fruit or, in conifers, the shoots that bear the leaves.
2. A tubular projection from a flower, usually from the base of a perianth segment. Well developed spurs are seen in columbine (*Aquilegia vulgaris*) and larkspur (*Consolida ambigua*).

squash A microscopical preparation in which the material is flattened before examination. It is commonly used in the study of chromosomes. Some tissues, e.g. anthers, may be squashed directly while others, e.g. root tips, must first be softened. This is commonly achieved by adding a cellulase enzyme.

stachyose A nonreducing tetrasaccharide made up of two galactose units and one unit each of glucose and fructose. It is found in many of the Labiatae, especially the tubers of *Stachys tuberifera*, and in some of the Leguminosae.

staining The treatment of biological specimens with dyes (stains) to colour part of the structure so as to make details more clearly visible through a microscope. Most laboratory stains are synthetic organic dyestuffs, which are absorbed by or bind to particular types of tissue. They are applied by immersing the specimen in a solution of the dye. *Nonvital staining* is the colouring of dead tissue; *vital staining* is the staining of living tissue without harming or killing the cells.

Most stains are organic salts consisting of a positive and negative ion. In *acid stains* the colour comes from an organic anion (negative ion). Such stains (e.g. eosin) tend to colour the cytoplasm of cells. In *basic stains* the colour comes from an organic cation (positive ion). Such stains, e.g. haematoxylin, tend to colour the nuclei. *Neutral stains* are mixtures of acid and basic stains and are used to stain both nucleus and cytoplasm. An example is Leishman's stain, which is made by mixing the acid stain eosin with the basic stain methylene blue in alcohol. Materials in cells can be described as *acidophilic* if they are receptive to acid dyes; *basophilic* if receptive to basic dyes; and *neutrophilic* if receptive to neutral dyes. Certain dyes need an additional substance (a *mordant*) to bind them to the tissue. The combination of the dyestuff and the mordant is called a *lake*. Iron alum, for instance, is used as a mordant for haematoxylin. In its absence, the tissue will not take up the stain.

Various complex techniques are used for staining different types of tissue for light microscopy. *Counterstaining* (or *double staining*) is the application of two stains in sequence so as to colour different parts of the specimen. For example, safranin is often counterstained with light green. Overstaining occurs when too much stain is taken up by the specimen. In such cases, part of the stain is removed by a solvent – a process known as *differentiation*.

By analogy with stains for the light microscope, *electron stains* are substances used with specimens for electron microscopy. They act by hindering the transmission of electrons (i.e. they are 'electron dense'). Examples are lead citrate, uranyl acetate (UA), and phosphotungstic acid (PTA). *See also* histochemistry, metachromatic stain, negative staining, permanent stain, temporary stain, vital stain.

stalk cell One of two cells, the other being the *body cell, formed by division of the generative cell in the male gametophyte of certain gymnosperms.

stamen The male reproductive organ of the flowering plant, many of which together make up the *androecium. It is a highly modified microsporophyll. A typical stamen is differentiated into *anther, *filament, and *connective. However in the very primitive flowering plants, e.g. *Magnolia* and *Degeneria*, distinct anthers and filaments are absent and the stamen is relatively broad and slightly flattened. In such cases the pollen sacs are borne on the surface of the microsporophyll. The more primitive flowering plants usually possess numerous separate stamens, as seen in the Ranunculaceae. The more advanced families have fewer stamens, which are often fused in some way. For example, there are usually four stamens in flowers of the Scrophulariaceae. Where the number of stamens is reduced they are strategically placed to maximize the chances of successful *pollination.

Various terms are used to describe the position of the stamens in the flower. If situated opposite the petals they are termed *antipetalous* and if opposite the sepals, *antisepalous*. If inserted on the corolla they are *epipetalous* and if inserted on the calyx, *episepalous*. If the stamens are exerted they are described as *phanerantherous* and if enclosed, *cryptantherous*. When there are two whorls of stamens, they are termed *diplostemonous* if the outer whorl is opposite the sepals and the inner whorl opposite the petals, and *obdiplostemonous* where the opposite occurs.

Similarly, terms used to describe stamen arrangement include *didymous*, when the stamens are in two equal pairs, and *didynamous*, when in two unequal pairs. *Tetradynamous* describes the arrangement where there are four long and two short stamens.
Compare carpel.

staminate flower A flower possessing male parts (stamens) but no female parts, as in the male flowers of holly (*Ilex aquifolium*). *Compare* pistillate flower. *See also* monoecious, dioecious.

staminode A sterile stamen. It may remain rudimentary or partially developed, as in figworts (*Scrophularia*), or become highly specialized and modified, as in *Iris*, where it forms a conspicuous part of the flower.

standard 1. The broad often erect petal at the top or posterior side of the flower of plants belonging to the subfamily Papilionoideae of the Leguminosae (i.e. 'pea' flowers). *See also* keel, wing.
2. Any of the three erect twisted inner petals of an *Iris* flower. *See also* fall.

standard deviation The average magnitude of deviations from the centre of a normal curve, obtained by squaring all the deviations, calculating their mean, and then finding the square root of the mean. It differs from the *mean deviation in that it removes difficulties introduced by sign, i.e. it does not matter which side of the centre of the curve any particular deviation is situated. The value so estimated, σ, is the point of maximum slope either side of the central line of the normal curve. If it were possible to make an infinitely large number of observations then the mean deviation would be as good a way of finding the true value of σ as the standard deviation. However with a limited number of samples, as always occurs in practice, the standard deviation provides a better estimate. Often the estimated standard deviation is represented by the Latin letter s while the true standard deviation is represented by its Greek equivalent, σ.

standing crop *See* biomass.

starch (amylum) The most abundant and important reserve polysaccharide in plants. Starch is an early end product of the photosynthetic reduction of carbon dioxide in chloroplasts: in the dark the starch thus formed is rapidly broken down to sucrose and transported to other organs.
The bulk of the starch in plants is found in storage organs. Commercial extraction of starch is from such organs. Common sources are the roots of cassava (*Manihot esculenta*), which yield tapioca, the rhizomes of arrowroot (*Maranta arundinacea*), the stem pith of the sago palms (species of *Metroxylon*, *Arenga*, and certain other genera), the tubers of potato, and the grains of various cereals, especially maize, wheat, rice, and sorghum.
Starch consists of two structurally different fractions, *amylose and *amylopectin. The relative amounts of amylose and amylopectin in a given starch sample depend on the species of plant from which the starch was obtained, amylose usually making up 20–30% of the starch granules. The shape of the starch grains formed by different species also differ (*see* hilum) and this variation has been used in taxonomic work.
Starch is synthesized from ADP-glucose by the enzyme starch synthetase. A branching enzyme, known as *Q enzyme*, is responsible for formation of the $\alpha(1-6)$ bonds in amylopectin. Starch breakdown is catalysed by a class of enzymes known as amylases. These enzymes break down starch to the disaccharide *maltose, which is then further degraded by the disaccharidase maltase. Amylase activity is high during periods of rapid growth such as seed germination or sprouting of tubers.

starch sheath *See* endodermis.

starch-statolith hypothesis A possible explanation for the way plants perceive gravity. Most gravity-sensitive organs contain specialized cells called *statoliths*, in their cytoplasm. The statoliths gradually fall to the lower side of the cell under the influence of gravity. The settling out of these grains in reoriented cells is said to provide an internal stimulus, which is transmitted to the growing region of the organ so that appropriate amendments to the growth pattern can be made. These growth movements restore the original position of the cells and hence of the whole plant part. *See also* geotropism.

star sclereid *See* astrosclereid.

statistics The mathematical analysis of experimental results by various methods that are designed to extract the maximum amount of useful and relevant information while at the same time taking into account the effects and potential errors produced by unknown or uncontrollable variables. The end result of any particular statistical treatment should be to reduce a mass of often incomprehensible raw data to a few figures that give an accurate summary of the results and usually an indication of how often such results could be expected to occur by pure chance. The application of statistical considerations to experimental design aims to ensure that results obtained from a necessarily limited sample can be seen to have general application.

statocyte Any gravity-sensitive cell that contains *statoliths.

statolith Any inclusion in the cytoplasm, such as a starch grain, that moves in response to gravity and is believed to be involved in the geotropic response. *See* starch-statolith hypothesis.

stearic acid An eighteen-carbon saturated fatty acid having the formula $CH_3(CH_2)_{16}COOH$. Stearic acid is formed by elongation of the smaller palmitic acid. It occurs in many lipids and is also an intermediate in the synthesis of *oleic, *linoleic, and *linolenic acids.

stele (vascular cylinder) The central core of the stems and roots of vascular plants, comprising the vascular tissue, ground tissue, such as pith and medullary rays, and the pericycle. The pericycle is considered to represent the outermost layer of the stele at the boundary with the *cortex. The many different types of stele are thought to reflect the direction of the evolution of the vascular system in plants (*see* illustration). Different types of stele may be found in different regions of the same plant. In dicotyledons the root characteristically has a solid central core of vascular tissue (*see* protostele), making it better able to resist pulling stresses, while the stem has a ring of discrete vascular bundles (*see* eustele) and is better able to resist bending stresses. *See also* atactostele, dictyostele, polystely, siphonostele, solenostele.

stem The part of the plant axis that is usually above ground and bears the leaves, reproductive parts, and buds. The terms *shoot* and stem are virtually synonymous except that an underground stem, such as a corm or rhizome, would rarely be described as a shoot. The vascular system of the stem conducts water and nutrients from the roots to the aer-

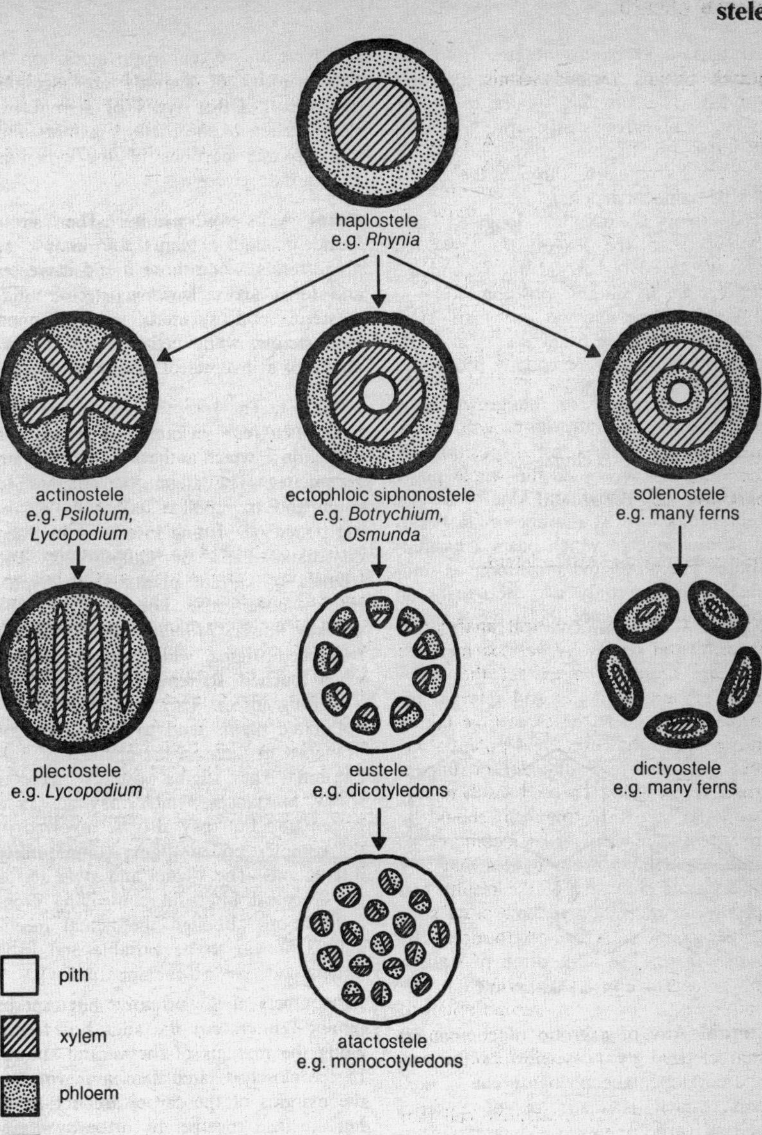

Proposed scheme of evolution of the various stelar types.

ial parts and photosynthates from the leaves to any regions where they are needed. The vascular tissues of higher plant stems are usually arranged as a cylinder (as in most dicotyledons) or dispersed irregularly through the cortex (as in monocotyledons).

The stem is usually elongated and branched so the leaves are separated and displayed to best advantage and the flowers are in suitable positions for pollination. There may be one main stem or trunk with side branches, as in many trees, or a number of equally prominent stems. *Compare* root.

steppe Eurasian *grassland.

stereid *See* prosenchyma.

sterigma (*pl.* sterigmata) One of usually four projections at the apex of a basidium, each one of which bears a basidiospore. The cultivated mushroom is unusual in having only two sterigmata on each basidium, hence the name *Agaricus bisporus*.

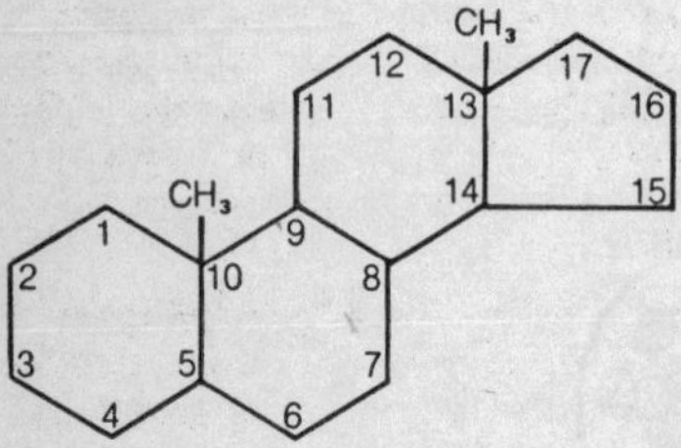

The steroid skeleton.

steroid Any of a group of compounds characterized by possession of the perhydrocyclopenteneophenanthrene skeleton, which is made up of a series of four carbon rings (*see* diagram). Steroids may be distinguished from certain terpenoids with a similar skeleton by their possession of methyl groups at carbons 10 and 13. All natural steroids also have an oxygen group at carbon 3. The majority of steroids are alcohols (*see* sterol). Other types of steroid include certain of the pseudoalkaloids and the aglycone portions of the saponins and cardiac glycosides.

sterol A *steroid alcohol. The sterols found in higher plants are known as phytosterols, while those found in yeasts and fungi are called mycosterols. Stigmasterol and sitosterol are common phytosterols, while ergosterol is an example of a mycosterol. *See also* saponin.

stigma 1. The receptive tip of the carpel, which receives pollen at pollination and on which the pollen grain germinates. The stigma is adapted to catch and trap pollen, either by combing pollen off visiting insects or by various hairs, flaps, or sculpturings. The stigmas of certain plants show haptotropic movements. For example, the monkey flower (*Mimulus guttatus*) has a two-lobed stigma, which closes together when touched, so removing pollen from a visiting insect. The stigmas of wind-pollinated plants tend to be feathery or branched to increase the chances of pollination. The stigma usually secretes sticky substances, which may act as a pollen trap but may also be involved in the complex pollen/stigma compatibility interactions. The stigma and style act as physiological filters in controlling cross fertilization, although the actual mechanisms appear to be variable and little understood (*see* self-incompatibility).

In members of the primitive angiosperm family Winteraceae the stigma is found along the margins of the ventral suture. In the closely related *Degeneria vitiensis* the margins of the carpel are not fused but are held together by interwoven papillae, which form the stigmatic surface. Studies of such plants have thrown light on the evolution of angiospermy.

2. *See* eyespot.

stilt root *See* prop root.

stimulus Any influence that, however received and whether acting directly or indirectly on an organism, gives rise to some form of response within that organism. A stimulus may arise externally, as do light, gravity, and chemical stimuli, or it may be generated within the plant itself as in the case of *autonomic movements.

stinkhorns *See* Phallales.

stinking smut *See* bunt.

stipe A stalk, especially: **1.** The part of the thallus of certain brown algae, e.g. *Laminaria*, joining the lamina to the holdfast.
2. The stalk of a mushroom or toadstool.

stipule One of a pair of leaflike structures, spines, glands, or scales at the leaf base or along a petiole. Stipules are believed to protect the developing leaf in the embryonic state. They may occasionally be modified to enclose a bud, as in beeches (*Fagus*). Large green stipules are seen in the garden pea (*Pisum sativum*) and replace photosynthetic tissues lost by the modification of the leaflets into tendrils.

stock 1. (rootstock) A plant onto which shoots or buds (scions) of another plant are grafted. The stock affects the size, vigour, and time of flowering and fruiting, of the scion, and specially bred rootstocks are selected accordingly. Where necessary rootstocks may be chosen that confer frost hardiness or resistance to certain soil-borne pathogens. *See* grafting.
2. A maintained culture of a particular organism, the origins and characteristics of which are known.

stolon A long branch that is unable to support its own weight and consequently bends down to the ground. Where nodes on the stolon touch the soil a new plant may develop from the axillary bud. Examples of stolons are the long shoots of currants and gooseberries (*Ribes*). Often ordinary shoots will behave like stolons if pegged to the ground, which is the basis of the layering method of vegetative propagation. *See also* runner.

stoma (stomate, *pl.* stomata) A pore in the epidermis of aerial parts of a plant, providing a means for gaseous exchange between the internal tissues and the atmosphere. Each stomatal pore is surrounded by a pair of *guard cells. The term is often used to refer to the epidermal pore with its associated guard cells. Stomata are very numerous, often more so on the lower surface of leaves than the upper surface.

stomach insecticide Any insecticide that must be taken into the alimentary canal to be effective. Stomach poisons are often inorganic compounds containing such elements as arsenic, mercury, or fluorine.

stomate *See* stoma.

stomium The point at which rupture occurs in a sporangium or pollen sac to release the spores or pollen. It is typically an area of thin-walled cells that rupture on the drying out of the surrounding tissue. The stomium is found under the annulus in ferns and between the pairs of pollen sacs in seed plants.

stone cell *See* brachysclereid.

stoneworts *See* Charophyta.

stooling The cutting back of a tree or shrub to ground level to encourage new growth from the base.

storied (storeyed) Describing the axial cells and rays of wood that are arranged in horizontal series as viewed in tangential section.

strain Any group of very similar or identical individuals, such as a pure line, clone, physiological race, or mating strain.

stratification 1. The arrangement of some factor in layers or strata. Thus vegetation in a woodland consists of an upper layer, the tree canopy, and a lower layer of herbs at ground level, often with one or more intermediate layers of shrubs and saplings. Other factors, such as temperature and light, show stratification, this being especially important in the study of aquatic ecosystems.
2. The practice of placing seeds between layers of moist sand or peat and exposing them to low temperatures, usually simply by leaving them outside during the winter. The treatment is necessary for those seeds that require a period of chilling before they will germinate. It is thought that low temperatures may block the action of a germination inhibitor. *Compare* vernalization.

strengthening tissue *See* mechanical tissue.

strobilus (cone) **1.** A well defined group of closely packed sporophylls bearing sporangia arranged around a central axis, as found in the gymnosperms and in certain lower vascular plants (e.g. Lycopodiales, Selaginellales, and Equisetales). In conifers it constitutes the commonly recognized 'cone' and comprises spirally arranged ovules and lignified ovuliferous scales.
2. Any of various cone-like structures seen in angiosperms, e.g. the aggregate fruits of alders (*Alnus*) and hops (*Humulus*).

stroma **1.** The hydrophilic proteinaceous matrix of *chloroplasts. It contains the enzymes and reagents necessary for the dark reactions of photosynthesis.
2. (*pl.* stromata) A solid mass of plectenchyma that may bear perithecia either on its surface, as in *Nectria cinnabarina* (coral-spot fungus), or embedded within the tissue with only the ostioles appearing at the surface, as in *Xylaria hypoxylon* (candle-snuff fungus).

stromatolites Structures found preserved in many rocks, especially carbonate rocks, that were formed by the activities of algae. They are among the oldest organic structures to have been recognized. Stromatolites have been identified in the gunflint cherts of Ontario, which are about 2000 million years old, and in Precambrian rocks in Australia and Africa some 3000 million years old. Often they consist of a number of white concentric rings, the outer of which may be up to a metre across. Microscopic examination has shown these rings to be composed of numerous blue-green algae that have been preserved due to their ability to secrete calcium carbonate and form large stony cushion-like masses. Stromatolite formation may be observed today in certain coastal zones, tropical streams, and by mineral springs where blue-green algae can flourish due to the absence of grazing invertebrate animals.

structural gene A length of DNA that codes for an enzyme or other protein. It is equivalent to a *cistron. *See also* regulator gene, operator gene, operon.

style The sterile portion of the carpel between the ovary and the stigma, which may be elongated or feathery, especially in wind-pollinated species, so that the stigma is presented in an effective place for pollination. In primitive forms the styles of individual carpels are separate. In more advanced forms they tend to be fused. In some plants, notably traveller's joy (*Clematis vitalba*) and the pasque flower (*Pulsatilla vulgaris*), the style elongates after pollination and remains attached to the fruit. This is an adaptation to promote wind dispersal of the seed.

subclimax vegetation *See* climax.

suberin A fatty acid polyester found in the cell walls of the endodermis (*see* Casparian strip) and of bark. It renders the tissue resistant to decay and entry of water.

sublittoral 1. (neritic) Describing a narrow zone in the sea or ocean that extends from low-tide mark (the edge of the littoral zone) to the edge of the continental shelf. The depth increases to about 200 m. The water is well oxygenated and light penetrates to the bottom. The temperature and salinity are fairly constant except near the shore line. In shallower regions sessile algae such as *Ulva* and *Laminaria* are found.
2. Describing the area of a lake or pond extending from the edge of the area occupied by rooted aquatic plants to about 10 m, i.e. to a depth below which the water becomes considerably colder.
3. Describing organisms that live in the sublittoral zone.

subsere *See* sere.

subsidiary cell (accessory cell) One of a group of morphologically differentiated epidermal cells immediately surrounding the *guard cells. Stomatal complexes lacking subsidiary cells are termed *anomocytic*. The number, arrangement, and development of subsidiary cells are often useful taxonomic characters. *See also* mesogenous, perigenous, haplocheilic, syndetocheilic.

subsoil *See* soil profile.

subspecies The rank subordinate to species in the taxonomic hierarchy. The category has been variously defined, a widely accepted definition being a population of several biotypes forming a more or less regional form of a species. The term subspecies is used when two or more populations are separated in some way (e.g. ecologically or morphologically) throughout their range. However they are not usually genetically isolated. There can therefore be a continuous intergrading of subspecies, thus making their delimitation more arbitrary than that of the species. Generally, if 90% or more of a group of infraspecific individuals are recognizably distinct from another similar group, then each may be ranked as subspecies. This is often referred to as the '90% rule'. Subspecies may share many similar attributes but retain essential differences, thus indicating that they are merely regional representatives of one species, sharing a common origin. The abbreviation 'subsp.' or 'ssp.' is used to indicate a subspecies, e.g. *Daucus carota* subsp. *gummifer*. The name of the subspecies that includes the type is always the same as the specific epithet.

substrate The molecule or molecules on which an enzyme exerts its catalytic action. An enzyme may be specific for only one substrate, e.g. aspartase, which catalyses the interconversion of aspartic and fumaric acids, and is strictly specific for the L-isomer of aspartic acid. Other enzymes can act on a range of substrates; alkaline phosphatase can hydrolyse many esters of phosphoric acid. The substrate reflects the structure of the enzyme active site in that the substrate must have a binding group, by which it can bind to the *active site, and a susceptible bond at which the enzyme can attack the substrate. These groups must have complementary structures in the active site.

subulate Shaped like an awl, i.e. narrow, pointed, and more or less flattened, e.g. the leaves of *Subularia aquatica* (awlwort).

succession A series of changes in the composition of the plant and animal life of an area beginning with the colonization of bare rock or soil by plants such as algae and lichens (a *pioneer community*) and ending with a stable *climax community, which is in equilibrium with the environment. Such a *primary succession* may take a number of years to complete. A *secondary succession* can occur in places where the original vegetation has been destroyed, for example on a burned heathland or woodland, and can be completed in a shorter time. *See also* sere.

succinic acid A four-carbon dicarboxylic acid with the formula HOOC $(CH_2)_2COOH$. Succinate is an intermediate in the TCA cycle, formed from succinyl CoA with concomitant formation of GTP. It is oxidized to fumarate by succinate dehydrogenase, a flavoprotein. In the reaction the FAD prosthetic group of succinate dehydrogenase is reduced to $FADH_2$. Succinate is also a product of the glyoxylate cycle, and succinyl CoA is an important precursor in porphyrin synthesis.

succubous Describing the leaf arrangement in leafy liverworts where the front edges of the leaves lie below the back edges of the leaves in front. *Compare* incubous.

succulent A plant that lives in places where water is either in short supply (physical drought), e.g. deserts or sand dunes, or where there is plenty of water but it is not easily obtainable (physiological drought), e.g. in salt marshes and mudflats. A succulent plant conserves water by storing it in large parenchyma cells in swollen stems and leaves. Many succulents reduce water loss by having rolled leaves, leaves reduced to spines, sunken stomata, etc. Some succulents conserve water by opening their stomata at night and closing them during the day (*see* crassulacean acid metabolism). Examples of succulents are desert cacti, many salt marsh plants, and *Sedum* species. *See also* halophyte, xerophyte.

sucker An adventitious shoot that develops from the root, often coming up some distance from the parent plant. Suckering shoots can be separated from the parent once they have developed their own root system. When suckers develop in grafted plants they are of the same constitution as the stock. If this is different from the scion, as is usually the case, the suckers should be removed before they affect the vigour of the scion.

sucrose (cane sugar, beet sugar, saccharose) A nonreducing disaccharide of glucose and fructose linked through a high-energy bond between carbon two of fructose and carbon one of glucose. Sucrose is the major transport sugar in higher plants. It is formed in the chloroplasts and transported through the phloem to other organs and tissues, where it is either metabolized for energy or utilized in the synthesis of reserve and structural polysaccharides. Present evidence indicates that in vivo synthesis of sucrose is a two-step reaction involving fructose 6-phosphate and glucose. These react to form sucrose 6-phosphate, which is then dephosphorylated by the enzyme sucrose phosphatase. The enzyme sucrose synthase can catalyse the direct manufacture of sucrose from UDP-glucose and fructose, but in the cell sucrose synthase is thought to be the major route of sucrose breakdown, allowing direct formation of *nucleoside diphosphate sugars from sucrose. The enzyme invertase also splits sucrose, to D-glucose and D-fructose, but its physiological significance is not clear. The mixture of D-glucose and D-fructose formed on hydrolysis by invertase is often termed *invert sugar* since the optical activity is inverted from dextrorotatory (sucrose solution) to laevorotatory (D-glucose and D-fructose mixture). Invert sugar is found in many fruits.

Table sugar is composed of sucrose crystals. Production is mostly from sugar cane (*Saccharum officinarum*), grown in the tropics, and sugar beet (*Beta vulgaris*), grown in temperate zones. Small amounts are obtained from other sources, e.g. sugar maple (*Acer saccharum*) and sweet sorghum (*Sorghum bicolor*).

suction pressure *See* diffusion pressure deficit.

Sudan stains Any of various temporary aniline stains used to colour fats and waxes. Examples are Sudan IV, Sudan blue, and Sudan black. They are used to stain cutinized and suberized tissues. Thus the Casparian band readily turns orange on staining with Sudan III or Sudan IV.

suffrutescent habit A growth form in which the plant is woody at the base but has herbaceous branches. It is seen in alpine willows.

sugar Any of the lower molecular weight carbohydrates, namely monosaccharides, smaller oligosaccharides, and derivatives of these. The term is sometimes used synonymously with monosaccharides; thus *sugar alcohols are reduction products of monosaccharides, while *sugar acids are monosaccharide oxidation products.

sugar acid An oxidized derivative of a monosaccharide. There are two biologically important types of sugar acid, the *aldonic* acids and the *uronic acids. In aldonic acids the aldehyde group of a monosaccharide is oxidized to a carboxylic acid. The aldonic acid of glucose, gluconic acid, is an intermediate in the pentose phosphate pathway. Ascorbic acid (vitamin C) is also an aldonic acid.

sugar alcohol A monosaccharide in which the aldehyde group is reduced to an alcohol. Thus D-glucose yields the sugar alcohol *sorbitol while the alcohol from mannose is *mannitol. Other important sugar alcohols are *glycerol, a central compound in lipid metabolism, and *inositol, an important intermediate in cell wall polysaccharide synthesis. Sugar alcohols, other than glycerol, are limited in nature to the plant kingdom.

sugar phosphate A phosphate derivative of a monosaccharide. Sugar phosphates are important intermediates in carbohydrate metabolism. The pentose phosphate ribulose 1,5-bisphosphate reacts with carbon dioxide in the first reaction of the Calvin cycle. All the subsequent steps of the Calvin cycle also involve sugar phosphates as do the reactions of glycolysis.

sulphur Symbol: S. A nonmetallic solid yellow element, atomic number 16, atomic weight 32.06. It is essential to growth, being found in the amino acids cystine and methionine, in the iron–sulphur proteins, e.g. ferredoxin, which are important in electron transport, and in *coenzyme A. Sulphur is also found in various secondary metabolites, e.g. the mustard-oil glycosides. Sulphur is absorbed from soil as the sulphate (SO_4^{2-}) ion. Various nitrogen and potassium fertilizers are applied as sulphates and sulphur deficiency is not commonly a problem.

Sulphur is used to control various pathogens and pests (*see* sulphur dust, dithiocarbamate fungicide). Sulphur dioxide, a by-product of many industrial processes, is an important atmospheric pollutant (*see also* acid rain).

sulphur bacteria *See* Beggiatoales, photosynthetic bacteria.

sulphur dust Powdered sulphur, used as an insecticide (especially against mites and scale insects) and as a fungicide, particularly in the control of mildew. It gives off sulphur dioxide when ignited and is thus often used as a *fumigant. It is also mixed with soil to kill soil-borne pathogens. Elemental sulphur, in combination with calcium hydroxide and water, forms *lime sulphur*, an orange-coloured liquid used to kill various mites and to control peach leaf curl, scab, and powdery mildew.

sun plant A plant that is able to flourish only in conditions of high light intensity. Sun plants usually have wide epidermal and palisade layers and numerous stomata on the lower surface of the leaf. Examples are the dominant trees of many woodland communities. *Compare* shade plant.

super gene (complex gene) A group of closely linked genes that determine different aspects of the same character and tend to act as a single functional unit. The S/s gene, which determines the 'pin' (s) and 'thrum' (S) style character of numerous *Primula* species, is an example. Besides controlling style length, it also has loci for stigma morphology, incompatibility reactions, pollen surface characteristics, and anther position. Very occasionally crossing over occurs between these loci resulting in a plant with both pin and thrum characteristics.

superior ovary *See* ovary.

supernumerary chromosome *See* B-chromosome.

supporting tissue *See* mechanical tissue.

surface area/volume ratio The ratio of the surface area of a cell, organ, or organism to its internal volume. The higher the ratio the greater the possibility for gaseous exchange and absorption of dissolved minerals. Thus the size of unicellular organisms is limited by the rate at which oxygen and carbon dioxide can diffuse into and out of the cell. The form of higher plants, especially the possession of numerous flat leaves, serves to increase the surface area/volume ratio. In this way a ratio of 30 cm^2 of surface to 1 cm^3 of tissues can be achieved. This ratio is greatly increased if intercellular spaces are taken into account.

surface tension A phenomenon apparent at the boundary of any liquid whereby an elastic film appears to be stretched over the surface. It is caused by the attractive forces between molecules at the surface of the liquid. The surface tension of water is particularly strong due to the orientation of the hydrogen bonds in the water molecule. This property slows down the vaporization of water and helps prevent plant desiccation. More importantly surface tension is responsible, with the strong adhesive forces of water, for *capillarity.

susceptibility The condition of a plant such that it is likely to succumb to attack by a pathogen. Extreme susceptibility, in which the host cells die soon after invasion by a pathogen, is termed *hypersensitivity and this reaction in fact confers considerable resistance. Susceptibility is often found where a crop cultivar has relied on one major-gene for resistance to a particular pathogen. If a mutant of the pathogen develops that can overcome this major-gene resistance then the cultivar is often seen to have no other means of preventing the invasion and spread of the pathogen. Susceptibility, like *resistance, varies according to environmental conditions. For example, excessive application of nitrogen fertilizer may lead to lavish growth that is more easily invaded by a pathogen.

suspended placentation *See* apical placentation.

suspension culture The system of growing single cells and small cell aggregates in a liquid growth medium that is kept agitated by means of bubbling, shaking, or stirring so the cells do not settle out. Microorganisms and cells derived from callus tissues may be grown in this way. Growth is maintained by providing continuous aeration and by either transferring portions of the suspension to fresh medium or replacing a part of the culture with fresh medium. Suspension cultures of plant cells derived from friable masses of callus show a similar growth curve to cultures of microorganisms in that there is a lag phase and a logarithmic phase of growth. Cells of many plant species can be induced to form embryoids (*see* embryogeny) in suspension culture, which, if removed from the culture and given appropriate conditions, will develop into complete plants.

suspensor The line of cells that differentiates from the proembryo by mitosis and anchors the embryo in the parental tissue. It also conducts nutrients to the embryo.

Svedberg unit *See* sedimentation coefficient.

swamp A region of vegetation that develops in stagnant or slow-flowing water, such as that around a lake margin. Initially there is a layer of peat supporting numerous reeds. When this builds up above the water level many other species become established, including sedges and yellow flag iris. Swamps may also develop on water-retentive clays in desert and semidesert areas.

Mangrove swamps are found along many tropical coasts. Mangroves are *halophytes and grow in a tangled mass to a moderate height. The stems and the exposed roots are covered with red algae and various lichens. Stilt roots anchor them in the soft mud and help to retain silt, thus gradually enabling other species to grow.

swarm cell In the *Myxomycota, an amoeba-like cell that is produced on germination of a spore. Swarm cells have two anterior whiplash flagella and lack a cell wall. Usually one, but occasionally up to four such cells emerge from one spore. They ingest food, reproduce by division, and then become gametes, in some instances becoming *myxamoebae first. The gametes fuse in pairs to form a zygote that develops into the *plasmodium. In adverse conditions the swarm cells can transform into microcysts.

syconium (syconus) A type of pseudocarp in which achenes are borne on the inside of a hollow receptacle. The fruit of figs (*Ficus*) is an example. *See illustration at* pseudocarp.

symbiosis An intimate relationship between two or more living organisms. It may be used in the narrow sense to mean only relationships in which all the partners benefit. In this sense it is synonymous with *mutualism. In its wider sense it covers other relationships, such as *parasitism and *commensalism.

sympatric Describing two or more populations that could interbreed but do not usually do so because of various differences, e.g. in time of flowering or type of pollinator. These populations may, through natural selection, become so distinct that they may be regarded as separate species (sympatric speciation).

Sympetalae *See* Asteridae.

sympetalous *See* gamopetalous.

symplast The continuum of cell protoplasts throughout the plant, linked by *plasmodesmata. Resistance to water flow through the symplast is far greater than through the *apoplast and the symplast is consequently only a secondary route for water movement.

symplastic growth A type of growth in which adjacent cells grow at the same rate so that the symplast is not disrupted. *Compare* intrusive growth, sliding growth.

symplesiomorphy *See* cladistics.

sympodial branching A type of *growth seen in some plants, e.g. elm and lime, in which the apical bud withers at the end of the growing season and growth is continued the following season by the lateral bud immediately below. *Compare* monopodial branching.

synangium A compound fruiting unit developed from the lateral fusion of individual sporangia, as seen in some ferns, e.g. *Marattia*, and gymnosperms, e.g. *Welwitschia*.

synapomorphy *See* cladistics.

synapsis The pairing of homologous chromosomes during *prophase of the first division of meiosis. The nature of the force that causes homologues to be drawn to each other is not known, but pairing is a very exact process, commencing with the pairing of the *centromeres and continuing with pairing of *chromomeres at various points along the length of the homologues. During the process, contraction of the chromosomes continues and they become closely coiled around each other to form bivalents.

syncarpous Describing a gynoecium in which the carpels are fused. The degree of fusion varies from only being joined at the ovary base (*semicarpous*), as in the Labiatae, to fusion of the ovary and style (*synstylovarious*), as in the Caryophyllaceae, to fusion of the whole carpel (*syncarpous*), as in the Cruciferae. This fusion may or may not persist in the fruit at dispersal.

The degree of fusion of the carpels is usually taken as an indication of the degree of specialization in the flower. *Compare* apocarpous.

syndetocheilic Describing a gymnosperm stomatal complex in which the subsidiary cells are derived from the same initial as the guard cells, as occurred in the Bennettitales. *Compare* haplocheilic. *See also* mesogenous.

syndiploidy A doubling of the chromosome number through a fault in the mitotic process. Commonly there is either failure of spindle formation at anaphase or failure to form a dividing wall between the two daughter cells. If such cells continue to divide normally an autopolyploid segment of tissue may arise in the plant. If this develops flowers, autopolyploid seeds may result.

synecology The study of the interactions between all the living organisms in a natural community, such as an oak wood or a pond, and the effect upon them of nonliving factors in the environment. *Compare* autecology.

synergidae (synergids) The two haploid nuclei at the micropylar end of the *embryo sac that do not participate in the fertilization process. Together with the egg nucleus they constitute the egg apparatus. Their function is not known and they abort soon after fertilization.

synergism The phenomenon whereby the effect of two substances acting together is greater than the sum of their individual effects. Combinations of different growth substances often act synergistically, for example, auxin and gibberellin combined bring about considerably greater internode elongation than either could individually.

syngamy *See* fertilization.

syngenesious Describing an androecium in which the anthers are fused, as in composite plants. *Compare* adelphous.

syntype Any one of two or more specimens or other material (descriptions, illustrations, etc.) designated by an author in the original publication of the name of a taxon. As a taxon can only have one type, a *lectotype can subsequently be chosen from a group of syntypes to serve as the nomenclatural *type.

systematics The scientific study and description of the variation in living organisms and the relationships that exist between them (the term is often used synonymously with *taxonomy). It is frequently preceded by a prefix, such as chemo- or bio-, denoting specialized fields of study that are valid in their own right. Several distinct phases in the development of systematics can be recognized. Pre-Darwinian systematics was primarily based on morphological and anatomical data. An understanding of natural selection and the behaviour of populations led to new interpretations and the inclusion of *phylogenetic considerations. Rapid developments in technology during recent decades have greatly aided the fields of chemosystematics and *numerical taxonomy.

systemic (translocated) Describing a chemical that is absorbed by a plant and transported throughout the tissues. Systemic fungicides and insecticides are applied to render plant tissues toxic to fungi or insects. Their effects are longer lasting than those of contact pesticides. Systemic herbicides, such as 2,4-D, 2,4,5-T, and MCPA, are used as selective weedkillers, killing most broad-leaved plants but leaving grasses and cereals unaffected. Such herbicides are also used to kill deeply rooted weeds.

T

2,4,5-T (2,4,5-trichlorophenoxyacetic acid) A synthetic *auxin of the *phenoxyacetic acid type with three substituted chlorine atoms. It has mainly proved of use as a selective weedkiller and as a defoliant. However, contamination with the highly toxic chemical dioxin has led to measures to restrict its use.

tachytelic *See* chronistics.

tactic movement *See* taxis.

taiga *See* forest.

take-all A disease of grasses and cereals caused by the fungus *Gaeumannomyces* (*Ophiobolus*) *graminis*. The pathogen survives in the soil on crop debris and can infect the crop at any stage. The dense mycelium on the roots and stem bases gives them a typical blackened appearance. If the plants survive until ear emergence the ears have little grain and appear bleached and are therefore called 'whiteheads'.

tangential **1.** Describing a longitudinal section cut down a cylindrical organ that does not pass through the centre of the organ. A tangential section through a woody stem is at a tangent to the growth rings and the growth-ring pattern appears paraboloid. *Compare* radial. **2.** *See* periclinal.

tannin (tannic acid) Any substance capable of precipitating the gelatin of animal hides as an insoluble compound, so changing the hide to leather, resistant to putrefaction. All tannins are obtained from plants and most are polyphenols. Common sources of tannin include tea (*Camellia sinensis*), sumac (*Rhus* spp.), and the bark and gallnuts of oak (*Quercus* spp.).

tapetum The food-rich layer of cells that surrounds the spore mother cells in vascular plants. In some plants it breaks down to form a fluid termed the *periplasmodium*, which is absorbed by the developing microspores. In others the tapetum remains intact until shortly before anther dehiscence and secretes substances into the locule.

Taphrinales An order of the *Hemiascomycetes containing many plant parasites, with most of the 105 species being placed in the genus *Taphrina*. There are three other genera. These fungi commonly induce hyperplasia, as in peach leaf curl, caused by *T. deformans*, and witches' broom of birch, caused by *T. betulina*. The Taphrinales form a limited dikaryotic mycelium that gives rise to asci. The ascospores reproduce asexually by budding before giving rise again to hyphae.

taproot A persistent robust primary root, often penetrating some depth below ground level and sometimes specialized for storage. Swollen taproots are produced by many biennial plants, e.g. carrot (*Daucus carota*).

tautonym A specific or infraspecific name that exactly repeats the generic name, e.g. *Magnolia magnolia*. This is not allowed in botanical nomenclature but is, however, acceptable in zoology.

Taxales An order of gymnosperms containing five extant genera (including *Taxus*, the yews) and about 20 species all in the family Taxaceae. They are often included within the order Coniferales but differ from conifers in having solitary ovules borne on short terminal shoots. The Taxales also have a distinct fossil record going back to the Triassic.

taxis (tactic movement, *pl.* taxes) A free directional locomotor movement exhibited by whole organisms that change their physical position in response to external stimuli. In the plant kingdom tactic movements are confined mainly to small, generally unicellular, aquatic plants, such as *Euglena*, which moves using a flagellum. In addition, certain reproductive cells and bacteria also show tactic movements. *Compare* tropism, nastic movements. *See also* phototaxis, chemotaxis.

taxometrics *See* numerical taxonomy.

taxon (*pl.* taxa) A named taxonomic group of any rank. Thus at the family level taxa may be represented by the Rosaceae and Labiatae, while *Rosa* and *Lamium* are examples of generic taxa. The term was coined to replace clumsy phrases such as taxonomic entity and taxonomic unit. Furthermore, the organisms contained within a rank (e.g. genus, order, or species) can also be referred to as taxa.

taxonomic characters Features, such as form, physiology, structure, and behaviour, that are assessed in isolation from the rest of the plant by taxonomists, in order to make comparisons and interpretations. It is important to distinguish between characters and character states. For example, leaf width may be a character, while leaves 4 mm wide are an expression of that character, i.e. its character state. Characters are often referred to as 'good' or 'bad' but this is strictly relative. Thus a good diagnostic character, such as compound leaves in a group of plants that mainly have entire leaves, would be bad for separating taxa in a group in which leaf divisions were either variable or often compound. *See also* weighting.

taxonomy The study of the principles and practices of classification. The term taxonomy is strictly applied to the study and description of variation in the natural world and the subsequent compilation of classifications. However, it is often used more loosely and includes the part of biological science referred to as *systematics. Taxonomy is a vast subject and various sections can be recognized within the discipline. In dealing with the flora of an area several phases can be recognized. The first phase is mainly concerned with identification and is sometimes referred to as exploratory or pioneer. Study of many tropical areas is still in this stage. Once material is better known and taxonomists have a good understanding of local and regional variation of the species it moves into the consolidation phase (the flora of Europe comes into this category). These two phases are jointly described by some as 'alpha' taxonomy. Once cytological or *biosystematic data are available these can be added to existing data. Taxonomy in which all available evidence is considered is described as the encyclopaedic phase or 'omega' taxonomy. Some authors make further distinctions between classical (mainly intuitive) and experimental approaches, the latter including biosystematic, chemosystematic, and numerical procedures.

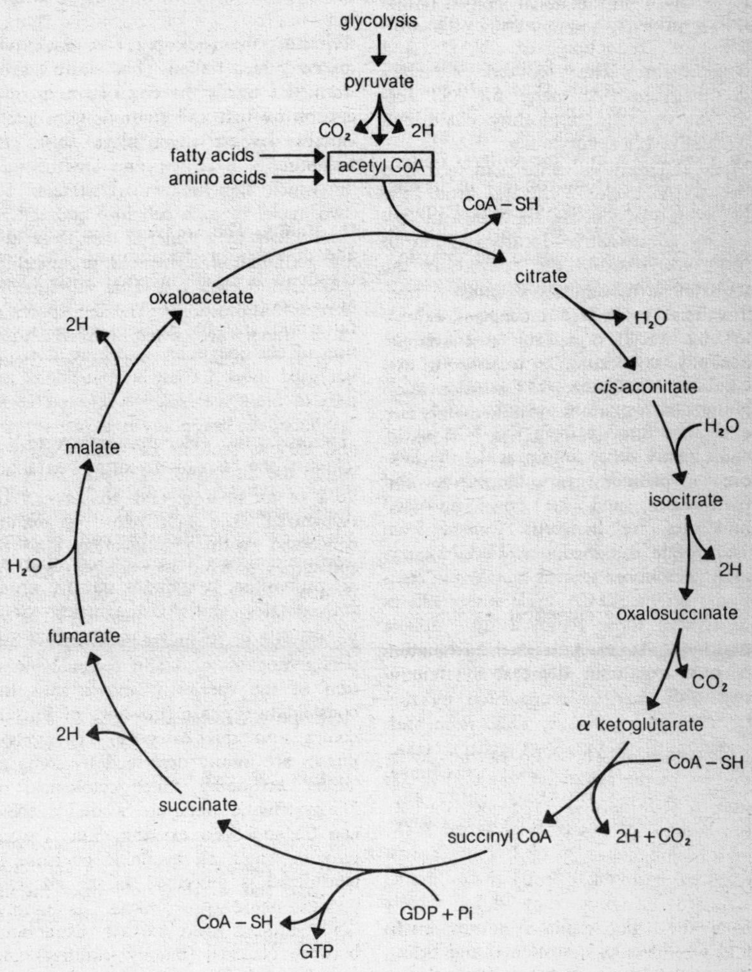

The TCA cycle.

TCA cycle (tricarboxylic acid cycle, citric acid cycle, Krebs' cycle) A cyclic sequence of reactions, found almost universally in aerobic organisms, in which the acetyl portion of acetyl CoA is oxidized to carbon dioxide and hydrogen ions. The energy released by this oxidation is principally conserved by the concomitant reduction of NAD and flavoprotein. The reduced cofactors eventually provide energy for ATP formation via the *respiratory chain and *oxidative phosphorylation.
Many different substrates can feed into the TCA cycle. Pyruvate, from the breakdown of glucose, feeds into the cycle by an oxidative decarboxylation to form acetyl CoA; the enzyme is the pyruvate dehydrogenase complex.
The TCA cycle is not simply catabolic in function. It is in fact an *amphibolic* pathway, combining both anabolic and catabolic functions. The amino acids glutamate, aspartate, alanine, and glycine arise directly from the TCA cycle, while many other amino acids, the hexose and pentose sugars, the purines and pyrimidines, and the porphyrin-based molecules are indirectly formed from TCA cycle intermediates. Certain enzymatic reactions, termed *anaplerotic* reactions, produce TCA cycle intermediates to replace those removed by synthetic pathways. An example is the formation of oxaloacetate by the carboxylation of pyruvate.

***t* distribution** The ratio of the deviation, d, to the estimated standard deviation, s. It is used in experiments where the number of observations is small (31 or less and hence 30 or less degrees of freedom) and consequently the estimated standard deviation may differ widely from the true standard deviation. To find whether any deviation is significant, its probability is looked up in 'Student's' t tables, which take the number of degrees of freedom (N) into account. The greater the value of N, the smaller the value of t for any given level of probability.

teleutosorus (telium) A sorus in which *teleutospores develop.

teleutospore (teliospore) A dark two-celled thick-walled binucleate spore formed towards the end of the growing season by rust and smut fungi. Teleutospores develop in a black sorus, the *teleutosorus* or *telium*, and are the form in which the fungus overwinters. The two nuclei in each cell fuse and the zygote nuclei then undergo meiosis so that on germination a haploid promycelium grows from each cell. This subsequently forms *basidiospores. The teliospores of smut fungi are often termed brand spores or chlamydospores. *See* heteroecious.

Teliomycetes (Hemibasidiomycetes) A class of the *Basidiomycotina containing those fungi in which a sorus (teleutosorus or telium) develops in place of a basidiocarp and produces resting spores (*teleutospores) that act as probasidia. It contains the two orders *Uredinales and *Ustilaginales. Some classifications include the Tremellales in this class.

teliospore *See* teleutospore.

telium *See* teleutosorus.

telocentric Describing a chromosome that has the centromere at, or very close to, one end so that only one chromosome arm is visible. *Compare* acrocentric, metacentric.

telome theory The proposition that certain plant organs are derived from modified reduced branch systems. Thus *megaphylls are believed to be derived by the flattening of a branch system and the subsequent development of ground tissue (webbing) between the branches. Some believe stamens may also have originated in this manner, rather than from microsporophylls.

telophase The stage in nuclear division following *anaphase, in which the separated chromatids or the separated homologous chromosomes of bivalents collect at the poles of the spindle and the nuclei of the daughter cells are formed. In *mitosis and the second division of *meiosis, the chromatids (now complete single-stranded chromosomes) become surrounded by vesicles as they reach the poles and these eventually merge to form the nuclear membrane. The chromosomes lengthen as they uncoil and the *chromatin assumes its interphase condition. Nucleoli reappear as they are formed at the nucleolar organizers.

At telophase of the first division of meiosis, the daughter haploid nuclei that form at the poles contain one member only of each homologous pair of chromosomes present in the original mother cell. Each chromosome is divided into two chromatids (genetic recombinants as a result of chiasma formation) held together by a *centromere. The duration of the following interphase is variable but there is no further DNA replication and the chromosomes remain in a relatively contracted condition. The two daughter haploid nuclei then enter the second division of meiosis.

As daughter nuclei become organized at the poles, a *cell plate begins to form in the cytoplasm in the equatorial region. This marks the beginning of cell wall formation. In meiosis a cell wall may not form at the end of the first division, the daughter cells being separated by a cell plate only.

TEM *See* electron microscope.

temperate phage A DNA phage that does not necessarily induce *lysogeny of an infected bacterial cell. Instead the phage DNA may be incorporated into the bacterial chromosome, where it replicates along with the bacterium. A bacterium thus infected is described as *lysogenic*.

temporary stain A type of stain used for immediate observation through the light microscope. Such stains often damage the section, or the colour of the stain fades after a short time. For example, when plant sections are stained with Schultze's solution and mounted in 50% glycerol, the stain eventually causes swelling and dissolves the walls. Some specimens can be mounted directly in the stain, for example, when using ruthenium red. *Compare* permanent stain. *See also* staining.

tendril A modified leaf, leaflet, branch, or inflorescence of a climbing plant that coils around suitable objects, such as other nearby plants, and helps support and elevate the plant. Examples are the branch tendrils of white bryony (*Bryonia dioica*), the leaf tendrils of yellow vetchling (*Lathyrus aphaca*), and the inflorescence tendrils of virginia creepers (*Parthenocissus*). The tendrils of some species terminate in disclike suckers while others, e.g. those of grapevines (*Vitis*), resemble climbing roots in being negatively phototropic and hence growing into dark cracks on the support.

tension wood *See* reaction wood.

tent pole In certain gymnosperms, an extension of the female gametophyte towards the pollen chamber. It is seen in *Ginkgo biloba* and is also characteristic of many fossil genera.

tepal An individual perianth part in flowers that have no distinct calyx and corolla, as occurs in most monocotyledons. In the tulip the tepals are all highly coloured.

teratology The study of malformations and abnormal growth. *See also* witches' broom, gall, leaf curl, club root.

terete Smooth, cylindrical, and tapering as, for example, a grass stem.

terminalization *See* prophase.

terpene *See* terpenoid.

terpenoid Any of a large class of compounds derived from multiples of the unsaturated hydrocarbon isoprene. They include the *terpenes*, which are all hydrocarbons. Terpenoids based on one, two, three, or four *isoprene subunits are called hemiterpenoids (e.g. apiose), monoterpenoids (e.g. camphor), sesquiterpenoids (e.g. zingiberine), and diterpenoids (e.g. phytol) respectively.

The terpenoids are a very diverse class of compounds. They may be cyclic, e.g. cannabidiol, or acyclic, e.g. phytol, a precursor of chlorophyll. Many give the characteristic odour or flavour to a plant oil, e.g. limonene (lemon oil), pinene (pine oil), and menthol (mint oil). The fat-soluble vitamins A, E, and K are terpenoids as are the ubiquinones and plastoquinones and the growth substances gibberellic acid and abscisic acid. The carotenoids (tetraterpenoids) and sporopollenin, gutta-percha, and natural rubber (polyterpenoids) are further examples of this group.

Terpenoid distribution is considered potentially valuable in taxonomic work. For example, various changes have been suggested at the tribe level in the Compositae following work on the occurrence of the bitter-tasting sesquiterpene lactones. In the genus *Pinus*, the distribution of turpentines has been used to justify placement of *P. jeffreyi* in the Macrocarpae. Traditionally it was placed in the Australes on the basis of its resemblance to *P. ponderosa*, a typical member of the Australes.

terra rossa A type of intrazonal soil rich in lime (calcimorphic) from the underlying limestone. It is a clayey soil with a low humus content and its bright red colour results from large amounts of iron oxides. It may be found in regions of Spain, southern France, and southern Italy. *Compare* rendzina.

Tertiary The first period of the *Cenozoic era from about 65 to 2 million years ago. It began with the Palaeocene epoch, which was followed by the Eocene, Oligocene, Miocene, and Pliocene (Pleiocene) epochs. The climate was tropical to the end of the Oligocene, becoming cooler and drier towards the end of the period as the oceans contracted and the present mountain ranges were formed. Abundant plant fossils have been found, many showing similarity to modern-day plants. Changes in the distribution of the flora can be related to changes in climate throughout the era. Siliceous remains of diatoms, the diatomaceous earths (Kieselguhr) are widespread, blue-green algae contributed to the formation of oil shales, green and red algae contributed to marine limestone deposits, and the fossil stoneworts contributed to the formation of freshwater limestone beds. The Gnetales appeared and modern genera of the Coniferales are represented. By the end of the period, the angiosperms were firmly established with about 260 families. This was accompanied by the evolution of insects. Grasses became more abundant in the Miocene, possibly because of the change to a drier climate. *See* geological time scale.

testa The protective outer covering of a seed, derived from the *integuments of the ovule after fertilization. In primitive forms the outer integument may remain fleshy and the inner become lignified to form a *sarcotesta* and *schlerotesta* respectively. In more advanced forms both the integuments fuse and become hard and dry except for a small unthickened area at the *micropyle, which facilitates radicle emergence at seed germination. In some species the outer surface of the testa becomes covered with mucilage, hairs, or fibres to aid seed dispersal. The fibrous nature of the cotton seed has long been exploited commercially.

test cross A cross between an individual of uncertain genetic constitution and a homozygous recessive. The latter is often a parent, in which case the test cross is a specific form of backcross. The purpose of a test cross is to ascertain the unknown genotype. For example, if T (tall peas) is dominant to t (short peas) then any tall peas (T_) could be either T$\underline{\text{T}}$ or T$\underline{\text{t}}$. If T_ were test crossed to tt and all the offspring were tall then the second allele could be assumed to be T. Alternatively, if there were any short (tt) peas among the offspring the second allele must be t.

test of significance A way of finding whether the results obtained from an experiment differ from the hypothesis the experiment was constructed to test because of sampling error or because the original hypothesis is invalid.

tetrad A group of four cells formed by meiosis. All four cells may survive to form spores, as occurs in the formation of pollen grains. Alternatively some of the cells abort, as in the angiosperm ovule, where one cell develops into the embryo sac and the remainder abort. The four cells of the tetrad may remain joined together. This provides a useful system for the study of the nature and extent of recombination, a procedure termed *tetrad analysis*. The shape of the tetrad, e.g. whether it is linear, tetrahedral, tetragonal, or rhomboidal can be a useful taxonomic character and is a major factor determining pollen morphology. *See also* triradiate scar.

tetrad analysis The genetic analysis of the four cells resulting from a meiotic division in order to gain information on the nature and extent of recombination. Tetrad analysis is only possible when the meiotic products remain together in groups (tetrads). This does occur in certain lower plants notably ascomycete fungi, e.g. *Neurospora*, and some bryophytes, e.g. the liverwort *Sphaerocarpus*.

tetradynamous Having four long stamens and two short stamens, as do the flowers of plants in the family Cruciferae.

tetramerous (4-merous) Describing flowers in which the parts of each whorl are inserted in fours, or multiples of four, as in the willowherbs (*Epilobium*) and members of the Cruciferae.

tetraploid An organism or cell with four times the haploid number of chromosomes. The fusion of diploid gametes will result in the formation of a tetraploid. Such gametes initially arise because of the failure of homologous chromosomes to separate at meiosis. Alternatively a tetraploid segment of a plant may arise because of the failure of chromatids to separate at mitosis. If this segment develops reproductive structures then diploid gametes will be formed. *See* autopolyploidy, allopolyploidy.

tetrasomic *See* aneuploidy.

tetrasporangium A sporangium in which four spores (the *tetraspores*) are formed. It is seen in some of the red algae, e.g. *Polysiphonia*.

thalamus *See* receptacle.

Thallophyta A former division containing all nonanimal organisms without differentiated stems, leaves, and roots, i.e. the *algae, *fungi, *lichens, and *bacteria. It was a feature of older classifications but has largely been replaced by more detailed systems that recognize the fundamental differences between these groups. In some recent classifications the Thallophyta (or Thallobionta) is retained as a subkingdom of the Plantae in distinction to the *Embryophyta.

thallus A plant body that is not differentiated into true leaves, stems, and roots. It is often a flattened structure, e.g. the gametophyte generation of the liverwort *Pellia*.

thermonasty A *nastic movement in response to a change in temperature. For example, when subjected to a temperature increase of between 5 and 10°C, *Crocus* flowers open in a few minutes due to a relative increase in growth rate on the inner side of the petals.

thermoperiodic response The response of a plant to a diurnal fluctuation in temperature. For example, growth of tomato plants is best when a certain day/night temperature regime is experienced. Meristematic activity and extension growth are often affected by thermoperiodicity.

thermophilic Describing microorganisms that require high temperatures (45–65°C) for growth. Such organisms are found in rotting vegetable matter and hot springs. *Compare* mesophilic, psychrophilic.

therophyte An *annual or *ephemeral plant that survives unfavourable conditions in the form of a seed. Therophytes are common on cultivated land and in desert regions where perennial plants cannot establish themselves. *See* Raunkiaer system of classification.

thiamin (thiamine, vitamin B_1) A substituted pyrimidine joined through a methylene bridge to a substituted thiazole. In its active form, thiamin pyrophosphate, it acts as a coenzyme in reactions involving the transfer of aldehyde groups in carbohydrate metabolism. Thiamin is synthesized by plants but not by some microorganisms or most vertebrates, which thus require it in the diet.

thigmotaxis A *taxis in which the movement is parallel to lines of stress in the substrate. It is seen, for example, in bacteria of the order *Myxobacteriales.

thigmotropism *See* haptotropism.

thin-layer chromatography A widely used, fairly fast chromatographic technique for separating the components of mixtures. A glass plate is covered with a thin layer of cellulose, silica gel, or alumina. This represents the solid stationary phase. A line is scratched on the base of the plate and a small spot of the mixture is applied to the base line using either a wire loop or a capillary tube. The plate is then suspended in a suitable solvent (the mobile phase), which rises up the plate by capillary action, carrying and separating the mixture into its constituents. The remainder of the procedure is similar to that used in *paper chromatography.

Thiobacteriales Formerly, an order containing the sulphur bacteria and the photosynthetic bacteria. Bacteria once placed in this order are now placed in either the Beggiatoales or Pseudomonadales.

thiourea A chemical, with the formula NH_2CSNH_2, widely used to stimulate germination in seeds that have a light requirement. It is believed to act by increasing gibberellin levels.

thorn A sharply pointed woody structure formed from a modified reduced branch and connected to the vascular system of the plant. Examples are those of the hawthorn (*Crataegus monogyna*) and sloe or blackthorn (*Prunus spinosa*). *Compare* spine, prickle.

thorn forest *See* forest.

threonine A polar amino acid with the formula $CH_3CH(OH)CH(NH_2)COOH$ (*see illustration at* amino acid). The biosynthetic pathway of threonine follows that of methionine from aspartic acid to the formation of homogentisic acid, at which point the two pathways diverge. Breakdown occurs via *glycine and acetyl CoA. Isoleucine is synthesized from threonine.

thrum *See* heterostyly.

thylakoid Any one of the layers making up the grana in *chloroplasts. It consists of a channel surrounded by a pigmented membrane.

thymine A pyrimidine base, characteristically present in DNA, and absent from RNA. In DNA it pairs specifically with adenine in the complementary strand, so that the thymine:adenine base ratio is 1:1. Thymine is linked to other bases of the same strand through a sugar-phosphodiester backbone, the nucleotide of thymine being called *thymidine* (thymine + deoxyribose sugar + phosphate). Thymine is more fully described as 5-methyl-2,4-dioxypyrimidine and is derived from sugars and amino acids. *Compare* uracil.

thyrse A complex densely branched inflorescence the individual branches of which are dichasia. An example is that of the horse chestnut (*Aesculus hippocastanum*).

tiller A shoot that develops from axillary or adventitious buds at the base of a stem, often in response to injury of the main stem. Tillering is seen when the trunk of a tree is lopped, as in the practice of coppicing. It is also characteristic of the growth of grasses, giving the tufted appearance of many species.

timber line (Waldgrenze) The more or less clearly defined region at high altitudes or latitudes beyond which normal dense tree growth does not occur. The trees growing between the timber line and the *tree line tend to be dwarfed and deformed and are termed *elfin forest.

tinsel flagellum *See* flimmer flagellum.

tissue culture The growth of isolated plant or animal cells or small pieces of tissue under controlled conditions in a sterile growth medium. The medium is designed to meet the requirements of the tissue involved. Experiments varying media composition have yielded information on the nutritional needs of particular cells, this not being readily ascertained in an entire organism. Whole plants can often be regenerated from small segments of tissue or even single cells. This feature has been used commercially in the propagation of various plants, notably orchids. The regeneration of plants from meristem explants has been used to free crops of virus infection since the virus particles do not penetrate meristematic tissues. Tissue culture also has various applications in plant breeding work. For example, crosses can be made between normally incompatible plants if the embryo is 'rescued' and grown in culture before abortion can take place. *See also* anther culture, protoplast fusion, suspension culture, callus.

toadstool *See* mushroom.

tocopherol (vitamin E) A terpenoid-like substance found in high concentrations in certain seeds, notably cereals. It is believed to prevent the oxidation of lipids and so prolong seed viability. In a few species applications of tocopherol have been shown to replace the vernalization requirement. Deficiency in certain animals appears to cause infertility. *See* vitamin.

toluidine blue A general purpose alkaline aniline stain commonly used to give an even coloration to the specimen and distinguish it from the embedding material. It is frequently used to stain the

1 μm sections cut on ultramicrotomes as an adjunct to transmission electron microscope studies.

tomentose Describing a surface densely covered in short hairs, such as that of the leaves of hoary mullein (*Verbascum pulverulentum*).

tonoplast The membrane separating the cell *vacuole from the protoplasm. It has the same structure as the plasma membrane and mediates between the protoplasm and the vacuolar sap.

torus 1. The thickened central part of the pit membrane in the *bordered pits of many gymnosperms, consisting of primary cell wall material. The torus is believed to act as a valve, blocking the pit when there is an inequality of pressure on either side of the pit, as is the case when an adjacent tracheary element is damaged.
2. *See* receptacle.

totipotency The capacity, exhibited by certain types of isolated differentiated plant cell, to regenerate whole plants. The phenomenon is seen as evidence for the theory that all nucleated plant cells possess all the genes necessary to direct the formation of a complete plant. To realize this potential the cell must be removed from the inhibiting influence of the rest of the plant body and given the appropriate stimuli, namely the correct balance of nutrients and growth substances. *See also* competence.

toxin A poisonous substance, usually of microbial origin, that stimulates the production of antitoxins in an animal body. Many bacterial diseases are due to the release of bacterial toxins. These may be *endotoxins*, which are formed within the bacterium and released on the death and disintegration of the bacterial cell, or *exotoxins*, which are secreted through the bacterial cell wall. The exotoxins, many of which are highly active enzymes, generally have more severe ef-

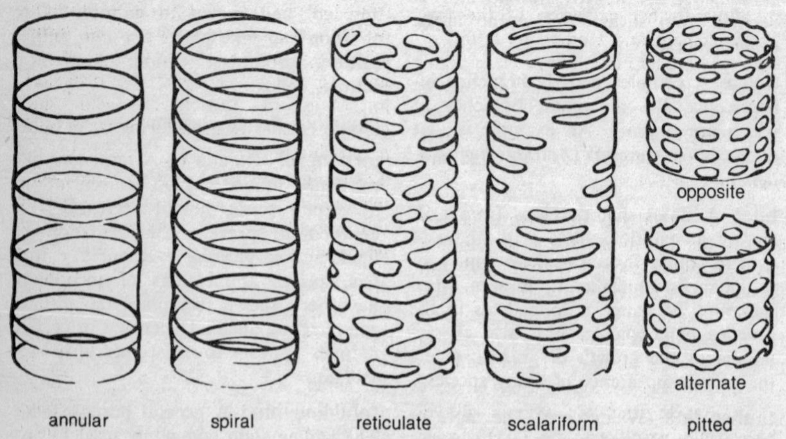

Types of secondary wall thickening in tracheary elements.

fects. Several fungi, e.g. death cap (*Amanita phalloides*) and ergot (*Claviceps purpurea*), also produce toxins. *Aflatoxins*, produced by the mould fungus *Aspergillus flavus*, can cause severe liver damage. Higher plant toxins are less common, an example being ricin, an albumin in castor oil (*Ricinus communis*) seeds, that causes agglutination of red blood cells.

TPN *See* NADP.

trabant *See* satellite.

trabecula A barlike structure extending across a lumen or lacuna. For example, in some plants extensions of the secondary wall thickening form trabeculae across the lumina of xylem tracheids.

trace element *See* micronutrient.

tracer *See* isotopic tracer.

tracheary element A water-conducting cell in the xylem, i.e. a *vessel element or *tracheid. The secondary cell walls of the tracheary elements show various types of thickening (*see* illustration). Annular and spiral patterns of thickening allow further extension of the tracheary element and tend to be found in protoxylem. The other forms of thickening are seen in metaxylem. Although bryophytes do not posses xylem, some do possess water-conducting cells known as *hydroids. Some consider these are also tracheary elements.

tracheid A relatively primitive tracheary cell in the xylem of many plants that has a lignified secondary cell wall and usually lacks a living protoplast at maturity. Tracheids may be distinguished from *vessels as they lack perforation plates and are generally smaller in diameter and greater in length. The only perforation in tracheids are the pits in the secondary cell wall, which are most concentrated on the end walls. In gymnosperms (except Gnetales) and some primitive angiosperms that lack vessels, the tracheids are the only cells specialized to transport water up the plant. *Compare* fibre, fibre-tracheid.

Tracheophyta In classifications that consider the possession of vascular tissue of greater significance than the seed-bearing habit, a division containing all the *vascular plants. It is divided into the subdivisions *Psilopsida, *Lycopsida, *Sphenopsida, and *Pteropsida. The psilopsids are considered the most primitive as they lack true roots and a well defined phloem. These features are properly developed in the other subdivisions. Also the psilopsid gametophyte resembles the sporophyte in being dichotomously branched and possessing vascular tissue, while in the other subdivisions the gametophyte is comparatively reduced.

trama The central tissue in the gill of a fungus, consisting of loosely packed hyphae.

transamination The transfer of the amino group of an amino acid to a keto acid, usually α-ketoglutaric acid, to form glutamate and the keto acid derived from the original amino acid. Transamination reactions are catalysed by *transaminases*, such as aspartate transaminase, which catalyses the reversible formation of oxaloacetic acid and glutaric acid from aspartic acid and α-ketoglutaric acid.

Transamination is central to amino-acid metabolism as it is one of the steps in the biosynthesis and the degradation of nearly all amino acids.

***trans* arrangement** (repulsion) The situation in which an individual is heterozygous for two linked *genes (either two units of function or two mutations in the same functional unit), and the recessive (or mutant) allele of one gene is on the same chromosome as the dominant (or normal) allele of the other gene. The homologous chromosome will thus possess the dominant allele of the first gene and the recessive allele of the second gene. If the recessive alleles are in the same functional unit then their effects will show in the phenotype because there is no unchanged functional unit to mask them. If they are in different functional units, then the phenotype will be normal. This is the basis of the **cis-trans* test. *Compare cis* arrangement.

transcription The part of protein synthesis that involves the formation of a complementary copy of the genetic code by *messenger RNA synthesis. Transcription occurs on the genetic material itself: principally nuclear DNA in eukaryotic organisms. *Compare* translation.

transduction The transfer of genes from one bacterium to another by the action of a temperate bacteriophage acting as a vector (transmitting agent). It occurs when a phage excises from a bacterial genome in a faulty manner, taking some of the bacterial genes with it and leaving some of its own behind. The transferred genes (the transducing material) are incorporated into the bacterial genome of a new host cell and result in the transduced cell showing a permanent genetic change. *Abortive transduction* is sometimes seen, in which the transferred genes do not become incorporated into the genome but remain as a plasmid in the cell. The recipient bacterium then only shows the new properties until the plasmid is lost from the cell. Transduction occurs naturally at low frequency but this can be increased substantially under laboratory conditions. It is an important tool in genetic engineering.

transect A line across an area along which a study can be made of the distribution and abundance of plant species. The line is marked out with a tape and its position indicated on a map of the area. A *profile transect* is made along a slope and the species present at various levels are recorded. The plants actually touching the tape may be recorded or *quadrats may be placed along the transect at regular intervals. A *belt transect* is a strip of ground, 1 m or 0.5 m wide, between two parallel transect lines. The species are recorded a section at a time. Scale diagrams, such as *profile diagrams* or *belt transect histograms*, can be constructed from the recorded results and the distribution of the species related to any habitat factors that have also been recorded.

transferase Any *enzyme that catalyses reactions in which a functional group is transferred from a donor to an acceptor molecule. Six classes of transferase are distinguished, including enzymes that transfer phosphate groups (phosphotransferases or *kinases) and enzymes that transfer acyl groups (e.g. phosphate acyltransferase, which catalyses the transfer of an acyl group from phosphate to coenzyme A).

transfer RNA (tRNA) A type of *RNA molecule that binds to amino acids in the cytoplasm and assists in their incorporation into polypeptide chains at ribosomes. The tRNA molecule is shaped like a clover leaf. Each amino acid binds only to a specific type of tRNA molecule. At one end of a tRNA molecule is a characteristic sequence of three bases, the *anticodon, which will temporarily pair with a complementary triplet, the *codon, in messenger RNA. At the same time the amino acid will detach from its tRNA and form a peptide bond with the amino acid that preceded it. By this means the polypeptide grows by the addition of amino acids one at a time. The amino acid that is incorporated will be determined by the specificity of an amino acid for a particular type of tRNA molecule and the specificity of pairing between the anticodon and codon. *See* translation.

transformation *See* bacterial transformation.

transfusion tissue The tissue immediately surrounding at least part of the vascular bundles in the leaves of gymnosperms, e.g. *Pinus*. Transfusion tissue is composed of tracheids (*transfusion tracheids*) with conspicuous *bordered pits, and parenchyma cells (*transfusion parenchyma*) containing tannin-like substances and sometimes starch. The main function of the transfusion tissue is believed to be the transport of materials between the vascular bundles and the mesophyll.

transition A mutation caused by substitution in the DNA of one purine base for the other, or one pyrimidine for the other. Hence adenine might replace guanine and vice versa, or thymine might replace cytosine and vice versa. It is thought that mutagens such as 5-bromouracil induce transitions. *Compare* transversion.

transition region **1.** The part of the plant body between the stem and the root in which there are intermediate arrangements of the tissues. *Vascular transition* takes place in this region, ensuring continuity of vascular tissues throughout the plant. *See* hypocotyl.
2. An *intercalary meristem located between the bladelike portion and the narrow stemlike portion of the plant body in the Laminariales (an order of brown algae).

translation The part of protein synthesis that involves decoding the sequence of triplets in messenger RNA, and the concurrent formation of a polypeptide based on that sequence. Translation occurs on ribosomes in the cytoplasm, the proteins being synthesized from the amino terminal. The process is highly endergonic requiring four ATP equivalents per amino acid residue. Four stages may be recognized: activation, initiation, elongation, and termination.
During activation amino acids are attached to small soluble *transfer RNA (tRNA) molecules in the cytosol. Each tRNA is specific for one amino acid. In initiation *messenger RNA (mRNA), transcribed from nuclear DNA, attaches to the ribosome. An initiating codon on the mRNA codes for a particular aminoacyl tRNA, which binds to the mRNA at the ribosome.
During elongation another aminoacyl tRNA now binds to the ribosome next to the first, and the two amino acids react to form a dipeptide on the second tRNA. The ribosome then moves along the mRNA, releasing the first tRNA molecule. A third aminoacyl tRNA now attaches to the ribosome and the process repeats, continuing until the full polypeptide chain is formed.
At termination, a termination codon on the mRNA causes the release of the completed protein from the final tRNA molecule. The last tRNA and the mRNA then dissociate from the ribosome. *Compare* transcription.

translocated *See* systemic.

translocation **1.** The conduction of soluble materials from one part of the plant to another. The process includes the movements of food substances in the phloem tubes, the transfer of growth substances from their point of production, and the upward flow of dissolved salts in the transpiration stream. Translocation of food materials is often preceded by enzyme action, which converts the substance concerned from an insoluble to a soluble form, as from starch to sugar. *See also* mass flow hypothesis, transpiration.
2. A chromosome mutation in which a chromosome segment has become detached and reattached to a different (nonhomologous) chromosome. The most commonly occurring translocations are known as *reciprocal translocations* or *interchanges*, which involve the mutual

exchange of segments from nonhomologous chromosomes. *See also* deletion, duplication, inversion.

transmission electron microscope *See* electron microscope.

transmitting tissue (conducting tissue) The specialized thin-walled tissue that constitutes the central part of the style in some angiosperms, through which the pollen tube grows down to the funiculus. It may be involved in self-incompatibility systems.

transpiration The loss of water by evaporation from a plant surface. Over 90% escapes through the open stomata, while about 5% is lost directly from the epidermal cells. It has been shown that although the combined area of stomatal pores is on average only 1–2% of the total leaf area, the amount of transpiration they allow is 90% of the transpiration that occurs from a water surface the same area as the leaf.

Transpiration rates are greatest when the leaf cells are fully turgid and when the external *relative humidity is low. Water forms a film around the mesophyll cells and evaporates into the substomatal chamber from where it diffuses into the air. The degree of opening of the stomata (stomatal resistance) is of prime importance in governing the rate of water loss. The width of the *boundary layer at the leaf surface is also important. In dry conditions, transpiration can cause wilting and so the plant may develop features such as waxy cuticles to minimize the problem. *See also* antitranspirant, cohesion theory.

transpiration stream The flow of water from the roots to the leaves via the xylem vessels that is caused by *transpiration at the leaf surface. *See also* cohesion theory.

transversion A mutation caused by the substitution of a purine base for a pyrimidine base (or vice versa) in the DNA. *Compare* transition.

trap crop Any crop planted in or around another crop to attract pests away from the more valuable crop. Trap crops are usually heavily treated with pesticides, burnt, or ploughed in.

traumatic acid A straight chain dicarboxylic acid, formula COOHCH:CH $(CH_2)_8COOH$. It has been isolated from green bean pods and is believed to act as a *wound hormone.

tree ferns *See* Filicales.

tree line (Baumgrenze) The more or less clearly defined region at high latitudes (the taiga/tundra boundary) or high altitudes (the subalpine/alpine boundary) beyond which trees do not grow. Occasionally inverted tree lines are seen in which trees do not grow below a certain line as, for example, in frost hollows. The limits to tree growth are presumably set by some climatic factor though it is by no means certain which particular factor is operating. Numerous theories have been advanced based on different climatic variables such as heat, light, or carbon dioxide deficiency, excessive wind or snow depth, etc. A recent theory suggests that, beyond the tree line, leaves become desiccated in winter because the summer growing season is not long enough to allow the leaves to mature fully and resist water stress. *Compare* timber line.

Tremellales An order of the *Hymenomycetes containing fungi that produce gelatinous basidiocarps, hence the common name gelatinous or jelly fungi. These number about 200 species (26 genera) and and are usually found growing on dead wood. They have characteristic rounded basidia divided by septa into four cells. Typical examples are witches' butter (*Exidia glandulosa*) and the yellow brain fungus (*Tremella mesenterica*). In some classifications the

Tremellales are included in the *Teliomycetes.

triacylglycerol (triglyceride) An ester of glycerol and three long-chain carboxylic acids or *fatty acids. Triacylglycerols may be triesters of either the same fatty acid, or more commonly of two or three different fatty acids (mixed triacylglycerols). The most common fatty acids found in triacylglycerols are palmitic, stearic, oleic, and linolenic acids.
Triacylglycerols are synthesized from glycerol phosphate and fatty acyl CoAs. The glycerokinase that phosphorylates glycerol is a soluble enzyme, but the other enzymes of triacylglycerol synthesis are associated with the microsomes. Breakdown of triacylglycerols to glycerol and free fatty acids is achieved by the action of *lipases.
Triacylglycerols are important energy storage molecules especially in seeds, many of which (e.g. rape, linseed, castor bean, coconut) are important commercially as sources of fats and oils. Fats and oils are solid and liquid triacylglycerols respectively.

Triassic (Trias) The first period of the Mesozoic era between about 225 and 195 million years ago. In the early Triassic, arid conditions prevailed though later the climate became mostly temperate, but variable. The early conditions have resulted in fossils of the period being generally rare. The fossil lycopsid *Pleuromonia* found in Triassic sandstone is thought to be an intermediate form between the Lepidodendrales and the Isoetales. Members of the sphenopsid order Sphenophyllales became extinct at the beginning of the period. The now extinct order of the gymnosperms, the Bennettitales, arose in this period and common fossils are *Bennettites* and *Williamsoniella*. Members of the extinct order Caytoniales are found and the Ginkgoales are well represented in the Upper Triassic (e.g. *Ginkgoites*). *See* geological time scale, Permo-Triassic.

tribe The rank subordinate to family but superior to genus in the taxonomic hierarchy. The term is applied to assemblages of similar genera within large families. The Latin names of tribes have the ending -eae. An example in the family Umbelliferae is the tribe Saniculeae, which embraces the genera *Eryngium*, *Astrantia*, and *Sanicula*. Similar tribes may be grouped together in subfamilies. Tribes may be split into subtribes, the Latin names of which have the ending -inae.

tricarboxylic acid cycle *See* TCA cycle.

2,4,5-trichlorophenoxyacetic acid *See* 2,4,5-T.

trichogyne A receptive, often hairlike, uni- or multicellular structure that projects from the female sex organ in some algae, ascomycetes, and lichens. It serves to attract and receive the male gamete or nucleus prior to fertilization.

trichome Any outgrowth, such as a root hair, from an epidermal cell. Trichomes are very varied in form and function, their morphology often yielding important taxonomic characters. They may be elongate, scalelike or peltate, glandular or nonglandular, and unicellular or multicellular.

trichothallic growth A form of growth seen in certain brown algae, e.g. *Ectocarpus* and *Cutleria*, in which cell division is restricted to well defined intercalary regions.

tricolpate Describing a pollen grain having three colpi, as is commonly found amongst most dicotyledon species.

trifoliate Describing a compound leaf having three leaflets arising from the same point, such as the clovers (*Trifolium*) and wood sorrels (*Oxalis*).

triglyceride The former name for a *triacylglycerol.

trihybrid An organism that is heterozygous for three genes, e.g. AaBbCc. Trihybrid crosses, especially of the type AaBbCc × aabbcc, are commonly used devices in chromosome mapping. By comparing the frequencies of the phenotypes, the chromosomal arrangements of the three genes A/a, B/b, and C/c can be deduced. This is technically easier than setting up three dihybrid crosses AaBb × aabb, AaCc × aacc, and BbCc × bbcc, which would only achieve the same result.

trilete *See* triradiate scar.

trimerous (3-merous) Describing flowers in which the parts of each whorl are inserted in threes, or multiples of three. This arrangement is characteristic of many monocotyledons.

triose phosphate A three-carbon phosphorylated sugar. Two commonly occurring triose phosphates are glyceraldehyde 3-phosphate and dihydroxyacetone phosphate (DHAP), which are both intermediates in the synthesis and breakdown of glucose. In glycolysis the triose phosphates are formed by the splitting of fructose bisphosphate, while in glucose synthesis they are formed from phosphoglyceroyl phosphate. Interconversion of these two triose phosphates is catalysed by the enzyme triose phosphate isomerase.

Triose phosphates are important in other metabolic processes. Glycerol is formed from DHAP, and glyceraldehyde 3-phosphate is an intermediate in the pentose phosphate pathway.

triphosphopyridine nucleotide *See* NADP.

tripinnate Describing a bipinnate leaf in which the secondary leaflets are further subdivided, as occurs in certain ferns.

triplet A sequence of three nucleotides on a DNA or messenger RNA molecule. Most of the 64 possible triplets (*see* genetic code) code for amino acids (*see* codon). However a few act as 'start' signals while others act as 'stop' signals to begin and end a polypeptide chain (*see* nonsense triplets).

triplet code hypothesis *See* genetic code.

triploid An organism, tissue, or cell that possesses three complete sets of chromosomes per nucleus. A triploid may arise from the fusion of a haploid and a diploid nucleus, or from the fusion of three haploid nuclei. Triploid organisms are normally infertile, since meiosis results in gametes that have more or less than the normal haploid number of chromosomes. The endosperm tissue of angiosperms is typically triploid (or pentaploid in some monocotyledons).

triradiate scar A scar on the surface of a spore that marks the point at which it was joined to the other three spores making up the tetrad. Triradiate scars are seen when the *tetrad shows tetrahedral symmetry. It is conspicuous because the spore wall is fairly thin at this pont while the rest of the wall is thickened with cutin. In the megaspores of *Selaginella kraussiana* the archegonia break through the megaspore wall at this point. Spores bearing a triradiate scar are termed *trilete*. *See* monolete.

trisomic An organism with an extra chromosome in addition to the normal complement, i.e. $2n+1$. Trisomy typically arises when a normal gamete (n) fuses with one containing an extra chromosome ($n+1$). The extra chromosome may be a simple copy of one of the chromosomes (primary trisomic), an *isochromosome (secondary trisomic), or may include parts of two nonhomologous chromosomes (tertiary trisomic). Trisomics often have abnormal phenotypes and are frequently less fertile than diploids. *See* aneuploidy.

tristichous Describing a form of alternate leaf arrangement in which successive leaves arise 120° around the circumference of the stem giving three vertical rows of leaves.

trivalent The temporary association of three homologous chromosomes in a triploid or trisomic organism. Such associations are observed between mid prophase and late metaphase of the first meiotic division.

tRNA *See* transfer RNA.

tropane alkaloids A group of alkaloids derived from the amino acid ornithine. They include hyoscyamine and its stereoisomer atropine, both solanaceous alkaloids. These are prepared from *Datura stramonium* (thorn apple) and *Atropa belladonna* (deadly nightshade) respectively. The cocaine alkaloids, derived from the leaves of *Erythroxylum coca* and *E. novagranatense* (coca), are also examples.

trophic level *See* food chain.

tropical rain forest *See* forest.

tropism (tropic movement) A directional growth movement of a plant or plant part that occurs in response to an external stimulus from a specific direction. Thus there are *positive* and *negative tropisms* depending on whether movement is towards or away from the source of stimulation. The movement or bending is accomplished by unequal rates of growth on the two sides of the organ, usually in response to increased auxin levels in the tissues. *Compare* nastic movements, taxis. *See* phototropism, geotropism, chemotropism; orthotropism, plagiotropism, diatropism.

tropophyte A plant that is adapted to survive in a climate where there are alternating wet and dry seasons, by having a resting phase during the dry season. For example, many trees in monsoon forests shed their leaves in the dry season.

true fungi *See* Eumycota.

truffles *See* Tuberales.

truncate Describing a leaf that is squared off at the apex.

tryptophan A nonpolar aromatic amino acid with the formula $C_8H_6NCH_2CH(NH_2)COOH$ (*see illustration at* amino acid). Tryptophan is synthesized via the shikimic acid pathway and broken down to alanine and acetyl CoA.
As with the other aromatic amino acids, tryptophan is the precursor of many aromatic compounds. Examples are the auxin indoleacetic acid and certain of the indole alkaloids, such as strychnine and yohimbine.

tube cell (tube nucleus) One of the cells contained within the microspore of gymnosperms, the others being the generative cell and the prothallial cell or cells. In *Cycas* it enters the haustorial pollen tube but its function is uncertain. In angiosperms the cell at the tip of the pollen tube that precedes the generative nuclei is sometimes termed the tube cell but it is more often called the *vegetative nucleus.

tube nucleus *See* tube cell, vegetative nucleus.

tuber A swollen part of a stem or root, usually modified for storage, and lasting for one year only, those of the succeeding year not arising from the old ones, nor bearing a position relative to them. Examples of such perennating organs are the stem tubers of potato (*Solanum tuberosum*) and the root tubers of *Dahlia*. Root tubers develop from adventitious roots (*compare* taproot). A stem tuber may be distinguished from a root tuber by the presence of buds or ‘eyes’.

Tuberales (truffles) An order of the *Discomycetes in which the ascocarps are formed underground and resemble tubers. It contains about 35 genera and 230 or so species, many of which are edible. Certain species are often found under particular trees, e.g. *Tuber melanosporum* (the European truffle), which grows under oaks. They are thus thought to be mycorrhizal.

tuberculate (verrucose) Having a surface covered with small warty projections as, for example, the rind of ridge cucumbers. Pollen covered in wartlike structures (verrucae), e.g. that of ivy (*Hedera helix*), is normally described as *verrucate*.

tumour-inducing principle A plasmid carried by the bacterium *Agrobacterium tumefaciens*, which causes crown gall disease in plants. The plasmid is necessary for the transformation of normal host tissue into tumour tissue. It is believed that this is brought about by the incorporation of the plasmid into the plant genome. Tissue removed from crown gall tumours can be grown in culture without the addition of auxin and cytokinin.

tundra A major regional community (*biome) in which the vegetation is poor, the few species being mainly lichens, mosses, heaths, sedges, grasses, and some herbaceous plants, but no trees. It is a region of cold *desert, the temperature rarely exceeding 10°C. The topsoil is frozen for about nine months of the year and the subsoil is subjected to permafrost. The plants are thus subjected both to extreme cold and physiological drought as the soil water is frozen. The topography, type of soil, degree of shelter, etc., give rise to differences in vegetation from one locality to another. Arctic tundra is found north of the tree line of North America and Eurasia in a band of varying width circling the Arctic Ocean. In the Antarctic, there are only small scattered areas of *arctic vegetation* consisting of mosses and lichens on some of the islands. *See also* alpine.

tunica *See* tunica-corpus theory.

tunica-corpus theory A concept of the organization and development of the *apical meristem, in which the meristematic region is differentiated into an outer peripheral layer or layers, termed the *tunica*, and an inner mass of cells, termed the *corpus*. The tunica is characterized by chiefly anticlinal divisions and the corpus mainly by periclinal divisions. The corpus gives rise to the interior part of the plant body and the tunica differentiates the outer layers including the epidermis. *Compare* histogen theory.

turgor The state existing in a plant cell when, due to the intake of water by osmosis, the protoplast exerts an outward pressure on the cell wall. When the cell is fully turgid, although the pressure is sufficient to make the wall bulge, the wall is strong enough to prevent more expansion and the further ingress of water. Turgidity is the main factor in maintaining rigidity and support in unlignified parts of the plant.

turgor potential *See* pressure potential.

turgor pressure In calculations of *osmotic pressure, the hydrostatic pressure exerted by the contents of the cell against the cell wall.

turion **1.** A type of perennating bud formed by certain aquatic plants, e.g. frogbit (*Hydrocharis morsus-ranae*), that is shed from the plant and lies dormant on the pond or river bed until the spring.
2. Any vegetative shoot or sucker.

two-dimensional analysis A technique used to separate mixtures of closely related molecules. One method involves two-dimensional paper chromatography, in which a chromatogram produced by a solvent running in one direction is then subjected to a second separation by placing the chromatogram at right angles in a second solvent. Another method involves separation by electrophoresis followed by paper chromatography.

tyloses Bladder-like ingrowths that protrude into the tracheary elements of older wood eventually causing blockage. They originate from adjacent parenchyma cells via paired pits in the cell walls. Tyloses often become filled with tannins, resins, gums, or various pigments, so giving the heartwood its characteristic darker colour. These substances also help to preserve and strengthen the wood, while some of the pigments are important commercially as dyes (e.g. haematoxylin). Tyloses are sometimes found in the vessels of herbaceous plants, in which their function is unclear. They may act to seal off damaged vessels.

type (nomenclatural type) The single element (e.g. illustration, specimen, etc.) on which the description associated with the original publication of a name was based. The type of a taxon can be a *holotype, *lectotype, or *neotype, as appropriate. The nomenclatural type of a species or infraspecific taxon of a vascular plant is usually a *herbarium specimen, but some species are typified by an illustration or description, in the absence of herbarium material. The names of some lower plants are based on type cultures, where the type is a living specimen. The type of a genus or infrageneric taxon (i.e. subgenus, section, series, etc.) is a designated species, while that of a family, subfamily, or tribe is a genus. The type of an order is a family, for example, the family Rosaceae is the type of the order Rosales.

Because of its importance, most type material is kept in different coloured folders, thereby enabling its rapid recovery in large collections.

tyrosine An aromatic polar amino acid with the formula $HOC_6H_4CH_2CH(NH_2)COOH$ (*see illustration at* amino acid). The biosynthesis of tyrosine is similar to that of *phenylalanine, diverging only in the last few reaction steps. Breakdown to acetyl CoA and fumaric acid occurs via phenylalanine. Like phenylalanine, tyrosine is a precursor of the isoquinoline alkaloids. It is also a precursor of various phenolic inhibitors, e.g. coumaric acid.

U

ubiquinone *See* coenzyme Q.

UDP *See* nucleoside diphosphate sugars.

Ulotrichales An order of the *Chlorophyceae containing simple unbranching filamentous algae (e.g. *Ulothrix*). Each cell of the filament contains one band-like chloroplast around its periphery and cell division is in one direction only. Asexual reproduction is by zoospores of various kinds while sexual reproduction varies from isogamy to the well developed oogamy seen in *Cylindrocapsa*. *See also* Ulvales.

ultracentrifuge *See* centrifuge.

ultramicrotome *See* microtome.

ultrastructure (fine structure) Structural details of cells below the limit of resolution of the light microscope and only revealed by the electron microscope.

ultraviolet microscope A microscope that uses ultraviolet radiation to illuminate the specimen and form the image. This increases the resolution because ultraviolet radiation has much shorter wavelengths (about 300 nm) than visible light, and resolution increases as wavelength decreases. Ultraviolet microscopy is difficult and complex and is rarely used since the development of the electron microscope.

Ulvales An order of the *Chlorophyceae containing pseudoparenchymatous thalloid algae (e.g. *Ulva*), characteristically found in marine and brackish waters. The sheetlike thallus is formed by division in two planes of a filament that initially resembles those of the *Ulotrichales. The Ulvales are consequently often placed in this order. There is an isomorphic alternation of generations in some species.

umbel A racemose inflorescence in which the flowers are borne on undivided pedicels originating from a common node on the main axis. The outermost flowers are borne on the longest pedicels so that the whole inflorescence is flat topped and gives the appearance of an umbrella. *Composite umbels* may be borne on branched pedicels so that numerous smaller umbels constitute the whole inflorescence. These inflorescences are typical of the Umbelliferae and provide landing platforms for pollinating insects. *See illustration at* racemose inflorescence.

Umbelliferae (Apiaceae) A dicotyledonous family containing about 2500–3000 species in about 300 genera and commonly known as the carrot family. Its members are distinguished by their characteristic inflorescence, the *umbel, a feature that made them one of the first natural families to be clearly recognized. Most umbellifers are herbaceous and have hollow internodes and often a characteristic odour. The fruit is dry and usually ridged and many are used as spices, such as caraway (*Carum carvi*) and coriander (*Coriandrum sativum*). Species used as herbs include parsley (*Petroselinum crispum*) and fennel (*Foeniculum vulgare*), while the carrot (*Daucus carota*) and parsnip (*Pastinaca sativa*) are important root crops.

uncoupling agent Any chemical that uncouples the process of respiration from that of phosphorylation. For example, in glycolysis the formation of 3-phosphoglycerate from 3-phosphoglyceroyl phosphate normally yields two molecules of ATP. However arsenate prevents the formation of ATP at this point. Arsenate also acts as an uncoupling agent in oxidative phosphorylation. Many other uncouplers of oxidative phosphorylation are known, most of which contain an aromatic ring and an acidic group. Examples are 2,4-dinitrophenol (DNP) and dicumarol. DNP has been used to investigate many aspects of plant physiology. For example, in research on phloem transport it has been shown that DNP inhibits translocation of assimilates. This implies there is some energy-requiring metabolic component in phloem transport and that the process is not purely due to osmotic potential and turgor pressure. DNP has also been used to investigate the theory that the respiratory climacteric that occurs during fruit ripening is due to the accumulation of natural uncoupling agents (uncoupling agents promote respiration as they cause a build-up of ADP).

undulipodia A collective name for *flagella and cilia, favoured by certain authorities because these organelles have the same (9+2) structure. The distinction between them has been an arbitrary one based on length and number.

unilocular sporangium A *sporangium that is not divided and comprises only one locule or compartment. Unilocular sporangia are seen in the sori of the brown alga *Laminaria*. *Compare* plurilocular sporangium.

universal veil A membrane that encloses the developing fruiting body of many basidiomycete fungi. It ruptures as the stalk elongates, the remnants forming the volva at the base of the stalk and the flecks of tissue seen on the upper surface of the cap.

uracil A pyrimidine base, characteristically present in RNA and absent from

DNA. The positions held by uracil in RNA are occupied by thymine in DNA; uracil is thus said to 'replace' thymine during RNA synthesis. Uracil, more properly called 2,4-dioxypyrimidine, is derived ultimately from sugars and amino acids.

uranyl acetate A widely used *electron stain suitable for high-resolution transmission electron microscopy. It stains nucleic acids and proteins thus enhancing the contrast obtained in observations of membranes and membrane-bound organelles.

Uredinales An order of the *Teliomycetes containing the *rust fungi, which number over 5000 species in about 126 genera. The teliospore is terminal, and the basidiospores develop on sterigmata and are actively discharged from the promycelium. *Compare* Ustilaginales.

uredospore (urediospore) A dikaryotic spore formed in a spore cluster (a *uredosorus* or *uredium*) by certain rust fungi. The uredospores of *Puccinia graminis* are formed on the main host, wheat, and are responsible for the characteristic rusty streaks of infected plants. On release they infect more wheat plants. *See* heteroecious.

uridine diphosphate *See* nucleoside diphosphate sugars.

uronic acids A class of sugar acids in which the carbon atom carrying the primary hydroxyl group (i.e. the CH_2OH end of the molecule) is oxidized to a carboxyl group (COOH). Uronic acids are biologically the most important group of the sugar acids and are common constituents of polysaccharides. For example, *glucuronic acid*, derived from glucose, is a common constituent of gums and mucilages. *Galacturonic acid* is a component of pectic substances while *mannuronic acid* is found in the seaweed gum *alginic acid.

Ustilaginales An order of the *Teliomycetes containing the *smut fungi, which number about 850 species in some 48 genera. The teliospore (the brand or smut spore) is not terminal and basidiospores develop directly on the promycelium rather than on sterigmata. The basidiospores are not actively discharged. *Compare* Uredinales.

utricle (bladder) **1.** An ovoid compartment lined with sensitive hairs that serves to trap and digest small animals. It is seen in species of *Utricularia* (bladderworts) and is a modification of the leaf.

2. An indehiscent bladder-like fruit formed by certain plants in the Chenopodiaceae and Amaranthaceae. It is a type of achene.

V

vacuolation *See* cell extension.

vacuole A fluid-filled cavity within the cytoplasm and separated from it by a membrane, the *tonoplast. In newly formed cells, when division has ceased, vacuoles are formed from small detached parts of the endoplasmic reticulum. As fluid (cell sap) accumulates they enlarge and coalesce to form a single vacuole, pushing the protoplasm against the cell wall. The surface area of the wall is increased, by stretching and the formation of additional material, to accomodate the increased volume of the cell. The vacuolar sap is a solution of organic and inorganic compounds. These may include sugars, soluble polysaccharides, soluble proteins, amino acids, carboxylic acids, red, blue, and purple anthocyanins, and mineral salts. Starch grains, oil droplets, and crystals of various kinds may also be present. These constituents of the sap probably represent metabolic by-products and reserve food material.

valine A nonpolar amino acid with the formula $(CH_3)_2CHCH(NH_2)COOH$ (*see illustration at* amino acid). Valine is synthesized from pyruvate via α-ketoisovalerate. The degradation of valine is extremely complex, leading eventually to proprionyl CoA and thence to succinyl CoA. Methionine, leucine, and isoleucine also degrade to proprionyl CoA; the last sequence of reactions to succinyl CoA is thus common to all four amino acids.

vallecular canal A longitudinal channel in the stem internode of *Equisetum* and some of its fossil relatives, positioned radially opposite a longitudinal furrow (valley) between the stem ridges. The vallecular canals are arranged roughly between the vascular bundles. *Compare* carinal canal.

valley bog *See* bog.

valvate Describing sepals and petals that meet but do not overlap in the bud. *Compare* imbricate.

valvule *See* palea.

variance Symbol V or σ^2. The square of the standard deviation. The mean square is the estimated variance.

variation The occurrence of differences between individuals. Such differences may be due to inherited (genetic) and environmental factors. Genetic variation is commonly due to recombination in sexually reproducing organisms, although the ultimate source of all genetic variation is mutation (*see* gene). Environmental variation may be caused by various factors, such as population density, nutritional status, light intensity, etc. A characteristic that is largely influenced by environmental factors is said to show phenotypic *plasticity and exhibit low *heritability. Variation may be due to differences in kind (*see* qualitative variation) or differences in degree (*see* quantitative variation).

variegation The occurrence of patches or streaks of different coloured tissues in a plant organ, usually a leaf or petal. It may be due to infection, particularly viral infection, mineral deficiency, or physiological or genetic differences between the cells. The variegated petals of certain tulip varieties (e.g. Rembrandt tulips) are due to a virus infection. The variegated leaves of *Coleus* are caused by groups of cells developing different pigment combinations. This is a form of somatic variegation and heritable differences between different coloured parts of the leaf are not evident. In contrast the variegated leaves of, for example, *Pelargonium* are due to genetic variation arising from mutation (*see* periclinal chimaera).

variety A rank subordinate to species but above the category form in the taxonomic hierarchy. Varieties are morphological variants, which may or may not have a clear geographical distribution. Sometimes they represent only a colour or habit phase. The variety of one author may be designated a subspecies or form by another. *See also* cultivar.

varve *See* varve dating.

varve dating A method used to determine the age of a particular sediment and of the fossils that it contains. A varve is a layer of sediment that has been deposited in a glacial lake by the melt waters in spring and summer. Because the particles brought down in the spring are much coarser than those deposited later in the year, annual layers can be distinguished. Varves were formed in the Pleistocene when the ice was retreating, and by counting them, the chronology for parts of northern Europe has been established. *See also* dendrochronology.

vascular bundle One of a number of strands of primary *vascular tissue constituting the *vascular system of the plant. Vascular bundles consist mainly of xylem and phloem, which may be separated by a *fascicular cambium. The relative position of xylem and phloem determines the type of bundle (*collateral, *bicollateral, *concentric, *amphicribal, or *amphivasal), and is often important taxonomically. The vascular bundles in a dictyostele are called *meristeles. *See also* eustele, atactostele.

vascular cambium A *lateral meristem, found in those vascular plants exhibiting secondary growth, that gives rise to *secondary xylem and *secondary phloem mostly by *periclinal cell divisions. The vascular cambium contains *fusiform initials, which give rise to the axial system, and *ray initials, which give rise to the radial (ray) system of the secondary tissues. *Compare* phellogen. *See also* cambium, fascicular cambium, interfascicular cambium.

vascular cylinder *See* stele.

vascular plant Any plant containing conducting tissue, i.e. xylem and phloem. Vascular plants are usually terrestrial or epiphytic and the sporophyte, which is the dominant generation, is differentiated into stem, leaves, and roots. They also differ from bryophytes and other nonvascular plants in possessing stomata. Fossil forms intermediate between the bryophytes and vascular plants have not been discovered and it seems probable that the two groups have evolved separately for a considerable time, possibly almost since the land was first colonized.

Depending on the relative importance attached to the possession of vascular tissue as compared to the production of seeds, the vascular plants may be classified in one division, the *Tracheophyta, or two, the *Pteridophyta and the *Spermatophyta.

vascular system The continuous network of *vascular tissue*, i.e. *xylem and *phloem, throughout a plant body.

vascular transition *See* transition region.

vector An agent that carries a pathogen. Strictly this includes wind, rainsplash, infected tools, etc. but more usually the term is applied to animal vectors and in particular to insects. Man can also be a vector – for example, a scientist examining an outbreak of disease may carry fungal spores on his clothing when moving to a healthy field. Insects are particularly important in the transmission of *virus and *mycoplasma diseases, the most common vectors being aphids, whiteflies, and leafhoppers. The insects acquire the pathogen while feeding on infected plants and transmit it when they move on to a healthy plant. Mites and nematodes are also vectors of virus diseases.

vegetative nucleus (vegetative cell) The large nucleus formed within the pollen grain of angiosperms along with one or two smaller generative nuclei. After germination of the pollen grain it migrates to the tip of the pollen tube where it may be termed the *tube nucleus*. It is thought to control the growth and development of the pollen tube and disintegrates when the pollen tube penetrates the nucellus.

vegetative reproduction (vegetative propagation) A form of asexual reproduction in which specialized multicellular organs formed by the parent become detached and generate new individuals. Such parts may include bulbs, corms, gemmae, rhizomes, stems, tubers, etc. The regenerative capacities of various plant organs have been exploited through a number of techniques used in agriculture and horticulture to multiply stocks. These methods include budding, layering, air-layering, cutting, and pegging, and may be enhanced by the careful control of microclimate or by growth regulators.

veil *See* universal veil, partial veil.

vein A *vascular bundle, or a group of closely associated bundles, in a leaf. In a leaf vein, the xylem is almost invariably positioned adaxially and the phloem abaxially, although there is sometimes an additional layer of adaxial phloem. Veins are sometimes surrounded by bundle sheaths of collenchyma, sclerenchyma, or parenchyma, which may extend to the leaf epidermis. The pattern formed by the veins in a leaf is called the *venation.

veination *See* venation.

velamen The multiple *epidermis of the aerial roots of many orchids, aroids, and other monocotyledons. It consists of densely packed cells that lack living protoplasts and have thickened walls. In wet weather these cells become filled with water but it is not certain whether or not the velamen performs an absorptive function.

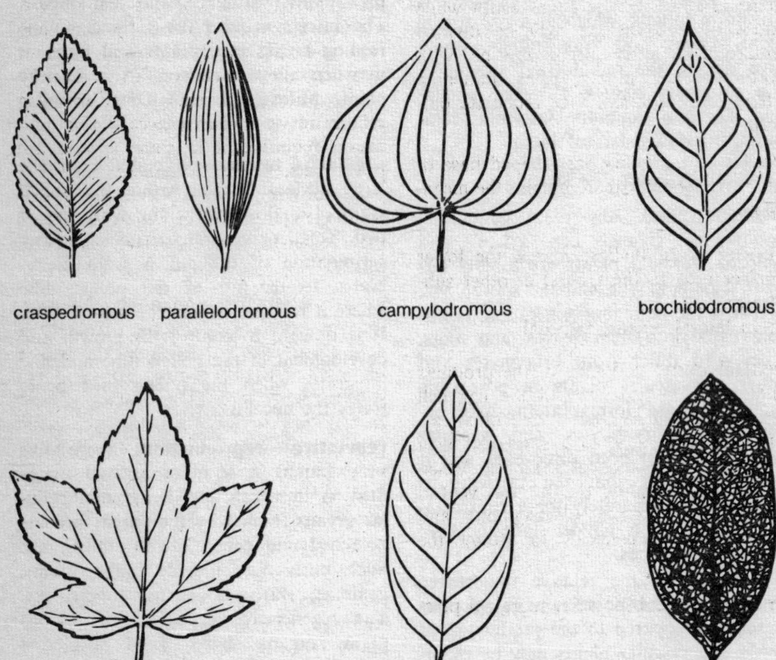

Some common forms of leaf venation.

veld (veldt, bushveld) South African savanna. *See* grassland.

velum 1. *See* annulus.
2. The flap of tissue that protects the sporangia in certain pteridophytes, e.g. *Isoetes*.

venation (veination) The pattern formed by the veins of a leaf, as viewed from above or below. The many different types of venation (*see* illustration) are useful diagnostically, especially in identifying fragmentary material. As a very general rule, the leaves of monocotyledons tend to have many parallel veins of more or less equal width while those of dicotyledons may be divided into one or a few primary veins, with secondary and tertiary veins branching off giving a net-veined or reticulate pattern.

venter The swollen flask-shaped base of an *archegonium that contains the megaspore.

ventral 1. In thallose plants, the lower surface next to the ground or other substrate.
2. In lateral organs, *adaxial.

ventral canal cell One of the products that is formed along with the egg cell when the primary ventral cell of an archegonium divides. It has no cell wall and lies at the base of the neck. When the archegonium is mature the ventral canal cell becomes mucilaginous and may produce chemicals to attract the male gametes.

ventral placentation *See* marginal placentation.

ventral suture The line of fusion where the margins of the megasporophyll join to form the characteristic tubular shape of the ovary as found in angiosperms. This is often one of the first lines of weakness along which dehiscence occurs at fruit ripening.

vernalin A hypothetical plant growth substance that, it has been suggested, is formed in meristematic regions of a plant subjected to cold. Experiments that apparently show this substance can be transmitted to other plants by grafting have probably failed to separate the effects of photoperiod from those of vernalization. Thus the transmitted flower stimulus could be *florigen rather than vernalin. Other experiments suggest that, if such a substance is formed, it is only transmitted to other cells by cell division. Thus if one apex on a plant is locally vernalized, the other apices remain unvernalized. It is probable that there is no one substance formed by vernalization and that the biochemical basis of vernalization is different in different cold-requiring species. Thus applying gibberellin to seeds replaces the vernalization requirement in some species but not others. Other substances that have partly or completely replaced the vernalization requirement in different species include auxin, kinetin, RNA, and vitamin E.

vernalization The promotion of flowering by exposure of young plants to a cold treatment. For example, the winter varieties of wheat, barley, oats, and rye will normally only flower in early summer if they were sown before the onset of winter. However in areas experiencing very harsh winters, as in the Soviet Union, this may not be possible. The plants are thus given an artificial cold treatment and planted in the spring. *See* vernalin.

vernation (ptyxis) The pattern of folding and rolling of leaves in a bud. Many types of vernation are recognized (*see* illustration), which may be useful diagnostically. The leaf margins may be rolled forwards to the upper side of the leaf (*involute*) or rolled backwards to the adaxial surface (*revolute*). The margin of one side may be wrapped around the other (*convolute*) or each leaf may be folded in a U-shape, enclosing the next youngest leaf (*conduplicate*). There may be many longitudinal folds in the lamina (*plicate*), as in certain palms, or the lamina may be rolled from apex to base (*circinate*), as seen in most ferns. *Compare* aestivation.

verrucose *See* tuberculate.

versatile Describing an anther that is joined to the filament about half way along its length and can move relatively freely. *Compare* basifixed, dorsifixed.

verticillaster (false whorl) The arrangement, resembling a complete whorl of flowers, that results when two dichasia form opposite each other at a node on an extended rachis. It is seen in members of the Labiatae, for example the white dead-nettle (*Lamium album*).

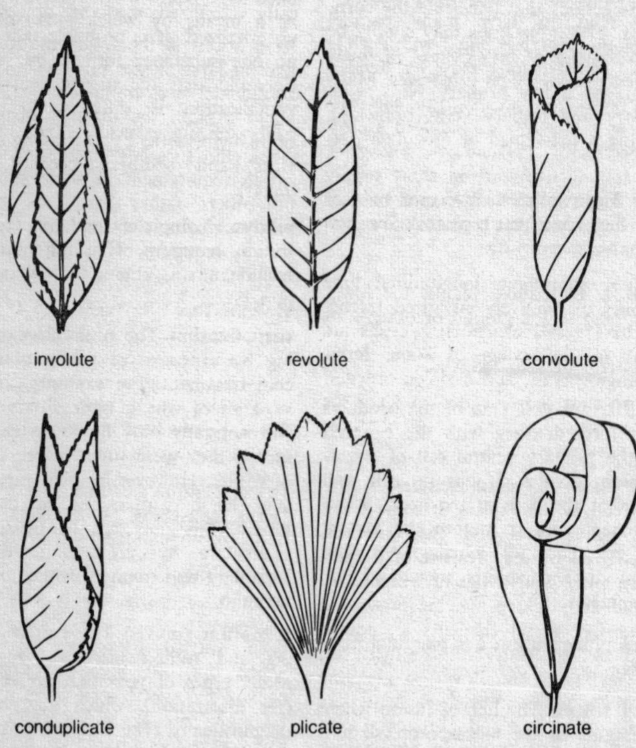

Types of vernation.

vesicle A general term for a cavity within the cytoplasm that is surrounded by a membrane. Vesicles vary considerably in shape, being tubular, spherical, discoid, ovoid, etc. They may contain particles, e.g. *endocytotic particles, or fluids, e.g. secretory products from the *golgi apparatus. The membrane isolates the contents from the cytoplasm.

vesicular-arbuscular mycorrhiza A form of *endotrophic mycorrhiza in which the fungus lives between the cells of the cortex and forms temporary hyphal projections that penetrate the cortical cells. The projections may simply be swollen vesicles or may consist of finely branched masses called *arbuscules*. Many grasses and crop plants form such associations and the fungi involved are often species of *Pythium* or *Endogone*. When soil nutrients are in short supply plants forming such mycorrhizae fare better than uninfected plants. *See also* ectotrophic mycorrhiza.

vessel A continuous longitudinal tube composed of relatively advanced tracheary cells (*vessel elements). Vessels are present in the xylem of some ferns, most angiosperms, and a single order of gymnosperms, the Gnetales, and are the main water-transporting cells of these plants. *Compare* tracheid.

vessel element (vessel member) A tracheary cell in the *xylem of some ferns, most angiosperms, and the Gnetales, characterized by the presence of a *perforation plate and a lignified secondary cell wall and lacking a living protoplast at maturity. Vessel elements are joined end to end to form *vessels, their end walls having broken down to give the perforation plates. Vessel elements are generally shorter and broader than tracheids and also differ in having perforation plates and often different types of wall pitting.

vibrio Any comma-shaped bacterium. *Compare* bacillus, coccus, spirillum.

vicariance The splitting up of an original biota into several isolated biotas (vicariants) by past geological or climatic events. The vicariants then develop independently and different species evolve. The process is presented as an alternative to the traditional theories of speciation by dispersal that have previously been used to explain biogeographical patterns. The existence of areas that contain many species different from but related to species in another distant area is taken by many as evidence for vicariance. One major event that is postulated as a means by which vicariants could have arisen is the fragmentation of the ancient land masses into the present-day continents (*see* continental drift). This could explain the disjunct distributions of many genera, e.g. *Liriodendron* (tulip trees) and southern beeches (*Nothofagus*).

villous Having a shaggy appearance due to a covering of long soft curly trichomes.

violaxanthin A *xanthophyll pigment that absorbs light in the blue region of the spectrum. Violaxanthin may in certain circumstances serve as a precursor of abscisic acid. It may also, with riboflavin, be involved in the photooxidation of endogenous auxins.

virion The inert extracellular phase of a *virus, consisting of a strand of DNA or RNA surrounded by a protein coat.

viroid An extremely small infectious agent consisting solely of RNA with no enclosing coat or capsid. Viroids have been isolated from various plants in which they are able to replicate and cause characteristic disease symptoms. Examples are potato spindle tuber viroid, chrysanthemum chlorotic mottle viroid, hop stunt viroid, and avocado sunblotch viroid.

virulence (pathogenicity) The capacity of a pathogen to cause disease. A pathogen often exists in a number of *physiological races, which, with respect to one particular crop cultivar, may be either virulent or avirulent. The release of resistant crop cultivars imposes a selection pressure on the pathogen to develop new genes for virulence. It has been shown there is a gene-for-gene relationship between crop resistance and virulence of the pathogen.

virus A small infectious agent that is only able to replicate by modifying the genetic machinery of living host cells. Outside the host cell a virus consists of DNA or RNA surrounded by a protein shell (*see* capsid). In this inert state a given virus has a characteristic size and shape (e.g. polyhedral, spherical, rod-shaped, etc.). Some of the simpler viruses, e.g. tobacco mosaic virus (TMV), can be crystallized.
Approximately 400 plant viruses are known, most of which are single-stranded RNA viruses. Some, e.g. TMV and cucumber mosaic virus, have a wide host range while others are limited to a few species. Viruses can produce a variety of symptoms, e.g. mosaics, leaf spots, and deformed growth of certain organs. The broken flower colours of certain ornamentals, e.g. Rembrandt tulips, are due to virus infection. Some viruses are symptomless though they may still markedly reduce yield. Viruses are transmitted by vectors and by infected seed and pollen. Control is by using virus-free seed or by breeding for hypersensitivity. Viruses are generally not found in meristematic shoot tips and virus-free stocks of certain species can be obtained by tissue culture of meristem explants. *Cross protection is successful in some crops.
A virus that infects bacteria is termed a *bacteriophage.

vital stain A stain used to dye living tissue without harming or killing the cells. Examples include trypan blue and vital red. *Intravital* staining involves the injection of a stain into an organism, some of the living cells taking up the dye. *Supravital* staining involves the removal of living tissue from a multicellular organism and its subsequent staining. *See also* staining.

vitamin Any compound, essential in trace amounts for the normal functioning of an organism, but not synthesized by some heterotrophs, which thus need to obtain it from plants or microorganisms. All vitamins with the exceptions of vitamins C (*ascorbic acid), E (*tocopherol), and K, function as coenzymes. They can be divided into two groups, the water-soluble vitamins, which contain all the B-group vitamins (*see* thiamin, riboflavin, nicotinic acid, pantothenic acid, pyridoxine, biotin, folic acid, cyanocobalamin) and vitamin C, and the fat-soluble vitamins, including vitamins A, D, E, and K.

vitamin A An isoprenoid compound important in many aspects of animal growth, an early symptom of deficiency being nightblindness. It is not present in plants but carotene pigments when ingested are cleaved into two molecules of vitamin A.

vitamin B_1 *See* thiamin.

vitamin B_2 *See* riboflavin.

vitamin B_6 *See* pyridoxine.

vitamin B_{12} *See* cyanocobalamin.

vitamin C *See* ascorbic acid.

vitamin E *See* tocopherol.

vitamin K A fat-soluble quinone found in most plants and many microorganisms. It is believed to play a part in the transfer of electrons from photosystem II to photosystem I in photosynthesis. In animals deficiency affects the normal blood-clotting mechanism.

vitta A resin canal or oil cavity. Vittae are often found in the fruits of plants in the Umbelliferae. The number and position (whether between or in the primary ridges of the fruit) are important diagnostic characters in the family.

vivipary 1. The differentiation of young plants or bulbils at the floral axils, instead of flowers. This is seen, for instance, in viviparous fescue (*Festuca vivipara*) and in the spider plant (*Chlorophytum comosum*).
2. The premature germination of seeds or spores in situ on the maternal plant before they have been released. This occurs in the mangroves where seeds may develop into sizeable seedlings before they are shed from the parent tree.

volunteer A plant that has grown from self-sown seed. Volunteers may be important inoculum sources for some diseases.

volva The cuplike structure that encircles the base of the fruiting bodies of many basidiomycete fungi. It is part of the remnants of the universal veil.

Volvocales An order of the *Chlorophyceae containing motile unicellular (e.g. *Chlamydomonas*), colonial (e.g. *Gonium*), and coenobial (e.g. *Volvox*) algae. Members of the Volvocales have one chloroplast in each cell, often containing an eye spot. Usually asexual reproduction is by zoospores while sexual reproduction varies from isogamy to oogamy. In some classifications the unicellular forms are placed in a separate order, the Chlamydomonadales.

W

Waldgrenze *See* timber line.

Wallace effect The idea, proposed by the naturalist A. R. Wallace, that reproductive barriers within a species may be developed and subsequently improved by selection. If there are a number of optimum phenotypes favoured by selection, intermediate organisms less fitted to the environment will be eliminated by natural selection. Thus hybrids between such phenotypes will be at a disadvantage and hybridization will be selected against.

wall pressure The force exerted upon cell contents by the cell wall. It is equal and opposite to the turgor pressure.

Warburg effect The inhibition of carbon dioxide assimilation and photosynthesis by atmospheric oxygen, described by O. Warburg in 1920. This phenomenon was later discovered to be due to *photorespiration.

water culture *See* hydroponics.

water moulds *See* Saprolegniales.

water potential Symbol Ψ. A measure of the energy available in an aqueous solution to cause the migration of water molecules across *semipermeable membranes during osmosis. Values of water potential cannot be calculated absolutely but are highest in pure water, which for convenience is given the value zero, and fall with increasing solute concentration. Water always tends to move from areas of high (less negative) to areas of low (more negative) potential. This principle governs water conduction in plants. Water potential is the sum of *osmotic potential, *pressure potential, and *matric potential. The term is replacing the concept of *osmotic pressure and differs from osmotic pressure in taking capillary and imbibitional forces into account.

wax A mixture of esters of higher fatty acids with higher monohydric alcohols or sterols. Waxes may also contain odd-carbon alkanes, long-chain monoketones, β-diketones, β-hydroxyketones, and secondary alcohols.
Waxes are important components of the waxy cuticle covering the stems, leaves, flowers, and fruits of most plants. They are manufactured as oily droplets in epidermal cells, from which they migrate to

the outer surface of the plant via tiny canalculi in the cell wall, and crystallize as rods and platelets. Their pattern of deposition is sometimes used as a micromorphological taxonomic character below the genus level.

The function of waxes is not fully understood although it seems likely that they are involved in water balance. Plant waxes are obtained on a commercial scale from the leaves of the carnauba palm (*Copernica prunifera*) and from the stems of *Euphorbia antisyphilitica.*

weed Any plant growing where it is not wanted. Many weeds are adapted to exploit disturbed areas of land and, unless measures are taken to prevent it, weeds are the first plants to establish themselves on cleared patches of ground. Many weeds are annual plants that produce large quantities of seed and often pass through several generations in a single year. An example of such a weed is shepherd's purse (*Capsella bursa-pastoris*). Weeds, especially those closely related to crop plants, may also harbour various pests and diseases. For example, the roots of the black nightshade (*Solanum nigrum*) are often infected by the potato cyst eelworm. *See also* rogue.

weedkiller *See* herbicide.

weighting In taxonomy, the assigning of greater or lesser importance to one character, as compared to another character, according to its known or assumed value. The initial choice of which characters to use in a classification is in itself a positive weighting process, termed *selection weighting.* Characters considered unreliable and consequently rejected, perhaps because they show too much environmental variation, are said to be given *residual* or *rejection* weighting. Once the characters have been selected some may be ascribed additional importance if, for example, they are known to be good diagnostically in other groups. When such weighting is applied before the classification is drawn up it is termed *a priori* weighting. An example would be the placing together of two taxa that share the one character chromosome number but differ in a number of other characters. Alternatively a classification may be constructed in which all the chosen characters are considered of equal importance; this is the normal procedure in *numerical taxonomy. When the classification has been erected it will then be apparent which characters correlate well with other characters. Such characters are given *a posteriori* weighting. This second procedure provides a method for obtaining unbiased correlations of visible morphological characters with other characters, e.g. chemical, cytological, or genetic characters, that are not immediately apparent.

Weismannism The theory of the continuity of the germ plasm proposed by A. Weismann in 1886. It opposed the idea that acquired characteristics could be inherited. Weismann distinguished the body of the organism (the *soma*) from the reproductive cells (the *germ plasm*) and stated that it was the germ cells alone that affected inheritance and not the soma. He suggested that the germ plasm was set aside during early development and was not affected by subsequent changes in the soma. Weismann also formulated a theory of inheritance based on the behaviour of chromosomes. His ideas led to a rejection of *Lamarckism that has continued to the present time.

Welwitschiales *See* Gnetales.

wet rot A plant disease in which there is disintegration of tissues and release of cell fluids. Brown rot of stored fruits, such as plums and apples, is caused by fungi of the genus *Sclerotinia.* Wet rot of structural timber is caused by various fungi including *Coniophora puteana* and *Poria vaillantii.*

wetting agent A chemical, such as detergent or soap, that lowers the surface tension of water. Wetting agents are added to fungicide, herbicide, and insecticide sprays to increase the area of the spray droplets in contact with the plant surface or insect body.

whiplash flagellum (acronematic flagellum) A threadlike projection arising from motile algal and fungal cells that has a smooth surface. Internally the fibrillar structure is typical of the *flagella and cilia of motile eukaryotic cells.

white alkali soil *See* solonchak.

whorled Describing a form of leaf arrangement in which three or more leaves arise at each node, as in the bedstraws (family Rubiaceae).

wild-type The common form of a gene or organism in natural (wild) populations. Wild-type genes are typically dominant. They are usually designated '+'.

wilt A plant disease characterized by wilting. Wilting often occurs in the advanced stages of root diseases when water uptake becomes inadequate. However the term is usually applied to diseases in which wilting occurs in the absence of marked root damage. Such wilting may occur either because the vascular tissues are blocked or because water is being withdrawn by parasitic plants, such as witchweeds (*Striga*) and broomrapes (*Orobranche*). Blockage of vascular tissues is caused by various fungi and bacteria. It may be due either to the physical presence of vast numbers of microorganisms or to substances, such as gums and tyloses, that the host forms in response to invasion. Examples of wilt diseases are Dutch elm disease, caused by the fungus *Ceratocystis ulmi*, and wilts of potato, tobacco, and banana, caused by the bacterium *Pseudomonas solanacearum*.

wind pollination *See* anemophily.

wing 1. Either of the two narrow lateral petals of a 'pea' flower. *See also* keel, standard.
2. The membranous outgrowth of certain fruits, e.g. the samara.
3. A flange running down a stem or stalk as, for example, seen along the stems of the hairy vetchling (*Lathyrus hirsutus*).

witches' broom A dense mass of deformed twigs, often resembling a bird's nest, caused by the host response to infection by certain insects, mites, viruses, fungi, or parasitic plants. Witches' broom of cacao is caused by the fungus *Marasmius perniciosus* and can result in almost total loss of yield as pods are also infected. Witches' broom of birch (*Betula*) is caused by the fungus *Taphrina betulina*.

wood *See* xylem, secondary xylem.

wood sugar *See* xylose.

woody perennial *See* perennial.

wound cork A layer of *phellem that forms over a damaged part of the plant. It prevents desiccation of the underlying tissues and the entry of pathogens. *See also* callus.

wound hormone Any substance produced by damaged cells that diffuses into nearby undamaged cells and there stimulates meristematic activity, resulting in the formation of a protective callus. Auxins, gibberellins, and certain products of wounding have been implicated in the wounding reaction. *See also* traumatic acid.

wound wood *See* callus.

wracks *See* Fucales.

X

xanthophyll Any of a class of oxygenated hydrocarbons derived from the carotenes. Xanthophylls function mainly as photosynthetic accessory pigments, and are found in many different plant species. However the xanthophylls fucoxanthin and peridinin of the brown algae are of especial interest, because they are the primary light absorbing pigments for these organisms. They absorb light at frequencies where chlorophyll has only poor absorption and transfer the absorbed energy to chlorophyll.

Xanthophyta (yellow-green algae) A division consisting mainly of freshwater and terrestrial *algae, including unicellular, colonial (palmelloid), filamentous, and siphonaceous forms. They are characterized by having disclike yellow-green plastids (due to the large amounts of carotenoids present) and by accumulating oil and leucosin rather than starch. The motile stages have two unequal flagella (one flimmer and one smooth flagellum) hence the older name Heterokontae. The cell wall is divided into two halves but this character is not obvious except in the filamentous forms, where disruption results in a number of H-shaped fragments.

xenia The modification of the form of a fruit or seed by the pollen due to its effect, through double fertilization, on the nature of the endosperm. For example, in maize the endosperm may show a variety of colours depending on the origin of the pollen.

xerophyte A plant that is adapted to living in dry conditions caused either by lack of soil water or by heat or wind bringing about excessive transpiration. Many xerophytes are found in deserts, on sand dunes, and on exposed moors and heaths. Some xerophytes, e.g. the *succulent desert cacti, store water in swollen stems and leaves. Many species reduce the rate of transpiration by having permanently rolled leaves, e.g. cross-leaved heath (*Erica tetralix*) or by having leaves that are rolled in dry weather, e.g. marram grass (*Ammophila arenaria*). Some have hairy leaves to trap moist air, e.g. great mullein (*Verbascum thapsus*) while others have stomata sunken into grooves producing pockets of moist air, e.g. *Pinus*. The leaves of many species are leathery with a thick cuticle and epidermis to reduce cuticular transpiration. Some species have stomata that close during the day and open at night (*see* crassulacean acid metabolism). *Compare* hydrophyte, mesophyte. *See also* xerosere.

xerosere A pioneer plant community that develops in a dry region. A *lithosere* develops on bare rock, beginning with simple plant forms, such as lichens. Soil building and accumulation of organic matter continue so that xerophytic herbs develop followed by *mesophytes, such as hardy herbaceous plants and even larger woody plants. A psammosere develops on bare *sand dunes and a *halosere on coastal mud flats and salt marshes. *See also* sere.

x-ray diffraction analysis A technique used to determine the three-dimensional pattern of large molecules such as proteins or DNA. A beam of x-rays is fired at a crystal and the resulting diffraction pattern is recorded on a photographic plate behind the crystal. The crystal is rotated slightly and is subjected to more x-rays. By repeating this, the three-dimensional pattern can be determined. It was work of this nature that showed DNA to be a regular double helix.

xylan A polysaccharide in which the major monosaccharide subunit is xylose. Xylans are abundant components of the *hemicelluloses. In monocotyledonous plants the dominant hemicellulose is an arabinoxylan, in which arabinose side chains are attached to a backbone of xylose residues.

xylem (wood) Vascular tissue whose principal function is the upward translocation of water and solutes. It is composed mainly of vessels, tracheids, fibre-tracheids, libriform fibres, and parenchyma cells. It should be noted, however, that all these cell types may not be present in any one wood sample. Wood anatomy is often very important taxonomically, the presence or absence of the various cell types and their distribution within the xylem being important diagnostic characters. The xylem occurs in association with, and usually internal to, the phloem. *See also* primary xylem, secondary xylem.

xylose (wood sugar) An aldopentose sugar (*see illustration at* aldose) widely found in plants, particularly in woody tissues. It occurs mainly in its polymerized form *xylan. It is also a constituent of the rare disaccharide primeverose.

Y

yeasts *See* Endomycetales, Blastomycetes.

yellow-green algae *See* Xanthophyta.

yellowing *See* chlorosis.

yellows 1. A plant disease, caused by mineral deficiency or a virus or mycoplasma, in which there is yellowing of the foliage. Yellowing is a common symptom when there is a deficiency of elements important in chlorophyll production, namely, iron, magnesium, manganese, nitrogen, or sulphur. Barley yellow dwarf and beet yellows are caused by viruses and coconut lethal yellowing is caused by a mycoplasma. *See also* deficiency disease, virus, Mycoplasmatales. **2.** A disease of cabbage caused by *Fusarium conglutinans.*

Z

zeatin (6-(4-hydroxy-3-methyl but-2-enyl)aminopurine) The first plant growth substance of the *cytokinin type to be isolated from plant tissues. It was obtained from the kernels of sweet corn (*Zea mays*) hence its name. Subsequently other cytokinins similar in structure to zeatin (i.e. an adenine molecule substituted with an isoprenoid derivative) have been isolated. Some of these are found as minor bases in transfer RNA molecules (*see* IPA).

zein A low-molecular-weight simple protein found as a major storage protein in maize grains. *See* prolamine.

zinc Symbol: Zn. A metal element, atomic number 30, atomic weight 65.38, needed in trace amounts for successful plant growth. Zinc ions are required as cofactors by certain enzymes, e.g. carboxypeptidase, carbonic anhydrase, and alcohol dehydrogenase. It has been suggested that zinc may play a role in the mechanism of action of the growth substance ethylene. A common symptom of deficiency is leaf mottling.

zonal soil *See* soil.

zoogloea *See* capsule.

zoosporangium A *sporangium that produces *zoospores.

zoospore A motile usually naked spore with one or more flagella. It may be produced either by a zygote or a zoosporangium and is dependent on water for dispersal. In some species, e.g. the green alga *Ulothrix*, two types of zoospore may be developed. In *Ulothrix* large macrozoospores with four flagella and small microzoospores with two flagella are seen. Zoospores may encyst under adverse conditions. *Compare* aplanospore.

Zygnemaphyceae (Conjugatophyceae) A class of the *Chlorophyta whose members are distinguished by their distinctively symmetrical cells, anatomically complex plastids, and the lack of free-swimming forms. In addition, they reproduce by *conjugation. It contains freshwater filamentous forms (e.g. *Spirogyra*, *Zygnema*) and unicellular forms, the *desmids*. The desmids are divided into the *saccoderm* desmids, commonly found in acidic pools and peat bogs, and the *placoderm* or true desmids, which are a major component of the phytoplankton. The cells of placoderm desmids are characteristically split into two virtually identical halves, joined by a narrow bridge at the centre. The cell wall is highly ornamented and indented.

zygomorphy *See* bilateral symmetry.

Zygomycetes A class of the *Zygomycotina containing both saprobic fungi and many that are parasitic, particularly on arthropods. It contains about 515 species in some 85 genera. There are two orders: the *Mucorales; and the Entomophthorales, most of which are insect parasites.

Zygomycotina A subdivision of the *Eumycota containing fungi with a mycelial thallus, characteristically with aseptate hyphae. It contains about 620 species in some 115 genera. Its members do not produce motile cells and asexual reproduction is by nonmotile aplanospores. Sexual reproduction leads to the formation of a resting spore (*zygospore). The Zygomycotina contains the two classes *Zygomycetes and Trichomycetes.

zygospore A thick-walled zygote that is formed by the fusion of *isogamous gametes. It is characteristic of organisms that reproduce by conjugation, e.g. fungi of the *Mucorales and algae of the *Zygnemaphyceae. The thickened walls enable it to survive adverse environmental conditions. *Compare* oospore.

zygote The product of the fusion of two gametes, before it has undergone mitosis or meiosis. In lower algae and fungi it may retain the motile nature of the gametes. More usually the zygote is immobile and may develop a thickened resistant wall to form a *zygospore. In higher plants the zygote is typically protected by maternal tissue, and divides immediately after fertilization forming a *proembryo. *See also* sexual reproduction.

zygotene *See* prophase.

zymase The mixture of enzymes, isolated from yeasts, that brings about *alcoholic fermentation. It includes pyruvate decarboxylase, which catalyses the formation of acetaldehyde from pyruvate, and alcohol dehydrogenase, which catalyses the reduction of acetaldehyde to ethanol. It also includes the enzymes of the glycolytic pathway.

APPENDIX

TABLE 1

The Meanings of Some Common Specific Epithets

Epithet	Meaning
acetosus	having an acid taste
aestivus, aestivalis	of summer
agrestis	of fields or cultivated land
alatus	winged
albus	white
alpestris, alpinus	of the Alps or high mountains
altissimus	very tall
altus	tall, high
amabilis	pleasing, lovely
angustatus	narrow, slender
arborescens	treelike
arenarius	of sandy places
argenteus	silvery
arundinaceus	reedlike
arvensis	of fields, especially ploughed fields
aureus	golden yellow
australis	southern
autumnalis	of autumn
borealis	northern
caeruleus	sky blue
caesius	blue grey
calcaratus	spurred
campanulatus	bell-like
campestris	of fields
candidus	shining white
canescens	hoary
capreolatus	having tendrils
carinatus	keeled
carneus	flesh coloured
carnosus	fleshy, succulent
castaneus	chestnut coloured
caudatus	tailed
cerasiferus	having cherry-like fruits
ceriferus	producing wax
cernuus	nodding, drooping
cinereus	ash grey
cirrhosus, cirrhatus	having tendrils
clavatus	club shaped
coccineus	scarlet
collinus	of hills
communis	common
comosus, comatus	tufted
concolor	uniform in colour
corniculatus, cornutus	having a hornlike appendage
costatus, costatalis	ribbed
crassus	thick, fleshy
crispus, crispatus	finely waved
cruentus	blood red
cyaneus	dark blue
demersus	growing under water
demissus	lowly, humble
discolor	not uniform in colour
dulcis	sweet
dumosus	bushy
echinatus	spiny
edulis	used for food
effusus	spread out thinly
elatus	tall
elodes	of marshes
ensatus	swordlike
esculentus	edible
ferrugineus	rust coloured
fistulosus	hollow and tubular
flabellatus	fanlike
flavus	pale yellow
flexuosus	bending alternately in opposite directions
floribundus	flowering profusely
fluitans	floating
fluviatilis	of rivers
foetidus	foul smelling
fontinalis, fontanus	of springs
fulgens, fulgidus	bright, shining
fulvus	yellow brown
furcatus	forked
gelidus	of cold regions
glutinosus	sticky
gracilis	slender

graveolens	strong smelling
griseus	pearl grey
hederaceus	ivy-like
hepaticus	liver coloured
hibernus, hiemalis	of winter
hirsutus	hairy
horridus	very bristly
hortensis	of a garden
humilis	dwarf
hystrix	bristly
incanus	grey, hoary
incarnatus	flesh coloured
impudicus	shameless, immodest
indicus	of India
infundibuliformis	funnel shaped
insignis	outstanding
integrifolius	having entire leaves
italicus	of Italy
junceus	rushlike
lacustris	of lakes or ponds
laevigatus	smooth, polished
lanatus, lanosus	woolly
latifolius	having broad leaves
leucanthus	having white flowers
limosus	of muddy places
lineatus	marked with parallel lines
littoralis	of the seashore
lividus	lead coloured
lunatus	half-moon shaped
luridus	dull yellow
luteus	deep yellow
maculatus	spotted, blotched
meridionalis	southern; flowers opening around midday
mollis	softly hairy
montanus, monticolus	of the mountains
moschatus	musk smelling
muralis	of walls
nanus	dwarf
natans	floating on or under water
nemoralis, nemorosus	of shade or woodlands
nervosus	having conspicuous veins
niger	black
nivalis, niveus	snow white
nudus	naked
nutans	nodding, hanging
occidentalis	western, American
officinalis	having medical use
oleraceus	vegetable crop
orientalis	eastern, Asian
paludosus, palustris	of bogs, marshes, or swamps
pannosus	densely hairy
parvus	small
patens	spreading
petraeus	growing among rocks
plenus, pleniformis	full, double
praecox	developing early
prasinus	bright green
oratensis	of meadows
pulcher	beautiful
pumilus	dwarf
pusillus	very small; weak
ramosus	branched
reniformis	kidney shaped
repens, reptans	creeping, prostrate
riparius	growing by rivers or streams
rivularis, rivalis	growing by streams or brooks
roseus	pink
rostratus	beaked
rubellus, rubens, rufus	reddish
ruber	red
ruderalis	growing in rubbish
rupestris	growing on rocks
sativus	cultivated
saxatilis	growing among rocks
scaber	rough
scandens	climbing
sempervirens	evergreen
sericeus	silky
serotinus	late
setaceus, setasus	bristly
sinensis	of China
somniferus	sleep inducing

speciosus	good looking
spectabilis	showy
squamatus	scaly
squarrosus	having overlapping leaves with outward-projecting tips
sylvaticus, sylvestris	of woods
tenellus	delicate
tenuis	slender
terrestris	of dry ground
tinctorius	used for dyeing
umbrosus	of shade
uncinatus	hooked
usitassimus, utilis	useful
velutinus	velvety
ventricosus	inflated, especially unevenly so
vernalis, vernus	of spring
versicolor	variously coloured
vescus	small; edible
virens, viridus	green
vulgaris	common

THE LIBRARY
THE COLLEGE
SWINDON

TABLE 2

The Systematic Names of some Organic Compounds

acetaldehyde	ethanal	α-ketoglutaric acid	1-oxybutanedioic acid
acetic acid	ethanoic acid	lactic acid	2-hydroxypropanoic acid
acetone	propanone	lauric acid	dodecanoic acid
alanine	2-aminopropanoic acid	maleic acid	*cis*-butenedioic acid
aspartic acid	aminobutanedioic acid	malic acid	2-hydroxybutanedioic acid
catechol	benzene-1, 2-diol	malonic acid	propanedioic acid
cinnamic acid	3-phenylpropenoic acid	mercaptans	thiols
citric acid	2-hydroxypropane-1, 2, 3-tricarboxylic acid	myristic acid	tetradecanoic acid
ethylene	ethene	oleic acid	*cis*-octadec-9-enoic acid
ethylene glycol	ethane-1, 2-diol	oxalic acid	ethanedioic acid
fatty acids	alkanoic acids	oxaloacetic acid	2-oxybutanedioic acid
formaldehyde	methanal	oxalosuccinic acid	1-oxypropane-1, 2, 3-tricarboxylic acid
fumaric acid	*trans*-butenedioic acid	palmitic acid	hexadecanoic acid
glutamic acid	2-aminopentanedioic acid	phloroglucinol	benzene-1, 3, 5-triol
glycerol	propane-1, 2, 3-triol	pyruvic acid	2-oxypropanoic acid
glycine	aminoethanoic acid	stearic acid	octadecanoic acid
glycols	diols	succinic acid	butanedioic acid
hydroquinone	benzene-1, 4-diol	toluene	methylbenzene
isoprene	methylbuta-1, 3-diene	*o*-xylene	1, 2-dimethylbenzene